SUBSEA VALVES AND ACTUATORS FOR THE OIL AND GAS INDUSTRY

SUBSEA VALVES AND ACTUATORS FOR THE OIL AND GAS INDUSTRY

KARAN SOTOODEH

ELSEVIER

Gulf Professional Publishing
An imprint of Elsevier

Gulf Professional Publishing is an imprint of Elsevier
50 Hampshire Street, 5th Floor, Cambridge, MA 02139, United States
The Boulevard, Langford Lane, Kidlington, Oxford, OX5 1GB, United Kingdom

Library of Congress Cataloging-in-Publication Data
A catalog record for this book is available from the Library of Congress

British Library Cataloguing-in-Publication Data
A catalogue record for this book is available from the British Library

ISBN: 978-0-323-90605-0

For information on all Gulf Professional publications
visit our website at https://www.elsevier.com/books-and-journals

Publisher: Joe Hayton
Senior Acquisitions Editor: Katie Hammon
Editorial Project Manager: Madeline Jones
Production Project Manager: Sojan P. Pazhayattil
Cover Designer: Victoria Pearson

Typeset by SPi Global, India

Contents

Acknowledgment

I would like to express my gratitude to my partner, Ms. Tamara Zhunussova, for her great support. In addition, many thanks to Galperti Engineering and Biffi companies for their support for writing this book.

CHAPTER ONE

Introduction to the subsea sector of the oil and gas industry

Contents

1.1 Introduction and history

The term "subsea" in simple language refers to underwater locations such as seas and oceans which are known as a sector of the offshore oil and gas industry. Many oil and gas resources are located under the water in offshore areas around the world. Americans began the exploration for oil and gas in the Pacific Ocean more than 70 years ago. In fact, no company took the risk to drill outside the borders of the land before 1947. The offshore oil and gas industry started in 1947 when the first oil well was completed by Kerr-McGee in just 4.6 m of water in the Gulf of Mexico. At the time, almost everyone believed that there was no oil in the Gulf or Mexico and that drilling in the Gulf would lead only to finding salt domes. Little equipment and few facilities were specially designed at that time for offshore oil exploration, drilling, and production. By the end of 1949, 11 oil and natural gas fields had been found in the Gulf of Mexico with 44 exploratory wells established according to the report of the National Ocean Industries Association. The offshore sector of oil and gas expanded afterward and the exploration and production

Subsea Valves and Actuators for the Oil and Gas Industry
https://doi.org/10.1016/B978-0-323-90605-0.00006-2

of oil and gas moved to deeper water depths. In 1961 the first subsea well was completed by Shell company. The first diver-less subsea completion was performed in 1967. The first North Sea subsea development project was named Ekofisk which was located at a depth of 220 ft (approximately 67 m) of the sea. One hundred and forty operational subsea wells existed worldwide by 1978. Shell Coulomb made the deepest subsea tieback at a depth of 7300 ft (approximately 2225 m) in 2004. By then, the number of operating subsea wells had reached 2404 worldwide. A joint venture of British Petroleum and Exxon Mobil started production of the Thunder Horse deep-water oil and gas field located 240 km off the west coast of New Orleans at a depth of approximately 1900 m. The Ormen Lange project operated by hydro, a Norwegian company, was the longest tieback at approximately 160 km.

Nowadays, the venture of exploration and production of oil and gas is moving to deep-water depths between 1 and 4 km. Subsea development is generally divided into two categories: shallow and deep-water depths. There is no clear definition of deep water but, in general, depths that are out of reach of a diver are called deep water. Based on this definition, depths in the range of 300 m and more can be considered deep may be 20 years ago. The definition of deep-water development is changing now since oil exploration venture is moving more and more to deeper water depths. Therefore the definition of deep-water depth could be changed now to depths of more than 1 km.

Many fixed platforms were used in the beginning for the development of offshore fields in shallow water depths. Fixed platforms are attached to the sea floor by legs made either from steel or concrete that rely on their own weight to be fixed to the floor. These platforms can be designed to be used for long-term (30+ years) oil and gas production and are suitable for installation on the field at a maximum of 300 m depth as illustrated in Fig. 1.1. A tension-leg platform TLP or extended tension leg platform is a floating structure

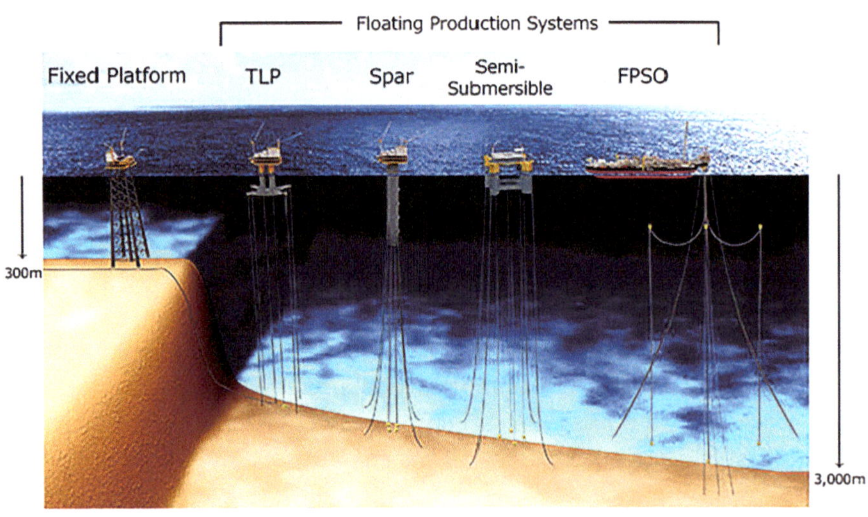

Fig. 1.1 Offshore field development/production solutions. *(Courtesy: MODEC/Springer.)*

that is moored (connected to the seabed by tendons). A TLP is a suitable field development platform for water depths between 300 and 1500 m. Spar platforms are floating platforms used for deep-water depths of more than 1.5 km; these platforms are anchored to the seabed. The robust design of the spar platform makes this option resistant to wind, waves, and currents. A spar platform contains a large diameter vertical cylinder supporting a deck as illustrated in Fig. 1.2. Semisubmersible platforms are buoyant and contain multilegged floating structures with a large deck. Offshore platforms are steel structures with legs directly connected to the sea floor and are only suitable for the development of a single oil well.

As the oil fields get deeper, oilfield development is changing and the use of ships called Floating Production Storage and Offloading (FPSO) vessels is becoming more popular. An FPSO receives the produced hydrocarbon from the subsea units and processes, stores, and finally transports it. Unlike a jacket platform which is connected to a single well, a FPSO can be connected to several wells. In fact, an FPSO may be used by offshore oil and gas operator companies for different purposes such as the production and processing of

Fig. 1.2 Spar platform. *(Courtesy: Offshore Technology.)*

petroleum and to store the finished products. Fig. 1.1 compares different field development solutions based on the depth of the field.

The main aim of subsea development is to select the optimum way in terms of safety and economical gain to extract the oil and gas from subsea oil fields. Subsea development is comprised of different activities such as exploration, drilling, completion, and production. Subsea technology is designed to meet very special and unique challenges and thus requires specialized engineering tasks.

The concept of subsea oilfield development was implemented in the early 1970s by installing subsea wellheads, Christmas trees, production components, and facilities such as manifolds, jumpers, and pipeline on the seabed. More detailed information about subsea production systems is provided in the next section of this chapter. The following parameters should be taken into account for subsea development:

- Water depth: shallow, deep, or intermediate;
- Stand alone or tieback: a subsea tieback is defined as an engineering solution that connects a new oilfield to an existing production field and is a popular trend in subsea development;
- Vertical or horizontal tree which affects the weight and figure of the tree;
- Hydraulic and chemical supply units;
- Subsea processing equipment and components;
- Field development architecture or configuration to provide a reliable, safe, and cost-effective arrangement;
- Artificial lift method.

1.1.1 Benefits of subsea engineering

A key benefit of subsea processing is that it improves oil and gas production and recovery from the reservoir. The other advantage of subsea processing is that it saves cost and increases ease of production by saving space on the topside facilities, whether platform or FPSO, and reduces the transfer of large volumes of untreated hydrocarbon from deep subsea depths to the surface. As an example, the separation of three phases of oil, gas, and water produced from the reservoir can be done subsea without any need to transfer all of these phases all the way to the topside. In addition, undesirable by-products of oil and gas such as carbon dioxide (CO_2) and hydrogen sulfide (H_2S) may be removed from the reservoir fluid (typically oil and gas) on the sea floor and there is no requirement to transfer them to the topside facilities through a flowline. A flowline, also called a pipeline, is defined as a segment of piping that transfers the fluid coming out from the wellhead to the topside facilities from the manifolds. An important point is that a flowline is connected to the topside facilities indirectly through risers (see Fig. 1.3). A riser is defined as a segment of pipe that connects an FPSO or a platform to a subsea system such as a flowline so this pipe (the riser) is extended between the seabed and the surface. Therefore both

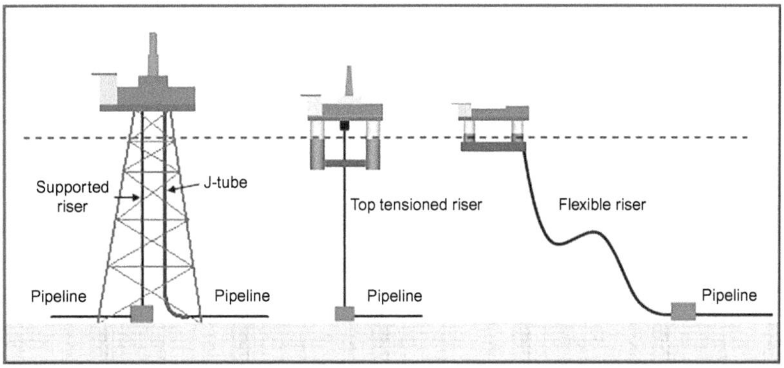

Fig. 1.3 Flowline and riser transportation system.

risers and flowlines (pipelines) are parts of a transportation system to bring the produced hydrocarbon to the topside facilities. Risers typically range in size from 3″ to 12″ in diameter and can be either flexible or rigid.

Subsea development can solve the flow assurance problems to a large extent. Multiphase flow, hydrate, and wax formation are operational problems associated with flow assurance and piping systems. Multiphase flow can have many different forms: one of the most severe forms is called slug flow. Slug flow refers to a condition in which subsea flowline or riser sections completely fill up with liquid and hold up the gas flow. Slugging has many different adverse consequences such as high loads and pressure drops in the piping system, corrosion, and erosion.

Another form of flow problem is called hydrate. This refers to solids like ice that can form in the natural gas at high pressure and low temperature. Formation of hydrate is very likely to occur during the multiphase flow movement of the three phases of oil, gas, and water or the two-phase oil and gas flow regime. However, hydrate can be formed in a single-phase gas after separation from the other two phases of water and oil. Thus a gas dehydration unit could be used on the platform of an FPSO to separate the saturated water from the gas to prevent hydrate formation through the injection of methanol or glycol into the gas service. Therefore separation of multiphase flow on the sea floor can significantly reduce the risk of hydrate formation in the gas service. Fig. 1.4 illustrates hydrate formation in a three-phase flow in a nonstraight piping system.

Crude oil can contain wax or paraffin waxes that can become deposited in the piping system especially at low temperature and reduce the piping flow capacity or block the piping. Injection of a wax inhibitor into the hydrocarbon in the subsea can mitigate the risk of piping blockage and flow reduction due to wax accumulation. Thus it can be concluded that subsea development and processing can reduce flow assurance problems to a large extent. Therefore subsea development can alter the strategy for the development of marginal fields. Marginal fields refer to fields that are not economical to

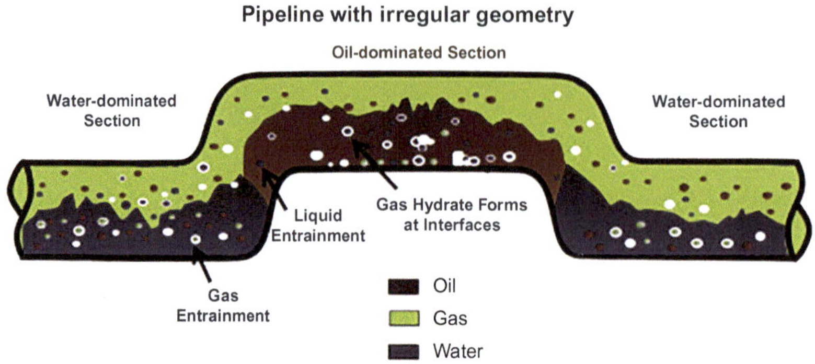

Fig. 1.4 Hydrate formation in three-phase fluid in a nonstraight piping. *(Courtesy: Elsevier.)*

develop and produce since they cannot provide acceptable revenue. Many fields once considered marginal can now be considered economical due to subsea development. Fig. 1.5 illustrates a subsea separation technology that leads to the separation of oil, gas, and water or just two phases of oil and gas. Subsea-produced oil and gas are transported to the topside platform separately through oil and gas flowlines.

The main challenges in designing a subsea system are that of dealing with high pressure and deep water. In conclusion, the benefits of subsea development and engineering are summarized as follows:

- Subsea engineering and processing may eliminate the need for having a platform or a ship;
- If a platform or ship is required, a less expensive one may be used;

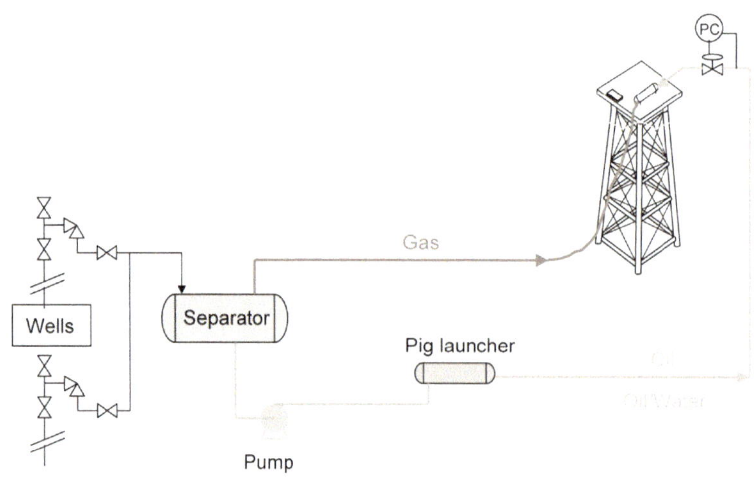

Fig. 1.5 Subsea separation and transportation. *(Courtesy: Springer.)*

- Oil and gas fields discovered in deeper and more remote areas are more economical to develop and produce;
- The oil recovery factor and production rate can be increased;
- Flow assurance improvement;
- Potential cost saving and improvement of project economic aspects.

1.2 Subsea production systems

Subsea production systems are located on the sea floor rather than on the surface. A subsea production system contains the whole process and all the facilities used for drilling, well completion, field development, and production. A typical subsea production system includes subsea completed wells, subsea wellheads and production trees, subsea manifolds, subsea tie-ins to connect the piping to another piping system or a facility including pipeline ends and inline structures, a flowline and subsea equipment such as a boosting pump, compressors, a chemical and hydraulic injection system, an umbilical and a production control system. Fig. 1.6 illustrates a typical subsea production system.

Some of the subsea production components can be categorized as "subsea processing" components. Processing refers to the treatment of the produced fluid from the reservoir. A petroleum reservoir or oil and gas reservoir is a subsurface pool of hydrocarbons contained in porous or fractured rock formations. Subsea processing involves different

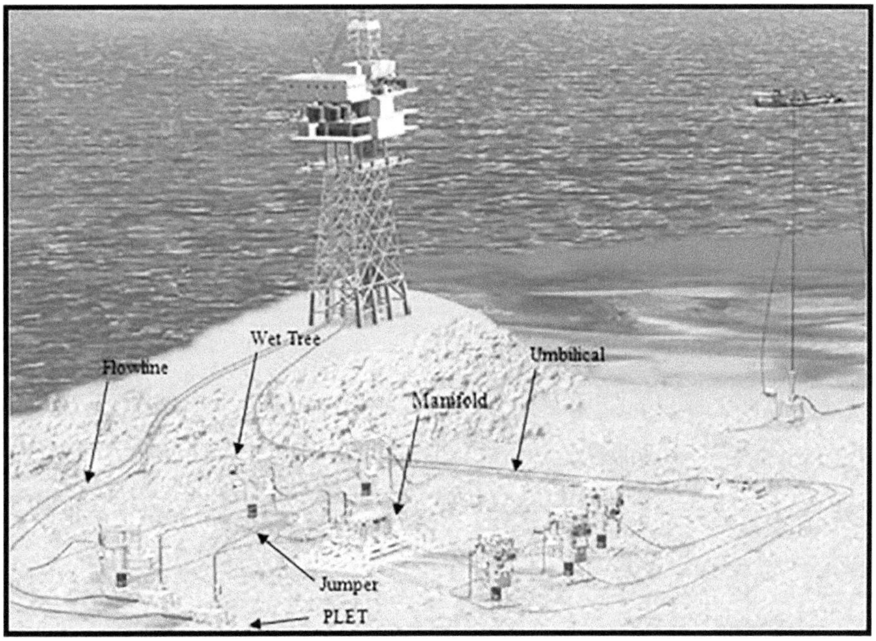

Fig. 1.6 Typical subsea production system. *(Courtesy: Elsevier.)*

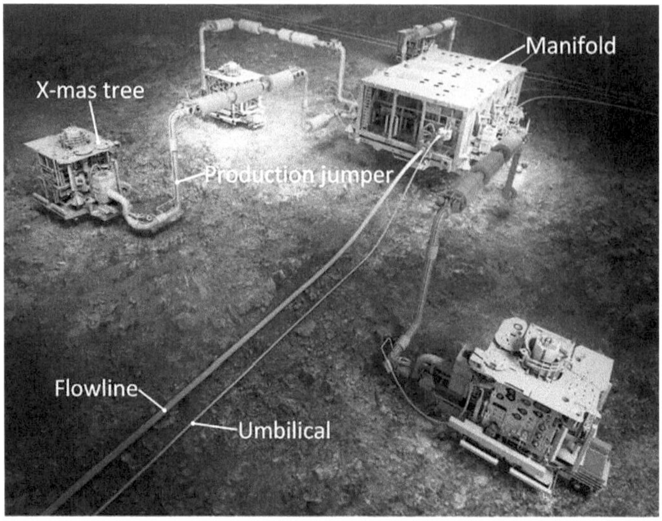

Fig. 1.7 Subsea production components and structures. *(Courtesy: Springer.)*

activities such as separation, pumping, compression, filtering, etc. which are explained later in this chapter.

Some of the main subsea production components and structures are highlighted in Fig. 1.7. Due to the high pressure and sometimes high temperature, as well as the harsh subsea environment and the external pressure from the seawater column, subsea components and equipment are exposed to high, complex, and critical load cases.

1.2.1 Wellhead and Christmas trees

A wellhead transfers the loads from the casing in the well and from the completed well to the seabed. It also provides a pressure barrier and access to the well bore and production tubing. A subsea wellhead provides structural and pressure integrity for the well casing and tubing. In short, a subsea wellhead can support casing, a tubing hanger, a production tree, and other drilling and completion components. Subsea wellheads are designed based on API 17D/ISO 13628-4 standards requirements. Different considerations should be considered for wellhead selection and design such as the loads transferred by the tubing hanger, annular injection requirements such as gas lift or water injection, special material selection for subsea wellheads and instrumentation. Definitions for the casing and tubing hanger will be provided later in this chapter.

A subsea well includes downhole well completion components as well as subsea trees. Well completion refers to a process of making a well ready for production and injection into the well after drilling. In fact, a drilled well should be closed with a piping, fitting, and valve assembly known as a Christmas tree (see Fig. 1.8) to gather the reservoir fluid from the wellhead and direct the produced hydrocarbon to the production facilities.

Fig. 1.8 Subsea (wet) Christmas tree.

Thus a Christmas tree (or X-mass tree) is the interface between the completed well and the production facilities and structures, i.e., a connection between the wellhead and the flowline or manifold. It should be noted that the function of a Christmas tree is more than a connection between the completed or finished well and production facilities. Other functions of a subsea Christmas tree are summarized as follows:

- A subsea Christmas tree is a pressure-containing barrier between the well and the subsea environment;
- It provides control of the produced fluid through a choke valve;
- It allows injection of water or gas to the well for improving hydrocarbon production also called "advanced oil recovery" process;
- Allows oil well intervention, i.e., maintenance, repair, and replacement tasks on the well and shutting down the well;
- Provides hydraulic interface for downhole hydraulically operated components such as subsurface safety valves;
- Provides electrical interface for downhole equipment and instrumentations that work with electricity;
- Provides access to the annulus for controlling the well and other tasks such as injection to the well.

Subsea wells can be split into two categories: satellite and clustered wells. A satellite well is a single well that shares a minimum number of facilities with other wells. When several wells are connected with common sharing facilities, they are considered clustered wells.

It is important to note that Christmas trees can be installed on land—these are called dry or surface trees—landed on the seabed; these are known as wet or subsea trees. Fig. 1.9 illustrates a dry or surface Christmas tree. Christmas trees can be vertical or

Fig. 1.9 Surface Christmas tree.

horizontal (see Fig. 1.8) with the possibility of access to the well bore from the top or side, respectively. The difference between a horizontal and vertical tree is weight, size, and configuration. The size of the well is equal to the nominal bore or internal diameter of the wellhead, normally measured in inches. As an example, 18 ¾″ is very common size for a well. Traditionally, 16 ¾″ wellbores are used for producing the hydrocarbons from the reservoir. The pressure rating of the tree could be one of the API 6A/API 17D standard pressure classes such as 5000 psi, 10,000 psi, 15,000 psi, or 20,000 psi. The temperature rating of a Christmas tree is based on the temperature classes given in API 6A/17D in the range of −60°C to 121°C. These temperature classes are defined in Chapter 2 of this book. It is important to consider that the temperature rating for a Christmas tree can be outside the temperature ranges given in the above-mentioned standards.

A subsea Christmas tree typically consists of the following components:

1. A tree connector that connects the Christmas tree to the wellhead;
2. The Christmas tree body, a large forging and pressure-containing component that contains the production flow path in the case of a single bore, or production flow path or annulus flow path in the case of a dual bore design and selection;
3. Tree valves that may be integral with the Christmas tree or bolted. These valves are located on the production bore in the case of single bore tree configuration or located on both the production and annulus in the case of dual bore Christmas tree configuration;
4. Control components, i.e., actuators installed on valves in the Christmas tree;
5. A choke used for regulating and reducing the pressure of fluid coming out of the well;
6. The tree cap which is used to blind off the wellhead. The cap on the tree should be protected against dropping objects;

7. The tubing hanger which is placed and sit in the wellhead and the tree is connected to the wellhead after completion. The tubing hanger transfers the loads from the casing to the sea floor. In addition, the tubing hanger provides structural support for the tubing and provides a pressure and fluid barrier. Fig. 1.10 illustrates a wellhead assembly including a tubing hanger. Different technical parameters should be considered for tubing hanger design and selection such as pressure rating, number of bores, and operational consideration such as loads.
8. Tubing is not a part of the wellhead but it is an important part of the well completion. Tubing cost can be as high as 20% of the wellhead cost. Failure of tubing can be very costly and lead to loss of production as well as injuries and loss of operation life. Load analysis and determining the mechanical strength of the tubing material are essential aspects of tubing design. Tubing is exposed to combinations of both internal and

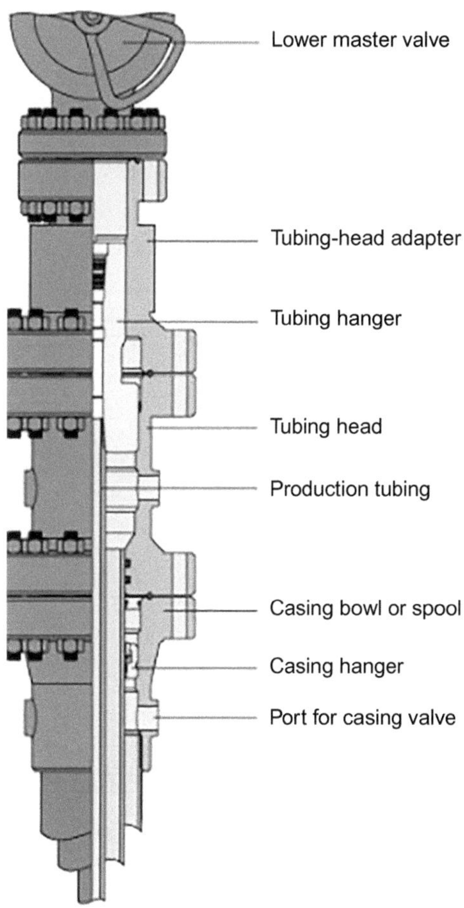

Fig. 1.10 Wellhead assembly including tubing hanger. *(Courtesy: Schlumberger.)*

external loads that can cause bursting and collapse of the tubing, respectively. The loads applied to the tubing are not limited to internal, external, and combination loads. Other factors create loads such as the weight of the tubing itself, buoyancy, thermal loads, buckling, etc. Fig. 1.11 shows how the combination effects of internal and external pressures affect tubing length and features.

9. A foundation is applicable for all subsea structures including the wellhead. A foundation is a template that withstands the weight and other loads of the wellhead.

The schematics on the left in Fig. 1.11 show a situation where internal pressure is higher than external pressure. The tubing is getting shortened and squeezed into a balloon shape. The schematic on the right side indicates a situation in which the external pressure exceeds the internal pressure. This condition increases the length of the tubing and reverse ballooning occurs.

Buckling of tubing is generated by compression forces that lead to changes in tubing shape as illustrated in Fig. 1.12.

Fig. 1.13 illustrates the installation of a subsea tree.

1.2.2 Wellhead and Christmas tree valves

Christmas tree valves are typically through conduit slab gate valves with ROV or actuator operation. ROV stands for Remotely Operated Vehicle which is explained in detail in Chapter 9. One of the advantages of through conduit gate valves is less restriction and

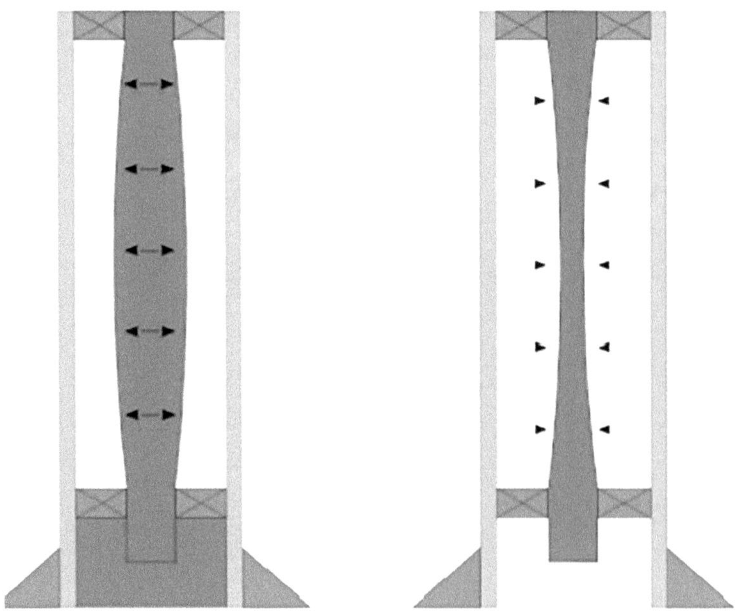

Fig. 1.11 Effect of internal and external pressure on the tubing.

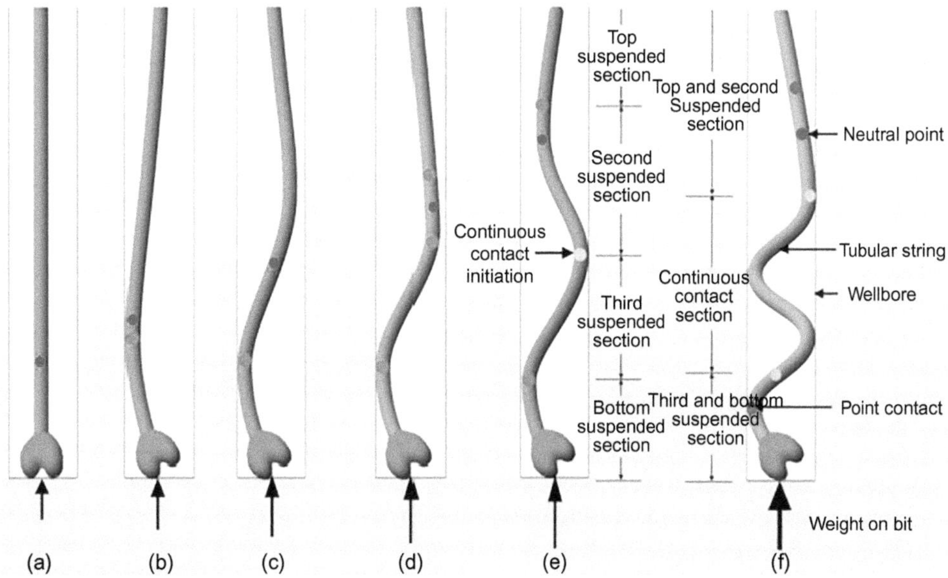

Fig. 1.12 Tubing buckling.

Fig. 1.13 Installation of subsea tree. *(Courtesy: Oceaneering.)*

flow drop creation during hydrocarbon production from the well. Slab gate valves are normally open for production fluid passage. More information about gate valves is available in Chapter 4. The actuators of gate valves are linear fail-safe closed or spring return type in almost all cases which are explained more in detail in Chapter 9.

Fig. 1.15 illustrates a dual bore vertical subsea tree with different valves. Dual bore wellhead configuration allows access to both tubing and the tubing and production casing annulus. An annulus is defined as an area between the tubing and the casing. The casing is a large diameter pipe that is assembled and inserted into the newly drilled well to protect the side of the well from collapse. Fig. 1.14 illustrates the different kinds of casing and production tubing.

Using a dual bore well is more reliable than single bore and it has some advantages. For example, the pressure between the production tubing and casing can be circulated and bled off. The main bore size can be standardized to 5″ but smaller sizes such as 3″ or 4″ can be used. The annulus bore can be standardized to a 2″ size. Each valve located on the dual bore Christmas tree is explained as follows:

Master valves located at the bottom of the Christmas tree control the flow of the wellbore. Since the function of the master valve is important, a couple of master valves are typically installed in the wellhead; one acts as a backup for the other one in case of leakage and for required maintenance. A smaller bore master valve is located on the annulus to control the flow in the annulus. The valve located on the vertical line in Fig. 1.15 above the lowest master valve is called underwater safety valve (USV) in case of location of wellhead under the water which is hydraulically actuated with a fail-safe close function to protect the production facilities if a failure or damage occurs in the well. This valve is connected to an emergency shut down (ESD) system and is closed rapidly when necessary. Swab valves installed on the top part of the Christmas tree provide access to the bore for intervention such as maintenance or cleaning of the well. A smaller swab gate valve is located on the annulus bore. Wing valves are located on the side of the Christmas tree. A production wing valve is used to isolate and control the production line and a kill wing

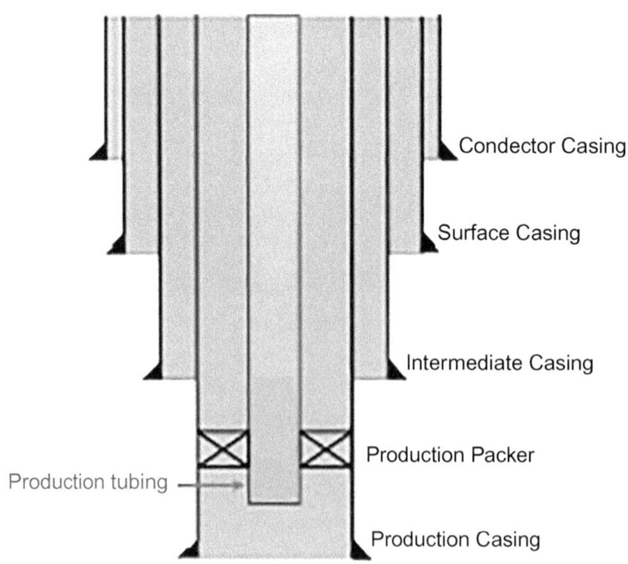

Fig. 1.14 Wellbore casing and production tubing.

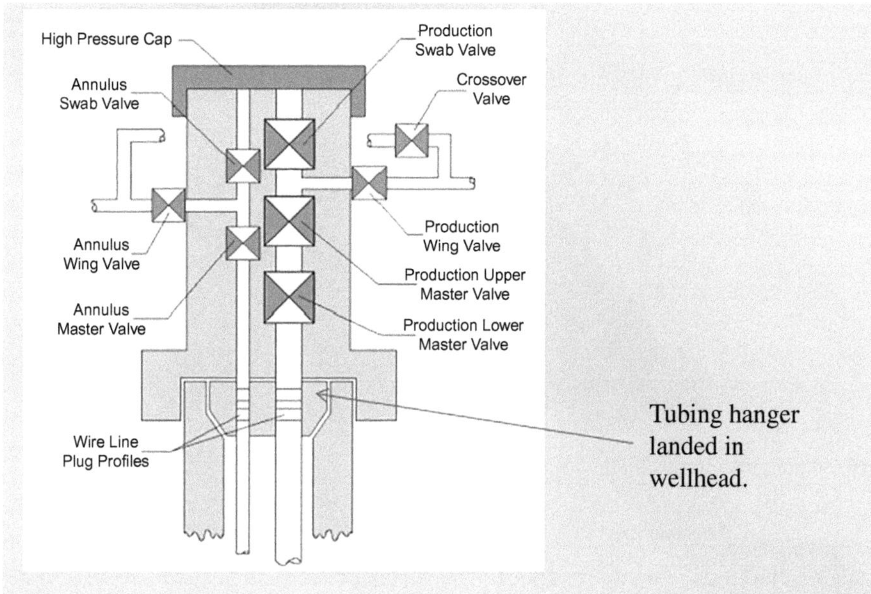

Fig. 1.15 Dual bore vertical tree.

valve located on the opposite side of the well is used for well treatment or control. Crossover valves allow communication between the annulus and production bore. Each bore of the well in dual bore well completion provides an independent barrier to flow.

1.2.3 Manifolds

Subsea manifolds (see Fig. 1.16), which are arrangements of piping, valves, connections, structures, and a foundation, are used in subsea production systems in order to receive, combine, and distribute hydrocarbon fluid and simplify the subsea system by reducing the quantity of piping such as pipelines and flowlines. Piping and valves are responsible for directing and controlling hydrocarbon flow and structures are used to protect and support the piping and valves located inside the manifold. Each piping header is typically connected to a flowline. The size of the piping depends on the Christmas tree and the size of the connected flowline. Foundations provide an interface between the seabed and manifold structure. The valves on the manifold direct or stop the fluid and can be either manual or actuated. Manually operated valves are operated with ROV. The main types of valves used in manifolds are through conduit gate (TCG), ball and axial flow, or nozzle check valves. In general, ball valves are selected for headers in sizes above 8″ and TCG valves are selected for branches or headers in 8″ and below. Small bore manually operated ball valves such as ½″ and ¾″ are used for chemical and hydraulic injection lines coming from the subsea distribution unit (SDU). Axial flow or nozzle check valves are nonreturn valves installed on the

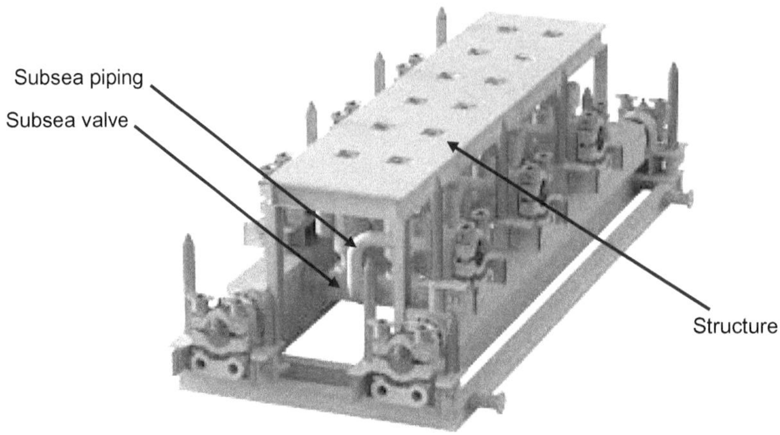

Fig. 1.16 Subsea manifolds.

chemical injection lines. The choice of an actuator for manifold valves may be either hydraulic or electrical. The actuators may be spring return, also called single acting, with both fail-safe close or fail-safe open functions; or double acting with a fail-as-is mode of operation. Double acting actuators are normally selected for larger valves in sizes such as 16″ or 22″ and higher pressure classes due to the compact design and high torque production. Most manifold valves are normally open. More information about actuators is available in Chapter 10. Manifold valves and actuators should be designed to withstand high pressure and operate under the hydrostatic seawater head. Installing a pressure balance or compensation system on the actuator can make it suitable for use in deep-water depths.

Fig. 1.17 illustrates the valve arrangement in a six-slot subsea manifold that includes three pieces of large bore 16″ ball valves on the headers, and a 7 1/16″slab gate valve on each branch connected on the branch piping in 8″ nominal pipe size:

The material of the piping and valves should be compatible with the produced hydrocarbon as well as the injected chemicals. Manifolds can collect the flow from different wells or inject water or gas into different wells for advanced oil recovery. Manifolds may have two, four, six, or eight slots. The number of slots is equal to the number of the wells connected to the manifold. As an example, six-slot manifolds have six branches that can be connected to six different wellheads. Fig. 1.18 illustrates a four-slot manifold connected to four wellheads through four branch lines. Fig. 1.19 illustrates a six-slot manifold connected to three wellheads from one side through three branch piping connections, and three more wellheads on the other side that are not visible in the picture.

1.2.4 PLET and PLEM

In addition to their other functions, subsea manifolds may also provide substructural support for more than one pipe-to-pipe connection or tie-in. Such manifolds are known as PLEM (Pipeline End Manifold). Therefore a PLEM is a subsea structure (a simple

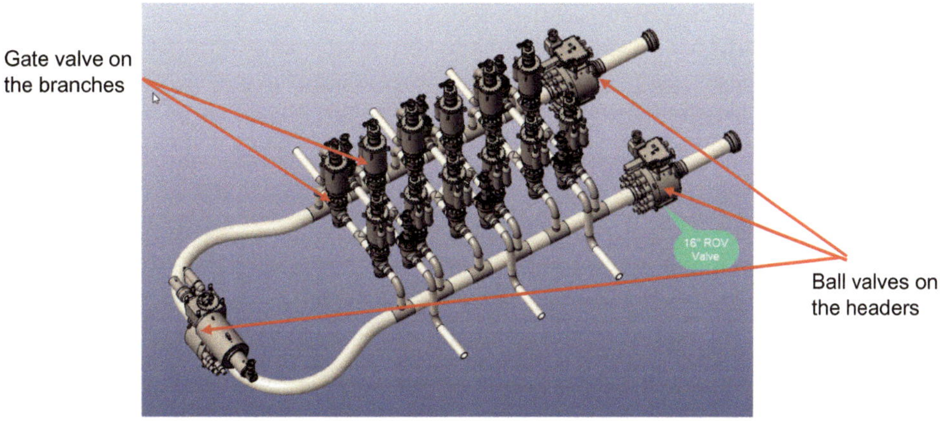

Fig. 1.17 Valve arrangement on a six-slot production manifold.

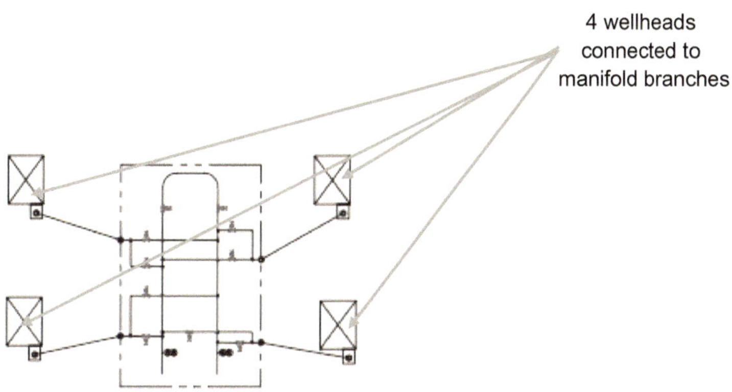

Fig. 1.18 A four-slot manifold schematic.

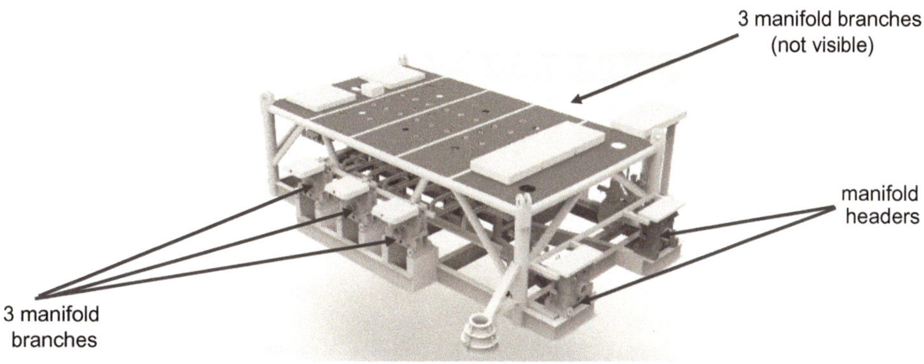

Fig. 1.19 A six-slot manifold.

manifold) placed at the end of a pipeline to connect the rigid pipeline with another structure or additional piping. As an example, subsea flowlines can be connected directly to a PLEM. In fact, pipe-to-pipe connections in subsea do not have any substructure for support, unlike connecting a pipe to another pipe inside a structure which is known as a tie-in structure. Therefore a PLET (Pipeline End Terminal) is required for a single pipe-to-pipe connection, e.g., connecting an export pipeline and a spool, or connecting a flowline to a manifold and riser. A PLET is known as a pipeline end or tie-in structure. In fact, pipeline or flowline cannot be connected directly to a manifold. Therefore the pipeline is connected to a PLET and a pipe spool or jumper is installed between the PLET and manifold as illustrated in Fig. 1.20. Fig. 1.22 illustrates a four-slot manifold connected to two flowlines through two PLETs. In addition, a PLET can be used to connect the flowlines to the risers. The flowline can be connected to the manifold in deep water with different types of connectors as follows:

- API 6A flanges with SBX or SRX gaskets (more information about API 6A flanges and gaskets is available in Chapter 2);
- Hubs and clamps according to API 16A (more information about these connections is available in Chapter 2);
- Proprietary mechanical connectors such as the collet style connectors illustrated in Fig. 1.21;
- Multibore connectors.

A pigging loop on the left side of the manifold in Fig. 1.22 provides the possibility of a pig (piping or pipeline injected gadget) running from the host facility such as a platform through one flowline all the way to the manifold header and returning back from the manifold to another flowline and finally returning to the host facility. Such a gadget may be run through the piping or pipeline for different purposes such as cleaning and inspection. Fig. 1.23 illustrates a pig used for cleaning the piping system.

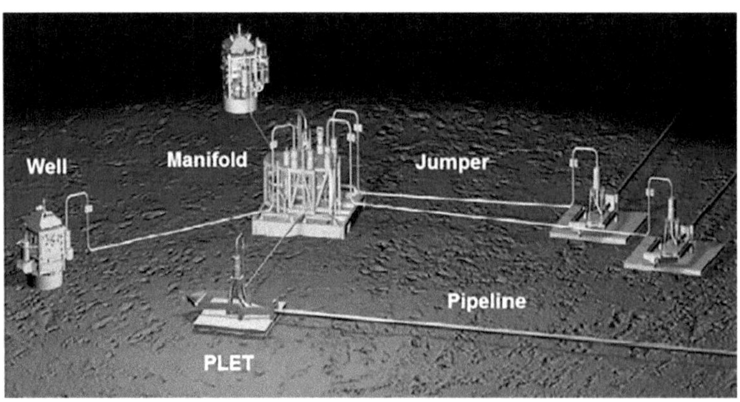

Fig. 1.20 Pipeline connection to manifold through a PLET tie-in.

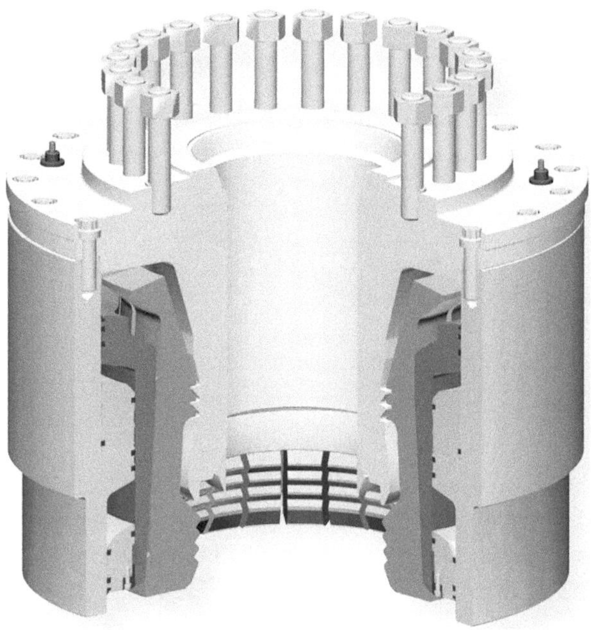

Fig. 1.21 Collet style connector. *(Courtesy: Cameron/Schlumberger.)*

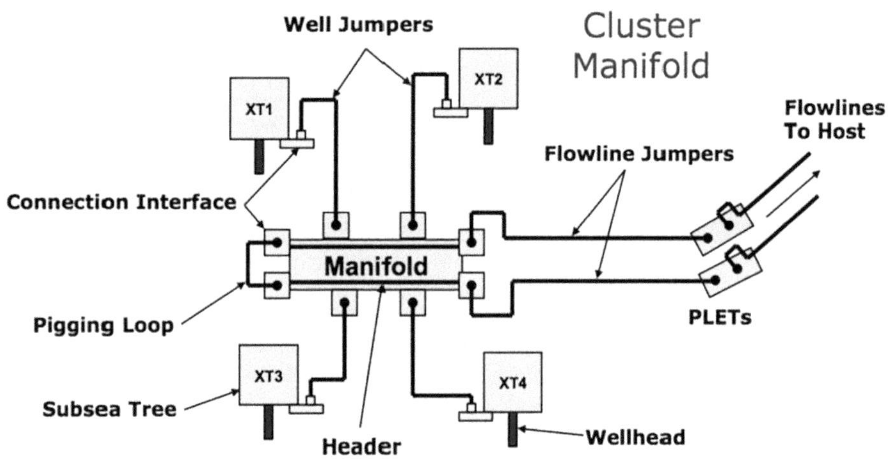

Fig. 1.22 A four-slot manifold and the connection to the PLETs. *(Courtesy: Subseapedia.)*

A pigging loop is normally required for export manifolds connected to flowlines and risers. Infield manifolds which are located upstream (before) from the export manifolds are connected to different wells. Let's clarify infield and export manifolds with one example.

Fig. 1.23 Pigging a pipe for cleaning purpose.

Example: Four pieces of 4-slot infield manifolds (refer to Fig. 1.18) are connected to 16 wellheads. In fact, each piece of infield manifold has four inlets and two outlets. Therefore eight flowlines should be connected to all four manifolds through eight PLETs and eight jumpers. The alternative and more economical solution is to use one eight-slot export manifold to connect all eight outlets of the infield manifold and reduce the number of flowlines to two instead of eight.

In conclusion, it is possible to summarize the functions of subsea manifolds as follows:
- Gathering the produced hydrocarbons from different wells;
- Distributing hydraulic fluid and chemicals through SDUs;
- Injecting water and gas into the wellhead to improve oil production and recovery;
- Providing substructural support for the connection of piping and spools;
- Providing an interface between production pipelines or flowlines and wells;
- Collecting the fluid from different flowlines and delivering it to an export pipeline;
- Facilitating the pigging of the subsea pipeline through its pig loop feature.

Fig. 1.24 illustrates a multipurpose manifold schematic with one production and two injection lines—one for gas and one for water injection to the reservoir for advanced oil recovery.

A jumper is a short pipe connector used to connect a manifold to a well, two manifolds to each other, or a manifold to a PLET. Simply speaking, a jumper is a short piece of pipe connector used to transport fluid between two subsea structures. Fig. 1.25 illustrates a subsea processing system with jumpers connecting the manifold to the wells and connecting a manifold to a PLET. Jumpers may be flexible or rigid. The distance and offset between two subsea structures affects jumper length and characteristics. The main route of the produced oil and gas in a subsea system could be from the tubing in the well to a Christmas tree, and then from the jumper to the manifolds, flowline, and risers.

The manifold concept does exist on the surface. Topside or surface manifolds are made from a connection of piping and fitting normally welded together. They are used

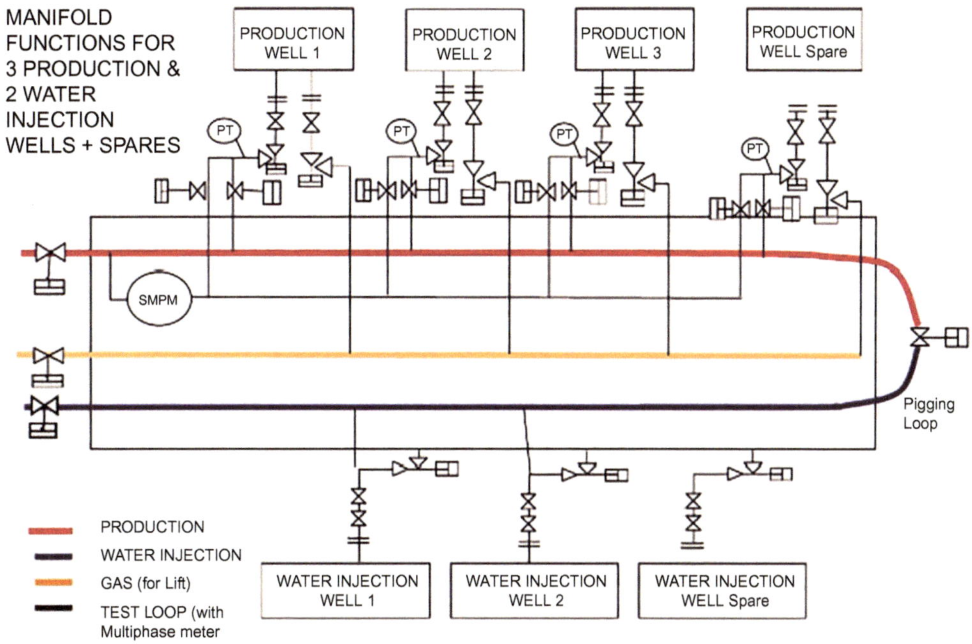

Fig. 1.24 A manifold for both production and injection purposes.

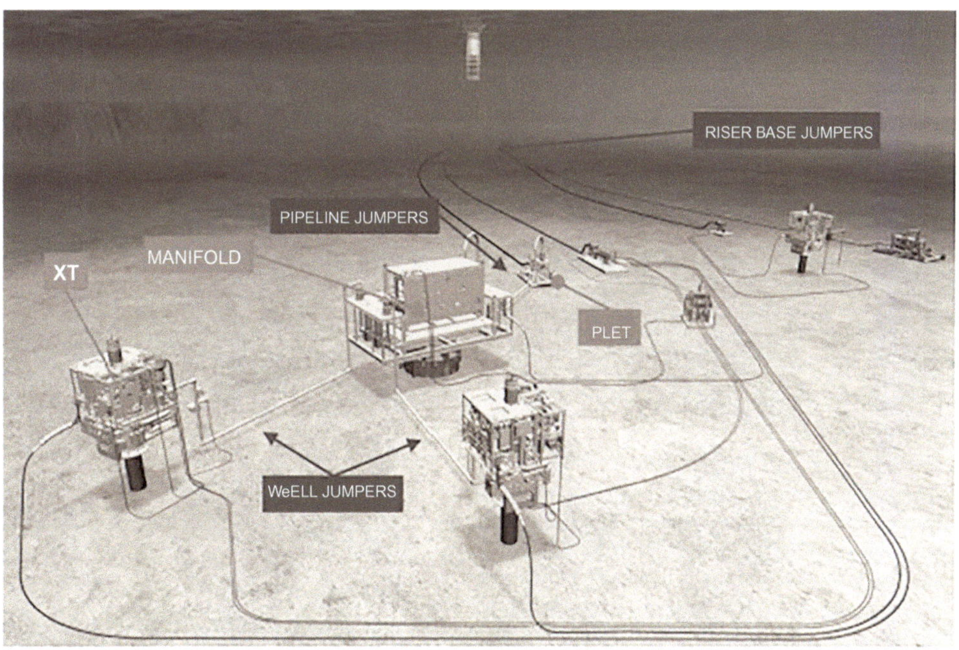

Fig. 1.25 Subsea processing system including jumpers.

to collect the produced hydrocarbon from different production wellheads or to inject water or gas from a single header to different wells located on different branches. Manifolds are designed to either merge multiple junctions into a single channel or divide one flowline into multiple outputs. Fig. 1.26 illustrates a surface manifold with a large header of 20″ pipe that collects the produced fluid from different wellheads in 12″ size. The header is fabricated by welding long length tees and the branch is blinded by blind hubs. Hubs and clamps are connections used in the offshore industry for large size piping above 2″ and high-pressure classes such as ASME pressure class 1500 and 2500, which are equal to a nominal pressure of 250 bar and 420 bar, respectively, to save space and weight compared to American Society of Mechanical Engineers (ASME) Standard flanges made according to ASME B16.5 and ASME B16.47. Alternatively, topside manifolds may be made from a piece of pipe for a header which is connected (welded) to different flange-olets used for branch connections. This type of manifold could be used when the branch size is relatively small such as 2″ or less. Fig. 1.27 illustrates a 6″ header pipe welded to 2″ branch flange-olets on a topside or surface manifold during the factory acceptance test (FAT). A flange-olet includes an olet plus a piece of pipe plus a flange. Olet fittings (also called branch connection fittings or outlet fittings) are fittings that provide an outlet from a larger pipe to a smaller one or one when there is large difference between the header and branch sizes. The main pipe onto which the branch connection is welded is usually called the run or header size. FAT is performed to make sure that there is no leakage from the manifold when it is pressurized at 1.5 times of the design pressure for 1 h. The headers and branches are blinded with flanges or hubs to isolate the manifold during the test.

Fig. 1.26 A surface manifold 20″ × 12″.

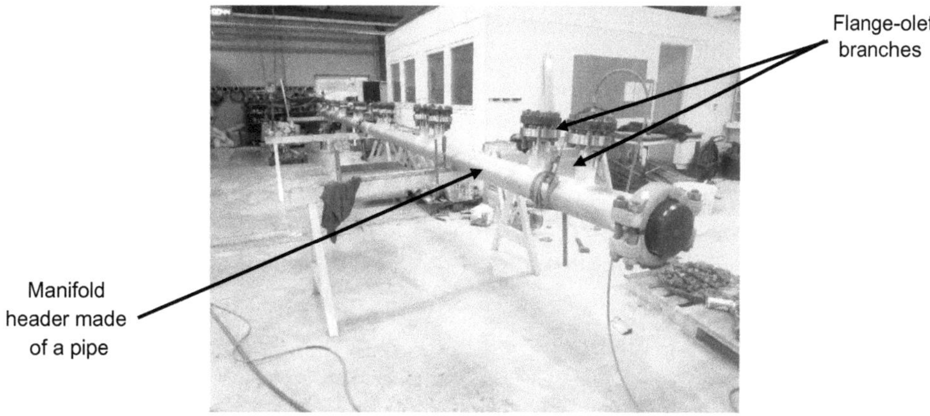

Fig. 1.27 A surface manifold 6″ × 2″.

1.2.5 Subsea structures and piping analysis

All subsea structures, foundation, and piping are subject to finite element analysis (FEA). Subsea structures and pipeline are modeled in software and all applicable loads are applied to the components to perform FEA. This analysis shows the effect of loads on the subsea components; structure and piping analysis is considered a key and essential part of both structure and piping design. Different parameters such as the geometry of the subsea components, material properties, applicable loads, and support conditions must be defined properly to facilitate accurate analysis. FEA is a computational tool for performing engineering analysis. This method is based on mesh generation technique to divide a complex problem into smaller elements. FEA simulates all of the different load scenarios that may apply for the pipeline or structure under analysis under different conditions.

1.2.6 Subsea distribution system (SDS)

Subsea distribution systems consist of a group of components and inline structures that provide communication and connection between the subsea control and topside. Through an SDS, hydraulic fluid, electricity, chemical injections, and communication are transferred between topside and subsea. Chemical injection and hydraulic fluid are distributed to the manifolds and wellheads through SDSs. Hydraulic fluid may be used for the operation of actuators installed on the valves as well as other subsea operations. An actuator is a mechanical device installed on a valve to control its operation (opening and closing). Chemicals could be wax inhibitor, scale inhibitor, annulus monitoring, Mono-Ethylene glycol (MEG), methanol, etc. Wax accumulation and inhibition in oil was explained earlier in this chapter. Scale inhibitors are a type of chemical injected into the well and production lines in order to prevent the formation of marine fouling in the form of bacteria and saturated salts that can be produced as a result of water injection. MEG is widely used in the oil and gas industry, especially in the subsea sector, to prevent hydrate formation in piping and other offshore facilities. It should be noted that the possibility of hydrate formation is high at subsea due to low-temperature application. Methanol has almost the same effect as glycol in oil and gas production where it is used to prevent hydrate formation and for deicing. Subsea equipment could be a separator, subsea pump, or compressor. Pumps are used to move the liquid forward and compressors are used to move the gas in the piping systems in oil and gas industry. An umbilical (Fig. 1.28) is an arrangement of tubing or piping as well as cables and hoses that transfer hydraulic and electric power to the subsea distribution unit or subsea umbilical termination assembly from the host facility such as a platform or ship. Electricity is required for subsea rotary equipment such as pumps and compressors as well as electrically operated actuators. Hydraulic fluid is used for the operation of subsea valves and other subsea components such as Christmas trees. Umbilical dimension can be up to 10″.

An SDS can be made of five different major components which are explained as follows:

- Umbilicals: As explained before, umbilicals contain hoses, tubes, and electrical cables as well as fiber optics to supply hydraulic fluid, electricity, and communication signals to the subsea components mainly on manifolds and Christmas trees. The main functions of a subsea production umbilical are as follows:
  1. To deliver hydraulic and electrical control signals to subsea control devices such as solenoid valves;
  2. To provide power for subsea valves and actuators;
  3. To transmit chemicals to the SDS unit or Christmas tree;

Fig. 1.28 Umbilical. *(Courtesy: Technip.)*

- A subsea umbilical termination assembly (SUTA) terminates the umbilical line and provides one or more connections for the transportation of the hydraulic, chemical, electrical, and fiber optic services illustrated in Fig. 1.29. This structure acts as a subsea connection or interface for the umbilical and a distribution center for the hydraulic fluid and chemicals. It should be noted that the umbilical is connected to the topside control system; the interface between the umbilical and the topside connection is known as a topside umbilical termination assembly (TUTA). A TUTA could be an enclosed unit with electrical power and communication cables, tubes, gauges, and valves for chemical and hydraulic supply.
- An umbilical termination head (illustrated in Fig. 1.30) is used for direct connection of an umbilical to a SUTA. This section could be made of different components such as a structural frame, plates, tubing, ROV connectors, etc.
- An SDU: Fig. 1.31 illustrates an SDU block diagram. An SDU is comprised of a hydraulic distribution unit such as the hydraulic power unit (HPU), pumps, strainers, and connected tubing or piping as well as electrical distribution modules and a communication distribution unit (CDU). The CDU should be a highly reliable communication distribution for fiber optic and other communication signals. Fig. 1.31 illustrates an SDU.
- Flying leads: connections between subsea structures such as manifolds, Christmas trees, UTUs, and SDUs. Flying leads are divided into hydraulic and electric with the abbreviation of HFL and ELF, respectively. As illustrated in Fig. 1.32, HFL and EFL are used in the transportation of hydraulic and electric power between the umbilical termination and SDU.

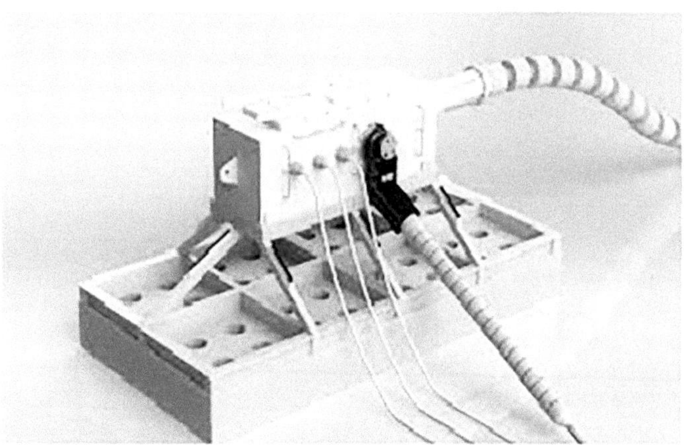

Fig. 1.29 Subsea umbilical termination unit (Offshore hydraulic innovation).

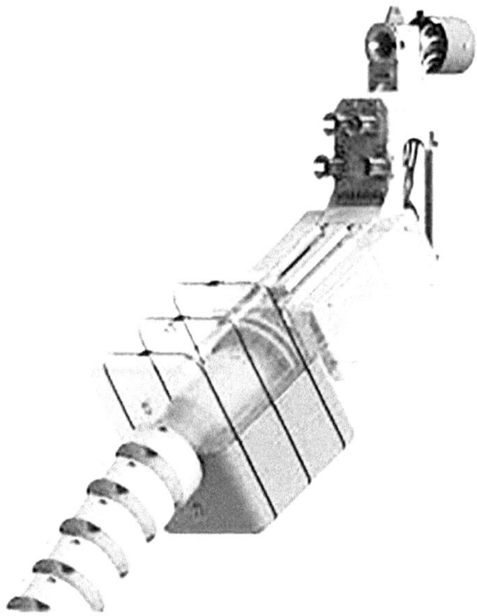

Fig. 1.30 Umbilical termination head (Offshore hydraulic innovation).

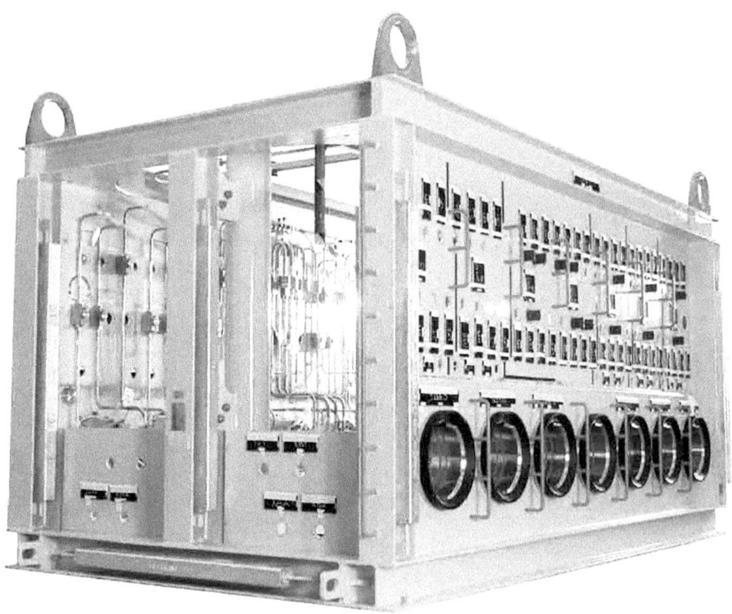

Fig. 1.31 Subsea distribution system. *(Courtesy: Schlumberger.)*

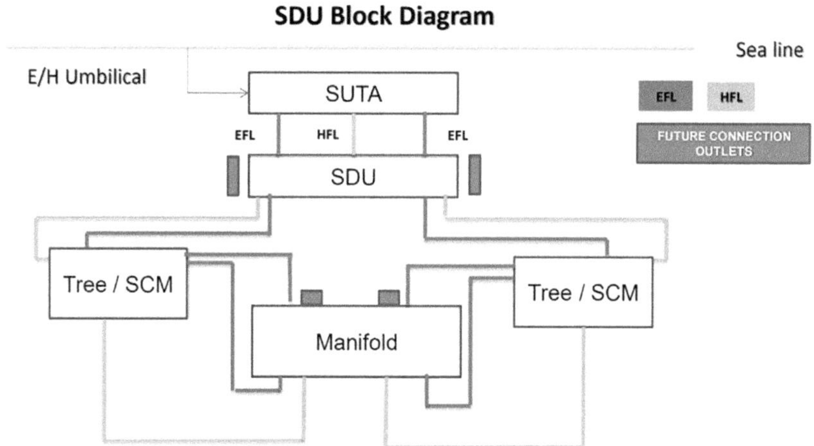

Fig. 1.32 SDU block diagram. *(Courtesy: Chess Subsea Engineering.)*

1.2.7 Subsea control system

The subsea control system is an essential part of the subsea development and processing system that can ensure the reliability of the subsea system during its design life. Subsea systems and equipment are mainly operated remotely as they are installed in very deep areas of the sea or ocean. Subsea control systems are covered by ISO Standard 13628-6, titled "Petroleum and natural gas industries—Design and operation of subsea production systems—part 6: Subsea production control system." Hydraulic and electric power are transmitted from topside facilities to subsea equipment for the safe operation of components such as valves and actuators, but they need to be controlled through a master control station (MCS) which is normally installed on a host facility such as a platform or FPSO that controls and receives data from the subsea equipment. In fact, an MCS provides an interface between the human operator and subsea equipment such as a subsea control module, sensors, etc. The operator can sit in the MSC and monitor and observe the status of the subsea equipment and components. An MSC consists of a computer and other electrical circuits and equipment.

In general, an MCS monitors and controls the following components and tasks:

1. Tree valves, which were explained earlier in this chapter;
2. Choke valves: Control valves are used to control (reduce) the pressure of the produced oil and gas. The choke valve kills the reservoir fluid pressure at the wellhead before it can be transported to other production units and facilities.
3. Sensors and transmitters: Different types of sensors are used for measuring and transmitting data about pressure, temperature, flow, level, moisture, and humidity to the MCS. For example, one important sensor and detector is a sand detector that can warn the operator about excessive sand production before severe damage occurs. Detecting sand in the produced fluid is important since it can prevent short- and long-term damage to components and equipment such as the choke, pipe, flowline, etc. In addition, increase

in sand production could be a warning of reservoir collapse. Sand erosion in subsea production systems can cause catastrophic failure. Sand accumulation at the bottom of the separators is known as a separator operational problem that should be predicted and managed properly. Separators will be explained later in this chapter. Sand can increase the wearing and erosion of the contact components and lead to flow assurance problems. Acoustic sand detectors are sensitive to noise in general, thus the accuracy of sand detectors can be affected by the noise created by a pump.

4. Warning and alarm announcement: a subsea control module (SCM) gives an alarm in case of any failure in the subsea system.
5. Emergency shut down (ESD): Some of the valves have an ESD function that should close down the process quickly to prevent an undesirable event in the case of failure. These valves are controlled and monitored through an MSC.
6. Well test management: A well test is performed to assess reservoir performance by evaluating parameters such as flow and pressure. A well test can provide information about the size, shape, and ability of the formation and reservoir to produce fluid. Formation is the name of the rock from which the oil, gas, and water are produced. Different electrical components and instruments such as a pressure gauge, wireline link, etc. are used to perform the test. All of these data will be transmitted to the MSC.
7. Hydrate control: Hydrate existence in subsea lines can be detected by sensors and the data transmitted to the MSC.
8. Data reporting and logging.

There are five different kinds of control system:
- Direct hydraulic;
- Piloted hydraulic;
- Sequenced hydraulic;
- Multiplex electrohydraulic;
- All-electric.

Direct hydraulic control system: the simplest remotely operated system for the control and monitoring of the subsea system. Each hydraulic actuator in this system is connected and monitored through a direct hydraulic control system as illustrated in Fig. 1.33. This system is made of the following major components:
- Hydraulic power unit (HPU): This is the heart and most essential part of each hydraulic system and consists of a hydraulic fluid storage tank or reservoir, a pump that moves the hydraulic fluid forward in the line or umbilical and a hydraulic accumulator. An accumulator is a pressure storage reservoir that is used to prevent hydraulic flow pulsation in the line by providing temporary oil supply to the line. An HPU contains instruments such as pressure regulators and pressure gauges, flow meters, and other devices. The hydraulic fluid should be cleaned according to the standard such as SAE-AS-4069.
- Control panel: The control panel, in general, manages, commands, directs, and regulates the behavior of the hydraulic system and the hydraulic actuator devices. The control panel can regulate and control the hydraulic flow to the actuator.

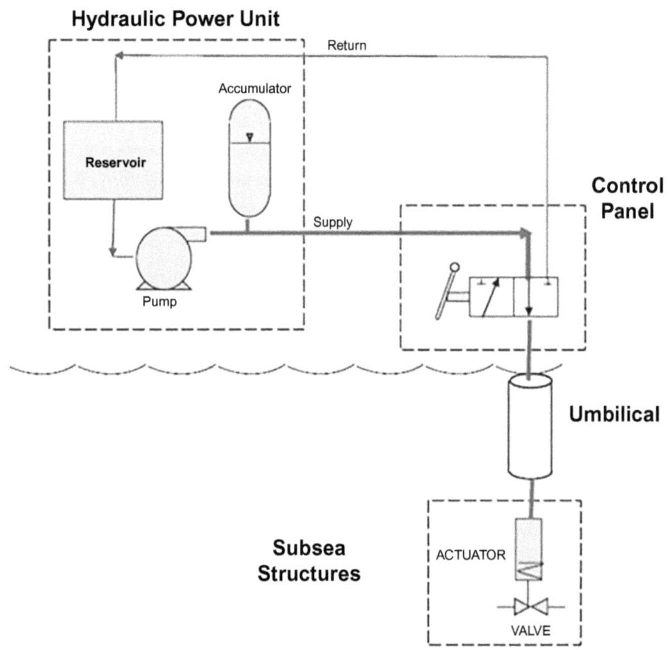

Fig. 1.33 Direct hydraulic control system. *(Courtesy: Elsevier.)*

- Umbilical: The hydraulic fluid is transmitted through umbilicals which were explained earlier in this chapter.

This type of control system can be reliable and relatively cheap, but it is very slow due to its large numbers of umbilicals.

Piloted hydraulic control system: The main characteristic of this system is to place the subsea control module (SCM) underwater as illustrated in Fig. 1.34. The subsea components work with hydraulic fluid and can be operated faster as compared to a direct hydraulic control system. However, the overall speed of operation is still slow. The other disadvantage of the system is that the number of umbilicals remains high. The reliability and accessibility of the system for maintenance is high since critical components such as the HPU are located topside.

Sequenced hydraulic control system: This system (see Fig. 1.35) consists of several sequentially operated valves and accumulators. This system has a better response time and operation speed compared to the two previous systems but it is still slow.

Multiplex electrohydraulic: The main addition to this system is an electrical power unit in the MCS that transmits electrical power to the subsea electronics module (SEM), which is a part of the subsea control module (SCM). The main advantage of this system is to reduce the number of umbilical and improve the response time and operation speed compared to the three previous systems. A detailed schematic of this system is provided in Fig. 1.36.

Two important modules of the multiplex electrohydraulic system located at the same level as the MCS in Fig. 1.36 are the electrical power unit (EPU) and the hydraulic power unit (HPU). The EPU provides electrical power for the SCMs. Power is supplied from

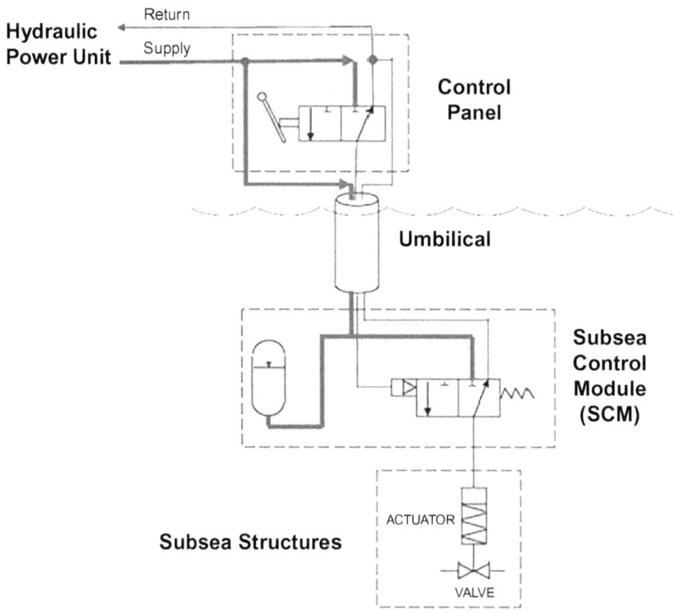

Fig. 1.34 Piloted hydraulic control system. *(Courtesy: Elsevier.)*

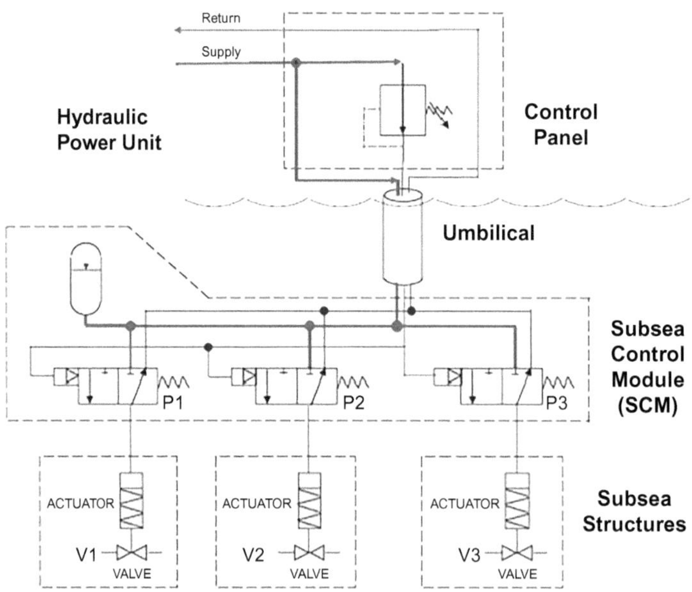

Fig. 1.35 Sequenced hydraulic control system. *(Courtesy: Elsevier.)*

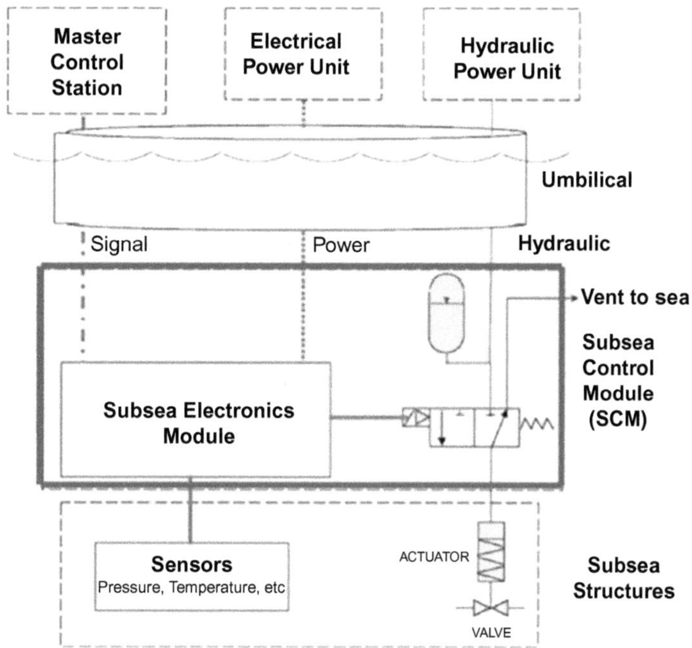

Fig. 1.36 Multiplex electrohydraulic. *(Courtesy: Elsevier.)*

the topside power supply unit. In addition, the EPU monitors the status of the dual power circuits in the umbilical. Filters and modems in the EPU provide a means of signal communication between the MSC and SCM. The HPU provides high- and low-pressure hydraulic control fluid to the subsea system as explained before. A hydraulic analysis should be done for the hydraulic system as follows:

- Field layout, such as the distance between wells and manifolds and the host facility;
- Number of hydraulic actuators installed on the valves;
- Type and size of hydraulic actuators (single acting or double acting);
- Pressure requirements such as high and low pressure;
- Actuator swept volume and operation time; swept volume is defined as the volume of oil which is displaced in an actuator through one length of stroke;
- Number of cycles (opening and closing) the actuators;
- Umbilical characteristics such as tube size and materials.

All electrical control system: An all electrical control system has no hydraulic component, so this solution is environmentally friendly and improves the response time of the system significantly. The next section explains about subsea control module as a main part of subsea control system.

An SCM (see Fig. 1.37) provides control, monitoring, intervention, and distribution of hydraulic and electrical power as well as communications. An SCM is an interface between the control system, HPU and EPU, as well as subsea equipment and components such as Christmas trees, valves and actuators, etc. An SCM can be installed on

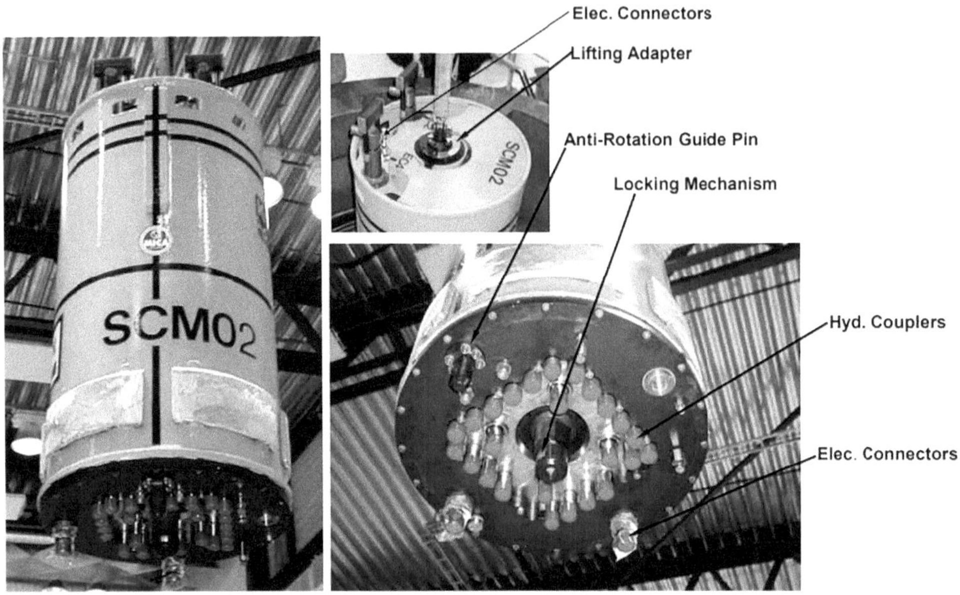

Fig. 1.37 Subsea control module. *(Courtesy: FMC Technologies.)*

the tree or manifold or other subsea structures and is replaceable with a running tool or ROV. The SCM is pressure compensated; it contains cylindrical housing, directional control valves, sensors, and subsea electronic modules. The base plate has all the hydraulic connections and required interfaces.

1.2.8 Subsea separators

Not all subsea development fields contain subsea separators. For those subsea fields with subsea separators, the separators are considered the heart of the processing system. A separator in oil and gas terminology is defined as a pressure vessel used for the separation of two- or three-phase fluids coming out of the wellhead. The concept of subsea separation is the same as topside and onshore separation. The subsea separator should be light and maintenance free or at least require only very minor maintenance during the design life of the plant. Gravity is the basic and traditional mechanism of separation in which oil, gas, and water are separated naturally due to their weight difference without adding any external assistance such as chemical injection or electrical shock to the fluid. For a three-phase separation, water is the heaviest phase so it is gathered at the bottom of the separator, gas is the most volatile and lightest so it will be gathered on the top, and oil is lighter than water and heavier than gas so it gathers in the middle. Fig. 1.38 illustrates a three-phase oil, gas, and water horizontal separator.

However, gravity separation is not always effective in separating the different phases of the produced fluid from each other. For example, water and oil emulsions cannot be

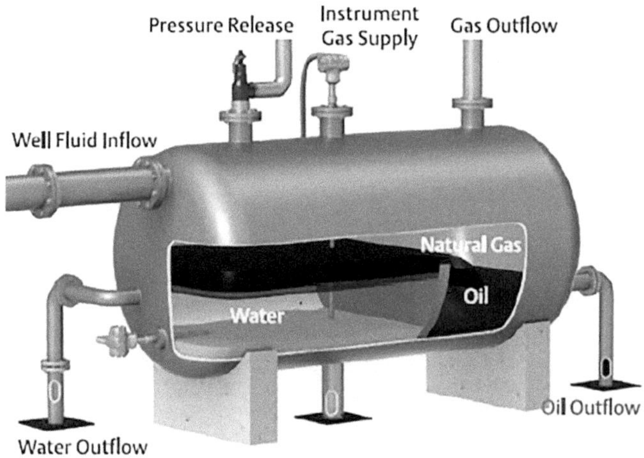

Typical Separator

Fig. 1.38 Gravity three-phase separator. *(Courtesy: Emerson Automation.)*

separated by gravity alone. An emulsion is defined as a mixture of two or more liquids that are unmixable. Different approaches such as using heat, electricity, or emulsifiers can be applied in order to dissolve and eliminate emulsions. An emulsifier can increase the kinetic energy between the water and oil molecules so that the water molecules join together with each other as do the oil molecules. After that, the separation of large water and oil molecules is easier by gravity force due to the greater weight difference between the water and oil molecules.

Separators in general are split into two categories: horizontal and vertical. Generally, horizontal separators are cheaper, more economical, and have a large area between the gas and liquid phase so they provide better separation. Foaming (see Fig. 1.39), which is one of the major separator operational problems, can happen more easily in vertical separators. On the other hand, horizontal separators provide more surface area for the gas phase and bubbles to escape. Foaming in separators occurs when small bubbles of gas become surrounded with a thin film of oil. Different reasons can cause foaming in separators such as pressure drop, impurities, high viscosity of oil, quick motion inside the separator, etc.

Sands that accumulate at the bottom of the separator are much easier to remove from horizontal separators. But vertical separators are more capable of handling motion and surge.

The advantages of subsea separation are summarized as follows:

- No requirement to transport sand and water to the topside facilities;
- Easier separation of crude oil subsea due to higher pressure subsea;
- Higher temperature results in lower viscosity of the oil and easier separation;
- Fewer flow assurance problems such as slugging;
- Probably less possibility of emulsion.

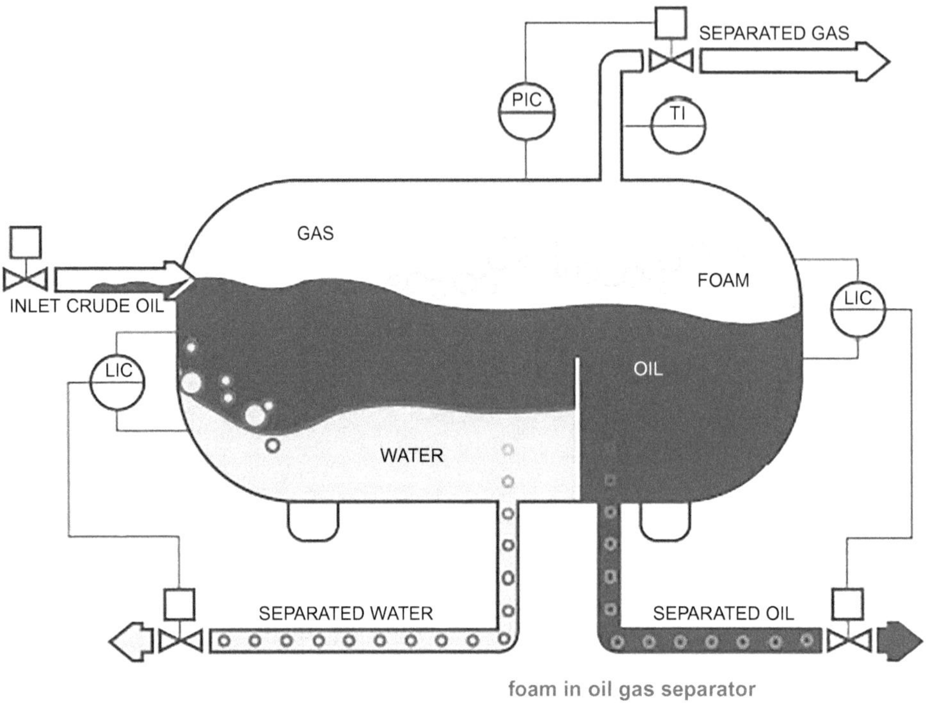

Fig. 1.39 Foam formation in a horizontal separator.

1.2.9 Subsea pumps

Subsea pumps (see Fig. 1.40) are divided into three different categories:

1. Electrical submersible pumps (ESPs) are utilized in the well for advanced oil recovery from the reservoir as an alternative method of water injection or gas lift. The reservoir pressure leads to the natural production of oil and gas from the well. However, the reservoir pressure drops after a while and the production rate from the well is reduced. Different methods for advanced oil recovery such as gas lift or water injection to the reservoir may be used to increase the oil recovery from the well. An ESP is typically installed inside the well. ESPs are suitable for multiphase flow as well as produced sand.
2. Subsea pumps can be used in order to pressurize the water to be injected into the well for advanced oil recovery.
3. Subsea pumps may be used after subsea separators to pump the produced oil to the host facility (see Fig. 1.5).

1.2.10 Subsea compressors

Subsea compressors are used either to inject gas into the reservoir for advanced oil recovery or in coordination with a subsea separator in order to pressurize and transport the produced gas from subsea to the host facilities. Fig. 1.41 illustrates a subsea compressor.

Fig. 1.40 Subsea pump.

Fig. 1.41 Subsea compressor.

1.3 Conclusion

Fig. 1.42 provides a very good overview of a subsea production system that includes the units and components summarized in this chapter.

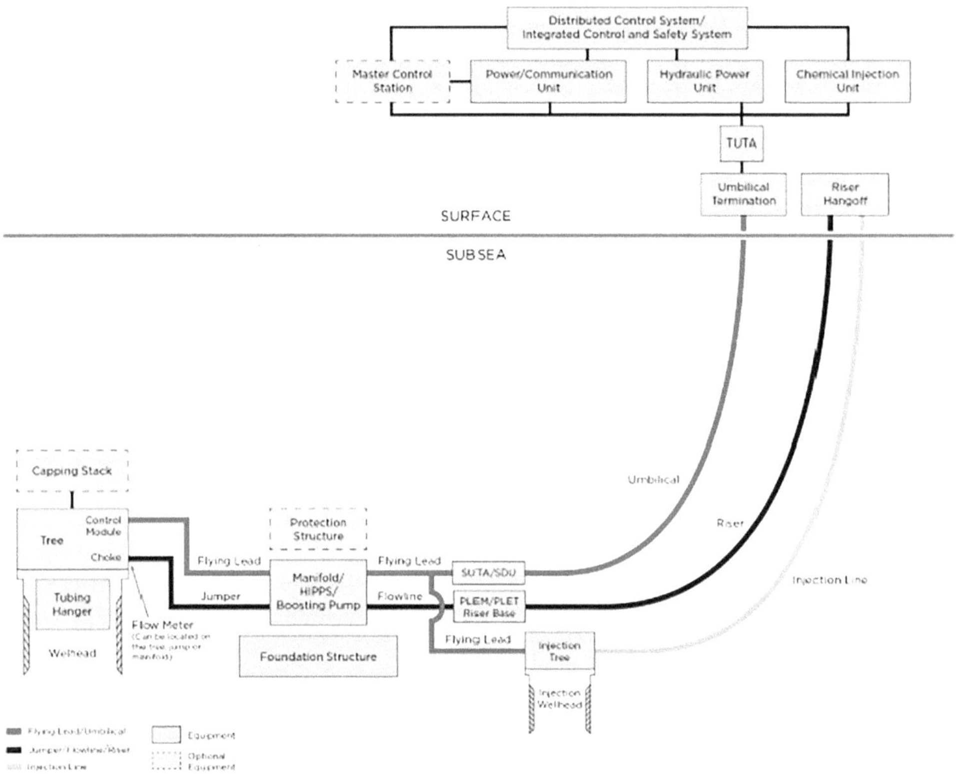

Fig. 1.42 Overview of a typical subsea production system. *(Courtesy: American Bureau of Shipping.)*

Further reading

FY. Bai, Q. Bai, Subsea Engineering Handbook, first ed., Elsevier, Atlanta, GA, 2012.

J.C. Berger, J.G. Cutler, Deepwater Reference Book, TotalFinaElf, 2000.

C.W. Burleson, Deep Challenge: The True Epic Story of Our Quest for Energy Beneath the Sea, Gulf Publishing Company, Houston, TX, 1999.

J. Praveen, M. Pathan, K. Ansari, Hyperbaric pressure testing of a subsea valve to validate deep water condition, Int. J. Mech. Prod. Eng. Res. Dev. 8 (2) (2018) 1011–1022.

S. Chakrabarti, Handbook of Offshore Engineering, first ed., Elsevier, 2005.

K. Sotoodeh, A review on subsea process and valve technology, J. Mar. Syst. Technol. 14 (2019) 210–219, https://doi.org/10.1007/s40868-019-00061-4.

K. Sotoodeh, Manifold technology in offshore industry, Am. J. Mar. Sci. 8 (1) (2020) 14–19, https://doi.org/10.12691/marine-8-1-3.

J.G. Speight, Subsea and Deep-Water Oil and Gas Science and Technology, first ed., Elsevier, 2015.

CHAPTER TWO

Subsea valve and actuator standards

Contents

Subsea Valves and Actuators for the Oil and Gas Industry
https://doi.org/10.1016/B978-0-323-90605-0.00001-3

2.1 An introduction to subsea valve and actuator standards

Subsea valves are designed and tested primarily based on the standards of the American Petroleum Institute (API) and the International Organization for Standardization (ISO). Each API standard used for designing and testing subsea valves and actuators has an equivalent ISO standard as follows:

1. API Spec.17D, "Design and operation of subsea production systems—Subsea wellhead and tree equipment," identical to ISO 13628-4.
2. API Spec. 6DSS, "Specification for subsea pipeline valves," identical to ISO 14723.
3. API Spec. 6A, "Wellhead and tree equipment," identical to ISO 10423.

The "SS" in API 6DSS means "standards for subsea." API 6DSS is very similar to the API 6D standard that is used for topside ball, through conduit gate, and axial valves in pipeline systems. Subsea actuators are also covered by the standards mentioned before with the exception of API 6DSS. API 6A and 17D are not limited to subsea valves and actuators and are applicable for other subsea components such as tees, tube hangers, casings, etc. Lately, however, the API and ISO standards in the list begun to diverge and no longer are identical.

The important point is that API 6A/ISO 10423 and API 17D/ISO 13628-4 are not limited to industrial valves; they cover the components and equipment listed in Table 2.1. API 6DSS/ISO 14723 only cover industrial valves.

In addition to the main subsea valve and actuator standards mentioned before, which are used for the design, manufacturing, testing, inspection, marking, and certification of valves and actuators, the following standards have to be considered for design, manufacturing, inspection, material consideration, and other aspects of the valves and actuators in the subsea sector of oil and gas industry by the engineers to improve the safety and reliability of the products and minimize the valve problems and failures.

ASME (American Society of Mechanical Engineers) BPVC (Boiler and Pressure Vessel Code) Sec VIII Div. 02 is titled "Rules for construction of pressure vessels" and covers rules for the design, manufacturing, testing, inspection, and certification of pressure vessels and boilers. A pressure vessel is defined as a closed container used to store hydrocarbon (oil and gas) at a pressure that is considerably higher than atmospheric pressure. Different types of pressure vessels are used for either storage or processing. Storage vessels could be storage tanks used for holding liquid or gas. Pressure vessels can perform different tasks like breaking down a product, combining products (towers), separating products (separators), or changing the temperature of a product (heat exchangers or boilers). Piping and valves are also considered pressure vessels. Pressure vessels can be either vertically or horizontally designed and installed. The main concept of ASME BPVC Sec VIII Div. 02 is "design by analysis," i.e., using stress analysis, finite element analysis (FEA), or finite

Table 2.1 Equipment and components covered by API 6A.

Wellhead equipment	Connectors and fittings	Casing and tubing hangers	Valves and chokes	Loose connectors	Other equipment
Casing Head Housing	Crossover Connectors	Mandrel Hangers	Single Completion Valves	Weld Neck Connectors	Actuators
Casing Head Spools	Tubing Head Adapters	Slip Hangers	Multiple Completion Valves	Blind Connectors	Ring Gaskets
Tubing Head Spools	Top Connectors		Actuated Valves	Threaded Connectors	
Crossover Spools	Tees and Crosses		Valves Prepared for Actuators	Adapter and Spacer Connectors	
Multistage Head Housing and Spools	Fluid Sampling Devices		Check Valves		
	Adapter and Spacer Spools		Chokes		
	Other End Connectors (Proprietary)		SSV/USV and Actuators for Offshore Services		Bull Plugs Running Tools Valve Removal Plugs Wear Bushings

Courtesy: Cameron.

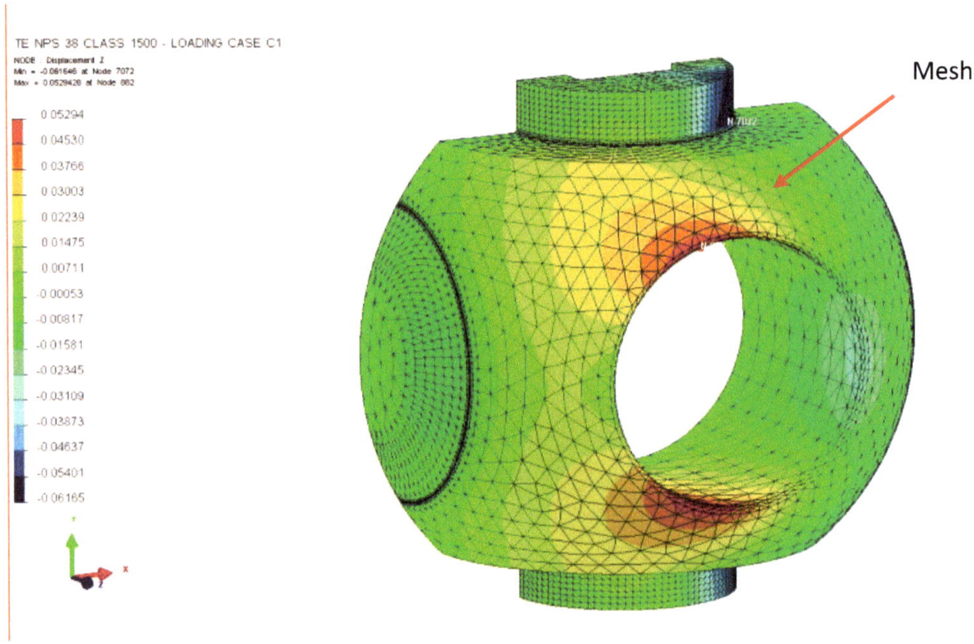

Fig. 2.1 FEA on half of the ball of a large ball valve.

element method (FEM) to demonstrate the accuracy and/or corrosiveness of the valve design. Fig. 2.1 illustrates the result of FEA implementation on half of the ball of a large ball valve. FEA simulates a physical phenomenon such as applied stress or flow of fluid through numerical mathematical technique. It should be borne in mind that FEA is not limited to load analysis and can be applied for dynamic flow analysis. But how does FEA work in practice? A 3D model of a component, such as an industrial valve, is transferred to or created in software such as Ansys. Applied loads and pressures such as fluid pressure, pipeline loads, etc. in different cases such as operation, test, and maintenance are created in the software. A large structure selected for FEA is divided into very small elements called mesh (illustrated in Fig. 2.1). After running the load simulation analysis, different colors appear for each mesh. Blue indicates the lowest level of stress concentration on the component. On the other hand, red mesh color indicates a high and critical level of stress concentration. The allowable stress values are given in ASME Sec. VIII Div. 02 and ASME Sec. II, "Materials." Allowable stress is defined as the maximum stress (axial, tensile, bending, compression) that a material can tolerate.

ASME B31.8, "Gas transmission and distribution piping system," covers gas pipelines, gas compressor stations, gas metering and measurement stations, etc. This design code describes a comprehensive solution for material, design, fabrication, assembly and erection, testing, and inspection of gas pipeline systems. This standard, like ASME

B16.25, is used for subsea pipeline valves since it provides some butt weld end arrangements (discussed in more detail later), as well as some solutions for welding together two different piping components with different thicknesses.

ASME B16.25 covers the butt weld ends preparation of piping components that are connected to the piping system through welding. Butt weld end connection is very common for subsea valves and meeting (connected) piping and is the most commonly used welding connection in the oil and gas industry. Circumferential welding and applying 100% radiography on the connected joints increases the welding joint efficiency to 100% and reduces the risk of leakage from the connection to zero. On the other hand, using flange connections increases the risk of leakage compared to using welding connection with a 100% radiography test. The valve body in butt weld end connections is welded to two forged manufactured pup pieces from each side by the valve supplier or a subcontractor. Welding the valve to the pup pieces is done prior to the shipment of the valve to the construction yard, where it is then welded to the pipeline. The forged pipe, or pup piece, is used to keep the fabrication weld heat input away from the valve internals, especially sealings. The pup piece and connected piping ending should be prepared according to ASME B16.25 standard (see Fig. 2.2). A root face of 1.6 mm on average is prepared close to the root of the welding, which is applied to prevent collapse of the welding. Two butt weld end connection arrangements are illustrated in Fig. 2.2. The one on the top is applicable to connections with a maximum thickness of 22 mm where the end of the piping connections is chamfered at 37.5 degrees on average. The butt weld

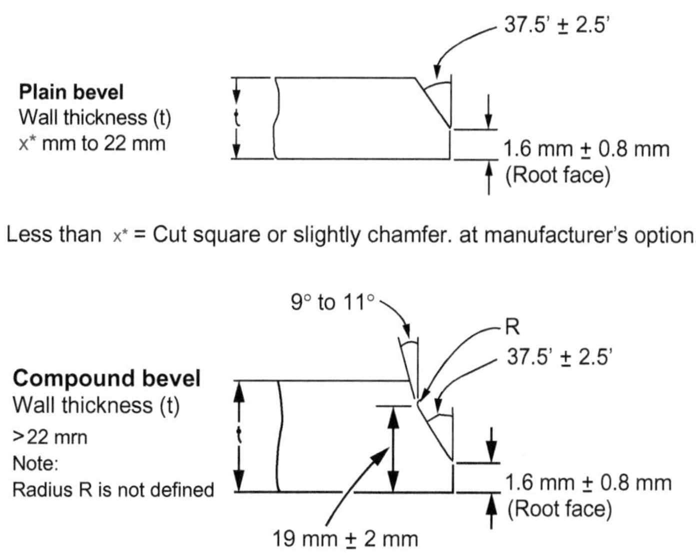

Fig. 2.2 Bevel end preparation according to ASME B16.25 standard.

end preparation for piping connections more than 22 mm thick requires two chamfered, the first one is 37.5 degrees till 19 mm thickness and then a chamfer at 10 degrees on average. Alternatively, narrow gap welding could be applied on butt weld end connections that include a chamfer at a smaller degree of 5–7; these can be welded more quickly and require less electrode (welding consumable). Although some concerns have been raised regarding the lack of fusion in narrow gap welding, this type of butt weld preparation and connection has been used successfully in the oil and gas industry.

API 6ACRA is titled "Age-hardened nickel-based alloys for oil and gas drilling and production equipment." This standard is known as a supplement to API 6A; it provides additional requirements for testing and materials (ASTM) for age-hardened hard nickel alloys such as Inconel 718 and 725 in regard to mechanical strength, chemical composition, heat treatment, testing, manufacturing route, etc. Age hardening, also called precipitation hardening, is a type of heat treatment used to increase the mechanical strength of the material.

API 6AV1 is titled "Specification for verification test of wellhead surface safety valves and underwater safety valves for offshore service." This standard is applicable for safety valves, including surface safety valves (SSV), underwater safety valves (USV), and boarding shutdown valves (BSDV) as per API 6A requirements. However, the end user may ask to apply this test for all the valves in the production line exposed to sand services in the project. Safety valves are discussed in more detail in Chapter 7. API 6A indicates a different qualification designation and performance requirement such as PR2F. This performance requirement defines the number of cycles for the qualification of the subsea valves, which is explained later in this chapter. The PR2F qualification level adds additional sand slurry testing as per the API 6AV1 requirement. Table 2.2 presents information related to performance requirements as well as sand slurry testing according to table 88 of API 6A, 21st edition. It should be noted that just 200 pressure-temperature cycles could be applied on subsea valves for design validation instead of the 200 operating and three thermal cycles provided in Table 2.2. In summary, the safety valve should not have any leakage during the seat or leakage test after 500 cycles of the valve in sand slurry fluid

Table 2.2 Summary of design validation requirements for safety valves based on API 6A.

	Performance requirements		Testing requirements	
Safety valve service class	**Operating cycles**	**Thermal cycles**	**Design validation testing**	**6AV1 sand slurry testing**
I	200	3	PR2F	Not required[a]
II	200	3	PR2F	API 6AV1 Class II
III	200	3	PR2F	API 6AV1 Class III

[a]Manufacturer may perform an API 6AV1 Class II test replacing the sand slurry with plain water as the test fluid.

with a differential pressure of 400 psi for 50 h. The details of the sand slurry test based on API6AV1 are explained in more detail in Chapter 7.

API 14A is a specification that addresses the requirements of subsurface safety valves (SSSV). These safety devices are fail-safe close flow controlling safety devices. The requirements for the installation and operation of SSSVs are covered in API 14B. Four classes of services were defined in older revisions of this standard. The service classes are defined as follows:

- Class 1 is a standard service that does not cover sandy services, services with a high amount of chloride or hydrogen sulfide that cause stress cracking corrosion or other services that can make a high amount of corrosion. Class 2 addresses sandy services; a SSSV is intended to be used to prevent valve failure due to sand. Class 3 addresses services that cause stress cracking corrosion due to hydrogen sulfide, mainly known as class 3S, and any other services such as chloride containing, known as 3C. Class 3 SSSVs should meet the requirements of Class 1 and 2. In addition, class 3 components should meet the requirements of NACE MR0175/ISO 15156, which provides guidelines for material selection in H_2S-containing environments (sour service) to prevent the cracking of materials and other failures associated with sour service such as blister, pit, etc. Failure of materials to be used in sour service is not only limited to loss of asset and production. A small concentration of hydrogen sulfide can kill humans. Hydrogen-induced cracking (HIC) is a type of sour corrosion that can easily cause carbon steel failure; this type of corrosion is addressed in TM0284. Class 4 addresses mass loss corrosion services. This class of SSSVs is intended to be used for the wells where the produced fluid can cause massive corrosion. Class 4 components should meet the requirements of both class 1 and 2. The material selection for class 4 depends on the well fluid service as well as the process conditions such as pressure, temperature, etc. There is no national or international standard that covers the materials in class 4. The material classes in API 14A have been replaced by validation grades in newer revisions of the standard, which are explained in detail in Chapter 7.

API 17H/ISO 13628-8 is a recommended practice (RP) that provides functional requirements and guidelines for remotely operated vehicle (ROV) interfaces in subsea production fields for the petroleum and natural gas industries. An ROV is a free-swimming submersible craft (see Fig. 2.3) used for the operation of subsea valves and actuators. These vehicles can be defined as state-of-the-art subsea (underwater) robots. There is a bucket or receptacle on the top of manual valves and actuators that can be operated with a torque tool carried by an ROV arm (see Fig. 2.4) for cycling (opening and closing) the valves and actuators. Interface data for the buckets, such as their dimensions, maximum input torque capacity, etc., are given in these two identical standards. More information about ROVs and ROV buckets is available in Chapter 8.

API 17N is titled "Recommended practice on subsea production system reliability, technical risk and integrity management." This RP aims to provide operators,

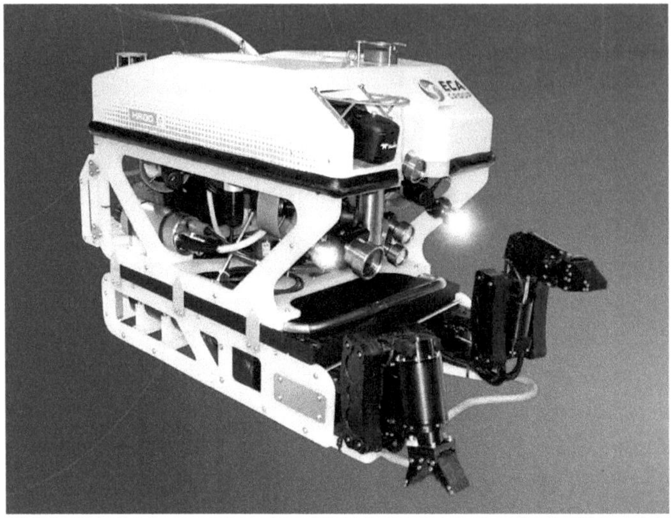

Fig. 2.3 Remotely operated vehicle (ROV).

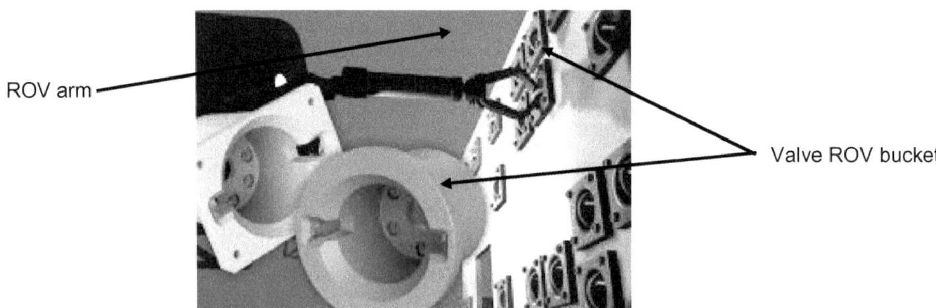

Fig. 2.4 Valve operation by a ROV.

contractors, and suppliers with guidance on the management and application of reliability management engineering techniques in subsea projects and operations within their scope of work and supply. Qualification of the new technology is very essential at an early stage such as design. The extent to which components such as valves and actuators are qualified for a specific application such as size, pressure rating, water depth, etc. should be formally evaluated to define the readiness level of the technology for the field through a term called "technology readiness level" (TRL). Eight levels of TRLs are defined in API 17N, from TRL0 to TRL7; they are defined as follows:

TRL0: Basic research: basic research and development (R&D) paper concept;

TRL1: Concept selection: proof of concept as a paper study or R&D experiments;

TRL2: Concept demonstration: experimental proof of concept using physical model tests;

TRL3: Prototype development: system function, performance and reliability tested;

TRL4: Product validation: preproduction system validated and environment tested;

TRL5: System integration testing: production system interface tested;

TRL6: System installed: production system installed and tested;

TRL7: System operation: production system field proven.

TRL4 requires the qualification of a product through 600 cycles, including 200 PR2 (pressure-temperature cycles), 200 hyperbaric tests, and 200 endurance cycle tests. Generally, TRL4 is the minimum acceptable qualification level for valves and actuators used in subsea projects. TRL7 means that the product has been used in the field during a past project without any failure record.

The other important term is TRC, which stands for technical risk criticality or technical risk categorization. TRC is common during all stages of a project such as feasibility study, front-end engineering, detail design, manufacturing, and installation. TRC is a formal assessment to be performed by quality control as well as technical departments through the following steps:

- Identifying technical uncertainties that can affect technical performance by considering five main risk categories: reliability, technology, architecture or configuration, environment, and organization.
- Providing a qualitative score for each risk item considering past experiences. These qualitative scores are split into four categories: A (very high), B (high), C (medium), or D (low).
- Adjust the level of TRC focus from the overall level to a detailed level.

API 17 Q is titled "Recommended practice on subsea equipment qualification." This RP provides suppliers, contractors, and operators with process level guidance to qualify equipment and components that are designed and selected for use in subsea applications. The document provides a high-level guideline only.

API 20 E is titled "Alloy and carbon steel bolting for use in petroleum and natural gas industries." This standard specifies requirements for the qualification, production, and documentation of carbon and low alloy steel bolting. Subsea bolting components (bolts and nuts) are subject to cathodic protection to prevent external corrosion due to contact with seawater. Therefore it is not common to use exotic bolting materials such as super duplex or Inconel 625 for subsea valves. This standard defines three different bolting specification levels (BSL): BSL1, BSL2, and BSL3. Each BSL defines the level of technical quality and the qualification level of the bolting. Higher BSL (BSL3) reflects higher quality bolting. In general, BSL1 bolting should satisfy the requirements given in the related ASTM standard such as ASTM A193, A194, A320. BSL2 requires additional testing compared to BSL1 and has more restricted requirements. BSL3 is the most restricted level with additional volumetric testing that is not required for BSL1 and BSL2. BSL1 is not proposed for subsea valve application.

API 20 F is titled "Corrosion resistant bolting for use in the petroleum and natural gas industries." This standard or specification specifies requirements for the qualification, production, and documentation of corrosion-resistant bolting used in the oil and gas industry. This standard is applicable for equipment and components that are covered by API standards. API 20 F introduces just two BSLs: BSL2 and BSL3. In fact, BSL1 is not applicable for corrosion-resistant bolting such as ASTM A453 Gr. 660 and API 6A718. ASTM A453 Gr. 660 is a nickel alloy bolting that contains 25% nickel and 15% chromium. API 6A718 covers Inconel 718 bolting, a hard nickel and chromium alloy bolting with high mechanical strength. Using this bolting material compared to Inconel 625 bolting, which has less mechanical strength, could lead to using fewer bolts for the body/bonnet or body pieces of a valve connection which could save some area on the body/bonnet of the valves.

DNV-RP-B401, "Cathodic protection design," is a RP to provide cathodic protection design and implementation by using aluminum or zinc for subsea components, including valves. Cathodic protection is a technique used to prevent the external corrosion of subsea components due to contact with seawater. Corrosion occurs when a metal loses an electron. Therefore giving electrons to a metal prevents corrosion in general. The main concept of cathodic protection is to connect a subsea component such as a valve to a less Noble material such as aluminum or magnesium that can be corroded easily. As a result, the less Noble material acts as an anode that is corroded and releases an electron to the cathode (valve or other subsea component) to mitigate corrosion.

DNV-RP-F112 is titled "Design of duplex stainless steel for subsea equipment exposed to cathodic protection." Historically, a small number of duplex and super duplex stainless steel subsea components such as piping and valves have failed due to hydrogen-induced stress cracking (HISC) corrosion. HISC corrosion has impacted the oil and gas industry, particularly the subsea sector, for the past 10–15 years. Much effort has been made recently to identify the failure mechanisms associated with HISC. As a result, Det Norske Veritas (DNV) has developed a guideline for the design of duplex components based on the DNV-RP-F112 standard to avoid HISC. HISC failure has mainly been attributed to a combination of stress coming from loads (e.g., tension in a pipeline) and an ingress of hydrogen formed at the steel surface due to cathodic protection, as well as the previously mentioned susceptible materials. Fig. 2.5 illustrates a combination of these three essential elements for HISC occurrence. DNV-RP-F112 provides limitations for stress and strain values for duplex and super duplex materials to prevent HISC. The gap in the DNV standard is that it considers only duplex and super duplex as materials susceptible to HISC. However, other research as well as industrial experience has proven that precipitation nickel alloys such as Inconel 718 and 725, which are widely used for subsea components such as valves, are at risk of HISC.

IEC 61508 is titled "Functional safety of electrical/electronic/programmable electronic safety-related systems." IEC stands for "International Electrotechnical

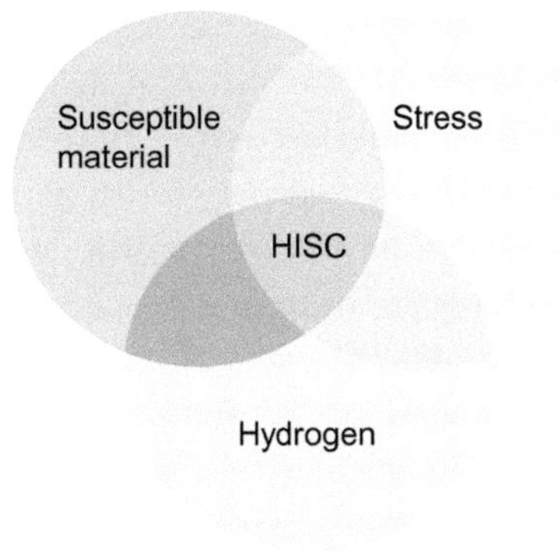

Fig. 2.5 Critical factors in HISC occurrence.

Commission." This standard addresses electrical, electronic, and programmable electronic devices and systems with safety functions such as actuated valves with emergency shutdown (ESD) functions and high integrity pressure protection systems (HIPPS). An ESD valve is an actuated valve that should be closed very quickly in the case of a hazardous event to protect the system, human lives, and the environment. HIPPS is a type of overpressure protection system in piping with high safety functionality requirements. IEC 61508 provides requirements to ensure that systems are designed and operated to maintain the required safety integrity level (SIL). Functional safety is a part of the overall safety of the component or equipment; it depends on the system or equipment to operate correctly in response to the input signals. SIL identifies the likelihood that a safety function will be performed satisfactorily. There are four levels of SIL defined in the standard: SIL1, SIL 2, SIL 3, and SIL 4. SIL1 has the lowest level of safety requirement and SIL4 has the highest level of safety with the least possibility of failure. SIL3 is a very common safety requirement for ESD and HIPPS actuated valves in subsea systems. Table 2.3, extracted from IEC 61508, provides the probabilities of failure for different SIL levels based on mode of operation.

Table 2.3 Probabilities of failure based on SIL.

Safety integrity level	Demand mode of operation (average probability of failure to perform its design function on demand—PFD)	Continuous/high demand mode of operation (probability of a dangerous failure per hour)
4	$\geq 10^{-5}$ to $<10^{-4}$	$\geq 10^{-9}$ to $<10^{-8}$
3	$\geq 10^{-4}$ to $<10^{-3}$	$\geq 10^{-8}$ to $<10^{-7}$
2	$\geq 10^{-3}$ to $<10^{-2}$	$\geq 10^{-7}$ to $<10^{-6}$
1	$\geq 10^{-2}$ to $<10^{-1}$	$\geq 10^{-6}$ to $<10^{-5}$

IEC 61508 requirements are not applicable for manual valves. Different dangerous failures in safety systems can be related to the incorrect specification of the system, random or systematic hardware failures, software errors, human errors, etc.

IEC 61511 is titled "Functional safety—Safety instrumented systems for the process industry sector." This standard provides practices in the engineering of the system to ensure the safety of an industrial process through the usage of safety instrumentation systems (SIS).

ISO 15156, "Petroleum and natural gas industries—Materials for use in H_2S-containing environments in oil and gas production," was explained earlier in this chapter. In general, materials used in wet H_2S (sour service) are subject to different types of cracking. Wet H_2S means that the presence of water is necessary for creating different sour corrosion types such as hydrogen-induced cracking (HIC), sulfide stress cracking (SSC), or stress-oriented hydrogen-induced cracking (SOHIC). These different corrosion mechanisms are explained in Chapter 12. ISO 15156 provides a guideline on the limitations of different material groups such as carbon steel, stainless steel, etc. in sour services and proposes hardness control and suitable heat treatment to mitigate corrosion.

ISO 5208, "Pressure testing of metallic valves," specifies the required pressure tests for valves, such as shell and seat tests, as well as allowable leakage rates in order to prove the structural integrity of the valves and the adequacy of the sealing mechanism. Different leakage rates such as A, B, C, D, etc. are defined in this standard. Leak rate A means zero leakage, which is normally applicable to shell test, as well as leak test from a soft (non-metallic) seat of a valve.

SAE-AS-4059 is titled "Aerospace fluid power—cleanliness classification for hydraulic fluids." Hydraulic fluid, which is used for hydraulic actuators and other applications, should be clean without particle contamination. Different cleanliness classes for hydraulic fluid are defined in this standard.

2.2 Reviewing API 6A and 17D and comparing them to API 6DSS

2.2.1 Size and dimension

Compact valve design and selection is important in the subsea sector of the oil and gas industry. The dimensions of manifold structures are largely dependent on valve dimensions. API 6DSS has a lot of similarities to the API 6D standard used for valves on platforms or ships and in refineries and petrochemical plants. Typical size ranges in the API 6DSS standard are 1, 1 ½, 2, 3, 4, 6, 8, 10, 12, 14, 16, 18, 20, 22, 24, 26, … 60 in. All sizes except for ½″, ¾″, and 1 ½″ are counting numbers. The valve size can also be called dimension nominal (DN). DN can be calculated by multiplying the size in inches (NPS) by 25. As an example, a 4″ valve size is equal to DN 100. The valve size in API 6DSS is normally equal to the connected piping size. Nominal sizes for the API 6A standard are normally 1 13/16″, 2 1/16″, 2 9/16″, 3 1/16″, 4 1/16″, 5 1/8″, 7

1/16″, 9″, 11″, 13 5/8″ which are not natural. The defined sizes in API 6A indicate the valve bore. As an example, an API 6A standard 7 1/16″ slab gate valve has a bore equal to 7.0625″ or 179.4 mm.

2.2.2 Bore dimension

Bore refers to the internal diameter of the valve. The valve could be full or reduced bore. A full-bore valve, according to API 6A, is defined as an oversized valve with a hole in the closure member (ball in a ball valve or gate in a through conduit gate valve) that is the same size as or larger than the connected piping, resulting in very low pressure loss (see Fig. 2.6). A reduced bore valve has a closure member with a smaller bore or flow passage compared to the connected piping. Fig. 2.7 illustrates a reduced bore ball valve; the reduction in bore can be observed from the pup piece connected to the valve. Subsea valves in many projects are welded to the piping with pup pieces rather than having flanged ends. Welding connection instead of flange connection has the advantage of zero leakage possibility of saving two flange connections to the valve.

Two examples of gate valves bore pattern evaluation have been reviewed as follows:

Example 2.1 A 7 1/16″ slab gate valve in pressure class 517 bar is connected to piping with a nominal pipe size (NPS) of 8″ in schedule 120 equal to 18.3 mm. The outside diameter of the pipe, according to ASME B36.10/36.19 standards, is 219.1 mm. The internal diameter of the pipe is shown in Figs. 2.8 and 2.9 and calculated through Formula (2.1):

Fig. 2.6 Full-bore subsea ball valve.

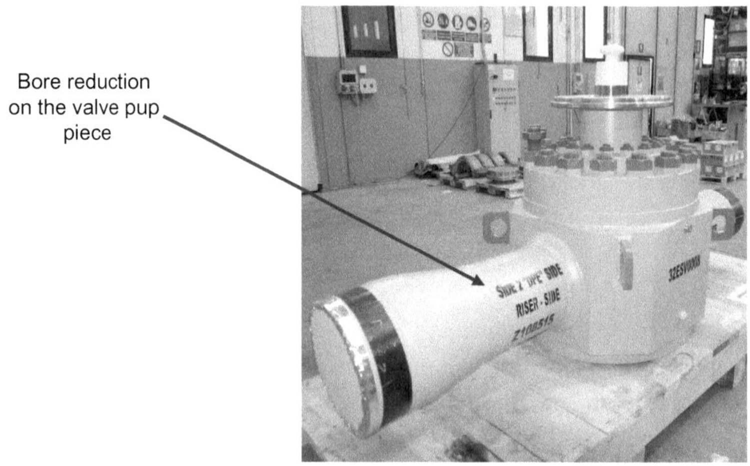

Fig. 2.7 Reduced bore subsea ball valve. *(Courtesy: Advanced Technology Valve.)*

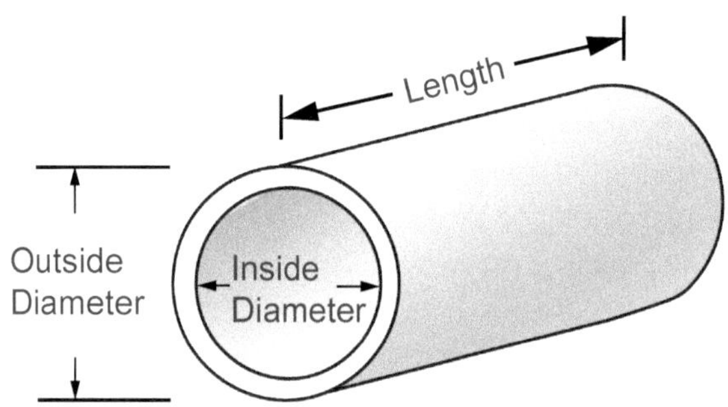

Fig. 2.8 Relationship between outside (OD) and inside diameter (ID) of the pipe.

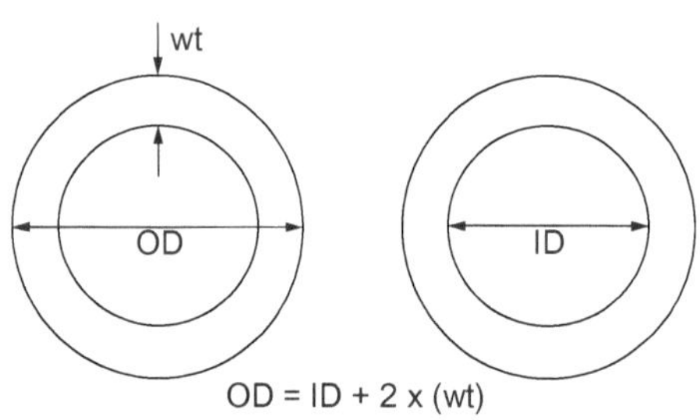

Fig. 2.9 Relationship between outside (OD) and inside diameter (ID) of the pipe.

Relationship between outside and inside diameter of a pipe

$$\mathrm{ID} = \mathrm{OD} - 2t\,(\text{thickness}) = 219.1 - 2{*}18.3 = 182.5\,\mathrm{mm} \tag{2.1}$$

The bore of the 7 1/16″ through conduit slab gate valve is equal to the size of the valve which is 7.0625 in. equal to 179.4 mm, which is smaller than the connected piping.

Example 2.2 A 2 1/16″ slab gate valve in pressure class 517 bar is connected to piping with a NPS of 2″ in schedule 80 equal to 5.5 mm. The outside diameter of the 2″ pipe according to the ASME B36.10/36.19 standards is 60.3 mm. The internal diameter of the pipe is calculated as follows:

$$\mathrm{ID} = \mathrm{OD} - 2t = 60.3 - 2*5.5 = 49.3\,\mathrm{mm}$$

The bore of the 2 1/16″ through conduit slab gate valve is equal to the size of the valve which is 2.0625 in. equal to 52.4 mm which is larger than the connected piping (full-bore valve). However, if the size of the valve connected to the 2″ pipe were changed to 1 13/16″, the bore of the valve would be equal to 1.8125 in. equal to 46.03 mm which is smaller than the connected pipe (reduced bore valve).

Although full-bore through conduit gate (TCG) valves maybe required by end users, the gate valve is not necessarily full bore. TCG valves are normally used in dirty (particle containing) services so using full bore may be preferred for higher flow capacity and ease of particle and sand movement through the bore of the valve and reducing the erosion rate and flow turbulence, which is achieved by selecting and using a larger port. A more comprehensive example for defining the bore type of gate valves is given in Table 2.4.

API 6DSS provides the minimum bore dimension for a full-bore valve based on the pressure class according to Table 2.5 extracted from API 6DSS. There is no limit for the maximum bore value looking at Table 2.5. According to the API 6DSS definition, a full-bore valve could have a bore that is bigger or smaller than the connecting pipe. Let's illustrate this fact with two examples.

Example 2.3 A 12″ valve in pressure class 1500 is connected to 12″ piping with schedule 100 equal to 21.44 mm. The outside diameter of a 12″ pipe according to ASME B36.10 or 36.19 for piping standard is equal to 323.9 mm. The inside diameter (ID) of the pipe can be calculated by deducting the double thickness ($2t$) value from the outside diameter (OD).

$$\mathrm{ID} = \mathrm{OD} - 2t = 323.9 - 2*21.44 = 281.02\,\mathrm{mm}$$

Table 2.4 Comparing the bore of API 6A slab gate valves with the meeting (connected) piping.

Nominal pipe size (in.)	Piping OD (mm)	Piping thickness (schedule 160)	Piping ID (mm)	Gate valve size (in.)	Gate valve bore (mm)	Valve bore
1	33.40	6.35	20.7	1 13/16	46	Full bore
2	60.33	8.74	42.85	1 13/16	46	Full bore
4	114.30	13.49	87.32	4 1/16	103.2	Full bore
6	168.28	18.26	131.76	5 1/8	130.2	Reduced bore
8	219.08	23.01	173.06	7 1/16	179.4	Full bore

Table 2.5 Minimum valve full-bore value according to API 6DSS.

		Minimum bore by class in. (mm)			
NPS	**DN**	**Class 150, 300, 600**	**Class 900**	**Class 1500**	**Class 2500**
1	25	1.00 (25)	1.00 (25)	1.00 (25)	1.00 (25)
1½	40	1.50 (38)	1.50 (38)	1.50 (38)	1.50 (38)
2	50	1.94 (49)	1.94 (49)	1.94 (49)	1.69 (42)
2½	65	2.44 (62)	2.44 (62)	2.44 (62)	2.06 (52)
3	80	2.94 (74)	2.94 (74)	2.94 (74)	2.44 (62)
4	100	3.94 (100)	3.94 (100)	3.94 (100)	3.44 (87)
6	150	5.94 (150)	5.94 (150)	5.69 (144)	5.19 (131)
8	200	7.94 (201)	7.94 (201)	7.56 (192)	7.06 (179)
10	250	9.94 (252)	9.94 (252)	9.44 (239)	8.81 (223)
12	300	11.94 (303)	11.94 (303)	11.31 (287)	10.44 (265)
14	350	13.19 (334)	12.69 (322)	12.44 (315)	11.50 (292)
16	400	15.19 (385)	14.69 (373)	14.19 (360)	13.13 (333)
18	450	17.19 (436)	16.69 (423)	16.00 (406)	14.75 (374)
20	500	19.19 (487)	18.56 (471)	17.88 (454)	16.50 (419)
22	550	21.19 (538)	20.56 (S22)	19.69 (500)	–
24	600	23.19 (589)	22.44 (570)	21.50 (546)	–
26	650	24.94 (633)	24.31 (617)	23.38 (594)	–
28	700	26.94 (684)	26.19 (665)	25.25 (641)	–
30	750	28.94 (735)	28.06 (712)	27.00 (686)	–
32	800	30.69 (779)	29.94 (760)	28.75 (730)	–
34	850	32.69 (830)	31.81 (808)	30.50 (775)	–
36	900	34.44 (874)	33.69 (855)	32.25 (819)	–
38	950	36.44 (925)	35.63 (904)	–	–
40	1000	38.44 (976)	37.63 (956)	–	–
42	1050	40.19 (1020)	39.83 (1006)	–	–
48	1200	45.94 (1166)	45.25 (1149)	–	–
54	1350	51.69 (1312)	–	–	–
56	1400	53.56 (1360)	–	–	–
60	1500	57.44 (1458)	–	–	–

The minimum bore for a 12″ valve in pressure class 1500 is equal to 287 mm which is larger than the connected piping ID which is equal to 281.02 mm.

Example 2.4 A 14″ valve in pressure class 2500 is connected to a 14″ pipe with schedule 120 equal to 27.79 mm. The outside diameter of a 14″ pipe according to ASME B36.10 or 36.19 for piping standards is equal to 355.6 mm.

$$ID_{pipe} = OD_{pipe} - 2t = 355.6 - 2*27.79 = 300.02 \text{ mm}$$

The minimum bore of a 14″ valve in pressure class 2500 is equal to 292 mm. In this case, the bore of a full-bore valve is less than the connected piping ID which is equal to 300.02 mm.

2.2.3 Pressure rating

API 6DSS covers pressure classes equal to 150 (PN 20 bar), 300 (PN 50 bar), 600 (PN 100 bar), 900 (PN 150 bar), 1500 (PN 250 bar), and 2500 (PN 420 bar); the same pressure classes are defined in the ASME B16.34 standard for valves with flanged, threaded, and welding ends. One of the weak points of the API 6DSS standard is that it does not cover any pressure above 420 bar equal to approximately 6090 psi (each bar is about to 14.5 psi). But many subsea valves have higher pressure ratings than 6090 psi such as 7500 psi, 10,000 psi, 15,000, and even 20,000 psi. API 6A covers pressure classes of 2000 psi (13.8 MPa or 138 bar), 3000 psi (20.7 MPa or 207 bar), 5000 psi (34.5 MPa or 345 bar), 10,000 psi (69 MPa or 690 bar), 15,000 psi (103.5 MPa or 1035 bar), and 20,000 psi (138 MPa or 1380 bar). Intermediate pressure ratings such as 7500 psi equal to 517 bar are common for subsea valves in manifolds to save weight, space, and cost rather than using pressure class 10,000 psi.

The other difference between API 6A and API 6DSS is that API 6A valves are designed to be operated at the maximum rated working pressures. Therefore the operation, design, and maximum rated pressure are equal in the API 6A standard. As an example, a subsea valve with a pressure rating of 690 bar has a design, operating, and maximum rated pressure of 690 bar. However, let's suppose that a subsea valve is designed based on API 6DSS with an operating pressure of 80 bar. In this case, the API 6A rule is not applicable so the design pressure (DP) could be 10% higher than the operating pressure, which gives a DP equal to 88 bar. Eighty-eight bar pressure is between class 300 equal to 50 bar and class 600 equal to 100 bar. Therefore the maximum rating for its working pressure is 100 bar.

The main question is how to determine the pressure rating of subsea valves. Subsea valves normally have a higher pressure rating compared to valves in other sectors of the oil and gas industry due to their installation at seawater depths. The seawater column head pressure must be taken into account during pressure rating selection.

Example 2.5 A subsea ball valve is installed at 1400 m water depth so the water head on the valve is equal to 140 bar. The valve internal fluid pressure at sea level is equal to 345 bar, so the fluid pressure at depth is equal to $345 + 140 = 485$ bar. There is no 485 bar in the API 6A and 17D standards; the minimum standard pressure rating is 690 bar. However, it is possible to use an intermediate pressure rating between 345 bar and 690 bar, namely 517 bar equal to 7500 psi. Using an intermediate pressure rating can reduce the weight and dimensions of the valve. External seawater pressure

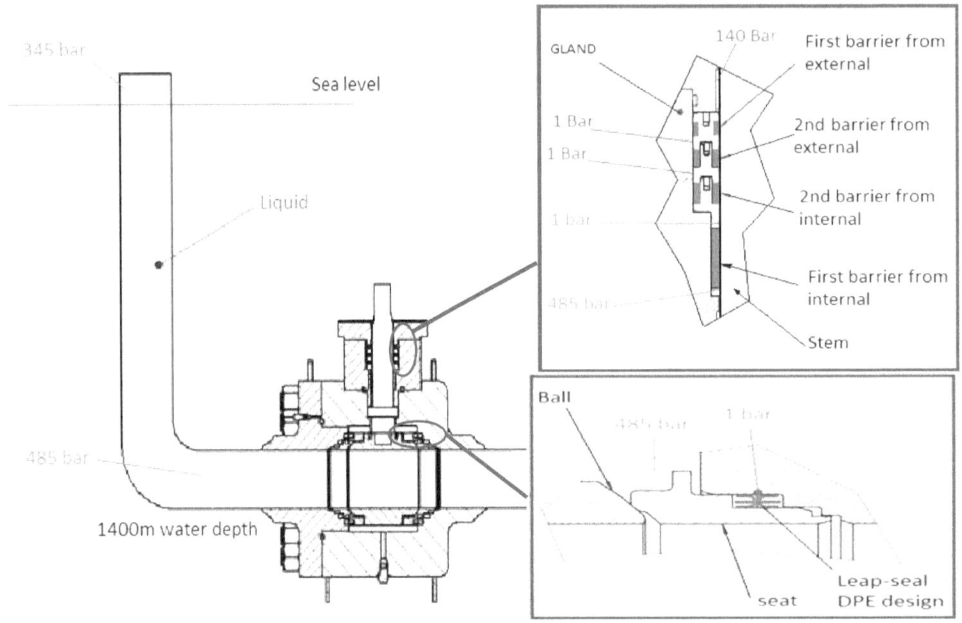

Fig. 2.10 Valve pressure rating determination and effect of internal and external pressure values on the valve sealing and seats. *(Courtesy: Valve World.)*

equal to 140 bar is applied to the top of the valve and the first barrier of external sealing. However, internal pressure equal to 485 bar is applied to the first sealing barrier from the internal pressure as well as the valve's seats. Fig. 2.10 illustrates how to calculate and select the pressure rating for a subsea valve.

The sealing of subsea valves is explained in detail in Chapter 13. There are two sets of sealing for subsea valves: internal pressure sealing, which is installed close to the valve internals to prevent the leakage of the fluid inside the valve to the subsea environment, and external sealing which is used to prevent leakage from seawater to the valve internals such as stem areas.

2.2.4 Product specification level (PSL)

PSL refers to the different levels of requirement for quality and quality control, qualification, testing, and documentation of materials. In addition, PSL affects the testing duration and requirements for subsea products such as valves. PSL level is defined in five product specification levels as per API 6A and API 17D. These five levels are PSL1, 2, 3, 3G, and 4. Some of the material qualities and testing affected by PSL are tensile, impact and hardness testing, dimensional inspection, visual examination, chemical analysis, material traceability, surface nondestructive testing such as magnetic particle or liquid penetrant test, welding and welding overlay nondestructive testing, etc. The other important consideration is that PSL is considered for pressure-containing and pressure-controlling valve parts such as the stem, ball or gate (closure member) and seats as well as the whole valve or actuator product. Pressure-containing parts are defined as valve parts whose failure to function will cause leakage of fluid to the environment. These parts are the body, bonnet, and stem at minimum. Pressure-controlling parts are those that affect the flow of fluid inside the valve such as the closure member and seats.

PSL1 is the minimum demand of API 6A for material quality control, testing, and requirement. This level of product quality is not normally used in subsea projects. PSL1 does not cover material traceability, chemical analysis, or nondestructive testing on the forged pressure-containing and pressure-controlling parts of the valves.

PSL2 has a higher level of material quality and is subjected to more strenuous testing compared to PSL1. It contains all the PSL1 requirements as minimum plus material traceability in some cases as well as chemical analysis.

PSL3 has a higher level of material quality and is subjected to more strenuous testing compared to PSL2. More precisely, more strenuous testing refers to additional time for both hydrotest seat and shell testing compared to PSL2. Table 2.6 provides the hydrostatic seat and shell test duration for subsea valves according to API 6A.

PSL3G, which is applicable for many subsea valves based on end users' requirements, has the same requirements as PSL 3 with additional gas test requirements for the (shell) body and seats. It is not common to specify PSL 3G for valve materials since material control and tests are covered by PSL 3 and not 3G. However, it could be the case that PSL3 products are marked with PSL 3G when the additional gas test is performed.

PSL4 has the highest level of material quality control; it has material qualities and control equal to PSL3 with additional requirements. A gas pressure test is also applicable for PSL 4.

Unlike API 6A and 17D, PSL is not defined/specified in the API 6DSS standard.

The minimum PSL level depends on the material class and pressure rating according to Table 2.7 extracted from API 6A. It should be noted that the required PSL level is normally given by the end user and is typically equal to PSL 3G. Material classes will be explained later in this chapter.

Table 2.6 Hydrostatic seat and shell test duration according to API 6A.

		PSL1	PSL2	PSL3	PSL3G	PSL4
Shell Test	Primary test duration	3 min	3 min	3 min	3 min	3 min
	Secondary test duration	3 min	3 min	15 min	15 min	15 min
Seat Test	Primary test duration	3 min	3 min	3 min	3 min	3 min
	Secondary test duration	3 min	3 min	15 min	3 min	15 min
	Tertiary test duration	NA	3 min	15 min	3 min	15 min

Table 2.7 Minimum PSL according to API 6A.

Material class	Rated working pressure 13.8 MPa (2000 psi)	20.7 MPa (3000 psi)	34.5 MPa (5000 psi)	69.0 MPa (10,000 psi)	103.5 MPa (15,000 psi)	138.0 MPa (20,000 psi)
AA, BB, CC	PSL 1	PSL 1	PSL 1	PSL 2	PSL 2	PSL 3
DD, EE, FF	PSL 1	PSL 1	PSL 1	PSL 2	PSL 3	PSL 3
HH, ZZ	PSL 3	PSL 3	PSL 3	PSL 3	PSL 3	PSL 4

Table 2.8 Minimum BSL for closure bolting.

	API 20E	API 20F
PSL 1	BSL-1	BSL-2
PSL 2	BSL-1	BSL-2
PSL 3	BSL-1	BSL-2
PSL 4	BSL-2 (bolt nominal diameter ≤ 2½ in.)	BSL-2 (bolt nominal diameter ≤ 2½ in.)
PSL 4	BSL-3 (bolt nominal diameter > 2½ in.)	BSL-3 (bolt nominal diameter > 2½ in.)

Courtesy: API 6A.

Table 2.9 Test requirements for the body, bonnet, and end connection of industrial valves based on API 6A based on different PSL.

Requirements	PSL1	PSL2	PSL3	PSL4
Tensile testing	Yes	Yes	Yes	Yes
Impact testing	Conditional	Conditional	Yes	Yes
Hardness testing	Yes	Yes	Yes	Yes
Dimensional verification	Yes	Yes	Yes	Yes
Traceability	–	Yes	Yes	Yes
Chemical analysis	–	Yes	Yes	Yes
Visual examination	Yes	Yes	–	–
Surface nondestructive test (NDT)	–	Yes	Yes	Yes
Weld NDT	–	Yes	Yes	Yes
Serialization	–	–	Yes	Yes
Volumetric NDT	–	–	Yes	Yes

Bolting Specification Level (BSL) was explained earlier in this chapter. The minimum BSL for closure bolting is based on PSL according to API 6A (refer to Table 2.8). However, the requirements for BSL given by the EPC contractor or end user overrule the standard requirements.

Table 2.9 summarizes the material tests for different PSLs as per API 6A requirements applicable for the body/bonnet and end connections of industrial valves.

Table 2.10 Test requirements for industrial valves as per API 6A.

Requirements	PSL1	PSL2	PSL3	PSL3G	PSL4
Drift test	Yes	Yes	Yes		Yes
Hydrostatic test	Yes	Yes	Yes, extended	Yes, extended	Yes, extended
Gas test	–	–	–	Yes	Yes
Assembly traceability	–	–	Yes	Yes	Yes
Serialization	–	Yes	Yes	Yes	Yes

Table 2.10 provides the level of quality control in terms of applicable tests for industrial valves made according to API 6A specification.

2.2.5 Material classes

API 6DSS does not define any material classes. However, API 6A defines seven material classes: AA, BB, CC, DD, EE, FF, and HH. The description of each material class is defined in Table 2.11, extracted from API 6A for the body and bonnet of the valve as well as the valve internals. The first three material classes (AA, BB, and CC) are selected for nonsour services with a relatively low concentration of hydrogen sulfide (H_2S). Sour service refers to H_2S-containing services; materials for such services should be selected as per the requirements given in the National Association of Corrosion Engineers (NACE) MR0175 and ISO 15156, "Petroleum and natural gas industries—Materials for use in H_2S-containing environments in oil and gas production." However, DD, EE, FF, and HH are used for sour H_2S-containing services.

Table 2.11 Material class definition in API 6A.

Material class		Body, bonnet, end, and outlet connectors	Mandrel hangers, valve bore sealing mechanisms, choke trim, and stems
AA	General service	Carbon or low alloy steel, or stainless steel or CRA[a]	Carbon or low alloy steel, or stainless steel or CRA[a]
BB	General service	Carbon or low alloy steel, or stainless steel or CRA[a]	Stainless steel or CRA[a]
CC	General service	Stainless steel or CRA[a]	Stainless steel or CRA[a]
DD	Sour service[b]	Carbon or low alloy steel or CRA[a,c]	Carbon or low alloy steel or CRA[a,c]
EE	Sour service[b]	Carbon or low alloy steel or CRA [a,c]	Stainless steel or CRA[a,c]
FF	Sour service[b]	Stainless steel or CRA[a,c]	Stainless steel or CRA[a,c]
HH	Sour service[b]	CRA[a,c,d]	CRA[a,c,d]

[a]CRA as defined in 3.1.22; NACE MR0175/ISO 15156 definition of CRA does not apply.
[b]As defined by NACE MR0175/ISO 15156.
[c]In accordance with NACE MR0175/ISO 15156.
[d]CRA required on retained fluid-wetted surfaces only; CRA cladding of low-alloy or stainless steel is permitted (see 7.5.1.2).

CRA stands for corrosion-resistant alloy. It generally refers to noncarbon and alloy materials such as different types of stainless steels like duplex and austenitic stainless steel as well as nickel alloys. But the definition of CRA is different in the API 6A standard. API 6A defines CRA material as a "nonferrous-based alloy in which any one or the sum of the specified amount of the elements titanium, nickel, cobalt, chromium, and molybdenum exceeds 50 % mass fraction." According to this definition, duplex and super duplex stainless steels are not categorized as CRAs.

Table 2.12 provides the chemical composition percentages for 22Cr duplex stainless steel. The maximum percentage of nickel, chromium, and molybdenum is 33%, which is less than 50%. Therefore duplex is not considered a CRA based on the API 6A definition. On the other hand, nickel alloys such as Inconel 625 are considered CRAs according to the API 6A definition. Table 2.13 provides the Inconel 625 chemical composition.

The minimum percentages of nickel, cobalt, chromium, and molybdenum for Inconel 625 are 58%, 0%, 20%, and 8%, respectively, which gives a total percentage of 86%.

The interpretation of Table 2.11 and the selection of material class as per API 6A does not seem an easy task, so some examples are provided in order to understand and select the material classes as per the API 6A standard based on industrial experiences.

Example 2.6 A 10″ ball valve has been selected and used for a manifold structure. The fluid service for the valve is a corrosive H_2S-containing hydrocarbon (gas service). The

Table 2.12 22Cr duplex chemical composition.

	C	Mn	Si	P	S	Cr	Mo	Ni	N
2205 (S31803)	0.03 max	2.0 max	1.0 max	0.03 max	0.02 max	min: 21.0 max: 23.0	min: 2.5 max: 3.5	min: 4.5 max: 6.5	min: 0.08 max: 0.20
2205 (S32205)	0.03 max	2.0 max	1.0 max	0.03 max	0.02 max	min: 22.0 max: 23.0	min: 3.0 max: 3.5	min: 4.5 max: 6.5	min: 0.14 max: 0.20

Table 2.13 Inconel 625 chemical composition.

	Ni	Cr	Fe	Al	C
Min %	58.0	20.0	–	–	–
Max %	–	23.0	5.0	0.4	0.1
	Mg	**Niobium**	**Mo**	**Co**	**Si**
Min %	–	3.15	8.0	–	–
Max %	0.5	4.15	10.0	1.0	0.5

Fig. 2.11 Inconel 625 weld overlay on the carbon steel body of a ball valve.

body of the valve is made in low-temperature carbon steel in ASTM A350 LF2 with a full Inconel 625 overlay in all sealing areas as well as the wetted parts. The reason for fully cladding carbon steel with Inconel 625 is to prevent sour corrosion. Fig. 2.11 illustrates Inconel 625 full cladding inside the body piece of a ball valve in carbon steel. The valve internals, including the ball, seats, and stem, should be in 25Cr super duplex material. Which material class should be selected for this valve?

Answer: The body of the valve is completely overlaid with Inconel 625 as a CRA material so we can ignore material classes DD and EE that have a carbon steel body in their descriptions. The remaining material class choices are FF and HH. Now we can look at the valve internals such as the seats and closure member, which are in 25Cr super duplex. Super duplex is not counted as a CRA as per the API 6A standard definition, so trim HH can be deleted. Therefore the correct material class for the valve in Example 2.6 is FF.

Example 2.7 A 5 1/8″ actuated TCG valve has been selected and used for a manifold structure branch. The fluid service for the valve is a corrosive H_2S-containing hydrocarbon (gas service). The body and bonnet of the valve are made in low-temperature carbon steel in ASTM A350 LF2 with a full Inconel 625 overlay of all sealing areas and wetted parts. The valve internals, including the gate (disk) and seats, are in Inconel 625. The valve stem should be in a material with the high mechanical strength of Inconel 718 or 725. Which material class should be selected for this valve?

Answer: In this case, the valve internals in Inconel 625, 725, and 718 are completely in CRA. The body of the valve is overlaid in Inconel 625 which is also CRA. Thus the correct material class is HH, which is the highest level. One should bear in mind that the material class is CC if the fluid is not H_2S containing.

Example 2.8 A 1 13/16″ TCG valve is selected on 2″ pipe with mono-ethylene glycol (MEG) fluid. MEG is injected into gas service as an inhibitor to prevent hydrate formation. Hydrates are ice-like solids that can form from the water or humidity in natural gas and cause flow assurance problems. MEG is not a corrosive fluid, so the body and bonnet of the valve are selected in carbon steel. The sealing areas and seat pocket of the valve are overlaid with Inconel 625, which is called partial cladding, to avoid crevice corrosion. In simple language, the valve fluid service can be trapped in the grooves and sealing areas in carbon steel and cause crevice corrosion. The valve internals, including the ball, seats, and stem, are in 25Cr super duplex. Which material class suits this valve, considering the assumption that the materials should be based on ISO 15156 for sour services?

Answer: The body and bonnet are in carbon steel, partially cladded with Inconel 625, and a carbon steel body suitable for sour service should be either DD or EE. The valve internals are in 25Cr super duplex as a type of stainless steel, so material class EE is suitable for this valve.

2.2.6 Temperature classes

Eight temperature classes are defined in the API 6A standard according to Table 2.14. These eight temperature classes cover temperature ranges from −60°C to 121°C. The design temperature in the project is not necessarily compatible with the temperature

Table 2.14 Temperature classes according to API 6A.

	Temperature range			
	°C		°F	
Temperature class	**min.**	**max.**	**min.**	**max.**
K	−60	82	−75	180
L	−46	82	−50	180
N	−46	60	−50	140
P	−29	82	−20	180
S	−18	60	0	140
T	−18	82	0	180
U	−18	121	0	250
V	2	121	35	250

Note: Minimum temperature is the lowest ambient temperature to which the equipment can be subjected. Maximum temperature is the highest temperature of the fluid that can directly contact the equipment.

classes given in Table 2.14. As an example, the design temperature of the valves could be from −29°C to 121°C, which is not completely covered by temperature class U. The main question is, how does temperature affect the valve? The simple answer is that a high temperature could affect material selection and strength. In general, higher temperatures could reduce the mechanical strength (allowable stress) of the material. A low-temperature range could affect material selection. As an example, the minimum design temperature of a carbon steel ASTM A105 is −29°C. Low alloy steel with 1.25% chromium and 0.5% molybdenum in forge material ASTM A182 F11 and 2.25% chromium and 1% molybdenum in forge material ASTM A182 F22 have a minimum design temperature of −29°C. Therefore both normal carbon steel and alloy steels in the above-mentioned grades cannot be used in a design or operating temperature less than −29°C. The alternative material could be 25Cr super duplex which can be used for a minimum temperature of −46°C. Alternative materials such as Inconel 625 should be used for the valve body and internals for a temperature value less than −46°C where 25Cr super duplex cannot be used.

Temperature range could affect the sealing material of the valves. All steel materials can be used for the maximum temperature values given in Table 2.14. NORSOK M-001 provides a maximum operating temperature limit of 100°C for 22Cr duplex and 110°C for 25Cr super duplex in an offshore environment. If duplex and super duplex are coated, these two materials can be used for higher temperature ranges above the limits given before. Subsea valves are coated so all exposed metallic materials for subsea valves can be used for the highest temperature limit which is 121°C as presented in Table 2.14. No temperature class is defined in the API 6DSS/ISO 14723 standards.

The temperature range for actuators does not necessarily follow from the temperature classes provided in Table 2.14. The maximum temperature range for an actuator should not be less than 65°C as per the API 6A requirement. The minimum temperature of the actuator can be equal to one of the minimum temperature classes provided in Table 2.14. Therefore the minimum temperature for the actuator could be 2°C or −18°C compatible with a minimum temperature value of class V or U, respectively. But it is possible to select another minimum temperature, for example −10°C, from the minimum temperature values given in the table for the actuators.

2.2.7 Performance requirements

Two levels of performance requirements (PR1 and PR2) are defined in Annex F of the API 6A standard for both valves and actuators. PR1 refers to 3 cycles and PR2 refers to 200 cycles for the qualification of valves and actuators. "Cycle" refers to the opening and closing actions of valves and actuators. PR2 is applicable for subsea valves and actuators; it includes 200 cycles during pressure and temperature fluctuations. Additionally, 200 hyperbaric test cycles and 200 endurance test cycles shall be applied to valves and actuators during the qualification tests. These three sets of qualification tests are explained in detail in Chapter 14. In fact, performance qualification testing should be performed on prototypes or the production model of the component made according to API 6A and/or API 17D standards to ensure that the performance requirement level is met for the design of the product under the specified pressure, temperature and load conditions, mechanical cycle, fluid service, etc.

2.3 Reviewing API 6DSS/ISO 14723

2.3.1 Valve types

API 6DSS covers different types of valves such as TCG valves, lubricated and non-lubricated plug valves, ball valves, swing and axial check valves, as well as axial on-off valves. It should be noted that all of the above-mentioned valves covered by API 6DSS/ISO 14723 are used in subsea except for plug valves. Plug valves maybe used in topside applications for very dirty particle-containing services. Plug valves are quarter-turn valves like ball valves; this means that the valve closure member is moved 90 degrees between open and closed positions. The closure member of a plug valve is a tapered or cylindrical plug. A lubricant plug valve is a type of plug valve in which a lubricant is used to provide sealing between the plug and the body. Fig. 2.12 illustrates a lubricant plug valve with two lubricant or sealant injection ports; port D provides stem sealing between the stem and body of the valve, and port F provides sealing between the plug and body on the downstream side. Chapter 3 explains ball valves in detail. Chapter 4 explains TCG valves such as slab, single and double expanding valves in detail. Swing and axial check valves are explained in Chapter 5. Axial valves are discussed in Chapter 8.

2.3.2 Valve bore

Two types of bore or valve internal diameter are explained in API 6DSS standard: full bore and reduced bore. The bore definitions and limitations were explained earlier in this chapter. Reduced bore valves, which have less bore area compared to end connections, could be the first choice if the valve could supply enough flow without any flow assurance problem. It is important to consider that many subsea valves are used in sandy services, so full bore is usually required for ease of sand movement and flushing the valve. Reduced bore or port is a cheaper choice than full bore, but is subject to higher wearing and provides less flow. The main question is, what would be the bore size of a reduced bore valve based on API 6DSS?

- For sizes from NPS 4″ (DN 100) to 12″ (DN 300), one size below the nominal size of the valve according to Table 2.5.
- For sizes from NPS 14″ (DN 400) to 24″ (DN 600), two sizes below the nominal size of the valve according to Table 2.5.

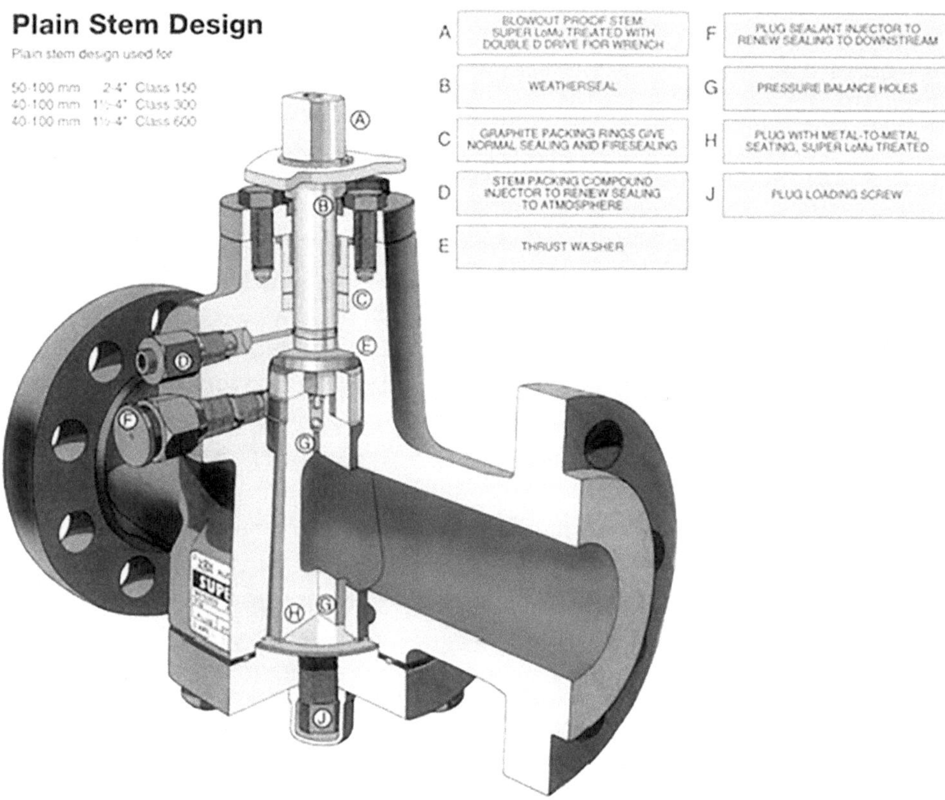

Fig. 2.12 Lubricated plug valve. *(Courtesy: M.M.K Engineering.)*

Note: According to the API 6DSS standard, the minimum size for bore reduction is 4″. A reduced bore 4″ valve is shown as 4″ × 3″, which means that the ending of the valves is compatible and equally sized with 4″ and the bore of the valve is compatible with 3″. However, according to industrial experience with small bore subsea valves, it is possible to have bore reduction in smaller sizes than 4″ such as 3″ × 2″, 1″ × ¾″, and ¾″ × ½″.

Example 2.9 A 20″ class 1500 subsea ball valve is selected as a reduced bore. What would be the minimum bore for this valve according to the API 6DSS standard?

Answer: A two-size bore reduction equal to 16″ is allowed for this valve (refer to Table 2.5 from API 6DSS). The minimum bore dimension for a 16″ valve in class 1500 is 360 mm.

The important consideration is that check valves, axial valves, and plug valves are always considered reduced bore. Ball, TCG, and swing check valves could be either reduced or full bore.

2.3.3 Face-to-face dimensions

The face-to-face dimensions of valves given in the API 6DSS standard are according to ASME B16.10, "Face-to-face and end-to-end dimensions of valves." Fig. 2.13 illustrates the face-to-face dimensions (parameter A) for a valve with a raised face, and the end-to-end dimensions (parameter B) for a butt weld end connection. It should be noted that raised face flanges are used for low-pressure class piping such as class 150 (PN = 20) and 300 (PN = 50) which are not applicable for subsea valves. Subsea valves could be supplied with pup pieces on both ends suitable for butt welding from both sides in many cases. It should be noted that subsea valves with pup pieces on both ends could have manufacturer standard end-to-end dimensions and not follow any standard. The tolerances on face-to-face as well as end-to-end dimensions are ±1.5 mm for valve sizes 10″ and smaller and ±3 mm for valve sizes 12″ and larger.

Table 2.15 is extracted from the API 6DSS standard, which provides the face-to-face and end-to-end dimensions of TCG valves in high-pressure classes of 1500 and 2500.

It should be noted that API 6A provides the face-to-face dimensions for flanged end gate, plug and ball, and swing and lift check valves according to Tables 2.16–2.18 extracted from the API 6A standard. The tolerances are also given in each table.

2.3.4 Required information from the valve supplier

The following information related to the valve should be provided by the valve supplier according to this book recommendation:

- Flow capacity or coefficient of the valve in the form of C_V and K_V;
- All of the different torque or force values for valve operation;
- Maximum allowable stem torque;
- Number of turns to open and close manually operated valves.

The flow coefficient (C_V) of a valve is defined as the amount of water that flows through the valve in gallons/minute at 60°F with a pressure drop of 1 psi across the valve. Different valve manufacturers provide different flow coefficient values for the same size and pressure classes of valves. C_V is calculated through Formula (2.2):

C_V value calculation of the valves

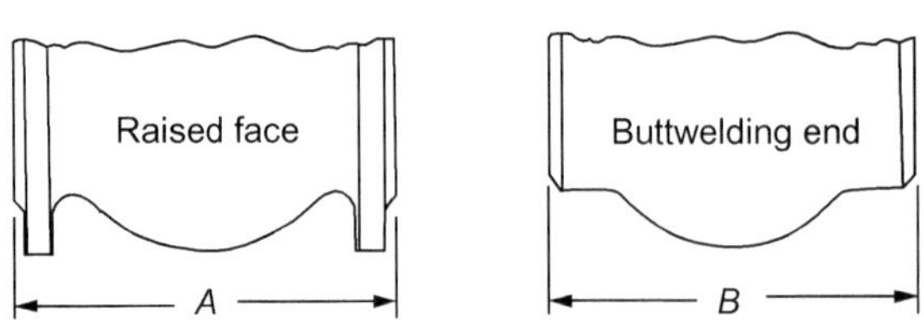

Fig. 2.13 Face-to-face and end-to-end dimension designation. *(Courtesy: ASME B16.10.)*

Table 2.15 Face-to-face and end-to-end dimensions of TCG valves in pressure classes of 1500 and 2500.

NPS	DN	Raised face *A* Class 1500	Welding end *B*	Ring joint *C*	Raised face *A* Class 2500	Welding end *B*	Ring joint *C*
2	50	14.50 (368)	14.50 (368)	14.63 (371)	17.75 (451)	17.75 (451)	17.88 (454)
2½	65	16.50 (419)	16.50 (419)	16.63 (422)	20.00 (508)	20.00 (508)	20.25 (514)
3	80	18.50 (470)	18.50 (470)	18.63 (473)	22.75 (578)	22.75 (578)	23.00 (584)
4	100	21.50 (546)	21.50 (546)	21.63 (549)	26.50 (673)	26.50 (673)	26.88 (683)
6	150	27.75 (705)	27.75 (705)	28.00 (711)	36.00 (914)	36.00 (914)	36.50 (927)
8	200	32.75 (832)	32.75 (832)	33.13 (841)	40.25 (1022)	40.25 (1022)	40.88 (1038)
10	250	39.00 (991)	39.00 (991)	39.38 (1000)	50.00 (1270)	50.00 (1270)	50.88 (1292)
12	300	44.50 (1130)	44.50 (1130)	45.13 (1146)	56.00 (1422)	56.00 (1422)	56.88 (1445)
14	350	49.50 (1257)	49.50 (1257)	50.25 (1276)	–	–	–
16	400	54.50 (1384)	54.50 (1384)	55.38 (1407)	–	–	–
18	450	60.50 (1537)	60.50 (1537)	61.38 (1559)	–	–	–
20	500	65.50 (1664)	65.50 (1664)	66.38 (1686)	–	–	–
22	550	–	–	–	–	–	–
24	60	76.50 (1943)	76.50 (1943)	77.63 (1972)	–	–	–

Dimension in inches (millimeters).
For Class 2500:
[a] Through conduit valves shall be 26.00 in. (660 mm).
[b] Through conduit valves shall be 32.00 in. (813 mm).
Tolerance: $\pm^1/_{16}$ in. (1.59 mm) on valve sizes 10 in. and smaller $\pm^1/_8$ in. (3.18 mm) on valve sizes 12 in. and larger.
Courtesy: API 6DSS.

Table 2.16 Face-to-face dimensions of full-bore TCG valves (in inches).

Nominal size (in.)	**Bore**	***L*, face-to-face length (in.)**					
	B	**2000 psi**	**3000 psi**	**5000 psi**	**10,000 psi**	**15,000 psi**	**20,000 psi**
Tolerance >	**+0.03/−0**	**±0.06**	**±0.06**	**±0.06**	**±0.06**	**±0.06**	**±0.06**
$1^{13}/_{16}$	1.81	–	–	–	18.25	18.00	21.00
$2^{1}/_{16} \times {}^{13}/_{16}$	1.81	11.62	14.62	14.62	–	–	–
$2^{1}/_{16}$	2.06	11.62	14.62	14.62	20.50	19.00	23.00
$2^{9}/_{16}$	2.56	13.12	16.62	16.62	22.25	21.00	28.50
$3^{1}/_{16}$	3.06	–	–	–	24.38	23.56	30.50
$3^{1}/_{8}$	3.12	14.12	17.12	18.62	–	–	–
$3^{1}/_{8} \times 3^{3}/_{16}$	3.19	14.12	17.12	18.62	–	–	–
$4^{1}/_{16}$	4.06	17.12	20.12	21.62	26.38	29.00	38.00
$4^{1}/_{16} \times 4^{1}/_{8}$	4.12	17.12	20.12	21.62	–	–	–
$4^{1}/_{16} \times 4^{1}/_{4}$	4.25	17.12	20.12	21.62	–	–	–
$5^{1}/_{8}$[a]	5.12[a]	22.12	24.12	28.62	29.00	35.00	–
$7^{1}/_{16} \times 5^{1}/_{8}$	5.12[a]	–	–	29.00	–	–	–
$7^{1}/_{16} \times 6$	6.00	22.12	24.12	29.00	–	–	–
$7^{1}/_{16} \times 6^{1}/_{8}$	6.12	–	–	29.00	–	–	–
$7^{1}/_{16} \times 6^{3}/_{8}$	6.38	22.12	24.12	29.00	35.00	41.00	PMR
$7^{1}/_{16} \times 6^{5}/_{8}$	6.62	22.12	24.12	29.00	–	–	–
$7^{1}/_{16}$	7.06	26.12	28.12	32.00	35.00	41.00	PMR
$7^{1}/_{16} \times 7^{1}/_{8}$	7.12[a]	26.12	28.12	32.00	–	–	–
9	9.00	–	–	41.00	PMR	–	–
11	11.00	–	–	PMR	–	–	–

[a]Bore tolerance for $5^{1}/_{8}$ in. size is +0.04/−0.
Courtesy: API 6A.

$$C_V = Q\sqrt{\frac{SG}{\Delta P}} \tag{2.2}$$

where:

Q is the flow rate in US gallon per minute;

SG is the specific gravity of the fluid, which is defined as the ratio of the fluid density to the water density. The specific gravity value of the water is equal to 1.

ΔP is the pressure differential across the valve in psi.

Torque is defined as a measure of force that can cause an object to rotate around an axis. In fact, torque is the rotational equivalent of linear force. Different existing torques and definitions are explained in Chapter 10. Torque values are associated with both valves and actuators. Actuator torque values are higher than valve torque values due to safety factor requirements. Safety factor is defined as the ratio of the torque actuator to the relevant valve torque which could be a minimum of 1.5 or 2 as examples. The required safety factor is information that should be provided by the valve purchaser and/or end

Table 2.17 Face-to-face dimensions of plug and ball valves (in inches).
(a) Flanged full-bore plug valves

Nominal size (in.)	Bore	*L*, face-to-face length					
	B	2000 psi	3000 psi	5000 psi	10,000 psi	15,000 psi	20,000 psi
Tolerance >	+0.03/−0	±0.06	±0.06	±0.06	±0.06	±0.06	±0.06
$1^{13}/_{16}$	1.81	–			18.25	18.00	21.00
$2^{1}/_{16}$	2.06	13.12	15.12	15.50	20.50	19.00	23.00
$2^{9}/_{16}$	2.56	15.12	17.12	18.00	22.25	21.00	28.50
$3^{1}/_{16}$	3.06	–	–	–	24.38	23.56	30.50
$3^{1}/_{6}$	3.12	17.62	18.62	20.75	–	–	–
$3^{1}/_{6} \times 3^{3}/_{16}$	3.19	17.62	18.62	20.75	–	–	–
$4^{1}/_{16}$	4.06	20.12	22.12	24.75	26.38	29.00	38.00
$4^{1}/_{16} \times 4^{1}/_{6}$	4.12	20.12	22.12	24.75	–	–	–
$4^{1}/_{16} \times 4^{1}/_{4}$	4.25	20.12	22.12	24.75	–	–	–
$5^{1}/_{6}$	5.12	25.12	26.12	–	29.00	35.00	–
$7^{1}/_{16} \times 6$	6.00	28.62	30.12	–	–	–	–
$7^{1}/_{16} \times 6^{3}/_{6}$	6.38	–	–	–	35.00	41.00	PMR
$7^{1}/_{16}$	7.06	29.12	31.62	38.50	35.00	41.00	PMR
$7^{1}/_{16} \times 7^{1}/_{6}$	7.12	29.12	31.62	38.50	–	–	–
9	–	–	–	–	PMR	–	–

(b) Flanged full-bore and reduced-opening ball valves

Nominal size (in.)	Bore	*L*, face-to-face length					
	B	2000 psi	3000 psi	5000 psi	10,000 psi	15,000 psi [a]	20,000 psi [a]
Tolerance >	+0.03/−0	±0.06	±0.06	±0.06	±0.06	±0.06	±0.06
$1^{13}/_{16}$	1.81	–	–	–	18.25	18.00	21.00
$2^{1}/_{16}$	2.06	11.62	14.62	14.62	20.50	19.00	23.00
$2^{9}/_{16}$	2.56	13.12	16.62	18.62	22.25	21.00	28.50
$3^{1}/_{16}$	3.06	–	–	–	24.38	23.56	30.50
$3^{1}/_{6}$	3.12	14.12	15.12	18.62	–	–	–
$4^{1}/_{16}$	4.06	17.12	18.12	21.62	26.38	29.00	–
$5^{1}/_{6}$	5.12	–	–	–	29.00	35.00	–
$7^{1}/_{16} \times 6$	6.00	22.12	24.12	29.00	–	–	–
$7^{1}/_{16} \times 6^{3}/_{6}$	6.38	–	–	–	35.00	41.00	–
$7^{1}/_{16}$	7.06	–	–	–	35.00	41.00	PMR
9	9.00	–	–	–	PMR	–	–

[a]Applies to full bore only.
Dimensions in inches.
Courtesy: API 6A.

user. In addition, the drive train of the valve should have a maximum allowable stem torque (MAST) higher than the actuator torque or force. Drive train, as per API 6DSS, is defined as "all parts of the valve between the valve operator and closure member that transmit the loads, including the valve closure member and excluding the valve operator." Fig. 2.14 illustrates a ball valve and drive train.

Table 2.18 Face-to-face dimensions of a swing and lift check valve (in inches).

(b) Regular and full-opening flanged swing and lift check valves, 2000 psi, 3000 psi, and 5000 psi RWP

Nominal size (in.)	***L*, short pattern face-to-face length**			***L*, long pattern face-to-face length**	
Tolerance>	**2000 psi ±0.06**	**3000 psi ±0.06**	**5000 psi ±0.06**	**3000 psi ±0.06**	**5000 psi ±0.06**
$2^{1}/_{16}$	11.62	14.62	14.62	–	
$2^{9}/_{16}$	13.12	16.62	16.62	–	–
$3^{1}/_{3}$	14.12	15.12	18.62	17.12	–
$4^{1}/_{16}$	17.12	18.12	21.62	20.12	–
$7^{1}/_{16}$	22.12	24.12	28.00	–	29.00
9	26.12	29.12	33.12	–	–
11	31.12	33.12	39.36	–	–

(c) Regular and full-opening flanged swing and lift check valves, 10,000 psi, 15,000 psi, and 20,000 psi RWP

Nominal size (in.)	***L*, face-to-face length (in.)**		
Tolerance>	**10,000 psi ±0.06**	**15,000 psi ±0.06**	**20,000 psi ±0.06**
$1^{13}/_{16}$	18.25	18.00	21.00
$2^{1}/_{16}$	20.50	19.00	23.00
$2^{9}/_{16}$	22.25	21.00	26.50
$3^{1}/_{16}$	24.38	23.56	30.50
$4^{1}/_{16}$	28.38	29.00	–
$5^{1}/_{6}$	29.00	–	–
$7^{1}/_{16}$	35.00	–	–

Courtesy: API 6A.

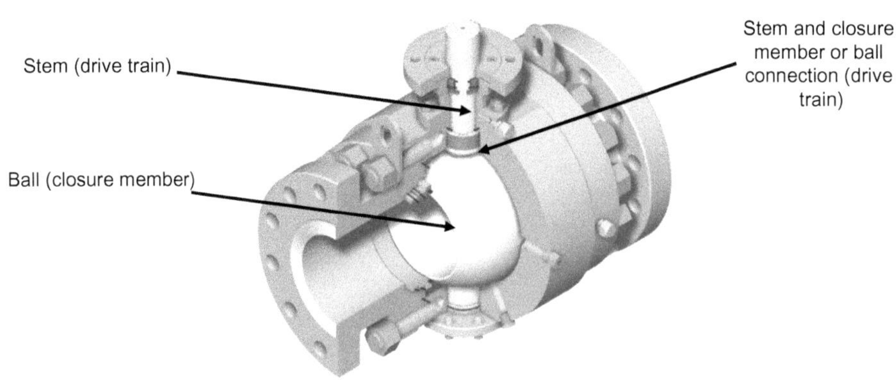

Fig. 2.14 A ball valve drive train.

MAST, as per API 6DSS, is defined as the "maximum torque/thrust that it is permissible to apply to the valve drive train without risk of damage, as defined by the valve manufacturer." MAST should always be higher than actuator torque to prevent damage to and failure of the drive train of the valve.

2.3.5 Required information from the purchaser or end user

The valve purchaser should provide some of the information related to the valves and actuators such as minimum and maximum operating/design temperature, valve working pressure, depth of water, valve type, required opening and closing time of the valve, orientation of the valve and actuator, ROV interface and classes, minimum and maximum actuator hydraulic supply pressure, type of actuator and failure mode of operation, piggability of the valves, etc. Piggable valves should normally have an internal bore diameter equal to that of the connected pipe with some degree of tolerance such as 2% or 3% to facilitate the passage of the pig through the valve and prevent the pig from getting stuck inside the valve.

Example 2.10 A 22″ piggable ball valve in a pressure class of 7500 psi equal to 517 bar is connected to 22″ piping with a schedule of 120. What is the bore size of this valve?

Answer: Refer to the ASME B36.10 standard; the outside diameter (OD) of a 22″ pipe is equal to 559 mm and the wall thickness of the 22″ schedule 120 is equal to 41.3 mm. Now it is possible to calculate the internal diameter (ID) of the pipe:

$$\text{ID} = \text{OD} - 2t = 559 - 2 \times (41.3) = 476.4\,\text{mm}$$

Thus the bore of the valve should be 476.4 mm with $\pm 3\% = 476.4 \pm 14.3 \rightarrow$ (minimum bore = 462.1 mm; normal bore = 476.4 mm; maximum bore = 490.7 mm).

2.3.6 Valve endings

Flanged and butt weld end connections are covered in the API 6DSS standard. In addition, hub and clamp end connections as well as compact flanges can be used for subsea valves. Compact flanges are covered by the NORSOK L-005 standard. Fig. 2.15 compares the size and dimensions of a standard ASME flange, which is the largest one on the right with a NORSOK L-005 compact flange and hub and clamp connection.

ASME flanges are not common for subsea applications, mainly due to their pressure limitation to 420 bar, so API 6A flanges (e.g., type BX) may be used for subsea valves. A BX flange as per API 6A means a ring joint and raised face flange as illustrated in Fig. 2.16.

API 6A type B flanges are just ring type joint (RTJ) without any raised face. A comparison between API 6A type B and BX is shown in Fig. 2.17. The picture on the left shows the RTJ face of an API 6A type B flange. The picture on the right shows a typical API 6A BX flange face. Different types of gaskets such as R, RX, and BX, illustrated in Figs. 2.18 and 2.19, can be used with the compatible flange face.

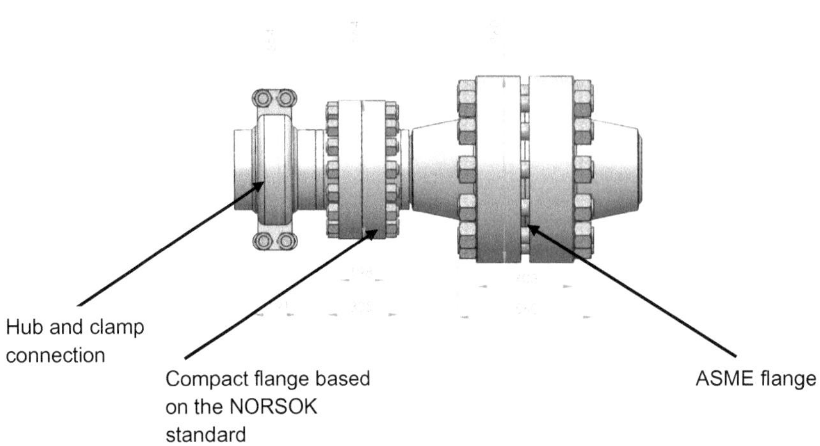

Fig. 2.15 Size and dimension comparison between ASME flanges, compact flanges, and hub and clamp connections. *(Courtesy: Tekflow.)*

Fig. 2.16 API 6A BX flange face.

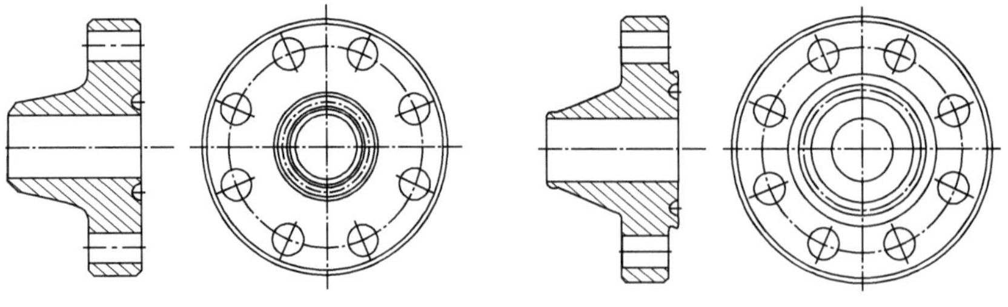

Fig. 2.17 API 6A B and BX flange face comparison.

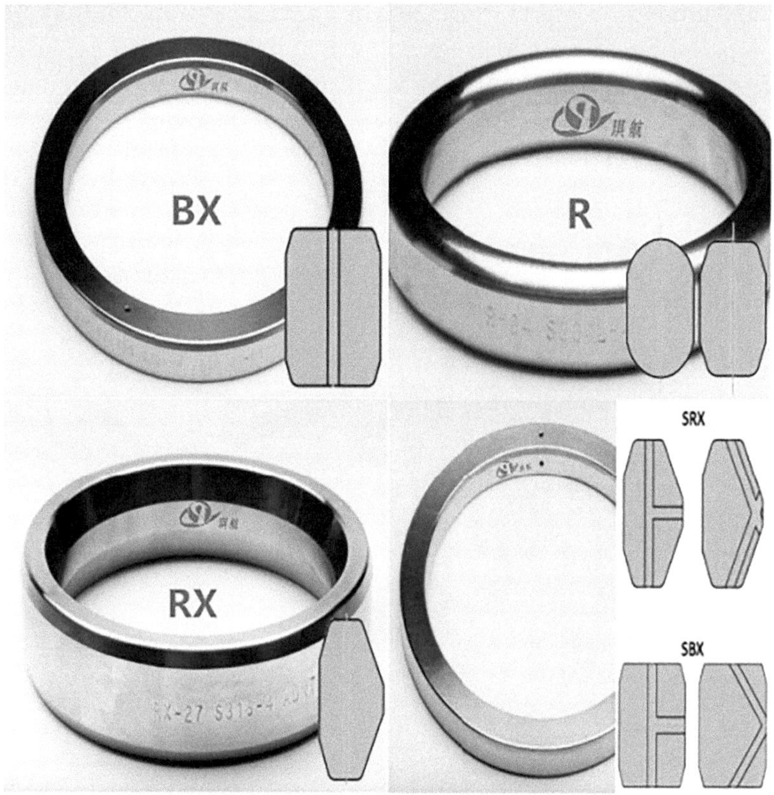

Fig. 2.18 Different RTJ gasket types.

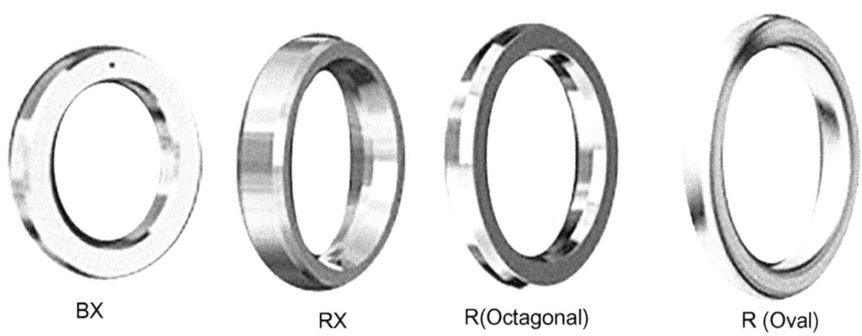

Fig. 2.19 Different RTJ gasket types.

SBX and SRX gaskets are similar to BX and RX gaskets, respectively, but the "S" prefix indicates that they have an intersecting vent hole allowing less metal-to-metal contact pressure between the gasket and flange groove and better sealing of the gasket and flange.

Fig. 2.20 illustrates a compact flange designed according to the NORSOK standard.

The compact flanges illustrated in Fig. 2.21 can be up to 60% more compact and up to 80% lighter compared to conventional ASME flanges. But the reasons for using compact flange end connections are not limited to space and weight saving. Other advantages of compact flanges are mentioned as follows:

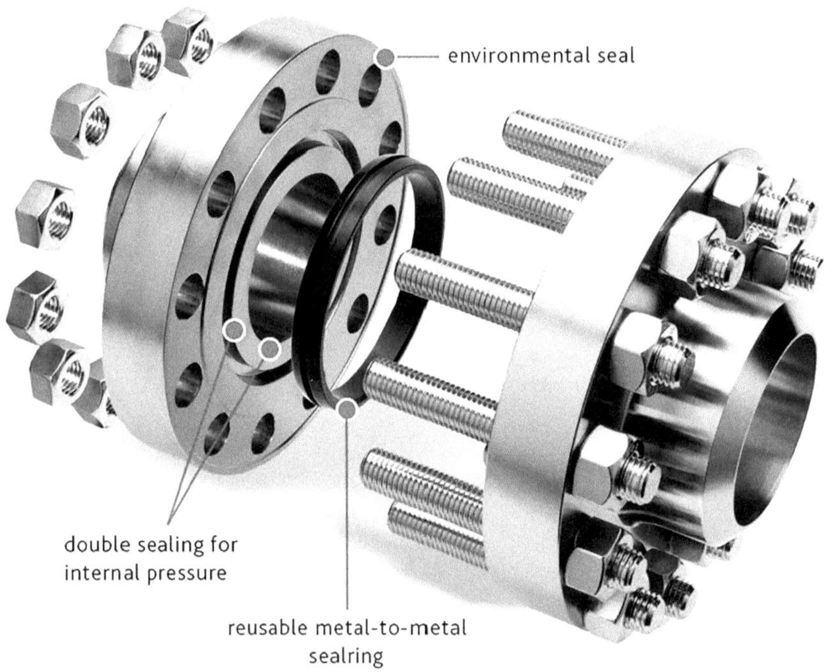

Fig. 2.20 Compact flange as per NORSOK standard. *(Courtesy: Vector.)*

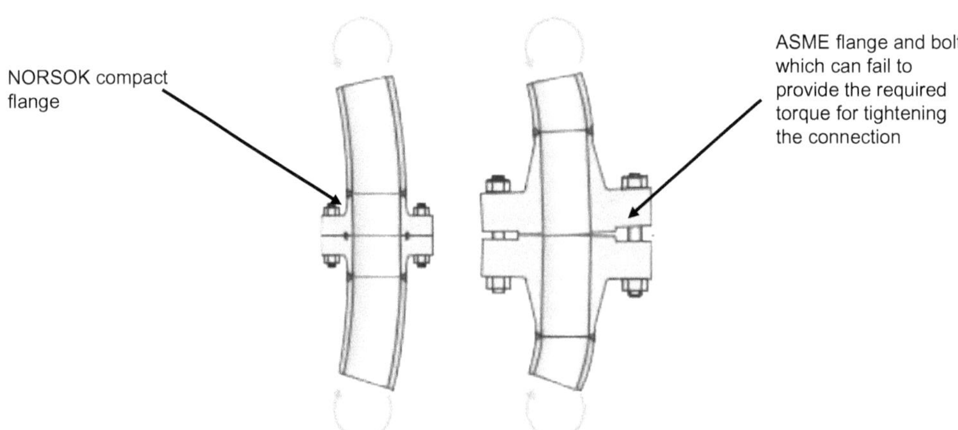

Fig. 2.21 Comparing ASME and NORSOK flanges exposed to bending dynamic load.

- Compact flanges have great load resistance as illustrated in Fig. 2.21.
- The full bore inside the flange prevents pressure drop and flow turbulence.
- The sealing (seal ring) sits between two compact flanges and is thus not exposed to any internal or external fluid as illustrated in Fig. 2.22, although ASME flange sealings can be in contact with the internal fluid and external environment.
- Suitable for high-pressure and high-temperature applications. ASME flanges can be used up to and including a pressure class of 2500 equal to a nominal pressure of 420 bar. However, compact flanges can be used up to and including a pressure class of 20,000 psi.
- They are easy and quick to install.
- The seal ring can be reused after the flange pieces have been dismantled.

Hub and clamp connections, also called mechanical joints, include two pieces of hubs and one clamp that are assembled together with four bolts and eight nuts as illustrated in Fig. 2.23. This type of hub and clamp connection is common for topside.

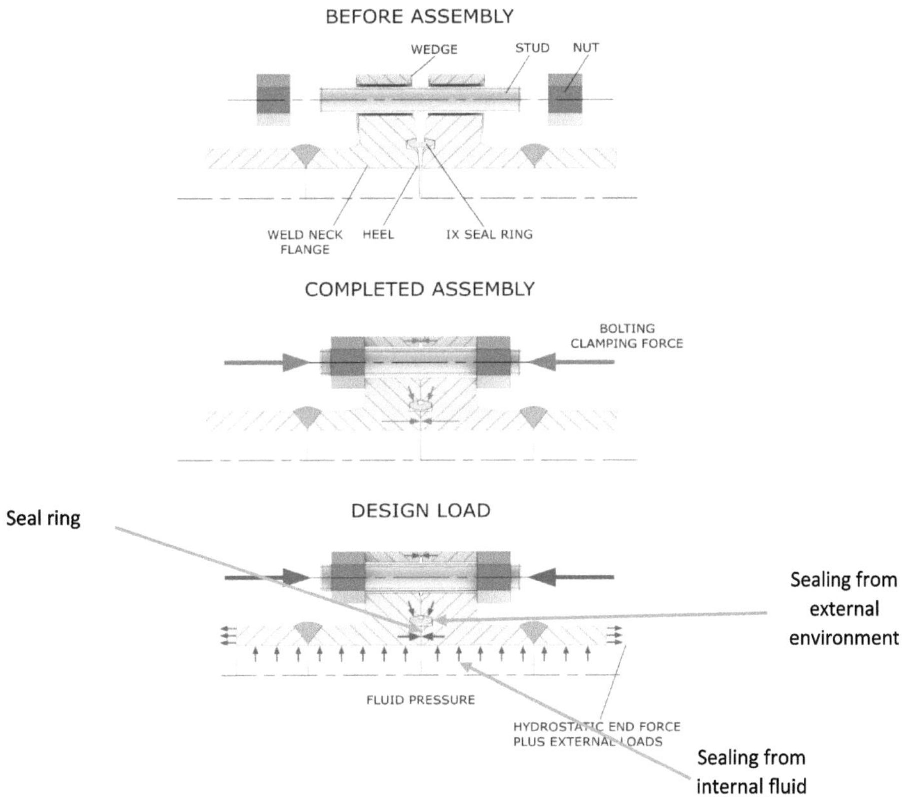

Fig. 2.22 Sealing protection of a compact flange. *(Courtesy: TP Connectors.)*

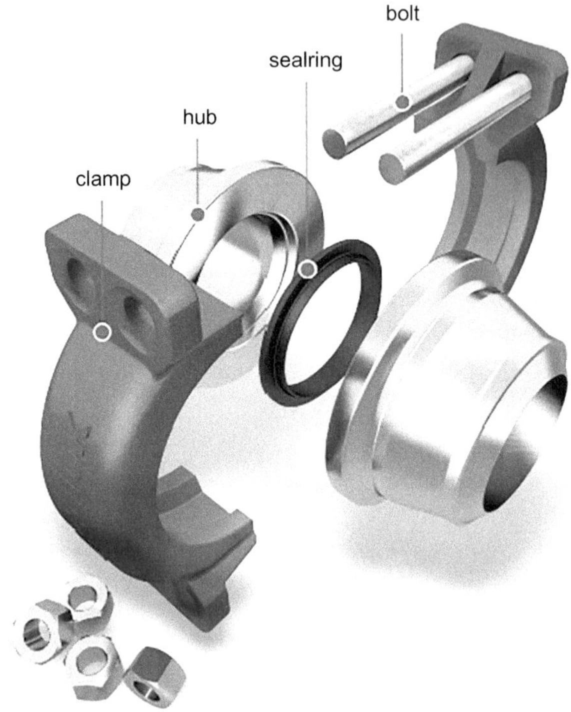

Fig. 2.23 Hub and clamp connection. *(Courtesy: Freudenberg Oil and Gas Technologies.)*

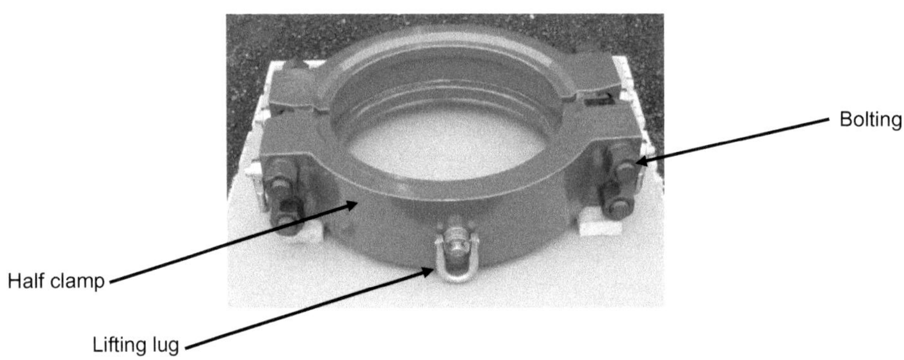

Fig. 2.24 Clamp connection as per API 16A.

Fig. 2.24 illustrates an API 16A subsea clamp connector in 2 9/16″ and 20,000 psi. API 16A defines the requirements for the design, testing, and other aspects of drill-through equipment including spools, connectors, adapters, etc. Fig. 2.25 illustrates a subsea clamp connector developed by Vector.

Fig. 2.26 illustrates a 7 1/16″ slab gate valve with hub end connection for subsea application.

Fig. 2.25 Subsea clamp connector. *(Courtesy: Vector.)*

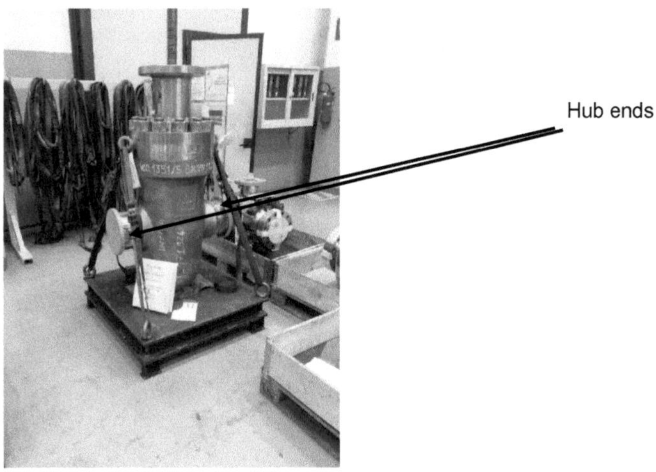

Fig. 2.26 Hub end connection slab gate valve.

2.3.7 Valve ports and connections

All ports for subsea valves are used for testing purposes and should be seal welded after finishing the test. Sealant injection ports are common for valves in sectors of the oil and gas industry other than subsea to inject grease or sealant as a part of emergency repair action prior to the main maintenance program. There are normally two sealant injection ports for the valves: one on the stem and the other on the seat. Fig. 2.27 illustrates sealant injection on the seat of a ball valve. A sealant injection port contains a check valve that

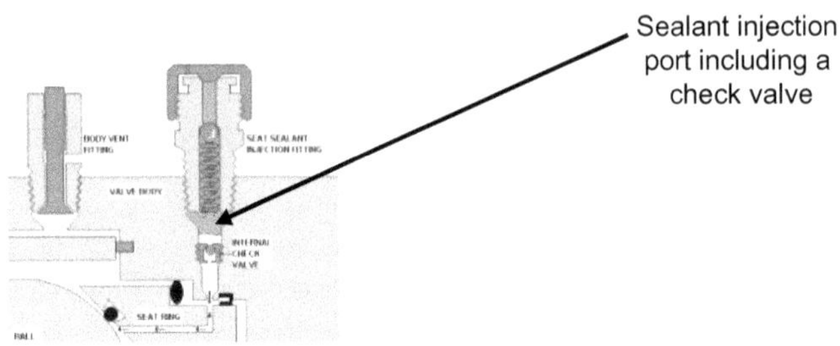

Fig. 2.27 Sealant injection on the seat of a ball valve.

gets opened when the sealant is injected and prevents the release of the internal fluid to the environment. Subsea valves are made to be maintenance free, so no sealant injection is required for them.

2.3.8 Operator auxiliaries and interfaces

All of the valve operators should have a visible position indicator on the top section of the valve and operator assembly showing both open and closed positions of the valve closure member. It is important that the position indicator be protected from marine growth by selecting suitable antifouling materials. More information about position indicators is given in Chapter 9 for manual valves and Chapter 10 for actuated valves.

In addition, travel stops should be provided on the valves or valve operators to adjust the travel of the closure member. Travel stops for a gate valve are normally located inside the valve. However, the end travel stops for ball valves are normally located on the actuators. Fig. 2.28 illustrates a slab gate valve disk that stops when it reaches the bottom of the

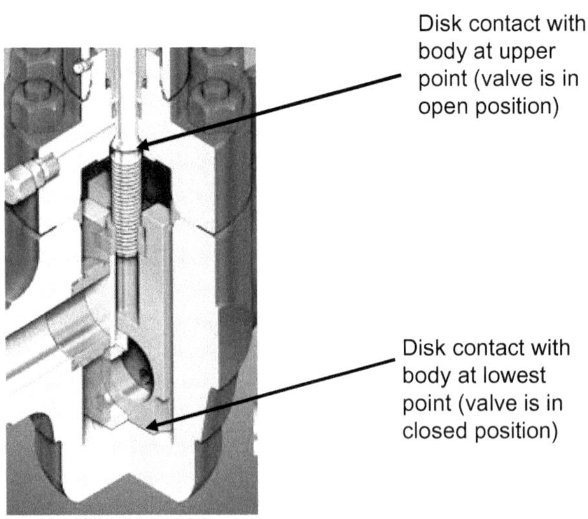

Fig. 2.28 Slab gate valve disk position.

valve for closing action and stops at the top due to the contact of the disk with the inner bonnet area, also called "back seat bushing," to provide full opening of the valve. The interface between the valve and actuator is very important. Proper seals such as O-rings can be used to seal the top work area of the valve to the bottom part of the actuator. In addition, dual pins could be used (as an example) for the proper alignment and connection of the valves and actuators. The purchaser should specify in an engineering document such as valve and actuator data sheet whether the actuators are fixed to the valves or retrievable. The ROV interface and class should also be provided by the purchaser.

2.3.9 Lifting points and requirements

Information about lifting points is available in Chapter 10.

2.3.10 Hyperbaric testing

When it comes to pressure testing for qualification or acceptance, hyperbaric testing (Fig. 2.29) is applicable for subsea valves to simulate deep-water conditions such as immersion in seawater and pressure in a chamber. This test is done before production as part of a qualification test for design validation, or before offshore installation as a part of a factory acceptance test (FAT). Hyperbaric testing improves the safety and reliability of valves, reducing possible downtime and failure risks, and can save significant costs during production or before placing the valves at a great depth for operation. It should be noted that hyperbaric testing may be replaced by FEA in some cases. As an example, let's suppose that a 5 1/8″ slab gate valve in a pressure class of 517 has been qualified through qualifications such as a hyperbaric test by the valve manufacturer. The qualification of a valve of one size could qualify a valve one size less and one size more than the qualified valve in

Fig. 2.29 Hyperbaric test for a subsea valve and actuator.

the same product family, i.e., of the same design, material, and pressure class. In the case of scaling, there is no need to perform a qualification test, but the valve supplier can apply FEA for validation of the scaled valve. As illustrated in Fig. 2.29, the valve and actuator shall be assembled together during the qualification tests, including the hyperbaric test in the hyperbaric chamber. However, if the hyperbaric chamber does not have enough space for the assembly of the valve and actuator, a hyperbaric test can be performed separately on the valve and actuator.

2.3.11 Material

It is recommended to have minimum EN 10204 type 3.1 for the pressure-containing and pressure-controlling parts of subsea valves. This degree of test certification allows the full traceability of components via heat number which is marked on the component and mentioned in the test certificate. API 6DSS provides the minimum material requirements that should be mentioned in the test report as follows:

- Material grade, which can be the relevant ASTM grade;
- Chemical properties, i.e., a list of the compounds in the material as well as their percentage;
- Heat treatment of the material;
- Mechanical properties such as yield, tensile;

Other material requirements are optional as follows:

- Carbon equivalent (CE);
- Harness test;
- Charpy impact test results. The Charpy impact test, also known as the Charpy V-notch test, is a standardized high strain rate test which determines the amount of energy absorbed by a material during fracture. Absorbed energy is a measure of the material's notch toughness.

It should be noted that forged materials are more common for subsea valves due to their higher quality. However, high quality casting such as centrifugal casting could be accepted for subsea valve components. Centrifugal casting is a process that delivers components with high material quality through a process in which molten metal is poured into a hot and spinning die. The die could be orientated horizontally or vertically.

2.3.12 Tests

Tests are divided into two main categories of FATs as well as qualification tests which are described in detail in Chapter 14.

2.3.13 Coating/painting

Coating and cathodic protection are two common approaches applied to subsea equipment including valves to prevent external corrosion due to contact with seawater.

Fig. 2.30 Top entry ball valve body welded to a pup piece and transition piece. *(Courtesy: Valve World.)*

The coating requirement and system should be defined by the valve and actuator purchaser. Coating or painting is applied to valve external parts such as the body, bonnet, and pup pieces. Valve pup pieces are pieces of forged piping welded to the body of the valve before valve assembly as illustrated in Fig. 2.30. The purpose of the pup piece is to provide some distance from the valves' internals and sealing from the heat input generated during the welding of the piping to the valve pup piece in the construction yard. In some cases, a transition piece should be welded between the pup piece and the body of the valve as a thickness convertor if the thickness differential between the valve and pup piece is more than 50% (refer to ASME B31.8 code). The coating systems for valves on topside platforms could be zinc-rich epoxy, phenolic epoxy, or thermal spray aluminum. However, these types of coating are not common for subsea valves.

The coating should be applied to the valve after the completion of the FAT; otherwise, any leakage from the valve body during the test could be hidden by the coating. The NORSOK M-501standard provides a requirement for the selection of coating materials, surface preparation, application procedures, and inspection of protective coating on offshore facilities. The aim of this NORSOK standard is to propose and obtain a coating system that is maintenance friendly, with minimal need for maintenance, easy application of the coating system, and Health, safety, and environmentally (HSE) friendly coating. Nine coating systems are defined in this standard. Table 2.19, extracted from the NORSOK standard, describes coating system 7. Coating systems 7B and 7C are applicable for submerged carbon and stainless steel materials that contain two component epoxies with a minimum of two layers of coating.

Table 2.19 Coating system for submerged materials as per NORSOK M-501 standard.
A.7 Coating system no. 7 (shall be prequalified)

Application (if not specified under others)	Surface preparation	Coating system (example)	Minimum DFT (μm)
Coating system no. 7A: Carbon and stainless steel in the splash zone	Cleanliness: ISO 8501-1 Sa 2½ Roughness: ISO 8503 Grade Medium	Two component epoxy or polyester based coating. Minimum number of coats: 2 Complete coating system:	600
Coating system no. 7B: Submerged carbon and stainless steel ≤50°C	G (50–85 μm. R_{y5})	Two component epoxy Minimum number of coats: 2 Complete coating system:	350
Coating system no. 7C: Submerged carbon and stainless steel >50°C		Two component epoxy Minimum number of coats: 2 Complete coating system:	350

Note 1: Stainless steel shall be blasted with nonmetallic and chloride free abrasive.
Note 2: The coating system shall always be used in combination with cathodic protection.
Note 3: For the splash zone, corrosion allowance in accordance with applicable regulatory requirements shall always be used in addition to the coating system, see NORSOK M-001.
Note 4: Antifouling may be required.
Note 5: Piping embedded in concrete shall be corrosion coated at least 300 mm into concrete.
Note 6: For the splash zone area, system no. 7A, the coating system shall also fulfill the prequalification requirements for coating system no. 1.

Example 2.11 A subsea valve with an operating temperature of −29°C to 121°C is operated by a rack and pinion actuator with an operating temperature of −10°C to 70°C. The valve is connected to the line via pup pieces and a butt weld connection. What coating systems can be proposed for the valve body, pup piece, actuator, ROV, and position indicator?

Answer: The valve and actuator have operating temperature values of more than 50°C, so NORSOK system 7C is proposed for the valve body, pup piece, and actuator. The important consideration is to avoid coating the end of the pup pieces since the endings of the pup pieces are going to be welded to the piping in the construction yard and the weld can damage the coating. Therefore 30 or 50 mm from the end of the pup piece should remain uncoated and will be coated in the yard after welding to the piping. A layer of primer could be added to the pup piece ends as a means of external corrosion protection before welding the pup piece to the pipe in the construction yard. The ROV and position indicator temperature is less than 50°C so coating system 7B is proposed. The other important consideration is the color of the coating. Different colors are proposed in Annex B of the NORSOK standard according to Table 2.20. Each

Table 2.20 Coating color as per NORSOK M-501.

Color	RAL-1K designation
White	RAL 9002 (Grauweiss)
Blue	RAL 5015 (Himmelblau)
Gray	RAL 7038 (Achalgrau)
Green	RAL 6002 (Laubgrün)
Red	RAL 3000 (Feuerrot)
Yellow	RAL 1004 (Goldgelb)
Orange	RAL 2004 (Reinorange)
Black	RAL 9017 (Verkehrsschwarz)

color corresponds to a RAL value. RAL is a European color matching system that defines colors for paint and coating. RAL 9002 which is a white color could be a choice for the valve body as well as the pup pieces and actuator. RAL 2004, an orange color (dark gray in print versions), is very common for ROVs; RAL 9017, a black color, can be selected for the position indicator. Fig. 2.31 illustrates a subsea slab gate valve with a hydraulic linear actuator in 1″ size and an ROV override. The ROV override is in orange color (dark gray in print versions), body of the valve as well as actuator and the pup pieces is in white color, and sings of "O" stands for open and "S" stands for shut for indicating the position of the valve are in black color.

Coating is not recommended for some parts of the valve such as the stem, bolting, flange faces (in case of flanged end valves), flange holes, and nut bearing areas. Bolt plugs (illustrated in Fig. 2.32) can be placed inside the flange bolt holes to protect the nut bearing areas and flange holes from the coating. Fig. 2.33 illustrates the uncoated areas of the flange such as the bolt holes and nut bearing areas.

2.3.14 Valve marking

Valve marking can be divided into two categories: valve information on the name plate and a low-stress die stamp on the body and bonnet of the valve. The name plate should be in austenitic stainless steel grade 316 which is the most resistant austenitic stainless steel for use in chloride and offshore environments. The name plates should be fixed to the body of the valve with stainless steel 316 rivets. Corrosion of the rivets leads to losing the name plate. Fig. 2.34 illustrates general corrosion on the rivets which are used to fix the name plate to the body of the modular valve. Although braided wire is proposed as an option to connect the name plate to the valve, this approach is more common for connecting temporary tag plates. API 6DSS provides the information that is required to be marked on the valve name plate, such as the name of the valve manufacturer, the valve type, design codes or standard, year of production, face-to-face dimensions, body material, bonnet, stem and trim, pressure and temperature rating, tag, and serial number. Trim refers to the

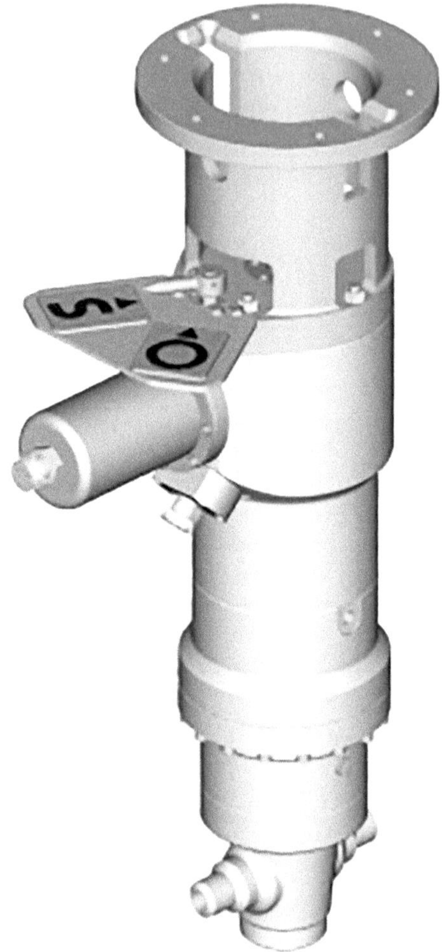

Fig. 2.31 Hydraulically operated 1″ gate valve.

Fig. 2.32 Bolt plugs.

Fig. 2.33 Uncoated nut bearing areas and flange bolt holes.

Fig. 2.34 Modular valves rivet corrosion in the offshore environment.

valves' internals which are in contact with the fluid service. Fig. 2.35 illustrates a valve tag plate sample provided by a valve manufacturer. Marking of the actuators is not required by the API 6DSS standard. Fig. 2.36 provides a sample of a tag plate on the actuator.

Some of the information should be engraved on the valves' body, bonnet, etc. As an example, the material grade should be engraved on the body of the valve as illustrated in

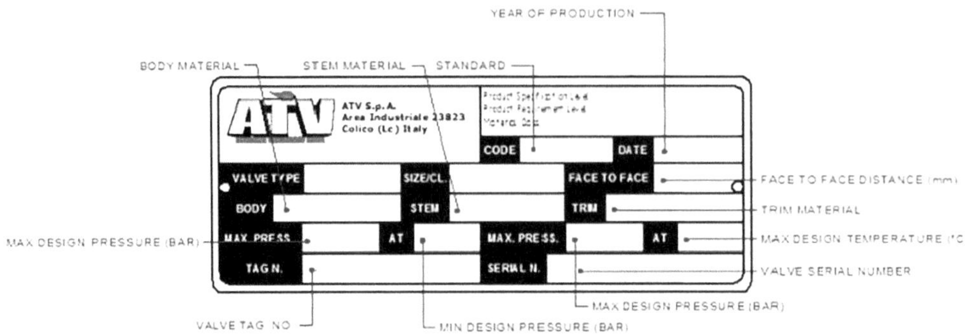

Fig. 2.35 Valve tag plate.

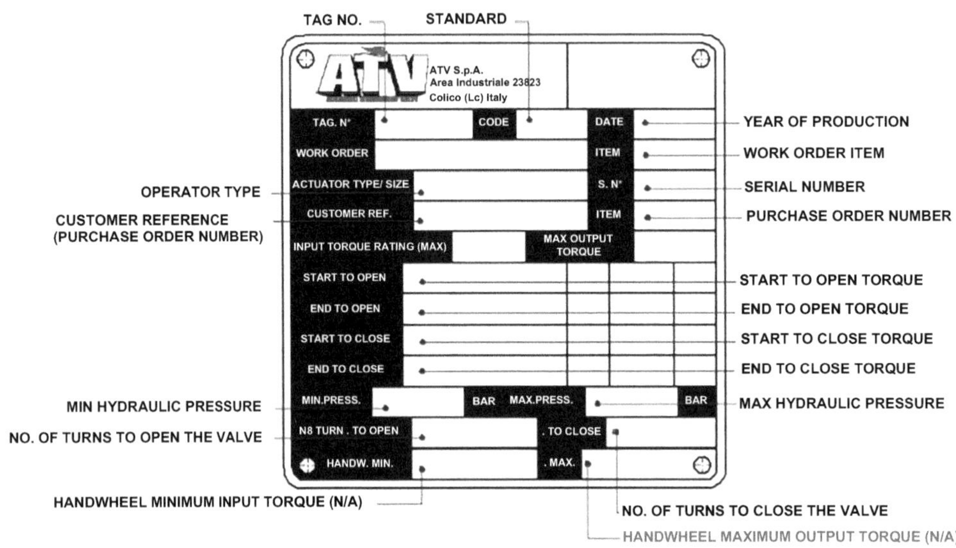

Fig. 2.36 Actuator tag plate.

Fig. 2.37. Ring type joint flange faces are sealed with a ring type joint gasket. Each ring groove has a number engraved on the outside area of the flange end body as per Fig. 2.38. The heat number should also be engraved on the body/bonnet as illustrated in Fig. 2.39. A heat number or heat lot is an identification number that is stamped on a material plate to prove that it meets industry quality standards, which require materials to be tested by the manufacturer. These numbers are used to identify production runs for quality control purposes. The heat number is also referred to in the material certificate for full traceability of the material. The sign of positive material identification (PMI) in Fig. 2.39 shows that the chemical composition of the component with the PMI stamp has been checked against the relevant material standard.

Fig. 2.37 Material grade A182 F51 duplex engraving on the check valve body.

Fig. 2.38 RTJ designation (R24) engraving on the outer body flange diameter.

In addition, the type of valve seat, i.e., self-relieving or double piston effect, should be marked on a separate tag plate on the valve as per Fig. 2.40 extracted from API 6DSS. The tag plate in Fig. 2.40 shows the seat of a double isolation and bleed type 2 (DIB2) valve with a self-relieving seat on the left side and a double piston effect seat on the right. The tag plate should be connected to the valve body through four holes and rivets.

Fig. 2.39 Heat number and material grade on the bonnet of a check valve.

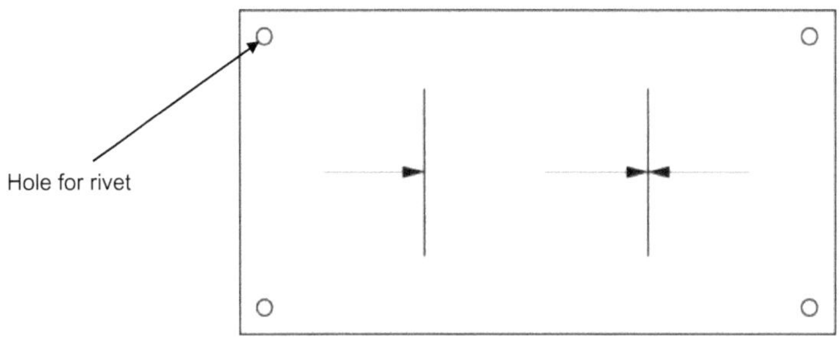

Fig. 2.40 Tag plate for a DIB2 valve with a self-relieving seat on the left and a double piston effect seat on the right as per API 6DSS.

2.3.15 Preparation for shipment

In preparation for shipment, the valve ends should be blinded off with plywood to protect the valve internals from the ingress of dust, humidity, moisture, foreign material, snow, etc. Fig. 2.41 illustrates a TCG valve blinded with plywood. Manually operated ball valves are shipped with the ball in open position, and manual gate valves are shipped with the gate or disk in closed position. The same rule is applicable for the shipment of ball and gate valves with double acting fail-as-is actuators. However, ball and gate valves with fail-safe closed actuators are shipped in closed position. Chapter 10 provides more detailed information about actuators and relevant terms such as single and double acting. Check valves are always shipped in closed position.

Fig. 2.41 Plywood for blinding the TCG valve.

2.3.16 Valve documentation

A list of valve documentations is provided by the valve purchaser in a document called the supplier document register list (SDRL). The main documents show interfaces with the piping layout and modeling on general arrangement drawings that provide all the dimensions for the valves and operators. Cross-section drawings normally provide a list of the valves' components and the material for each part. An inspection test plan (ITP) is a document that describes the plan for managing the quality control and assurance of a particular element of the construction work and provides information on the requirements, an overview of the method(s) to be used, and the responsibilities of relevant parties. This document should be provided in the first weeks after placing the order. Different material test reports and procedures such as a nondestructive test (NDT), PMI, etc. can be considered retained documents. Retained means that these documents should be kept by the valve manufacturer and submitted upon the request of the purchaser. Coating procedure is another important document that should be submitted by the valve manufacturer. The installation, operation, and maintenance (IOM) manual, the packing and preservation procedure, FAT procedure, declaration of conformity with the purchase order and standards, as well as many other documents should be generated and supplied by the valve manufacturer.

2.4 Review of JIP33 specification

2.4.1 Introduction

The International Association of Oil and Gas Producers (IOGP) released Joint Industry Program 33 (JIP33), a supplementary specification to API 6DSS for subsea pipeline valves, in August 2020. The overall aim of this supplement is to improve the level and content of API 6DSS and provide an optimum specification for the procurement of subsea industrial valves. The main aim of this section is to review the main changes proposed by JIP33 to API 6DSS.

2.4.2 Summary of the main changes

2.4.2.1 Valve types and configuration

Valve types

API 6DSS covers four types of valves: ball, TCG, axial, and plug valves. Plug valves are not used in subsea, so the JIP33 specification has removed them from the standard.

One-piece ball

JIP33 has added the requirement of having a one-piece solid ball, but this requirement seems unnecessary since the top ball valve suppliers already typically supply a one-piece ball. Fig. 2.42 illustrates a one-piece ball belonging to a ball valve.

Fig. 2.42 One-piece solid ball.

Valve bore

API 6DSS provides a minimum bore value (internal diameter) for each valve size and pressure class. However, the pressure classes in API 6DSS are normally lower than the subsea valve pressure classes. Therefore the minimum bore values given in API 6DSS are not applicable for subsea valves in most cases. API 6A provides a bore value equal to the size of the valve, which is more practical for the subsea gate valves used on Christmas trees and manifold branches. As an example, the bore of a 7 1/16″ gate valve in pressure class 10,000 psi (see Fig. 2.43) is typically 179.4 mm. API 6DSS does not cover 7 1/16″ size and 10,000 psi pressure class, so API 6DSS cannot be used as a reference for the bore of a gate valve in such a size and pressure class. The other example is about valves subject to pig running for cleaning or inspection purposes. These valves should have a bore equal to the internal diameter of the connected piping. Thus the bore of the valve does not follow the standard in the case of pig running. In the second example, none of the subsea

Fig. 2.43 7 1/16″ Class 10,000 subsea gate valve with highlighted bore value of 179.4 mm.

valve standards are good references for the internal diameter value of the valve. The valve data sheet is the best reference for the bore value of the valve as per JIP33.

2.4.2.2 Valve design

Design standards

The design and calculation of valves' pressure-containing parts, including the bolts, should be made with reference to recognized international standard such as ASME BPVC Section VIII Div. I or II, ASME B16.34, or certain EN standards as per API 6DSS. However, EN standards are not very common for the design and analysis of subsea valves, so JIP33 recommends limiting the standards to the above-mentioned ASME codes. Fig. 2.44 illustrates a deformation load analysis on the body and pup piece of a subsea ball valve according to ASME BPVC section VIII Div.02.

Bolting

JIP33 has added some additional requirements for valve bolting, summarized as follows:

- The bolting allowable stress is at maximum 70% of the bolt yield strength.
- Small size bolting less than 10 mm in diameter is not allowed.
- The bolt preloading for tightening should be according to API 6A or ASME PCC-1 or EN 1591 standards.
- Pressure-containing parts such as the body and bonnet should be jointed with stud bolts and nuts and not cap screws. Cap screws are allowed for nonpressure-containing components.

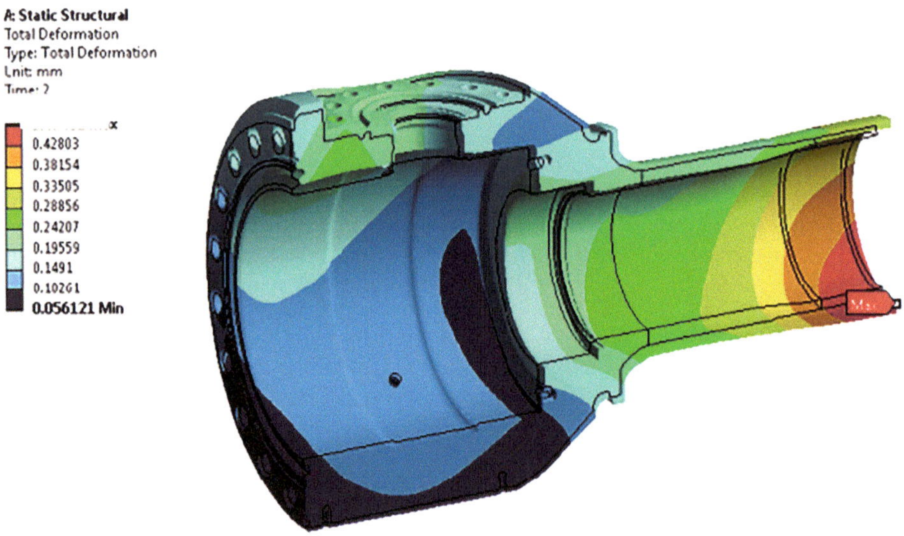

Fig. 2.44 Deformation load analysis on a subsea ball valve as per ASME BPVC.

Fig. 2.45 Body and bonnet joint through cap screw.

Fig. 2.46 Cap screws for the body and bonnet connection of a slab gate valve.

Fig. 2.45 illustrates the body and bonnet of a slab gate valve connected with cap screw bolting shown in Fig. 2.46 which does not follow the JIP33 recommendation.

Piggability

Pig stands for a piping injected gadget that is sent inside the pipeline and connected valves, mainly for cleaning or inspection purposes. JIP33 emphasizes that piggable valves and connected components such as transition pieces and pup pieces should allow the passage of the pig. For this purpose, the internal diameter of both the pup piece and transition piece (if any) as well as bore of the valve should be equal to the connected piping with some tolerances that are mentioned in the valve specifications.

Stem sealing

JIP33 mentions different stem sealings, including primary and secondary internal pressure sealings, as well as environmental valve stem sealing to prevent the ingress of seawater to the valve internals for production valves. JIP33 requires a test port for every single sealing to make individual stem sealing test possible during the validation (qualification) test. On the other hand, API 6DSS mentions that test ports for production valves are required if they are requested by the valve purchaser. The JIP33 approach, to provide a test port for each stem sealing, improves the safety and reliability of production valves. It is also recommended to test every single stem seal during qualification. The test ports for subsea valves are only used for testing the valves in the factory, so they should be sealed after test implementation. API 6DSS allows four different approaches for sealing the test port plugs such as National Pipe Tapered (NPT) threads, seal weld, flange end, or other welded connections. However, JIP33 states that "seal welding of the plugs shall be performed if specified." The recommendation of this writer is to apply seal welding on the test port plugs. Fig. 2.47 illustrates a test port plug in Inconel 625 material that blinded the test port. The red part (dark gray in print versions) shows Inconel 625 overlay seal welding.

Stem and seat injection ports

Subsea valves are designed for maintenance-free operation during their design life. Stem and seat injection points are required on the stem and seats of the valves for emergency maintenance and repair of the seat or stem sealing. API 6DSS allows using the injection ports if it is agreed between the valve supplier and purchaser. However, as stated in JIP33, there is no need to add injection points on subsea valves.

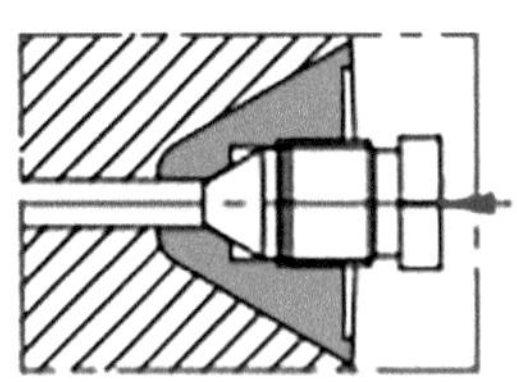

Fig. 2.47 Inconel 625 test port plug with seal welding of Inconel 625 in *red color* (dark gray in print versions).

Position indicator

JIP33 calls for position indicators that are mechanically attached to or integral with the valve. The other important consideration for the position indicator is to use antifouling material to prevent the growth of marine animals and bacteria.

Valve lifting

JIP33 requires the valve and actuator supplier to provide a sketch and clear instructions for the lifting and handling of the valves and actuators. A load verification and lifting load test should be performed for the valve and actuator lifting lugs. Fig. 2.48 shows a sketch for lifting a double acting subsea actuator with slings or chains. The load testing of the lifting lugs should be tested as per API 17D Annex K which is specified in JIP33.

Drive train safety factor

Valve drive trains are defined as all the valve components located between the operator and the closure member, including the closure member and excluding the operator, according to API 6DSS. The main components of the valve drive train are the stem and closure member. The mechanical strength of the stem is very important since it should withstand the loads coming from the valve operator. Therefore API 6DSS has asked for a minimum safety factor of two for stem design and calculations. This means that the stem of the valve should withstand at least twice the applied torque of the valve.

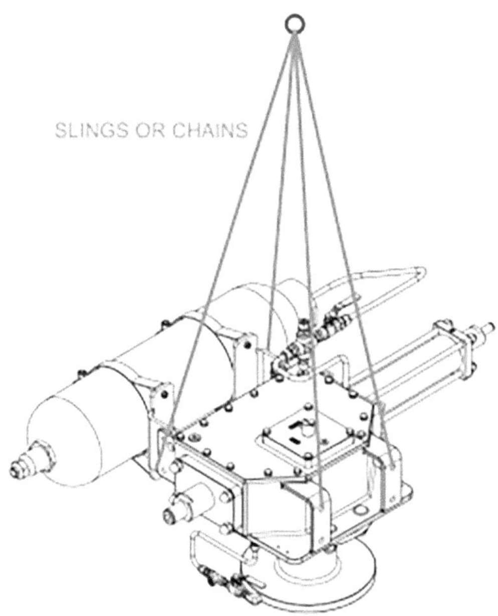

Fig. 2.48 A sketch for lifting a double acting actuator with slings or chains.

In other words, the maximum allowable stem torque should be at least twice the valve breakaway torque. JIP33 has increased this safety factor to a minimum of 2.5 of the valve torque.

Operator overpressure protection

API 6DSS specifies the means of pressure relief from the stem or bonnet of the valve to prevent overpressurizing the valve operator. JIP33 specifically puts emphasis on using a pressure relief valve for this purpose. In addition, the set pressure of the pressure relief valve should be stated on the valve general arrangement drawing, and the relief ports are required to be marked on the valve as per JIP33.

2.4.2.3 Material

Different material data sheets (MDS) for materials, such as duplex, super duplex, carbon steel, low alloy steel and nickel alloys, are provided in JIP33 Annex R, which proves information about the required chemical composition and mechanical strength of the materials, the required qualification tests, qualification of manufacturers and the process of producing the materials, etc. Table 2.21 presents a sample of MDS for a low alloy steel in ASTM A694 F60 grade which can be used for the body of the valve.

The MDSs proposed by JIP33 are not limited to the materials to be selected for valve components. Additionally, two MDSs are given in JIP33 for applying hard-facing alloys such as stellite or tungsten carbide on the valve internals. The intention of putting hard facing on the valve internals is to reduce erosion and galling. Galling is defined as metal-to-metal friction between two metaling surfaces that causes wearing. Galling could happen between the ball and seats of a ball valve due to opening and closing the valve. Thus both ball and seat surfaces or contact surfaces should be hard faced with tungsten carbide. Fig. 2.49 illustrates galling between two metallic surfaces in contact. Galling could happen between the bolt and nut as well as the stem bearing and stem. The stem bearing is a component installed around the stem of the valve to prevent side-by-side movement. JIP33 specifically focuses on galling prevention for bearings. The common approach in the valve industry is to apply a Teflon lining to the internal surface of the bearings to prevent them from galling due to friction with the stem. The emphasis of both API 6DSS and JIP33 is that corrosion allowance to compensate metal loss due to carbon dioxide in the hydrocarbon service should be considered only for noncorrosion-resistant alloys (CRA) such as carbon steel. Exotic or CRA materials are not corroded by carbon dioxide so they do not require any corrosion allowance. Annex R of JIP33 also requires an impact test. An impact test is performed on some of the materials to make sure that tested materials are compatible with the minimum temperature condition. As an example, low-temperature carbon steels are typically used and designed for a minimum design temperature of −46°C, so they are impact tested at −46°C.

Table 2.21 MDS sample for a low alloy steel as per JIP33 Annex R.

Material data sheet	**MDS no. F60**[a]			**Rev. 00**
Type of material: Carbon stool				
Product form	**Standard**	**Grade**	**Acceptance class**	**Supplementary requirement**
Forgings	**ASTM A694**	**F60**		
Scope	This MDS defines applicable options and/or requirements that supplement or amend the referenced standard specification and DNVGL-RP-0034.			
Metal Making	No additional requirement			
Manufacturing	Manufacturing and qualification of forgings shall be in accordance with DNVGL-RP-0034 SFC2.			
Chemical Composition	CE < 0.430%. All trace elements shall be reported.			
Heat Treatment	Forgings shall be delivered in quenched and tempered condition. A normalizing treatment may be used prior to quality heat treatment. If reheat treatment is required, it shall be justified, prior to carrying out the activity.			
Tensile Testing	No additional requirement.			
Impact Testing/Toughness testing	Test temperature shall be equal or lower than minimum design temperature of a valve but not more than −18°C.			
Micrographic Examination	No additional requirement.			
Extent of Testing	No additional requirement.			
Nondestructive Testing	No additional requirement.			
Repair of Defects	Weld repair shall not be permitted.			
Sour Service (additional metallurgical, manufacturing, testing and certification requirements)	No additional requirement.			
Marking	No additional requirement.			
Certification	The inspection documents shall be issued in accordance with ISO 10474/EN 10204 Type 3.1 and shall confirm compliance with this specification.			

[a]The supplementary suffix "S" shall be used to designate a material delivered in accordance with the MDS plus the additional supplementary requirements for sour service.

The additional requirement for soft or nonmetallic materials as per JIP33 is to qualify these materials as per NORSOK M-710. This standard provides a series of tests for the qualification of soft materials, which are typically used for the sealing of the valve in different areas such as the stem, body, seats, etc. in the offshore industry (see Fig. 2.50). The other important consideration required as per JIP33 is that the spring part, which is used in energized seals, should be in a CRA. The gas could ingress into the soft seals and make them fail through a phenomenon called "rapid gas compression" (RGC). This failure

Fig. 2.49 Galling on two metallic components.

Fig. 2.50 NORSOK M-710 soft material approval for the offshore industry. *(Courtesy: Ge Mao Rubber Industrial Co.)*

mechanism of soft seals could be riskier in high-pressure class gas services such as a nominal pressure of 100 bar equal to pressure class 600. However, JIP33 requires that RGC be considered for all of the seals inside valves exposed to gas service. The proper sealing property to prevent RGC is called antiexplosive decompression (AED).

Forging is the preferred choice of material manufacturing and shaping for subsea valves due to the high quality associated with the forging manufacturing process. However, both API 6DSS and JIP33 allow the use of casting according to API 20A, which addresses carbon steel, alloy steel, stainless steel, and nickel-based alloy casting for use in the petroleum and natural gas industry.

Carbon equivalent is an important parameter for carbon steel and low alloy steel; it depends on the different chemical percentage such as carbon, chromium, molybdenum, etc. A high carbon equivalent of more than 0.43% should be prevented according to both API 6DSS and JIP33 since it could reduce the weldability and increase the brittleness of the carbon and low alloy steel materials.

Subsea valve components in contact with seawater are protected against seawater corrosion by both coating and cathodic protection. Cathodic protection is done by connecting the valve to a sacrificial anode in aluminum, magnesium, or zinc that loses electrons easily and will be corroded in favor of the valve components. Components in specific materials such as duplex and super duplex that are subject to cathodic protection are at risk of HISC. Therefore the guideline provided in DNV-RP-F112 should be applied to duplex and super duplex materials subject to cathodic protection to prevent HISC. However, both API 6DSS and JIP33 include only 22Cr duplex for HISC analysis according to the DNV guideline. Fig. 2.51 illustrates a cathodic protection mechanism for a subsea pipeline.

2.4.2.4 Valve assembly

JIP33 requests that the following tasks to be performed before the valve assembly:

- Checking all the valve components to make sure that they are free from rust, damage, sharp edges, etc.

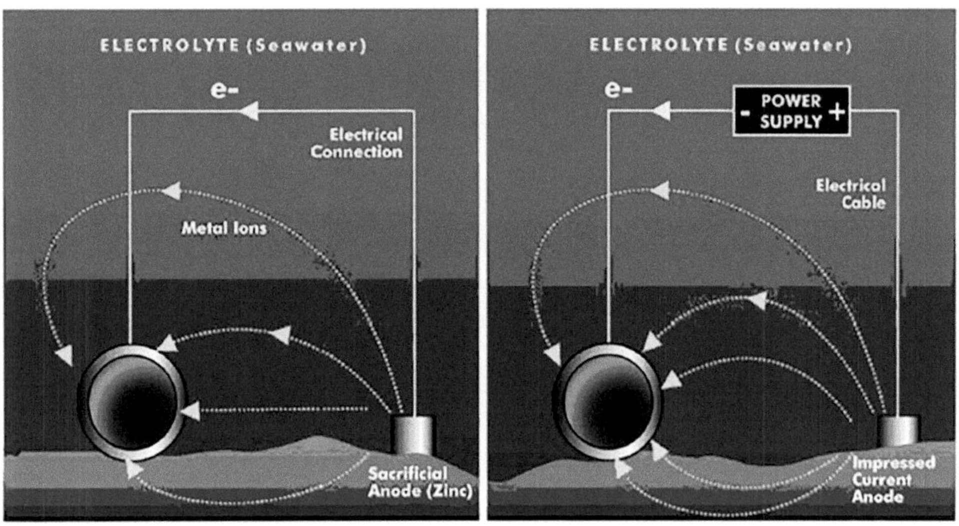

Fig. 2.51 Subsea pipe subject to cathodic protection. *(Courtesy: Wordpress.)*

- Nonmetallic sealings should not be reused after disassembling the valve and removing the sealing, and the seals should be replaced.
- Lubricant with a certain viscosity may be used for assembling the valve components. The lubricants should be removed after assembly and prior to the testing of the valves.

2.4.2.5 Valve testing

Testing is an essential part of the subsea valve manufacturing process to maintain safety and reliability and ensure that the valve can provide its required function without any need for maintenance over its design life. JIP33 provides a sequence for all the tests that should be performed as part of the FAT. Different types of tests, including pressure tests, function tests, drift test, electrical continuity measurement, etc., are listed in this sequence. Fig. 2.52 illustrates a ball valve installed on a test bench during the body and seat pressure tests. The table that lists the sequence of the tests contains different information such as the applicable section number, type of test medium (gas or water or air), the minimum duration of the tests, and the application of the test. Pressure tests are those that are applied to the body and seats of the valve to evaluate its sealability. A drift test is a

Fig. 2.52 Ball valve during the pressure test.

type of test that is performed to check that the bore or internal diameter of the valve has a specific diameter.

The important point that is added in JIP33 is that the primary sealing, including the stem seal, body and bonnet or body piece seals, should be tested during the shell test. In fact, if the test fluid during the shell test passes the primary seals and exits the relevant test port, the conclusion is that the primary seals do not function properly. The function test is performed to measure the required torque or force for opening and closing the valves as well as the speed of operation. JIP33 recommends applying the function test in two conditions: first while the valve is empty and second when the valve is under full differential pressure and the test media could be water. The test medium in the second case should be water with a corrosion inhibitor.

Further reading

H.E. Amaya, One subsea, A.D., Hydrogen embrittlement stress cracking (HISC) in the oil and gas industry, in: A Cameron & Schlumberger Company, Conference Paper, Materials Science Technology, Columbus, OH, 2015.

American Petroleum Institute (API) 6A, Specification for Wellhead and Tree Equipment, 21st ed., API, Washington, DC, 2018.

American Petroleum Institute (API) 17D, Design and Operation of subsea Production Systems, subsea Wellhead and Tree Equipment, 2nd ed., API, Washington, DC, 2011.

American Petroleum Institute (API) 6DSS, Specification for Subsea Pipeline Valves, third ed., API, Washington, DC, 2017.

American Petroleum Institute (API) 14A, Specification for Subsurface Safety Valve Equipment, 12th ed., API, Washington, DC, 2015.

American Petroleum Institute (API) 17H, Remotely Operated Tools and Interfaces on Subsea Production Systems, third ed., API, Washington, DC, 2019.

American Petroleum Institute (API) 17N, Recommended Practice on Subsea Production System Reliability, Technical Risk, and Integrity Management, second ed., API, Washington, DC, 2017.

American Petroleum Institute (API) 17Q, Recommended Practice on Subsea Equipment Qualification, second ed., API, Washington, DC, 2018.

American Petroleum Institute (API) 20E, Alloy and Carbon Steel Bolting for Use in the Petroleum and Natural Gas Industries, first ed., API, Washington, DC, 2012.

American Petroleum Institute (API) 20F, Corrosion Resistant Bolting for Use in the Petroleum and Natural Gas Industries, first ed., API, Washington, DC, 2015.

American Petroleum Institute (API) 6ACRA, Age Hardened Nickel-Based Alloys for Oil and Gas Drilling and Production Equipment, first ed., API, Washington, DC, 2015.

American Society of Mechanical Engineers (ASME), Design and Fabrication of Pressure Vessels. Boiler and Pressure Vessel Code, ASME Section VIII Div.02, New York, NY, 2012.

American Society of Mechanical Engineers (ASME), Buttwelding Ends. ASME B16.25, ASME, New York, NY, 2017.

American Society of Mechanical Engineers (ASME), Carbon, Alloy and Stainless-Steel Pipes. ASME B36.10/19, ASME, New York, NY, 2004.

American Society of Mechanical Engineers (ASME), Gas Transmission and Distribution Piping System. ASME B31.8, ASME, New York, NY, 2018.

A. Bahrami, P. Woollin, Hydrogen induced stress cracking of duplex stainless-steel subsea components, in: TWI Ltd. Conference paper, Offshore Mechanics and Arctic Engineering (OMAE) Conference, Shanghai, China, 2010.

Det Norske Veritas (DNV) RP-F-112, Design of Duplex Stainless-Steel Subsea Equipment Exposed to Cathodic Protection, Det Norske Veritas, Hovik, 2008.

International Association of Oil and Gas Products, Supplementary Specification to API Specification 6DSS Subsea Pipeline Valves, JIP33, [online] Available at: https://www.iogp-jip33.org/wp-content/uploads/2020/08/Supplementary-Specification-to-API-Specification-6DSS-Subsea-Pipeline-Valves-S-708v2020-08.pdf (Accessed 12 September 2020).

International Electrotechnical Commission (IEC), IEC 61508 Functional Safety of Electrical/Electronic/Programmable Electronic Safety-Related systems part 1: General Requirements, second ed., 2010. Geneva.

International Electrotechnical Commission (IEC), IEC 61508, Functional Safety of Electrical/Electronic/Programmable Electronic Safety-Related Systems, Part 4: Definitions and Abbreviations, second ed., 2010. Geneva.

International Electrotechnical Commission (IEC), IEC 61508, Functional Safety of Electrical/Electronic/Programmable Electronic Safety-Related Systems, Part 2: Requirements for Electrical/Electronic/Programmable Electronic Safety-Related Systems, second ed., 2010. Geneva.

International Electrotechnical Commission (IEC), IEC 61508, Functional Safety of Electrical/Electronic/Programmable Electronic Safety-Related Systems, Part 6: Guidelines on the Application of IEC 6158–2 and IEC 6158-3, second ed., 2010. Geneva.

International Organization of Standardization (ISO) 10423, Petroleum and Natural Gas Industries—Drilling and Production Equipment-Wellhead and Christmas Tree Equipment, fourth ed., 2009. Geneva.

International Organization of Standardization (ISO) 13628-4, Petroleum and Natural Gas Industries—Design and Operation of Subsea Production Systems-Part 4: Subsea Wellhead and Tree Equipment, second ed., 2010. Geneva.

International Organization of Standardization (ISO) 14723, Petroleum and Natural Gas Industries—Pipeline Transportation System—Subsea Pipeline Valves, second ed., 2009. Geneva.

International Organization of Standardization (ISO) 13628-8, Petroleum and Natural Gas Industries—Design and Operation of Subsea Production Systems-Part 8: Remotely Operated Vehicle (ROV) on Subsea Production Systems, first ed., 2006. Geneva.

International Organization of Standardization (ISO) 15156, Petroleum and Natural Gas Industries—Material for Use in H_2S Containing Environments in Oil and Gas Production, third ed., 2015. Geneva.

International Organization of Standardization (ISO) 5208, Industrial Valves-Pressure Testing of the Metallic Valves, fourth ed., 2015. Geneva.

Norsok, Surface Preparation and Protective Coating. Norsok M-501, sixth ed., 2012. Lysaker.

M. Solnordal, S. Wasberg, G. Heiberg, E. Odd, Hydrogen Induced Stress Cracking (HISC) and DNV-RP-F112, Det Norske Veritas, Hovik, 2009.

K. Sotoodeh, Valve preservation requirement, Valve World Mag. 21 (5) (2016) 53–56.

K. Sotoodeh, Valve failures, analysis and solutions, Valve World Mag. 23 (11) (2018) 48–52.

K. Sotoodeh, Welding a carbon steel body valve to 22Cr duplex pipe: a case study, Valve World Mag. 24 (5) (2019) 38–41.

K. Sotoodeh, Manifold technology in offshore industry. Am. J. Mar. Sci. 8 (1) (2020) 6–13, https://doi.org/10.12691/marine-8-1-2.

Workshop Agreement NORSOK M-WA-01, HISC Guideline for Duplex Stainless Steel. Revision 01, NORSOK, Lysaker, 2005.

CHAPTER THREE

Subsea ball valves

Contents

3.1 Introduction and applications

Ball valves are quarter turn valves, which means that they require a 90-degree rotation of the stem and ball-shaped closure member to move between open and closed positions. Subsea ball valves are not selected for wellheads or Christmas trees, while through conduit gate (TCG) valves have a long history of being used on the trees. But ball valves provide some advantages over gate valves such as less height, faster operation (i.e., possibility of quick opening and closing), less stem wear, less sensitivity of the valve operator to the depth, the possibility of being made in double block and bleed (DBB), etc. which are explained in detail in Chapter 4. The application of ball

Subsea Valves and Actuators for the Oil and Gas Industry
https://doi.org/10.1016/B978-0-323-90605-0.00016-5

valves is for manifolds and subsea distribution units (SDUs). SDUs are explained in more details in Chapter 1. Ball valves located on manifold headers are typically in 8″ size and larger. These valves can either be operated manually or by an actuator. Fig. 3.1 illustrates a 16″ ball valve in pressure class 7500 psi located on a subsea manifold and equipped with a symmetrical double acting rack and pinion hydraulic actuator. Fig. 3.2 illustrates a 10″ ball valve in class 7500 psi in the fabrication yard located on a subsea manifold header.

Ball valves for SDUs are typically small valves in ½″ or ¾″ with manual operation. Fig. 3.3 illustrates a ½″ ball valve in a pressure class of 10,000 psi that is operated manually by a remotely operated vehicle (ROV) by means of a low torque paddle installed on the top of the valve. These valves are located inside the SDU and supply chemicals such as wax and scale inhibitors to the manifold piping system. In addition, some of these valves supply hydraulic pressure equal to 5000 psi, known as "high pressure," to the actuators and some are used for delivering medium hydraulic pressure equal to 3000 psi to the actuators. Despite these differences, all of the ½″ valves that are used for chemical injection

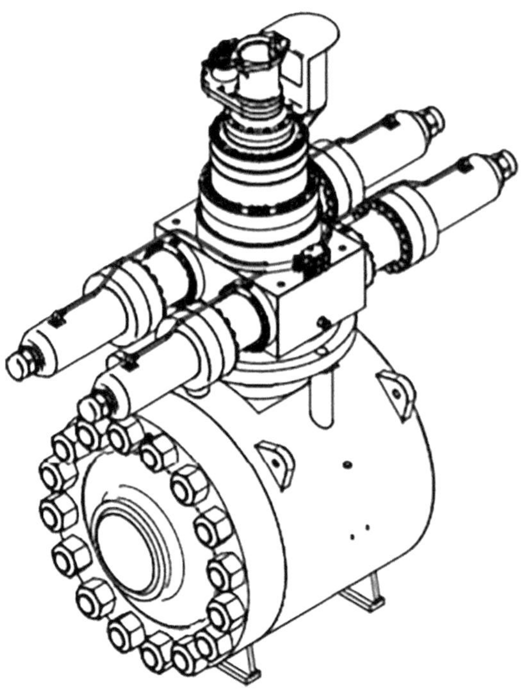

Fig. 3.1 16″ class 7500 ball valve with a symmetrical rack and pinion double acting hydraulic actuator located on a subsea manifold.

Fig. 3.2 10″ class 7500 ball valve with ROV override in the construction yard.

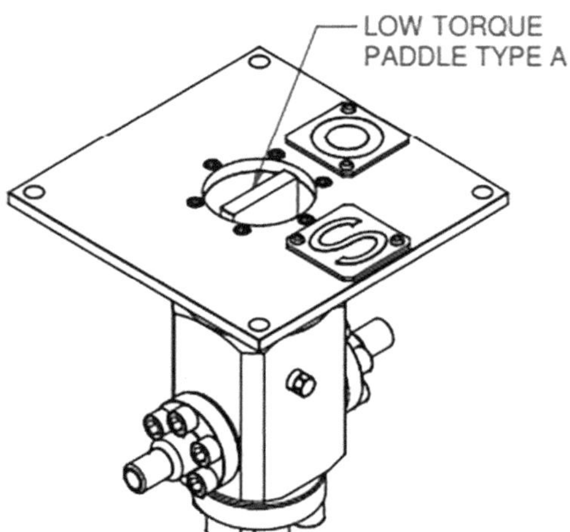

Fig. 3.3 ½″ ball valve class 10,000 psi with a low torque paddle on SDUs for supplying chemicals and hydraulic fluid.

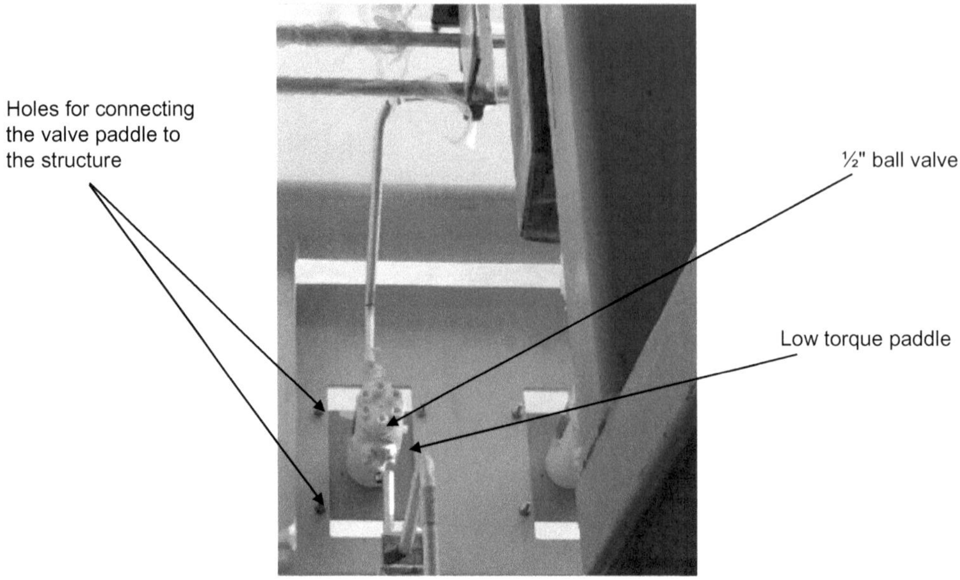

Fig. 3.4 ½″ ball valve with low torque paddle connected to the structure through the holes.

and hydraulic pressure are standardized to 10,000 psi. Standardization of the valves and piping leads to better control and management of the valves and reduces the number of necessary tests. The important point for the operation of a ½″ ball valve, such as the one illustrated in Fig. 3.3, is to have a very robust connection between the low torque paddle and the valve to prevent the separation of the paddle from the valve during ROV operation. Normally, the torque for operation of these valves should be low: less than 200 N m. The four holes on each corner of the low torque paddle are used to connect it to the structure as illustrated in Fig. 3.4. Fig. 3.5 illustrates all of the small bore valves located in one manifold; all are ½″ with a low torque paddle on the top. Fig. 3.6 illustrates a ¾″ ball valve with a ROV bucket located in an SDU for chemical and annulus monitoring.

In addition, ball valves can be used as subsea isolation valves (SSIVs) installed on the pipeline to protect the platform and personnel from an undesirable event in the subsea pipeline. More information about SSIVs is available in Chapter 7. Fig. 3.7 illustrates an SSIV actuated ball valve. In conclusion, subsea ball valves are located on the manifold, flowline, pipeline end manifold (PLEM), and pipeline end terminal (PLET), and can be used for different applications such as emergency shutdown for SSIVs, high pressure and high temperature as well as high integrity pressure or piping protection systems (HIPPSs).

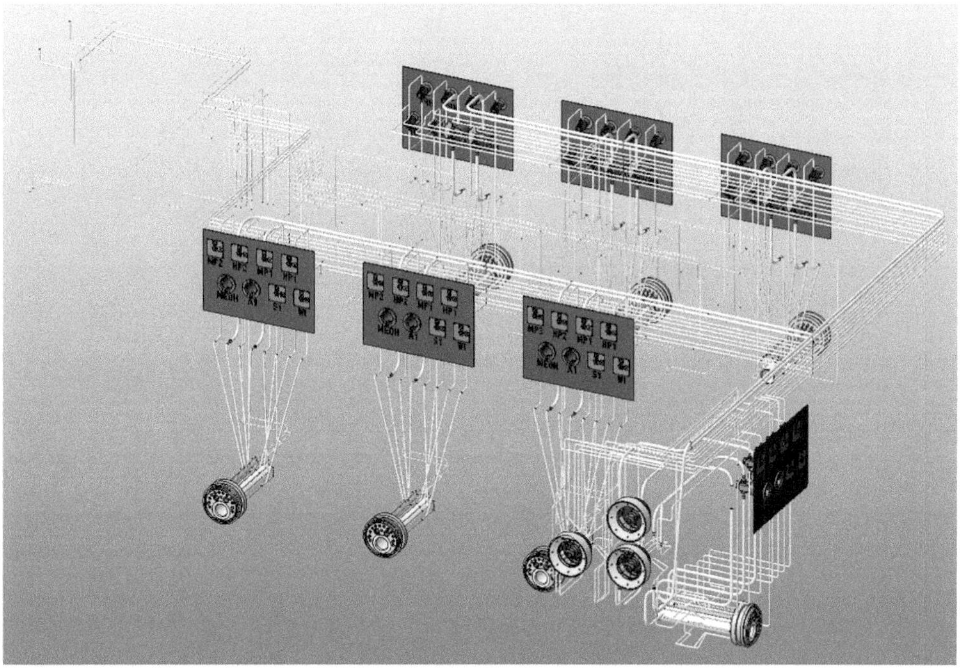

Fig. 3.5 All small bore ball valves in ½″ with a low torque paddle located on a manifold.

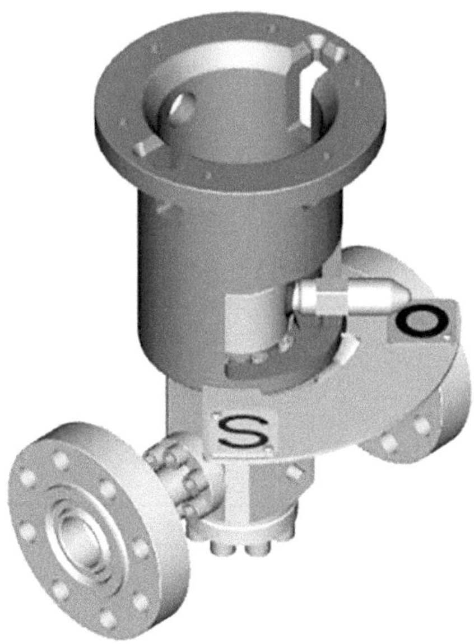

Fig. 3.6 3/4″ ball valve with ROV bucket for chemical supply located on an SDU.

Fig. 3.7 SSIV ball valve. *(Courtesy: Pinnacle.)*

3.2 Ball valve design features

3.2.1 Size

The size of ball valves do not follow the size ranges provided in the API 6A and API 17D standards in most cases. Ball valve size is typically equal to the nominal pipe size (NPS) and is selected mainly according to API 6DSS, for example: ½″, ¾″, 10″, 14″, 16″, 18″, 22″, 30″, etc. However, there are cases in which a ball valve could be sized according to API 6A sizes such as 9″ and 11″. As an example, Fig. 3.8 illustrates a DBB valve made of two ball valves in 9″ size as per API 6A; the bleed valve is 2″.

3.2.2 Pressure class

Subsea ball valves typically follow the pressure classes given in API 6A such as 5000, 10,000, 15,000, and 20,000 psi. Intermediate pressure classes such as 7500 psi can be used instead of 10,000 psi to reduce the weight, space, and cost of the valve and operator.

3.2.3 Temperature class

Temperature classes could be according to the classes given in API 6A or different. The minimum and maximum design temperature of the valve is given in the valve data sheets.

3.2.4 Bore

Ball valves could have full bore (see Fig. 3.9), reduced bore (see Fig. 3.10), or specific bore. If the ball valves are piggable, then the bore of the valve should be equal to the connected pipe. It is important to note that normally some tolerances exist for the pig

Fig. 3.8 Double block 9″ ball valve with a 2″ bleed valve between two block valves. *(Courtesy: Galperti Engineering.)*

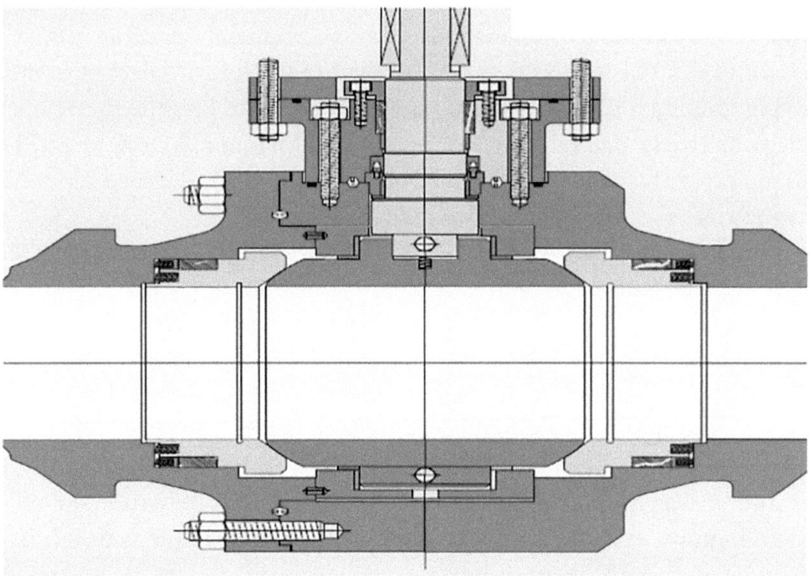

Fig. 3.9 Full bore ball valve.

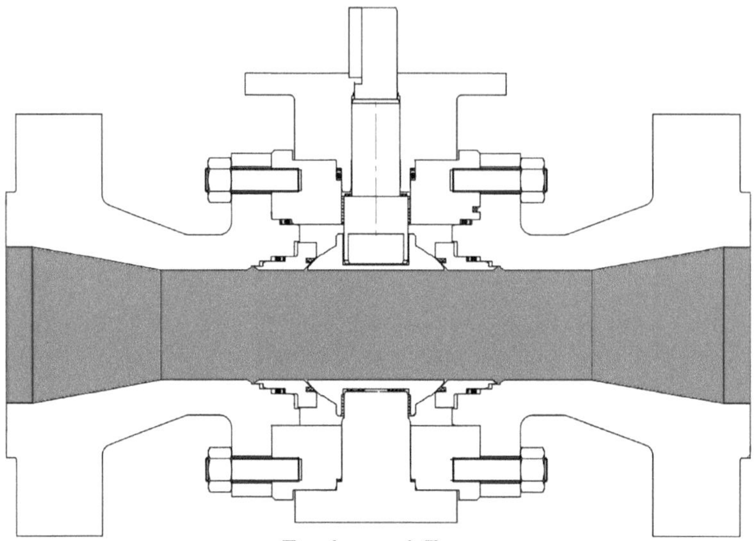

Fig. 3.10 Reduced bore ball valve. *(Courtesy: EOG Platform.)*

diameter, which could be 3% of the piping diameter. Therefore the bore of the valve is not exactly equal to the meeting pipe and some tolerances like ±3% may be allowed.

The preferred bore for ball valves in sizes above 2″ is reduced bore. Reduced bore selection makes the ball valve lighter, more compact, and less expensive. In addition, some saving in terms of the space, weight, and cost of the valve operator can be achieved. However, typically two questions should be asked before making the decision to go for a reduced bore ball valve. The first is related to flow assurance and pressure drop within the reduced bore valve. The second question is whether the line connected to the valve is subject to pig running for inspection or cleaning purposes. If pressure drops and reduced flow capacity due to the ball valve bore or port reduction is not an issue and the line is not subject to pig running, then the ball valve could be selected as reduced bore. API 6DSS has following rules regarding valve bore reduction:

For sizes 12″ and smaller, the bore could be one size less than the end connection size of the valve. As an example, a reduced bore ball valve located on a 12″ line could be selected as a 12″ end connection with a reduced bore equal to 10″. The dimension of this valve can be written as 12″ × 10″.

For sizes of 14″ and larger, the bore could be reduced into two sizes less than the end connection of the valve. As an example, a 14″ reduced bore ball valve connected to a 14″ line could be selected as a valve with 14″ ends and a bore reduction equal to 10″, which is two sizes smaller than 14″. The dimensions of this valve can be written as 14″ × 10″.

Based on industry experience, it is possible to use reduced bore subsea ball valves in small sizes such as 1″ × ¾″ as an example.

3.2.5 Design standards

Subsea ball valves are designed and tested mainly based on the standards of the American Petroleum Institute (API) and the International Organization for Standardization (ISO). Each API standard for subsea valve and actuator design and testing has an equivalent ISO standard as follows:

1. API Spec.17D, "Design and operation of subsea production systems—Subsea wellhead and tree equipment," identical to ISO 13628-4.
2. API Spec. 6DSS, "Specification for subsea pipeline valves," identical to ISO 14723.
3. API Spec. 6A, "Wellhead and tree equipment," identical to ISO 10423.

3.2.6 Seat design

Ball valve seats can be in soft thermoplastic materials or metal. Thermoplastic materials are explained in detail in Chapter 13. "Soft seat" has advantages such as reducing the valve weight and cost and increasing the possibility of having zero leakage from the valve's seats. However, soft materials have limitations such as temperature and pressure resistance, potential damage in sandy and particle-containing services, and damage from operating the valve against full differential pressure. Metal seats are more typical for subsea ball valves since they are more robust. A soft seat, as illustrated in Fig. 3.11, includes a metallic seat carrier and a seat insert in soft material. However, a metal seat for a ball valve includes just a metallic ring without any insert as shown in Fig. 3.12. The other important seat feature for subsea ball valves is a floating seat. Seats can be categorized as floating or fixed. Floating seats are energized with springs at the back of the seat. Typical self-relieving or double piston effect seats, which are explained in the next section, are applicable for floating seats.

The other important seat design figure is the choice of "self-relieving" seats, also called "unidirectional" or "single piston effect," versus "double piston effect seats." Fig. 3.13 illustrates a ball valve with one single piston effect seat on the left (self-reliving)

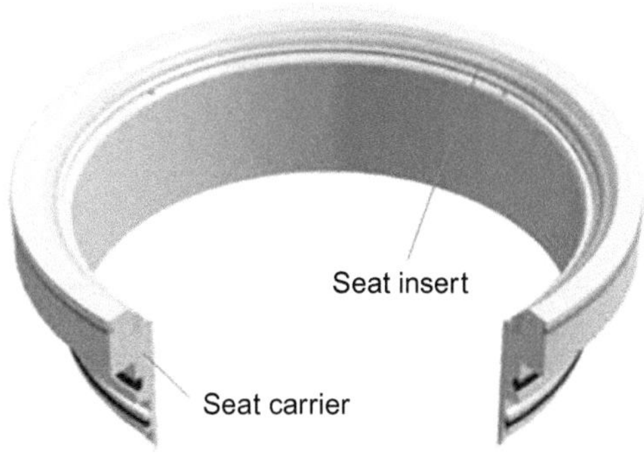

Fig. 3.11 A ball valve soft seat including a metallic seat carrier and soft seat insert.

Fig. 3.12 Metal seat of a ball valve.

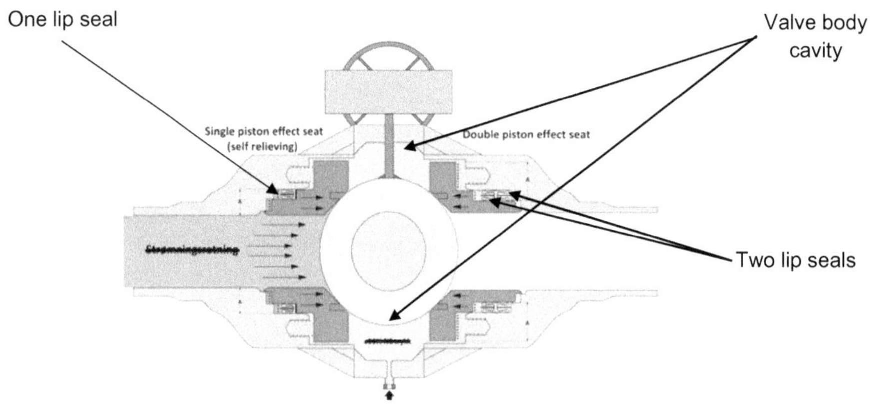

Fig. 3.13 A DIB2 ball valve with one single piston and one double piston effect seat. *(Courtesy: Valve World/Elsevier.)*

and one double piston effect seat on the right. The area in the green color (dark gray in print versions) between two seats and above the ball in yellow color (light gray in print versions) is called the body cavity of the ball valve. The ball valve in Fig. 3.13 has two different seat designs, called double isolation and bleed (DIB2). A self-relieving seat is illustrated on the left side of Fig. 3.13. This seat is unidirectional, meaning that it provides sealing only from the fluid in the red color (dark gray in print versions) on the left side of the valve in closed position. The fluid in red color (dark gray in print versions) can enter the cavity area when the valve is moving from closed to open position or vice versa. In that case, the self-relieving seat releases the fluid trapped in the cavity to the upstream side, shown in red (dark gray in print versions).

This valve is unidirectional. The flow direction should be marked on the body of the valve and on the tag plate to show the self-relief and double piston effect seat direction on the valves. Fig. 3.14 illustrates a double piston (DP) seat marking on a ball valve flange. The DP marked on the flange does not provide good visibility, so it would not be considered a good seat marking example. Alternatively, a tag plate, such as the one illustrated in Fig. 3.15, can be attached to the valve body with rivets to show the direction of each seat. The right-hand side with two arrows shows a DP effect seat, and the one with one arrow on the left side shows a self-relieving seat. Subsea isolation valves (SSIVs), which are explained in Chapter 7, are typically ball valves and are operated on the flowline to protect the platform against any undesirable failure subsea. These valves could be in DIB2, and the end user may prefer to have the double piston effect seat toward or on

Fig. 3.14 Double piston marking on the body of the valve. *(Courtesy: Valve World.)*

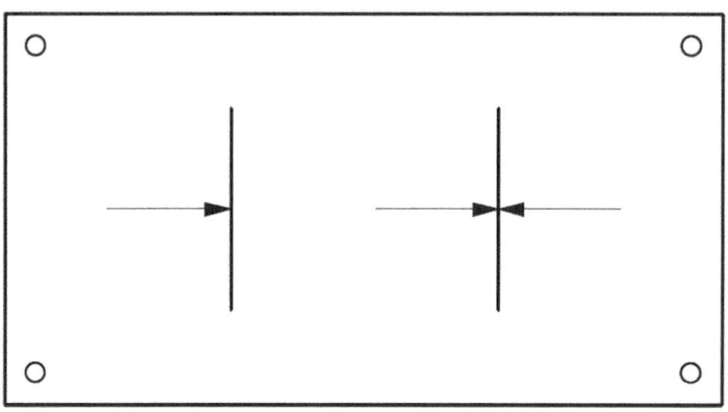

Fig. 3.15 Marking double piston effect and self-relieving on the body of a ball valve as per API 6D.

the side of the platform for higher safety. If the double piston effect seat were placed on the upstream side and the self-relieving seat were placed on the downstream side toward the platform, what could happen during valve shutdown in the case of an emergency for the protection of the platform? If the first seat barrier, which is a double piston effect seat, fails then the fluid can enter the valve cavity and the second seat barrier, which is a self-relieving seat, releases the fluid to the downstream pipeline and the platform. However, if a self-relieving seat is used as the primary barrier against the upstream pipeline pressure and *it* fails, the fluid that is leaked to the cavity of the valve would not leak further from the double piston effect seat toward the platform. Thus it is safer and more highly recommended to have the double piston effect seat on the downstream side, toward the platform, for SSIVs. The second question is why would DIB2, which is not bidirectional, be preferred over DIB1 with two double piston effect seats for better protection? There is probably no advantage in selecting two double piston effect seats compared to a situation in which DIB2 is used with a self-relieving seat upstream and a double piston effect seat is used downstream. DIB1 is more expensive than DIB2 and requires a pressure relief valve on the valve cavity in the case of liquid service. The accumulated pressure in the cavity of a DIB1 type ball valve cannot be released out through the double piston effect seats; also, the liquid, unlike gas, is not compressible. Thus a pressure relief valve should be installed on the cavity in the case of liquid service to release the trapped liquid in the cavity and prevent overpressurizing the fluid which can damage the valve body and other components. However, adding a pressure relief valve to the body of the valve could result in an extra hole in the body and thus a potential leakage point, which is a disadvantage. Fig. 3.16 illustrates a pressure relief valve on the body cavity of a valve.

According to API 6DSS, the pressure inside the cavity should be released by the self-relieving seat at a pressure not exceeding 33% above the design pressure. The valve

Fig. 3.16 Pressure relief valve on a valve body cavity.

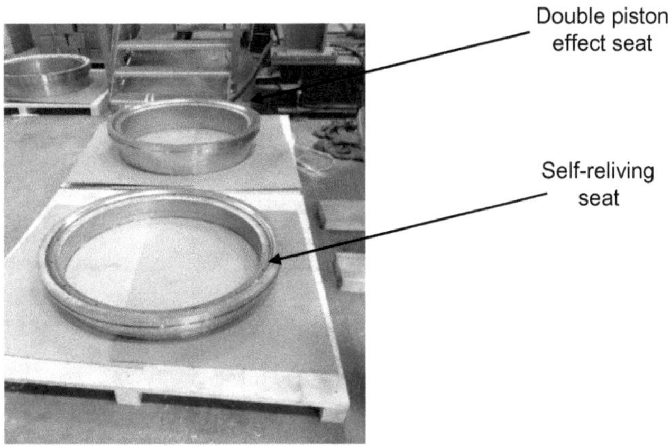

Fig. 3.17 Self reliving and double piston effect seats for a ball valve.

specification may require a higher safety margin such as 10% above the design pressure. Self-relieving seats are called unidirectional since they can provide sealing just in one direction. The design of a self-relieving seat has just one lip seal in the area between the seat and the body. A lip seal is also a unidirectional seal. The seat on the right-hand side of Fig. 3.13 is bidirectional and a double piston effect type. It contains two lip seals to provide sealing from both directions: one direction is from the cavity to the valve bore in white color and the other direction is from the white valve bore to the cavity. A double piston effect seat is larger than a single piston effect seat as illustrated in Fig. 3.17.

3.2.7 Top entry vs side entry

The internal parts of top entry ball valves may be accessed by removing the bonnet from the top of the valve after removing the bolts. This valve has the advantage of online maintenance without any requirement to remove the valve from the line. The advantage of being able to perform online maintenance is one of the main reasons for using top entry ball valves on the pipeline system topside and in other sectors of the oil and gas industry. However, it should be noted that subsea valves do not require any maintenance, so choosing top entry due to the possibility of online maintenance is not relevant in the subsea sector of the oil and gas industry. Top entry ball valves are typically welded to the piping system instead of attached via flange connections. Using welding end connections for top entry valves has some advantages, such as reducing the possibility of leakage from the valve and saving two flanges on the piping. Fig. 3.18 illustrates a subsea top entry ball valve welded to the meeting or connected piping from both ends with pup pieces. Fig. 3.19 illustrates an 8″ subsea ball valve with a design pressure of 233 bar in the fabrication yard. This valve weighs 2200 kg and will be used at a depth of 800 m.

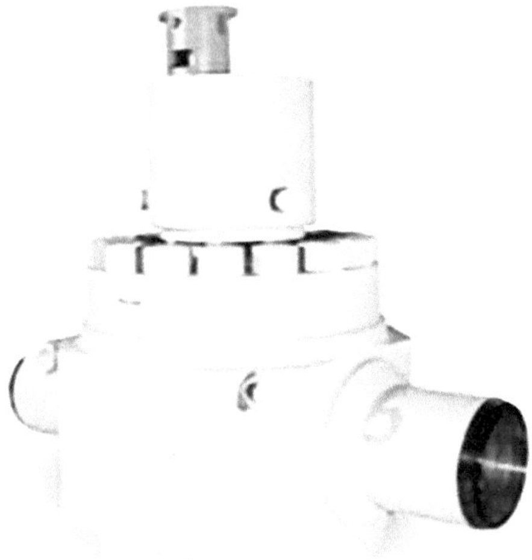

Fig. 3.18 Top entry subsea ball valve. *(Courtesy: Perar.)*

Fig. 3.19 Top entry subsea ball valve in the construction yard.

Side entry ball valves are bonnetless and can be made in two or three body pieces typically bolted together. A side entry ball valve is normally lighter than a top entry ball valve of the same size and pressure class. The bonnet of a top entry ball valve is usually heavy, which makes top entry ball valves relatively heavy compared to those with side entry design. The valves illustrated in Figs. 3.1 and 3.2 are side entry subsea ball valves.

Fig. 3.20 Side entry subsea ball valve. *(Courtesy: Viar Valvole.)*

Fig. 3.20 illustrates a subsea side entry ball valve welded to the pipeline through a pup piece. Side entry ball valves are flanged end in sectors of the oil and gas industry other than subsea. However, subsea side entry ball valves are typically welded to the piping as illustrated in Figs. 3.1, 3.2, and 3.20. The body piece, which is similar to a flange connection and integrated with a pup piece, is called body closure by some of the valve manufacturers.

In general, top entry ball valve design has the following advantages over side entry:

- Maintenance can be performed for this valve while it is online, without removing the valve from the piping. As discussed before, online maintenance possibility is not an advantage for subsea, since subsea valves are maintenance free.
- More flexibility in stem enlargement in case a more robust and thicker stem is required to improve the maximum allowable stem torque (MAST). MAST is defined as the maximum torque or force that the stem of the valve can withstand and bear without any breakage. There are typically two ways of improving MAST: the first is by upgrading the stem material to a material with higher mechanical strength, and the second is by increasing the stem diameter and making the stem thicker. The ball valve components, including the stem and the closure member (ball), should bear the loads produced by the actuator and ROV manual override. If a safety factor of two has been specified by the end user for manual operation as well as actuation of the ball valve, it means that the actuator and manual ROV produce torque values at least twice the torque required for valve operation under the most severe conditions. The most severe condition, as an example, could be referred to as opening the valve under full differential pressure across the valve. MAST should be equal to or higher than the highest torque value produced by the ROV and actuator.

- Top entry design can withstand higher load values from the pipeline compared to side entry design; Figs. 3.21–3.23 illustrate axial, bending, and torsion pipeline loads on a top entry ball valve located on a pipeline.

Fig. 3.24 shows the body of a ball valve with top entry design during welding to one of the pup pieces. The welding of the pup piece to the valve body is done before the complete valve assembly. One reason to weld the pup piece to the body of the valve before assembly is to avoid damage to the valve internals, especially the soft seals, from the heat produced during the welding.

After top and side entry design, the third choice for the body and bonnet design of the valves is fully welded, in which the body and bonnet are not bolted together. Fully

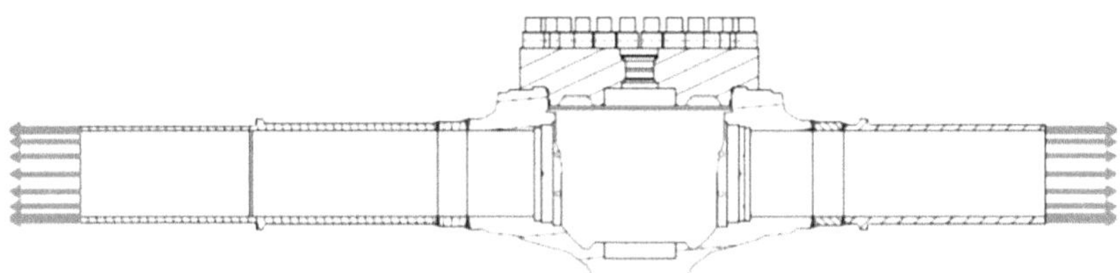

Fig. 3.21 Axial loads on a top entry ball valve. *(Courtesy: Valve World/Elsevier.)*

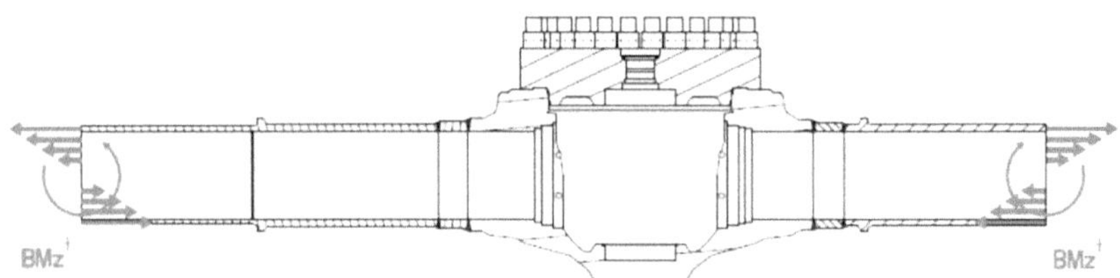

Fig. 3.22 Bending loads on a top entry ball valve. *(Courtesy: Valve World/Elsevier.)*

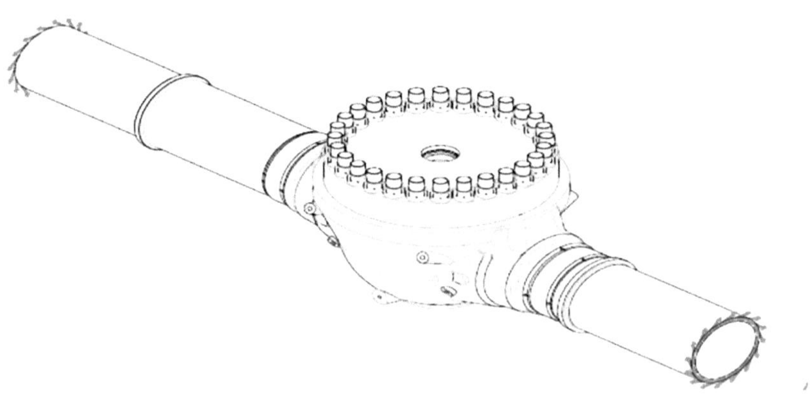

Fig. 3.23 Torsion loads on a top entry ball valve. *(Courtesy: Valve World/Elsevier.)*

Fig. 3.24 Welding the body of a top entry ball valve to a pup piece. *(Courtesy: Mir Valve.)*

welded valves have less possibility of leakage compared to side entry or top entry in which gaskets are used between two body or body/bonnet joints. Figs. 3.25 and 3.26 illustrate a fully welded ball valve for subsea application.

3.2.8 Floating vs trunnion ball

In general, floating ball valves are used for smaller size and lower pressure class ball valves. A floating ball valve does not have any support under the ball as illustrated in Fig. 3.27. Fig. 3.28 illustrates a trunnion mounted ball valve with trunnion support under the ball. Trunnion support can facilitate operating the ball valve more easily at a lower torque value. If a large size and high-pressure class ball valve (e.g., 8″, 7500 psi) is selected as floating, the operating torque would be very high. It is therefore common to select trunnion mounted ball valves in the subsea sector of oil and gas engineering. Fig. 3.29 shows a detailed view of a trunnion mounted ball valve. The trunnion could be in the shape of a plate or trunnion support.

3.2.9 Antiblowout stem

An antiblowout stem means that the stem pushing up to the outside should be prevented. The pressure inside the valve has a tendency to always push the stem up. Blowout of the stem could damage the operator who is working with the valve. The solution is to have a thicker area for the stem, which is stopped by a piece of valve body or bush extension as

Fig. 3.25 Fully welded ball valve for subsea. *(Courtesy: Mir Valve.)*

Fig. 3.26 Fully welded ball valve for subsea. *(Courtesy: Mir Valve.)*

illustrated in Fig. 3.30. Some valve manufacturers may not apply an antiblowout stem design like the one shown in Fig. 3.30 where the stem step is located at the bottom of the stem. This design could be affected by the seawater head pressure and could cause side movement on the stem which could be transferred to the ball and seat of the valve and disturb the sealing of the valve through the seats. Thus valve manufacturers may consider adding a stem shoulder or stem step to the upper part of the stem to prevent the effect of the seawater head on the valve sealing.

Fig. 3.27 Floating ball valve.

Fig. 3.28 Trunnion mounted ball valve.

3.2.10 End-to-end

Ball valves are typically supplied with two pup pieces on both ends, with end-to-end dimensions according to the manufacturer standard. The end-to-end and face-to-face length of ball valves for flange and welded ends are given for typical American Society of Mechanical Engineers (ASME) pressure classes and not API pressure classes.

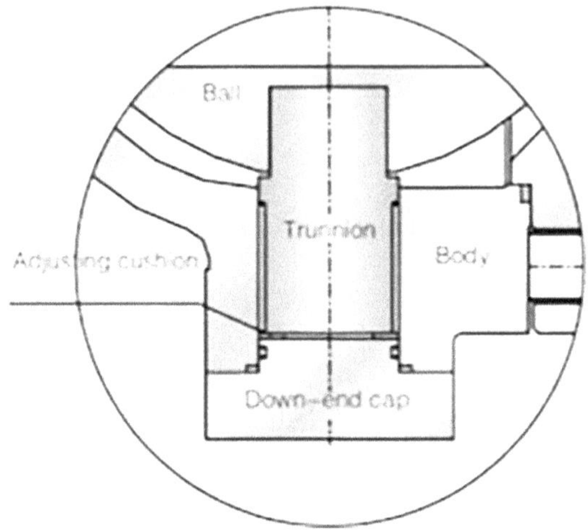

Fig. 3.29 Trunnion mounted support detail.

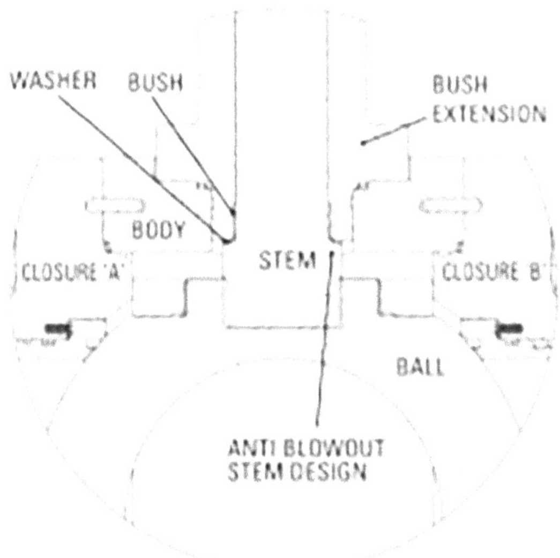

Fig. 3.30 Antiblowout stem design. *(Courtesy: CAM Industry.)*

3.3 Ball valve components

Fig. 3.31 illustrates a detailed drawing of a small bore ball valve in ¾″ and pressure class 10,000 psi. The aim of this section is to go through the different parts of a small size ball valve and provide a detailed review. The choices of materials are provided in the review.

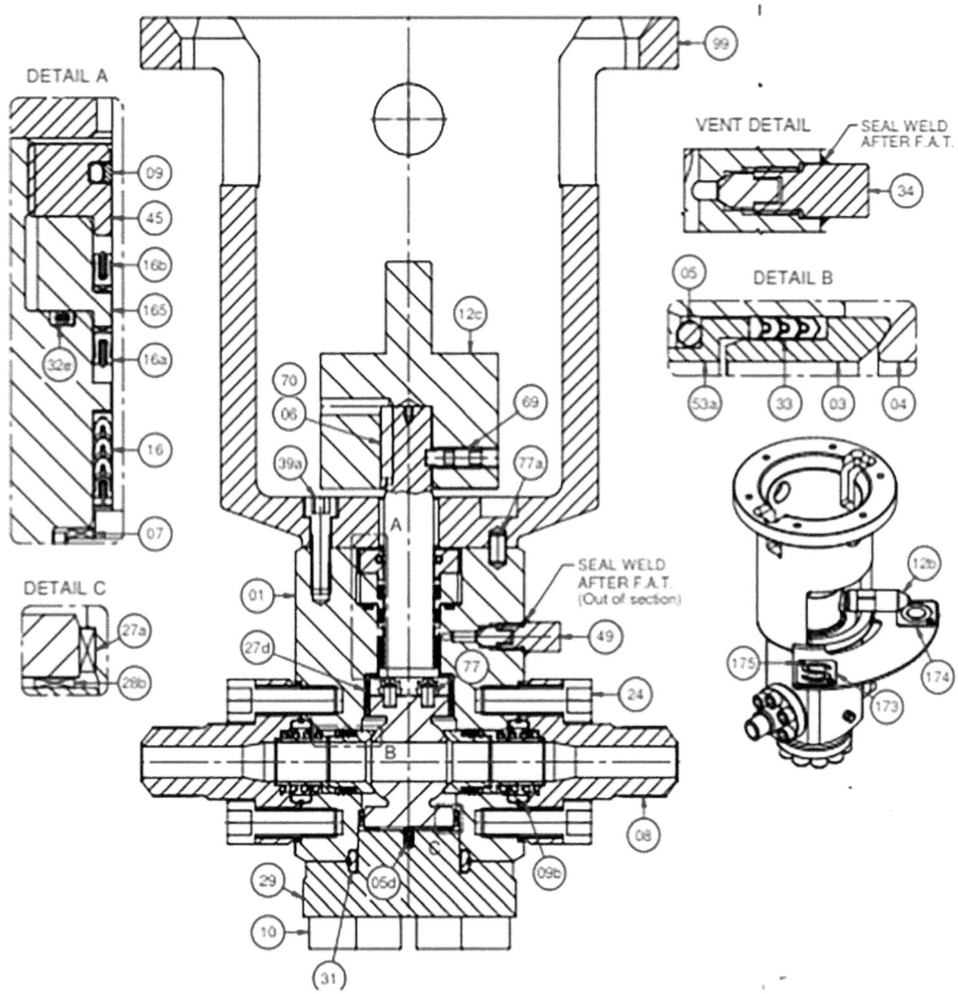

Fig. 3.31 Detailed drawing of a ¾″ pressure class 10,000 psi. *(Courtesy: Perar.)*

Item #1 is the body of the valve, which is the main pressure-containing part. The design of the body is important, since a failure of the body leads to leakage to the sea with highly negative, adverse consequences. The body of a such a small ball valve could not be in carbon steel with an overlay of Inconel 625. The first reason is that applying Inconel 625 overlay on a valve of such small size is challenging and difficult, and the second reason is the relatively high cost associated with placing an Inconel 625 overlay on carbon steel compared to body material in 25Cr super duplex. The body of the valve could be in 25Cr super duplex or Inconel 625 as an alternative. Item #8 is the body closure, which is integrated with the valve pup piece. The valve pup piece is typically integrated with the body or body closure of valves in small sizes such as less than 2″. The material of the body closure and integral pup piece should be same as the body of the ball valve, which could be 25Cr super duplex or Inconel 625 for example. Item #4 in detail B is the ball of the valve. The ball should be made in the same material as the body, or a material with higher corrosion resistance at minimum. The ball in this case could be in 25Cr super duplex or Inconel 625. Item #3 in detail B is the seat of the valve. Subsea ball valves typically have metal seats. The material of the seat could be in 25Cr super duplex or Inconel 625, i.e., the same as the ball material choices. However, tungsten carbide hard facing in a relatively thick layer such as 0.1 mm or higher should be applied on both the ball and seat contact surfaces to avoid galling on both the seats and the ball. Galling is a kind of wear caused by adhesion and friction between two sliding metallic surfaces. Item #5 is a spring located at the back of the seat to provide force for pushing the seat forward toward the ball. This type of seat is floating since it is energized through a spring. The material of the spring could be Elgiloy, which is a strong cobalt alloy suitable for seawater service. Item #5d is another spring material installed between the ball and body of the valve to provide electrical continuity between these two components. Providing electrical continuity assures the effectiveness of the cathodic protection in a subsea environment. Item #6 is the stem of the valve, which could be in 25Cr super duplex or Inconel 625, the same as the ball and seats material. The valve is small, so the operation torque for opening and closing this valve should not be high. This means that, most likely, there is no need to upgrade the stem material to a higher mechanical strength material such as a hard facing nickel alloy like Inconel 718 or 725. Item #6 together with Item #70 shows the stem key of the valve. The stem key of the valve shows the direction or position of the ball when the operator (gear or actuator) and the position indicators are off the valve. Fig. 3.32

Fig. 3.32 Top entry ball valve with a stem groove for placing the stem key.

illustrates a top entry ball valve with a bare stem (without gear box or actuator) containing a groove in the stem of the valve for the stem key. The groove is perpendicular to the flow passage, meaning that the valve closure member hole or ball hole is perpendicular to the flow passage, so the valve is in closed position.

Item #7 is called a thrust washer, which could be in Inconel 625 material or a double layer of Inconel 625 as an external layer and a PTFE (Teflon) internal layer. The thrust washer, also called a stem bottom bearing, is a type of bearing that provides resistance against axial load applied to the stem and prevents side movement of the stem. This presentation in turns leads to less friction and galling of the stem. Item #8 is the body closure of the valve which is integrated with the pup piece. The pup piece provides some distance between the pipe welding to the valve done in the construction yard to protect the valve internals, especially the seals. The length of the pup piece should be long enough to prevent the melting of the soft seals during welding in the yard. The closure member should have exactly the same material as the body of the valve. Item #9 is a seal ring which acts as the primary external pressure barrier to prevent seawater ingress to the valve stem area and the valve internals. Different materials, such as Buna N, hydrogenated Buna N, or a mix of PTFE (Teflon) and Buna N, could be selected. Item #9b is a metallic gasket used between body pieces for sealing. The gasket is normally a BX type as per API 6A and could be in 25Cr super duplex or Inconel 625 materials. Item #10 is bolting; these could be studs and nuts or cap screws made of low alloy steel bolts with zinc and nickel plating. The function of these bolts is to bolt the valve bottom part of the body, also called a trunnion, to the other body connections. Item 12b is a position indicator, which can be in 25Cr super duplex. Item #12c is an override or override nut, which is the part to be rotated by ROV for valve operation. The override is in 25Cr super duplex material. Item #16 is a V-pack, which provides the preliminary internal pressure stem sealing. The stem sealing could be made in three different material combinations, such as two thermoplastic parts made of PTFE and PEEK as well as Elgiloy springs. Item #16a is a lip seal made of PTFE and an Elgiloy spring; it acts as a secondary stem sealing against internal fluid. Item #16b is also a lip seal that acts as a secondary stem seal barrier against the external seawater. Item #24 are cap screws made of low alloy steel material or carbon steel with zinc and nickel plating. Item #27a and #27d are bearings in nickel 625 with an inner layer of PTFE to prevent side movement of the lower and upper parts of the stem. The inner PTFE layer inside the bearing is made to prevent galling between the stem and the bearing. Item #28b is another thrust washer, explained before. Item #29 is a trunnion in the same material as the body, such as 25Cr super duplex or Inconel 625. It provides a support for the ball, which leads to less torque requirement for valve operation. Item #31 is a metallic gasket, same as item #9b. Item #33 is a Chevron V-pack used for sealing between the seat and body of the valve. The material of item 33 is same as item #16. Item #34 is a plug, which could be in super duplex or Inconel 625 material; it is used for blinding the test port after performing the tests. The plug is seal welded after the test to prevent any possibility of being removed from the valve. Item #39a are cap screws which are used to connect the ROV bucket to the valve. These bolting pieces are in low alloy or carbon steel with zinc-nickel plating. Item #45 is a spacer that fills the area between the stem seals. The spacer could be in super duplex or Inconel 625 as two examples. Item #49 is another plug used for the test of the stem seals; it could have a different material than the body of the valve to prevent galling during plugging and unplugging actions. As an example, Inconel 625 plugs could be used for 25Cr super duplex body valves. Item #53a is a spring holder used to keep the spring (item #5) in place. The spring holder should be made of the same material as the seat of the valve. Item #69 is a set screw used to connect the override nut to the stem of the valve. In fact, the rotary movement of the torque tool is transferred to the override nut as well as the stem. The set screw inside the bucket for connecting the stem nut to the stem could be in super duplex stainless steel as an example. Item #99 is a ROV bucket or receptacle, which could be in normal carbon steel, low-temperature carbon steel, or a soft material like Devlon. Devlon is a thermoplastic material developed by James Walker that can be used for a ROV bucket or valve seat. Chapter 13 provides more information about soft materials, including thermoplastics and elastomers. The ROV interface should be designed according to the API 17H/ISO 13628-8 standard. Alternatively, subsea ball valves could have a handwheel for diver operation if installed in shallow water. Fig. 3.33 illustrates a subsea ball valve with side entry design and a handwheel for diver operation at a depth of 100 m. The body of the valve is in ASTM A694 F65 low alloy steel with an Inconel overlay and Inconel 625 trim. Alternatively, the ball of the valve could be in low alloy steel such as ASTM A694 F65 grade with an Inconel 625 overlay as illustrated in Fig. 3.34. The Inconel 625 weld overlay is called cladding. One of the main differences between subsea ball valves and ball valves in other sectors of the oil and gas industry is that an emergency sealant injection fitting is not installed on subsea valves. Sealant injection points are installed on the valve stem sealing and soft seats to expedite emergency repairs on the soft materials prior to the next maintenance program. However, subsea ball valves are maintenance free, so there is no need to install emergency sealant injection ports on them. This reduction in sealant injection ports leads to making fewer holes on the body of the valves which reduces the chance of leakage.

Items #173 and 174 are markers for indicating the position of the valve with the letters "S" and "O" indicating shut and open, respectively. These signs are made of antifouling materials to prevent marine growth and be visible and readable for the ROV during operation. Fig. 3.35 shows a detailed drawing of a large bare stem ball valve in 10″ and a pressure class of 10,000 psi equal to 690 bar. This valve has a side entry design and can be disassembled from the right side. Bare stem means that the valve is without any operator like a gear or actuator. This valve is manually operated by a ROV. Item #1 is the body of the valve, which is in carbon steel or low alloy steel ASTM A694 F60 or F65 and fully overlaid or cladded with Inconel 625 in 3 mm. Alternatively, super duplex can be selected for the body of the valve, but it would be more expensive. Item #2 is the closure

Fig. 3.33 Side entry ball valve with a handwheel for shallow water operation by a diver. *(Courtesy: Mir Valve.)*

Fig. 3.34 Inconel 625 cladding on the ball of a ball valve.

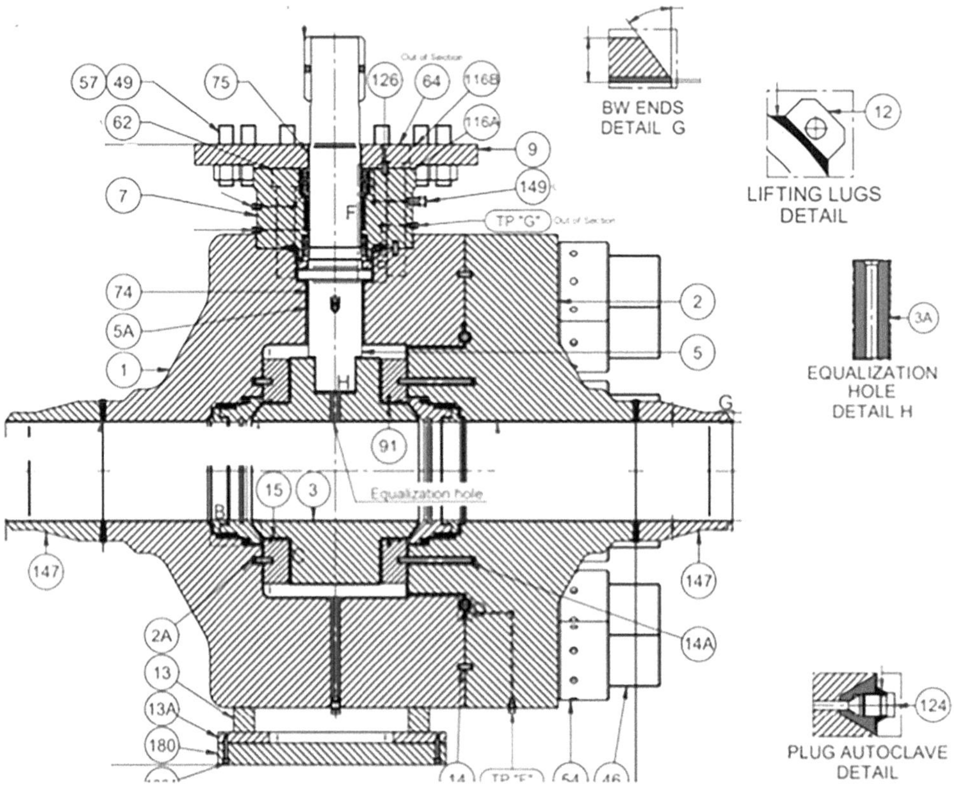

Fig. 3.35 Bare stem 10″ ball valve in pressure class 10,000 psi detailed drawing. *(Courtesy: ATV.)*

member, which is typically in the same material as the body, which is low alloy steel with three milometers of Inconel 625 overlay. Item #3 is the ball of the valve in low alloy steel with a full overlay of Inconel 625 like the ball illustrated in Fig. 3.34. The ball external surface is hard faced with tungsten carbide in a relatively thick layer of 0.2–0.3 mm. There is a hole inside the ball for pressure equalization, which means that the valve can be operated more easily with less torque. Some end users do not allow an equalizer hole in the ball to reduce operational torque. There is a rod (item #3A) in the equalizer hole in Inconel 625 material to prevent crevice corrosion inside the equalizer hole in the ball. Item #5 is the stem of the valve, which could be in super duplex or Inconel 718 or 725. The shoulder of the stem is located at the upper part, relative to the connection of the stem to the ball, to prevent valve sealing disturbance due to seawater head. More information about this type of antiblowout design is provided in Section 3.3. Item #140 is the stem key which is recommended to be same as the stem of the valve. The function and purpose of the stem key is explained earlier in Section 3.3. In some cases, the valve supplier may propose a different material for the stem key compared to the stem of the valve. As an example, the stem key could be in super duplex and the stem in Inconel 725. In such a case, the possibility of galvanic corrosion on the coupling of the stem and stem key should be evaluated. Item #7 is a body piece called a gland trunnion that is made of the same material as the body of the valve. Item #9 is called a top flange or adapter plate, which is an interface between the valve and operator. The top flange is normally made according to the dimensions or other requirements given in the ISO 5210 and ISO 5211 standards. Item #12 is a valve lifting lug welded to the body of the valve. Lifting lug welds are critical and should be tested as per a suitable nondestructive test (NDT) method; this could be a magnetic particle test or liquid penetration test. The valve lifting lug is in carbon steel and the lifting lug shown in the figure is made of plate. Item #13 and 13a are valves supports in carbon steel plates and could be 50 mm just as an example. Item #46 is the bolting that connects the body and body closure. The bolt is in low alloy steel with electroplated zinc and nickel for corrosion protection. All of the external parts of the valve, including the bolting, are exposed to cathodic protection. Item #74 is a stem bearing which is typically in Inconel 625 with a PTFE or Teflon lining to prevent side movement of the valve stem. The lower part of the ball is supported by a trunnion or a support plate which is item #15. Item #147 is the pup piece of the valve welded to the valve body. The pup piece end is given in detail "G" according to ASME B16.25. Alternatively, narrow gap welding can be selected for ending the pup piece to reduce the welding consumable and for faster fabrication. Figs. 3.36 and 3.37 illustrate narrow gap welding between the valve pup piece and the connected pipe. The pup piece material should be the same as that of the connected piping. Typical

Fig. 3.36 Narrow gap welding end preparation.

Fig. 3.37 Narrow gap welding end welding.

pup piece materials could be carbon steel or low alloy steel, carbon steel or low alloy steel with Inconel 625 cladding as well as super duplex. Figure #149 shows a pressure relief valve used to release fluid leaked from the stem sealing to the sea to prevent leakage from the stem area to the operator of the valve.

3.4 Quality control

Quality control is an important part of the manufacturing of subsea valves, including ball valves, to make sure that there is no defect or nonconformity in the products. Positive material identification (PMI) should be performed on the pressure-containing and pressure-controlling parts, at minimum, to make sure the material of these components is according to the project specification with the correct chemical composition. Fig. 3.38 illustrates PMI being performed on the stem of a ball valve to check its chemical compositions. The wrong material could be due to a mistake from the foundry or forge master as providers of the raw material, or could be due to a mismatching of components during the assembly and fabrication of the valve. Fig. 3.39 illustrates a nondestructive

Fig. 3.38 PMI performed on the stem of a ball valve. *(Courtesy: Hit Valve.)*

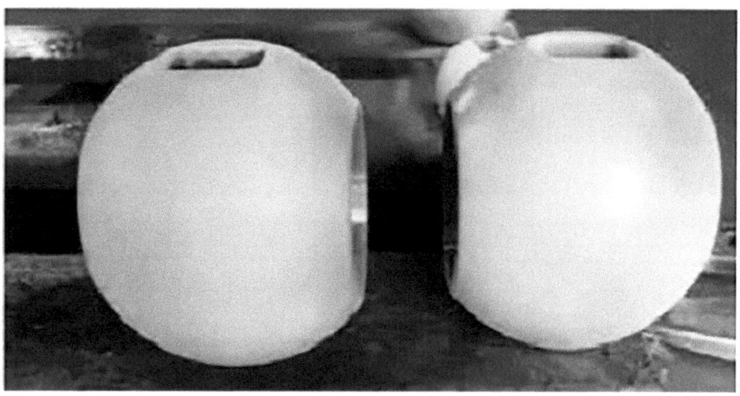

Fig. 3.39 Liquid penetration test on the balls. *(Courtesy: Hit Valve.)*

liquid penetration test on the balls to make sure that there is no defect. A dimension check is an important part of the quality control process and final inspection. Fig. 3.40 illustrates measuring the end flange connection of a ball valve. Fig. 3.41 illustrates measuring the roughness of the ball of a ball valve. Fig. 3.42 illustrates the end-to-end length measurement of a top entry ball valve after the pressure test. Typically, final inspection, including

Fig. 3.40 Dimension control and flange end measurement. *(Courtesy: Hit Valve.)*

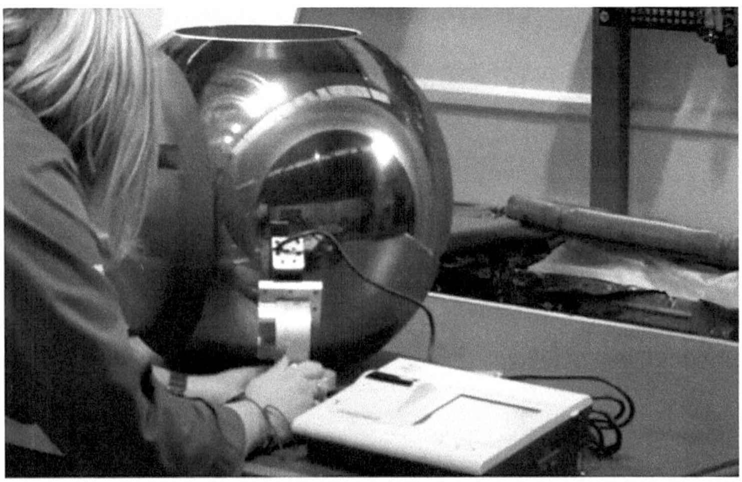

Fig. 3.41 Measuring the roughness of a ball valve. *(Courtesy: FCT.)*

the dimensional check, is done after the pressure tests. Fig. 3.43 illustrates the measurement of the pup piece diameter of a top entry ball valve. Fig. 3.44 illustrates bolting diameter measurement for a flanged body end ball valve.

3.5 Three-way ball valves

Three-way or 3-way ball valves are special valves with multiports, i.e., three or four ports, that are typically used for diverting or mixing the flow. Three-way ball valves may be selected and operated for subsea application. This type of valve could be designated with an "L"-port (see Fig. 3.32) or T-port depending on the application. The type of

Fig. 3.42 End-to-end length measurement of a top entry ball valve.

Fig. 3.43 Measurement of a valve pup piece diameter.

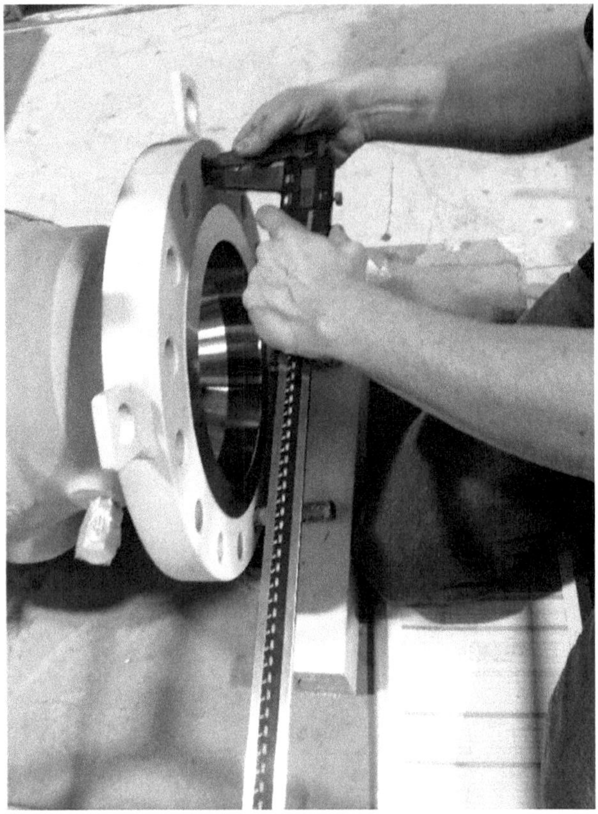

Fig. 3.44 Measurement of the bolt diameter of a body flange valve.

operation could be either manual via ROV or actuated. Based on industrial experience, it is better to use a ¾″ three-way ball valve to divert the flow in a subsea distribution unit (SDU) in a methanol fluid piping.

The L-port ball valve illustrated in Fig. 3.45 has a ball with the shape of an "L." The figure on the left shows Position #1, in which the flow in blue color (light gray in print versions) enters the valve from port B, moves 90 degrees to the left, and exits the valve from port A. The ball is rotated 90 degrees counterclockwise in position #2 to divert the flow from port A to port C. In this case, the flow enters the valve from port B, gets diverted to the right, and exits the valve through port C.

There are different flow patterns for a three-way ball valve in both L-port and T-port. In some rare cases a Y-port may be used. Each port design could have two or three positions as illustrated in Fig. 3.46. Three-way ball valves with two positions are called 90-degree turn; a valve with three positions is called 180-degree turn. Flow plan #9 in Fig. 3.46 is same as the L-port ball valve in two positions provided in Fig. 3.45.

Question: A three-way ball valve is required to divert the flow from point "A" to either point "B" or "C." The fluid is hydrocarbon and it moves from point "A" to

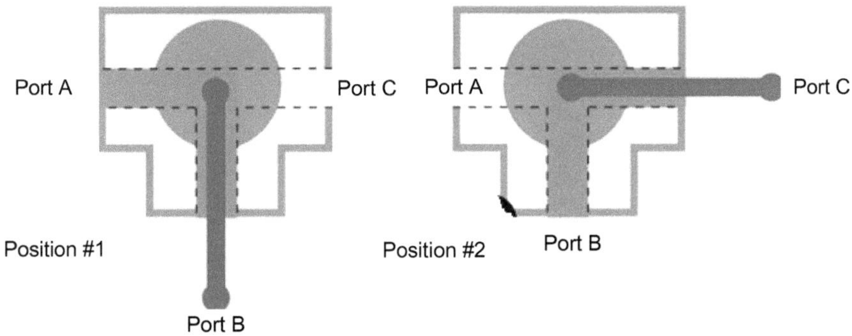

Fig. 3.45 L-port ball valve.

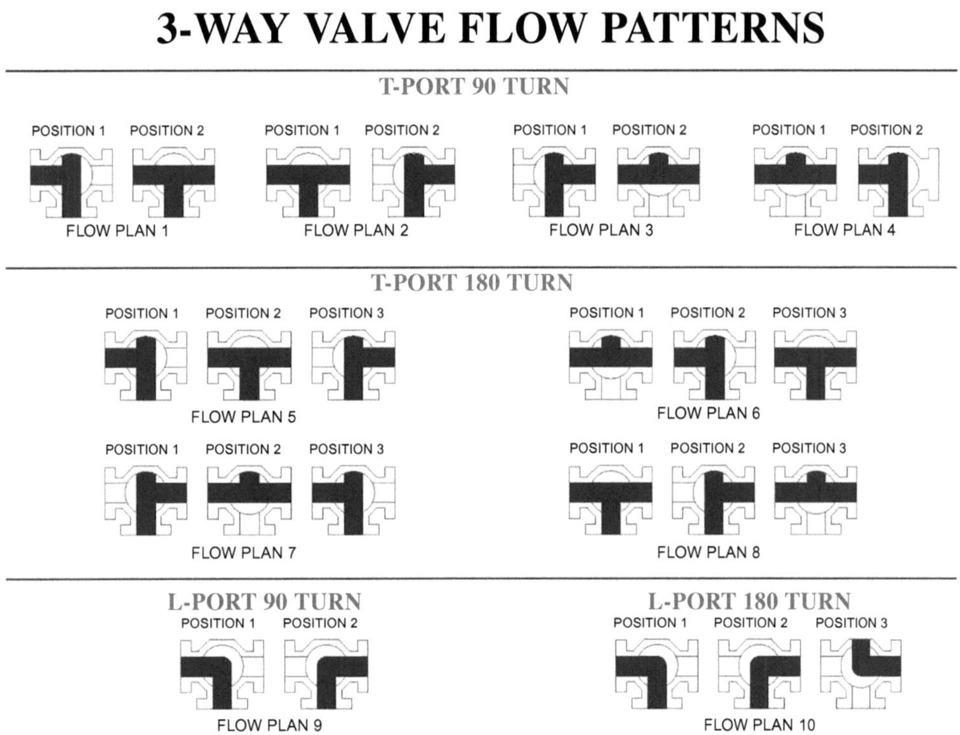

Fig. 3.46 L-port and T-port three-way ball valve flow patterns. *(Courtesy: Pipingnow.com.)*

"B" in a straight line in normal condition. The fluid should be diverted from point "A" to point "C" in the other position. In this case, the fluid undergoes a 90-degree rotation and moves downward. Which type of three-way ball valve is required for this application?

Answer: A T-port three-way ball valve is required, since an L-port cannot provide fluid transfer on a straight line. A valve with two positions should be sufficient for the

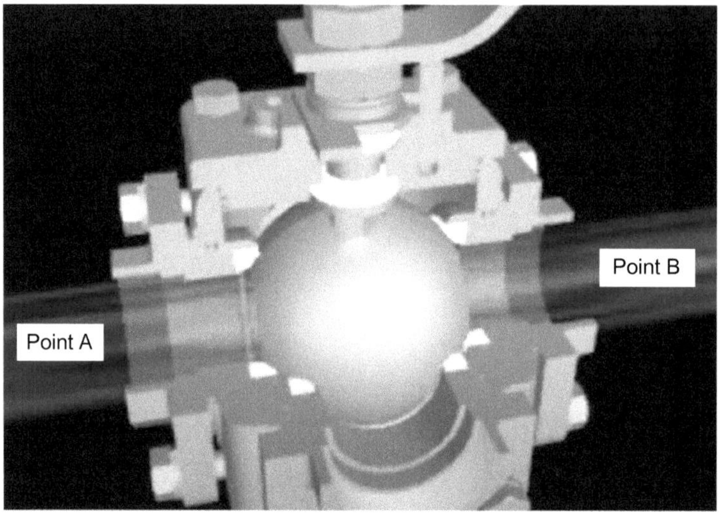

Fig. 3.47 T-port three-way ball valve with flow plan 4 (shown in Fig. 3.46 in position #1).

application since the diverting of the flow is done in two scenarios. The proposed flow plan for this application is flow plan #4. Fig. 3.47 illustrates a three-way ball valve in position #1, where the fluid is transferred between point A and point B.

The ball moves 90 degrees and transfers the fluid from point A to point C as per Fig. 3.48.

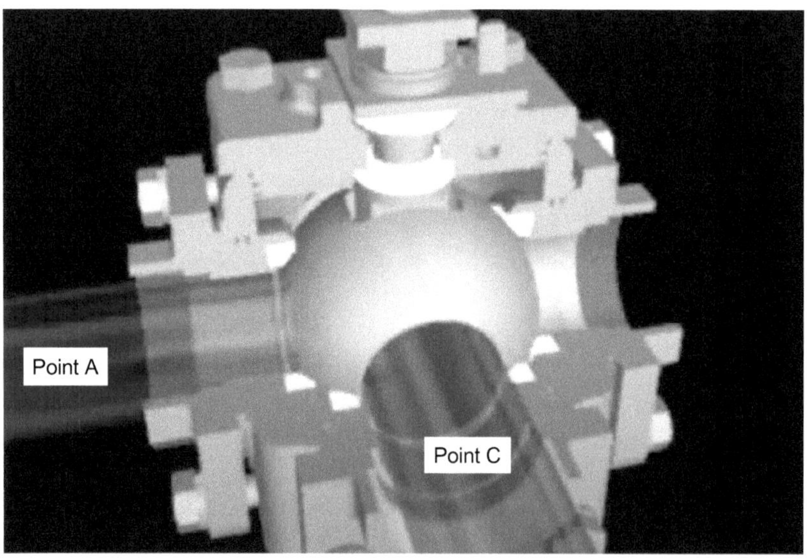

Fig. 3.48 T-port three-way ball valve with flow plan 4 (shown in Fig. 3.46 in position #2).

The number of ports could be more than three, as in a four-way ball valve or even higher; ports can be manufactured by the valve supplier upon the request of the purchaser. Four-way and other multibore valves are ideal for flow switching operations.

3.6 Modular valves

Two ball valves could be integrated in a single body with a needle valve as a bleed between; a valve of this design is called a modular valve. Some engineers call this type of valve a double block and bleed (DBB) valve, since double isolation can be achieved via the two balls and a needle valve between that acts as a bleeder. This type of valve could be used on chemical injection lines to provide a double barrier between the main process line and the chemical injection line. Two blockages by means of two ball valves provide extra safety and reliability; if one valve fails to provide isolation, the other could provide isolation as a backup. Fig. 3.49 illustrates a subsea modular valve that includes two small ball valves in sizes less than 2″ that are operated with ROV and a needle valve in between. The modular valve is side entry with the possibility of access to the valve internals through the flange connection on both sides of the valves. Fig. 3.50 illustrates another DBB valve in $2\frac{1}{16}''$ and a pressure class of 10k (10,000 psi) installed in 3300 m water depth. Reduced bore is the preferred choice for DBB valves to save weight, required space, and cost if there is no problem with pressure drop and flow assurance. The DBB valve in

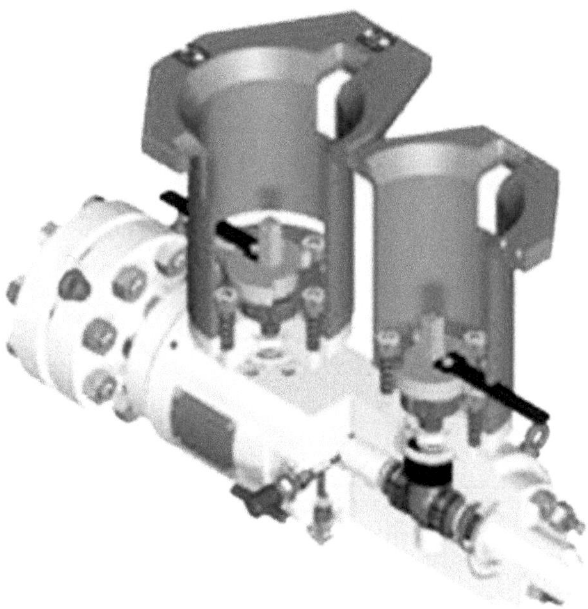

Fig. 3.49 Subsea modular valve. *(Courtesy: Pacson.)*

Fig. 3.50 Subsea modular valve. *(Courtesy: Galperti.)*

Fig. 3.50 is a reduced bore valve and both ball valves have an ROV bucket operator. A low torque paddle is used to operate the needle valve. Any accumulated fluid and pressure between the two balls is released through the needle valve. Needle valves, as a type of flow controlling valve, are covered in Chapter 8.

3.7 Ball valve case studies

3.7.1 Transferring the branching connections to the valve pup piece

The first case study involves a 22″ ball valve in a pressure class of 7500 psi equal to 517 bar located on a manifold header. The 22″ ball valve has a 2″ bypass valve, illustrated in Fig. 3.51, that is mainly used for pressure equalization across the 22″ to open the 22″ ball valve more easily during start-up. Two branches are required from the 22″ header; each

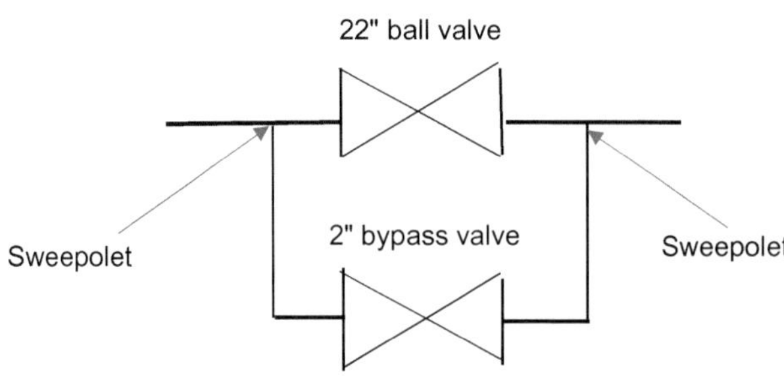

Fig. 3.51 A 22″ ball valve with 2″ bypass for pressure equalization.

Fig. 3.52 Sweepolet.

one is 2″. Two sweepolets in 22″ × 2″ are required to make a branch from both ends of the valve. Fig. 3.52 illustrates a sweepolet for the purpose of branching from the header when significant difference exist between the header and branch sizes. Welding the sweepolets to the valves' pup pieces instead of the connected piping could save the length of the piping system, including the valves. Sweepolet is a branching fitting with high stress resistance and high fatigue life compared to other olet family group fittings. The sweepolet could be fillet welded from the side on which it sits on the pipe (see Fig. 3.53); it could provide a butt weld connection from the other side. If the pup piece

Fig. 3.53 Sweepolet preparation for welding to the pipe.

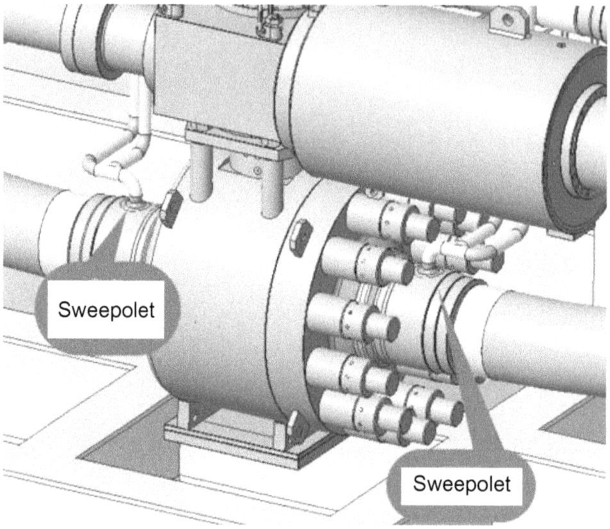

Fig. 3.54 Ball valve 3D model with sweepolets on the pup piece.

of the valve is in 22Cr duplex or 25Cr super duplex, hydrogen induced stress cracking (HISC) analysis and risk evaluation should be performed on the pup piece of the valve, including the weld joints of the sweepolet, to make sure that the pup piece plus welded sweepolet are not at risk of failure due to HISC corrosion. Fig. 3.54 provides a 3D model of a ball valve with sweepolets welded on both pup pieces.

In addition, the valve supplier may need to increase the length of the pup piece due to welding the sweepolets. In some cases, mono-ethylene glycol (MEG) could be injected to the piping and to save space, the MEG connection could be transferred to the valve pup piece. The other risk related to adding sweepolet welding on the valve pup pieces by the valve supplier instead of welding the sweepolet in the fabrication yard on the connected piping to the valve could be associated with the extra time required to delivering the valves by the valve supplier.

3.7.2 Rotation of the actuators

Ball valve actuators can be single acting spring return or double acting with two hydraulic cylinders. The actuators are typically installed parallel to the connected piping to the valves as well as valves' port. Thus the actuators could occupy a relatively large space sidewise. The type of actuator used for ball valves could be scotch and yoke or rack and pinion. More information about actuators is provided in Chapter 10. Fig. 3.55 illustrates a 16″ ball valve with a spring return actuator parallel to the routing of the connected piping and occupying a relatively large space sidewise. The actuator length or end-to-end length in this case is around 4 m and 20 cm. One solution is to change a rack and pinion

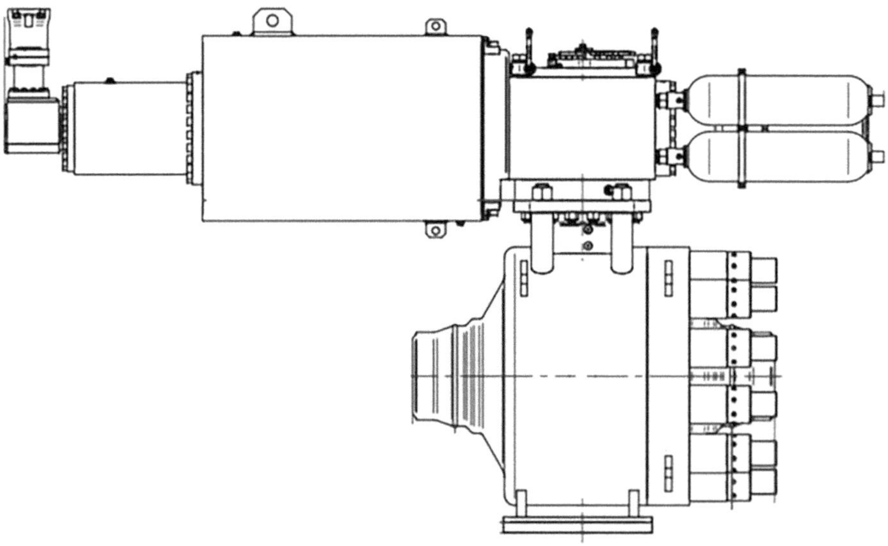

Fig. 3.55 A 16″ ball valve in pressure class 517 bar with a spring return actuator.

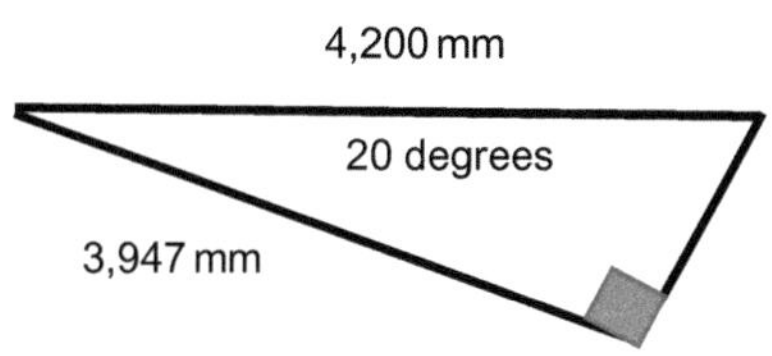

Fig. 3.56 A triangle showing the effect of 20-degree rotation on the actuator.

actuator to helical spline to save space sidewise. A helical spline actuator, like a linear actuator, stands vertically. More information about helical spline actuators is available in Chapter 10. The other solution is to rotate the actuator 20 or 30 degrees to save space sidewise. Fig. 3.56 illustrates a triangle showing the length of the actuator sidewise as well as the effect of a 20-degree rotation of the actuator. The side length of the actuator after the 20-degree rotation is calculated by multiplying 4200 mm by cosine 20 degrees which is equal to 3947 mm. Thus rotation of the actuator 20 degrees can save 253 mm of space sidewise. Fig. 3.57 illustrates a tilted rack and pinion actuator rotated 20 degrees so the actuator direction is not parallel to the port of the valve connected to the piping.

3.7.3 HIPPS application

HIPPS stands for high integrity pressure or piping protection system. The details of HIPPS are explained in Chapter 15. HIPPS is an application in which the valve should

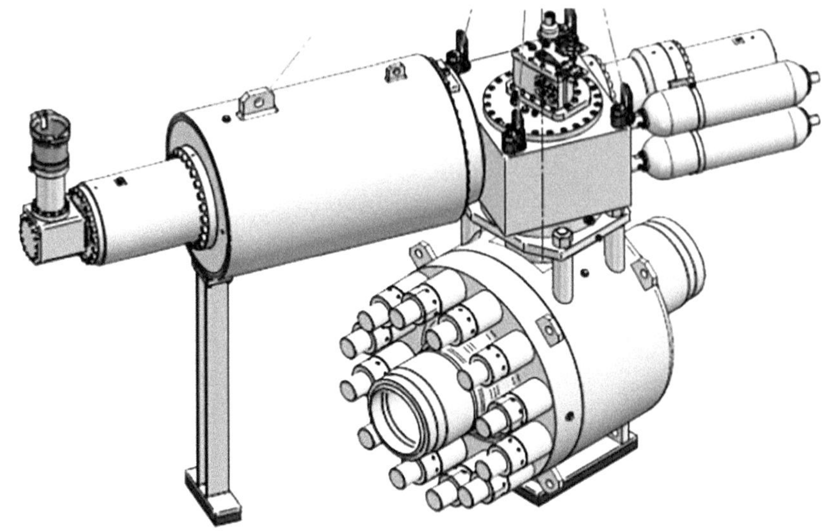

Fig. 3.57 Rack and pinion actuator tilted 20 degrees.

be closed and provide isolation in a very fast manner. Ball valves have quarter turn application so that they can be opened and closed quickly. Thus ball valves are a good choice for HIIPS application due to their fast operation. With that said, axial valves could be a more suitable choice of valve for HIPPS application, which is explained in Chapter 8.

3.8 Self-assessment questions

1. Which sentence provides the complete correct statement about subsea ball valves?
   A. Ball valves are just used in sizes above 8″ for subsea applications.
   B. Full bore ball valves have priority over reduced bore since they do not produce high pressure drop.
   C. Ball valves have some advantages over gate valves, such as less height, quicker opening and closing, less stem wear, and less sensitivity of operation due to water depth.
   D. Small ball valves are always operated via ROV.
   Answer: Option A is wrong since ball valves could be selected in small sizes like ½″ and ¾″ in SDUs. Option B is not correct; reduced bore ball valves have priority in terms of selection because they are lighter and more compact and also require a lighter and more compact actuator or manual operator in most cases. Option C is completely correct. Option D is not correct since small ball valves in shallow water depths could be operated by a diver.

2. Which design feature choice is applicable to ball valves?
   A. Antiblowout stem, and reduced or full bore.
   B. Seat single or double piston effect, and quality control.
   C. Backseat feature, and retrievable operator.
   D. Top or side entry, and pressure balance design.
   Answer: Option A is correct since an antiblowout stem as well as full bore or reduced bore is applicable to ball valves. Option B is wrong since quality control is not a design feature. Option C is wrong as well because a backseat feature is applicable to through conduit gate valves and not applicable to ball valves. Option D is wrong because pressure balance design is not a ball valve feature. Pressure balance design could be applicable for axial valves used for flow controlling; these valves are covered in Chapter 8.
3. Which sentence is correct about modular valves?
   A. Isolation ball valves inside the body of a modular valve can be used for flow regulation or throttling.
   B. Modular valves typically consist of three valves in one body in forged material.
   C. The needle valve inside a modular valve is used for isolation of the fluid.
   D. Modular valves for chemical injections could be in large sizes such as 6″.
   Answer: Option A is wrong because two ball valves inside a modular valve body are used for isolation of the fluid. Standard ball valves cannot be used for flow regulation since keeping the ball valve in partially open position could cause wear and damage to the valves' internals. Option B is correct since two ball valves and one needle valve are integrated in the body of a modular valve. Additionally, forged materials have a higher quality than cast materials, so forging is more common for subsea valves. However, investment casting or any high-quality casting could be selected for subsea valves as a material processing method. Option C is not correct since the main function of the needle inside a modular valve is not isolation. Isolation is achieved with two ball valves; the needle valve drains the fluid in the area between the two balls. Option D is wrong since valves 2″ and smaller are typical for chemical injection lines; a 6″ modular valve would not be common for a chemical injection line.
4. Which sentence is not correct about double piston effect seats and double isolation and bleed valves?
   A. Double piston effect seat dimension is typically different from that of a single piston or self-relieving seat.
   B. DIB2 is preferred over DIB1 since it can provide efficient double isolation at a lower cost and no need to have an extra hole on the body cavity in liquid service.
   C. There is no difference in sealing between the seat and body of the valve between self-relieving and double piston effect seats.
   D. DIB2 is not a bidirectional valve.

Answer: Option A is correct since a double piston effect seat is normally longer than a self-reliving seat. Option B is correct and some end users prefer DIB2 for the reasons mentioned. Option C is wrong since a double piston effect seat requires a two-lip seal or v-pack, while a single piston effect seat requires just one seal. Option D is correct since DIB2, as per the API 6DSS definition, is not bidirectional and contains one single piston and one double piston effect seat.

5. Which parameters make ball valves the better choice compared to gate valves?
   A. More compact end-to-end length and actuator.
   B. Less sensitivity to water depth, more compact in height, less stem-sealing friction.
   C. Possibility of having either side entry and top entry design.
   D. Lower cost.

   Answer: Option A is not correct since ball valves are not necessarily shorter in length compared to gate valves. Option B is correct and all the points mentioned in this option are advantages of a ball valve over a gate valve. Option C is a correct statement, but it does not provide any advantages for ball valves over gate valves, which are always top entry. Option D could be correct in many cases and ball valves could be cheaper than gate valves of the same size and pressure class, but it is not correct in all cases.
6. Which sentence is not correct about quality control and final inspection?
   A. A pressure test is typically done before final inspection.
   B. Packing and preservation control is part of final inspection.
   C. Quality control is always done after valve assembly.
   D. A nondestructive test on the valve components after machining is a part of quality control.

   Answer: Option A is correct. Option B is also correct, since packing and preservation are a part of the final inspection that is done after the pressure test. Final inspection could be done in a separate session from the factory acceptance test. Option C is wrong since quality control could be done on the valve components before assembly. Option D is correct since NDT is a part of quality control that is done after machining or welding.
7. Which item is correct in regard to differences between subsea ball valves and valves in other sectors of the oil and gas industry?
   A. Type of seats (self-relieving and double piston effect) is different from subsea ball valves with the ball valves in other sectors of oil and gas industry.
   B. Using emergency sealant injection ports on the stem and seats is different from subsea ball valves with the ball valves in other sectors of oil and gas industry.
   C. Lifting lug arrangement is different from subsea ball valves with the ball valves in other sectors of oil and gas industry.
   D. Possessing a pup piece is different from subsea ball valves with the ball valves in other sectors of oil and gas industry.

Answer: Option A is not correct since a self-relieving and double piston effect seats are applicable for ball valves in all sectors of the oil and gas industry. Option B is correct since subsea valves are maintenance free so, unlike ball valves in other sectors of the oil and gas industry, they do not require any sealant injection port on the seats or stem sealing. Option C is not correct because the lifting lug arrangement is the same for ball valves in subsea and nonsubsea sectors of the oil and gas industry. Option D is not correct since a pup piece is used for ball valves in general to keep the valve internals away from the heat produced during the welding of the valve to the pipe in the fabrication yard.

8. Which sentence is wrong about material selection for ball valves?

A. In the case of using a bellows stem seal, Inconel 725 is a good choice of material for stem.

B. The ball of a ball valve could be solid or weld overlay.

C. The use of low alloy bolts with zinc-nickel electroplate is common for subsea valves, including ball valves.

D. Inconel 625 is the best choice of stem material, especially in the case of high toque production by the operator of the valve.

Answer: Option A is correct and Inconel 725 is proposed for the metallic stem seal of the valves since it has high mechanical strength and fatigue resistance. More information about bellows stem sealing is provided in Chapter 13. Option B is correct since it is possible to choose a solid corrosion resistant alloy (CRA) material such as 25Cr super for the ball of the ball valve or make the ball in carbon steel or low alloy steel and apply Inconel 625 on the ball. Option C is correct; low alloy steel bolts as per API 20E are common for subsea valves subject to cathodic protection. The cathodic protection mechanism is explained in Chapter 12 of the book. Option D is not correct since Inconel 625 is not a hard nickel alloy and cannot provide a mechanical strength as high as hard nickel alloys such as Inconel 718 and 725.

Further reading

American Petroleum Institute (API) 6A, Specification for Wellhead and Tree Equipment, 21st ed., API, Washington, DC, 2018.

American Petroleum Institute (API) 17D, Design and Operation of Subsea Production Systems, Subsea Wellhead and Tree Equipment, second ed., API, Washington, DC, 2011.

American Petroleum Institute (API) 6DSS, Specification for Subsea Pipeline Valves, third ed., API, Washington, DC, 2017.

Y. Bai, Q. Bai, Subsea Engineering Handbook, first ed., Elsevier, Atlanta, GA, 2012.

D.R. Mefford, Deepwater Subsea Ball Valves, Cameron, 2014.

K. Sotoodeh, Top entry valves in offshore oil and gas, Valve Mag. (2019) [online] Available at: http://www.valvemagazine.com/web-only/categories/technical-topics/9717-top-entry-valves-in-offshore-oil-and-gas.html (Accessed 21 September 2020).

K. Sotoodeh, Valve failures analysis and solutions, Valve World Mag. 23 (11) (December 2018) 48–52.

K. Sotoodeh, Welding a carbon steel body valve to 22Cr duplex pipe: a case study, Valve World Mag. 24 (5) (2019) 38–41.

K. Sotoodeh, Manifold technology in offshore industry. Am. J. Mar. Sci. 8 (1) (2020) 6–13, https://doi.org/10.12691/marine-8-1-2.

CHAPTER FOUR

Subsea gate valves

Contents

4.1 Introduction

Historically, API 6A through conduit gate (TCG) valves have been used for subsea field development, especially for Christmas trees and wellheads. TCG valves have robust design and suitable for particle containing or dirty fluid services. Valves used for wellheads and Christmas trees are gate, typically TCG valves from 1 13/16″ to 7 1/16″ in size. TCG valves could be slab, single or double expanding types. Wedge gate valves (see Fig. 4.1), which are widely used in the topside sector of the offshore industry on platforms and ships and in refineries and petrochemical plants, are not used in the subsea sector since they are not suitable for operation in particle-containing applications. One main disadvantage of wedge gate valves is the possibility that particles could accumulate in the bottom section of the valve body where the gate (disk) sits during valve closure; these particles may cause

https://doi.org/10.1016/B978-0-323-90605-0.00012-8

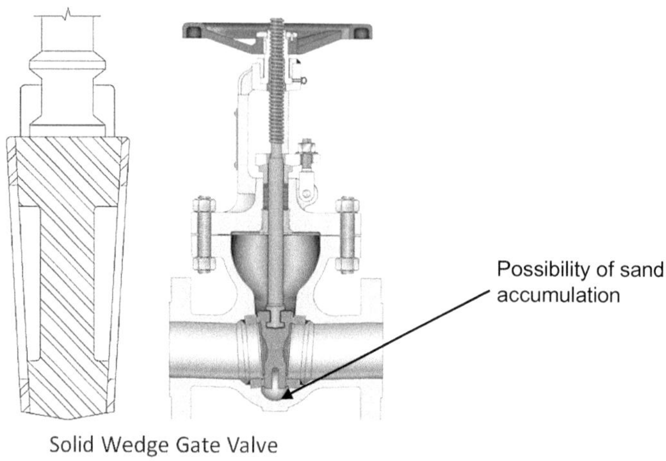

Fig. 4.1 Wedge gate valve.

damage to the disk during closure. Small wedge gate valves in ¾″ or 1″ are used in the nonsubsea sector of the oil and gas industry for venting and draining of the piping headers. Therefore gate valves for subsea use should be designed and considered with the erosive particle-containing fluid coming out from the reservoir and wellhead in mind.

TCG valves are common for the manifold branch piping connected to the wellheads. Unlike wedge gate valves, TCG valves are suitable for dirty (particle-containing) services. Fig. 4.2 illustrates a six-slot manifold, meaning that the manifold is connected to six

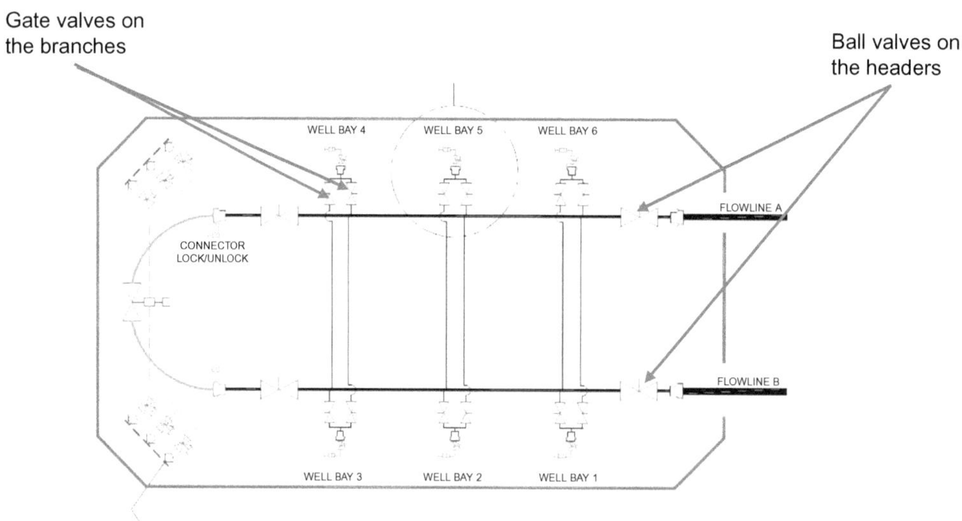

Fig. 4.2 Six-slot subsea production manifold process flow diagram.

wellheads. That is the reason why TCG valves are used on Christmas trees as well as manifold branch lines. The next section explains the different types of TCG valves.

4.2 Through conduit gate valve types

4.2.1 Slab gate valve

A slab gate valve is an on/off valve used in process services such as hydrocarbon (oil and gas) to stop and start the flow. It is not recommended to use a TCG valve for flow control, since keeping the gate in half open position leads to vibration, erosion, and cavitation in the valve. The opening and closing of the valve is performed through a linear movement of the stem that transfers to the gate (disk), also called the "closure member." The gate is provided with a fully circular port opening that can be in line with the flow in open position or block the flow in closed position. The gate moves or slides between two seats. The valve sealing in a standard design with two floating seats is achieved through metal-to-metal (MtM) contact between the downstream seat and the disk. In fact, the seats guarantee a correct continuity between the body and the gate. The disk is pushed through the fluid pressure to the downstream seat as illustrated in Fig. 4.3.

Therefore it would not be a communication path between the valve cavity area where P_3 pressure exists and the valve bore downstream side with P_2 pressure indication.

However, the upstream part of the valve with P_1 pressure indication is connected to the valve cavity. The pressure in the cavity releases to the upstream valve bore side as illustrated in Fig. 4.3. This standard type of slab gate design, from a sealing point of view, is called "pressure energized-upstream downstream sealing only." The sealing between the downstream seat and disk depends on spring force as well as fluid pressure. It is

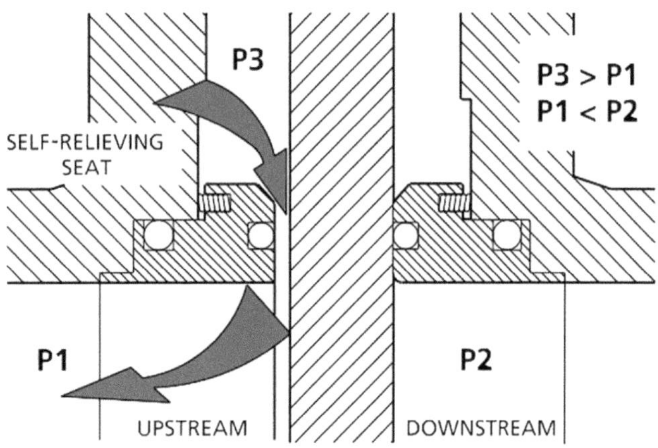

Fig. 4.3 Slab gate valve sealing on the downstream seat. *(Courtesy: Cameron.)*

important to make sure that the valve can provide acceptable sealing when the internal fluid service is at low pressure.

According to API 6DSS, the maximum cavity relief pressure shall be maximum 33% above the maximum rated or working pressure or 1.33 times of the maximum rated or working pressure. As an example, the cavity pressure shall be released to the valve bore at a pressure equal or less than 918 bar when a slab gate valve working pressure is 690 bar equal to 10,000 psi. If the valve seats were to fail to release the cavity excess pressure to the line, the cavity pressure could damage the valve and lead to operational problems. The springs that energize the seats are located behind the seats and play essential roles in the flexibility of movement of the seats to release the cavity pressure. Inconel X750 cannot be used for spring materials since this material has low resistance to seawater. Alternatively, Elgiloy, which is a cobalt alloy with approximately 40% cobalt, 20% chromium, 15% nickel, and 7% molybdenum, is proposed for spring material in subsea valves. The client or Engineering, Procurement, and Construction (EPC) contractor may ask for less than 33% maximum cavity excess pressure compared to the line (such as 10%) to achieve a higher safety factor in this regard. The other important consideration is that both seats of a slab gate valve in standard design should be floating type, meaning that the seats are energized with springs. The opposite design with a fixed seat is not proposed since the downstream seat cannot seal tightly. The standard slab gate valve design is bidirectional, which means that the valve can be rotated 180 degrees and connected to the piping. The standard design with two floating seats may be considered double isolation and bleed type 2 (DIB2) as per API 6DSS due to its tight sealing of the downstream seat. Both seats of a standard design slab gate valve are able to provide double isolation but not at the same time. The standard design of a slab gate valve cannot provide a double block and bleed (DBB) feature—refer to API 6DSS standard for subsea pipeline valves.

DBB, according to API 6DSS, refers to a valve with two sealing surfaces that provides a seal from the applied pressure from outside of the valve in closed position. Additionally, the DBB feature includes a cavity connection between the two seats for a vent or drain connection. Fig. 4.4 illustrates a DBB ball valve design according to API 6DSS. Refer to the right-hand side picture showing a DBB ball valve in closed position; the seats can provide proper sealing against the applied pressure from both the upstream and downstream sides of the valve.

DBB is not achievable for a standard design slab gate valve. However, DBB design is possible for a slab gate valve with two double piston effect seats, also called double isolation and bleed 1 (DIB1) as per the API 6DSS definition.

The concept of DIB1 sealing applicable for standard slab gate valves is given in Fig. 4.5. The fluid is flowing from the right side (highlighted in yellow; light gray in print versions) and presses the disk to the downstream seat. Therefore the fluid enters the cavity of the valve but the downstream seat prevents flow of the fluid to the downstream section (highlighted in green; dark gray in print versions).

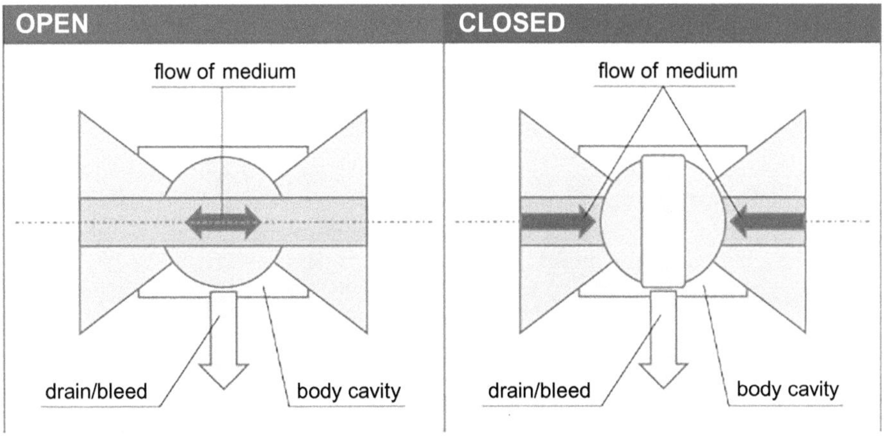

Fig. 4.4 DBB ball valve according to API 6DSS. *(Courtesy: MRC Global.)*

Fig. 4.6 illustrates a manually operated slab gate valve in 7 1/16″ size and a 517 bar pressure rating. The main parts (components) of slab gate valves are explained as follows.

4.2.1.1 Body/bonnet

The body and bonnet, illustrated in Fig. 4.6, are known as "pressure-containing parts." The definition of a pressure-containing part is given in API 6DSS, the standard for subsea pipeline valves, as "parts whose failures to function as intended results in a release of contained fluid into the environment and as a minimum includes bodies, end connections, bonnet also called cover as well as stem." The body and bonnet are bolted together and the bolting should also be considered pressure containing. The bolting should be in high mechanical strength materials such as B7 alloy steel to withstand the pressure and loads without any failure. The sealing between the body and bonnet of slab gate valves is achieved through a metallic ring type gasket as well as an O-ring or a lip seal (see Fig. 4.7). An O-ring or lip seal located on the external part provides sealing against leakage of seawater to the valve, so it can be called external sealing. The metallic gasket provides bidirectional sealing against leakage of external seawater fluid to the valve internal,

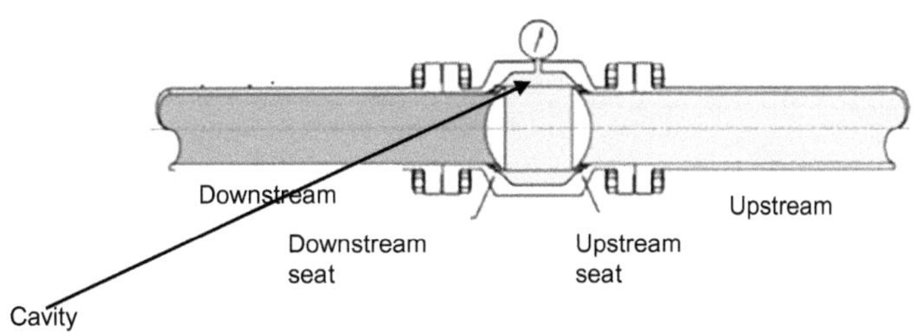

Fig. 4.5 DIB1 slab gate valve sealing concept.

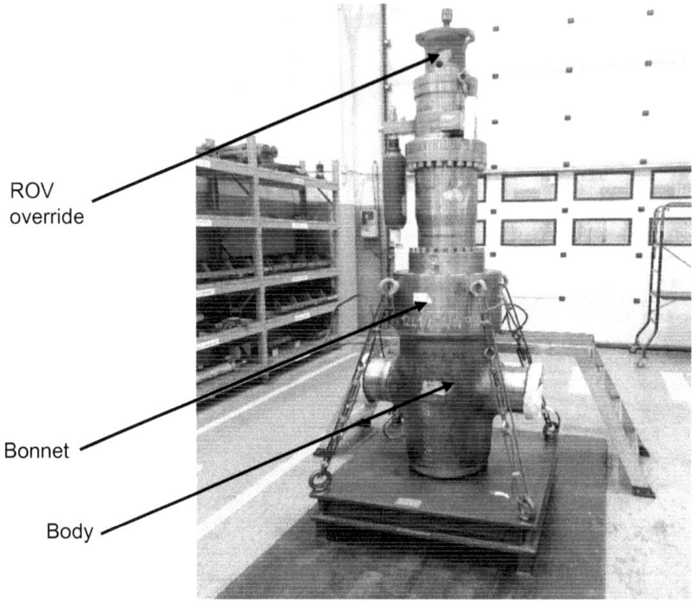

Fig. 4.6 A 7 1/16″ 517 bar slab gate valve.

and leakage of the valve internal fluid (typically hydrocarbon) to the seawater. The metallic gasket is made according to API 6A and is normally BX type. More information about API 6A flange faces and gaskets is provided in Chapter 2. The body/bonnet grooves where the gasket is placed shall be overlaid with Inconel 625 when carbon and low alloy steel materials are used for the body and bonnets. The port between the lip seal and metallic gasket shown in Fig. 4.7 is used for test purposes. In general, making holes in the body and bonnet of the valves should be minimized and should be limited to the test ports. All of the test ports are seal welded after finishing the factory acceptance test (FAT) so that there is no possibility of unplugging them. Therefore there is no vent and drain plug or flange connection on the cavity of subsea gate valves.

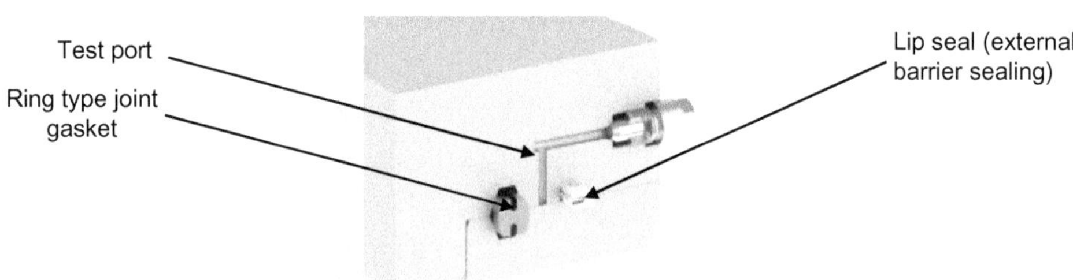

Fig. 4.7 Body and bonnet sealing mechanism for a gate valve. *(Courtesy: Petrol Valve.)*

Seats (explained in the next section) provide preliminary stem sealing. All TCG valves shall be provided with a secondary stem seal, also called backseat bushing. A backseat is a part of the valve inner bonnet area where the stem sits and prevents the flow of internal fluid to the stem sealing area. A conical area in the stem sits on the backseat area to provide fluid isolation of the stem sealing areas. The backseat area could be hard faced with tungsten carbide to minimize fretting or MtM contact corrosion as a result of its contact with the conical part of the valve stem. Fig. 4.8 illustrates a slab gate valve in open position. When the stem moves upward, the thicker part of the conical stem will sit on the backseat area and seal the stem sealing areas. The backseat could be subject to a test during the FAT, which is explained in Chapter 14. For TCG valves to be used topside, it is not recommended to do maintenance on the valve stem area, such as sealing replacement, during backseating position of the stem. Backseating position is when the conical part of the stem is attached to backseat area and provides secondary stem sealing. In fact, the backseat is just considered a disk travel stop at the top and is not recommended for stem sealing. Alternatively, an upper disk travel stop could be used inside the valve close to the backseat to provide a travel stop for the upper part of the disk. The gate valve bottom part acts as the lower disk travel stop. The other important gate valve component is a gland flange, which provides force on the sealing. This force causes expansion and sealing of the seals. The gland flange is used in connection with another component,

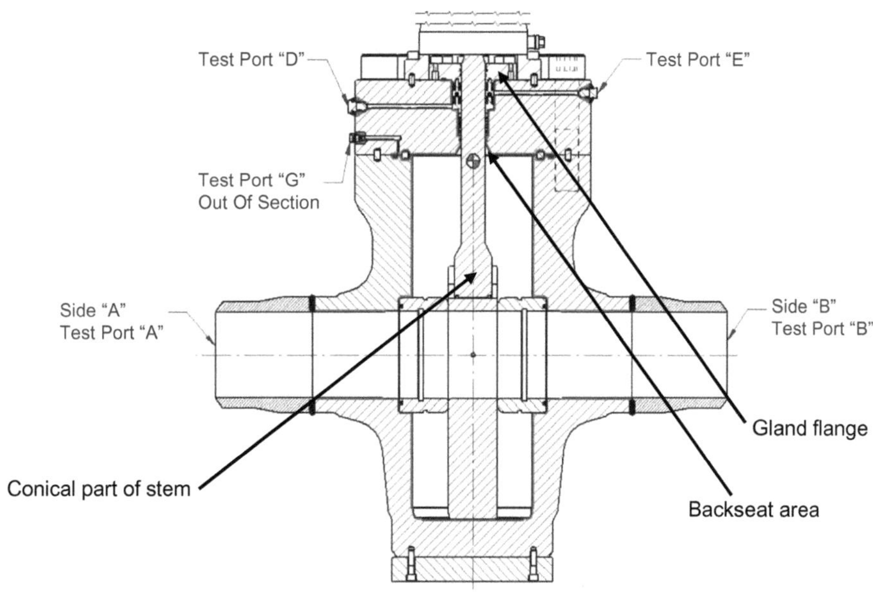

Fig. 4.8 A slab gate valve highlighting the backseat area, the conical part of the stem and the gland flange. *(Courtesy: Advanced Valve Technology.)*

called a gland, that is located below the gland flange. The gland and gland flange may be integrated in some design patterns.

Gate valves have a top entry design that provides access to the valve internals when the bonnet is removed. Top entry design has some advantages, such as online maintenance, meaning that there is no need to dismantle the valve from the piping for maintenance. However, this advantage is not relevant in a subsea context, since subsea valves are designed for no maintenance over their design life. Additionally, top entry valves are more resistant to piping loads. They have more flexibility for stem enlargement, especially in the case of actuation where a more robust stem may be required.

4.2.1.2 Gate (disk) and seats

The gate could be considered the most expensive part of a slab gate valve. It should be polished perfectly and coated with a thin layer of tungsten carbide to prevent wearing due to erosive fluid, and galling due to friction and contact with the seats. The connection between the gate and stem is known as a T-shoulder, which is illustrated in Fig. 4.9. A T-shoulder connection is a proper attachment between the stem and gate and cannot be disconnected due to fluid pressure or movement of the disk and stem. It also allows movement of the stem with the gate during operation.

The gate of a slab gate valve could be normal port or reverse port. Reverse port, as illustrated in Fig. 4.10, refers to a gate with the hole for fluid passage on the top part of the gate as close as possible to the valve operator. This type of port is used for fail-safe closed

Fig. 4.9 Disk and stem/disk connection in a slab gate valve.

Fig. 4.10 Reverse port gate.

(FSC) single acting spring return actuators. These actuators are opened with hydraulic fluid supply and closed through spring force. When hydraulic supply fluid enters a FSC linear actuator on the top of a gate valve, it pushes the stem down so the reverse port in the gate will be aligned with the bore of the valve, and the valve will open (see Fig. 4.12, left side). This situation is exactly opposite to the closing action of a wedge type gate valve, in which the gate closes the valve when the stem moves downward. When the hydraulic fluid stops in an FSC linear actuated slab gate valve, the stem moves upward and the bottom part of the gate closes the valve.

Normal port gate refers to a gate with the hole at the bottom (see Figs. 4.11 and 4.12, right-hand side). Normal port acts in exactly the opposite way from reverse port, and can be used for fail-safe open (FSO) or manually operated slab gate valves. It should be noted that FSO gate valves are not common in subsea applications.

Normal or standard port defines the position of the gate bore at the bottom located at a point that is furthest away from the valve operator or actuator that is used for a fail-safe open (FSO) gate valve. In an FSO slab gate valve with a normal port, the hydraulic fluid enters the actuator and pushes the stem downward so the valve closes. When the hydraulic fluid supply to the actuator stops, the stem moves upward and the valve opens (see Fig. 4.12, right-hand side).

25Cr super duplex, Inconel 625, and Inconel 725 are common materials for the gates and seats of subsea gate valves, and should be overlaid with tungsten carbide to prevent galling due to contact and friction of the gate (disk) with the seats, and to mitigate disk erosion due to contact with particle-containing fluid. One- or two-millimeter erosion

Fig. 4.11 Normal port.

allowance (as examples) can be added to the thickness of the disk to mitigate the erosion risk. Galling is defined as wear created due to adhesion and contact between two metallic surfaces.

The bore size of a slab gate valve depends on the size of the hole in the gate; as a general rule, the gate valve bore as per the API 6A/ISO 10423 standard is equal to the size of the valve. As an example, a 5 1/8″ slab gate valve as per API 6A has a bore equal to 5.125 in., which is approximately 130.2 mm. Full bore valves with a bore equal to or greater than the connected or meeting pipe are proposed for slab gate valves, since these valves are operated in particle-containing fluids. Having a full bore (FB) valve facilitates flushing the particles, minimizes erosion, and provides higher flow capacity across the valve. Table 4.1 compares TCG valve bores with the connected piping in schedule 160 which is a thick piping, and the result shows that the valve bore is always larger than

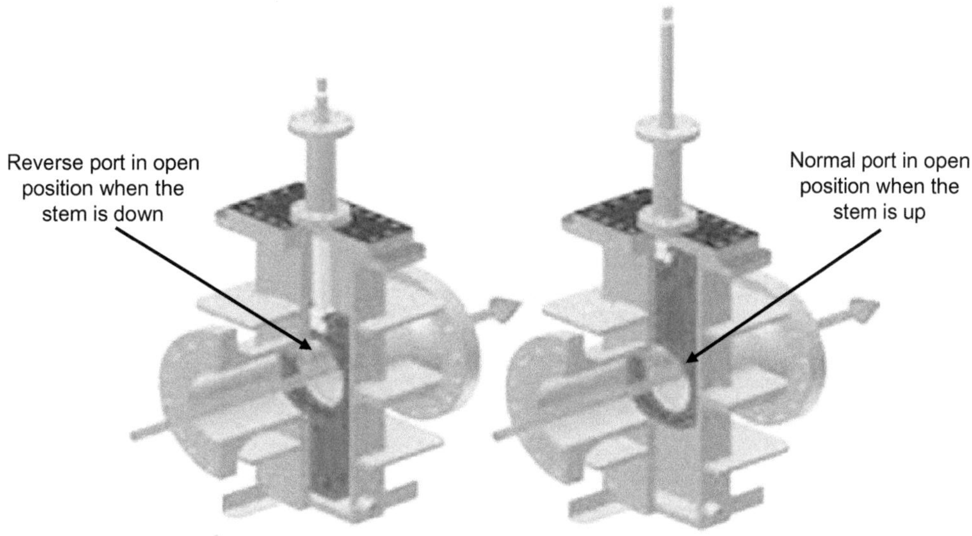

Fig. 4.12 Reverse and normal port gate opening. *(Courtesy: Cameron.)*

the piping internal diameter. However, the gate valve bore could be smaller than the connected piping, especially when the pipe has a lower schedule number and less thick. Schedule number which could be a number such as 40, 80, 160, or combination of letters such as XS or XXS is a designation for the piping thickness. As an example, a 1 13/16″ gate valve has a bore of 46 mm according to the API 6A standard. The connected pipe is 2″ with an outside diameter equal to 60.3 mm according to ASME B36.10/19. The schedule of the pipe is 80, equal to 5.5 mm, so the inside diameter of the pipe is 49.3 mm which is larger than the bore of the valve.

The other important consideration in designing the gate or disk is that the gate should move downward where some sand may accumulate at the bottom of the valve in the

Table 4.1 Comparing TCG valve bores with the internal diameter of the pipe in schedule 160.

Nominal pipe size (in.)	Standard valve bore as per API 6A standard	Pipe internal diameter for schedule 160
2	1 13/16″ or 2 1/16″	1.689″
3	2 9/16″	2.624″
4	4 1/16″	3.438″
6	5 1/8″	5.189″
8	7 1/16″	6.813″
10	9″	8.500″
12	11″	10.126″
16	13 5/8″	12.814″

(Courtesy: Cameron.)

cavity. Thus tapering the disk or gate bottom (see Fig. 4.13) facilitates moving the disk through the sand.

The other important feature for slab gate valves, especially those that are installed on a vertical line with the stem on the horizontal plain, is the use of disk guiding plates or a seat skirt to protect the seats from particles entering the seat, especially during the valve closure member travel, and to provide a guide for the linear movement of the disks during opening and closing of the valve. Fig. 4.14 illustrates seat arrangement, as well as disk and guiding plates for a slab gate valve.

The important consideration for guided plates is that they should be robust enough to withstand the weight of the disk, especially in large valve sizes. The guided rings could be welded to the body of the valve internally. Two seat retainers or seat rings, shown in red (dark gray in print versions) in Fig. 4.14, are located in the machined area in the valve body, called the seat pocket. The seats of gate valves are normally more robust and larger than those of ball valves, as illustrated in Fig. 4.15. Some end users may allow having a hole inside the gate (disk) for pressure equalizing between the upstream, downstream, and cavity of the valve, which leads to less force necessary for valve operation and allows for a more compact operator.

The seats are energized by springs and the sealing between the seats and body of the valve is assured by some sealing such as a lip seal, V-pack, or seat insert made of PEEK (soft material) as an example. Fig. 4.16 illustrates the seat of a slab gate valve with a PEEK seat insert for sealing between the body and the seat. PEEK is a thermoplastic sealing that can be used as a seat insert for ball and gate valves in process (hydrocarbon oil and gas)

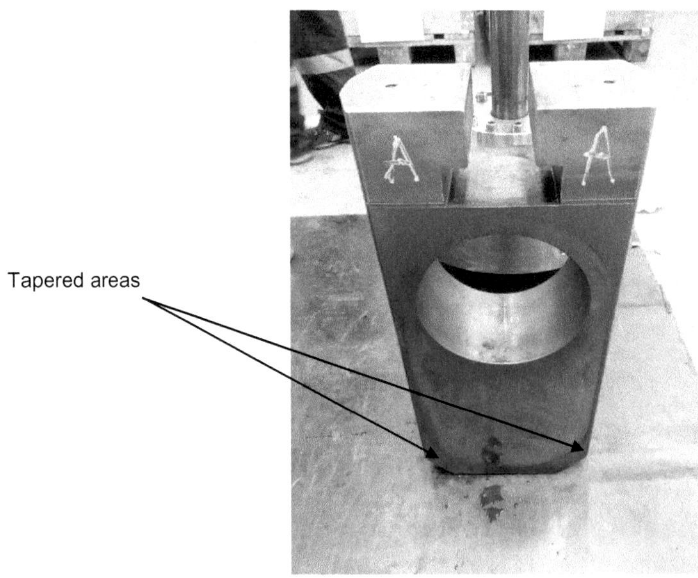

Fig. 4.13 Tapered gate valve disk.

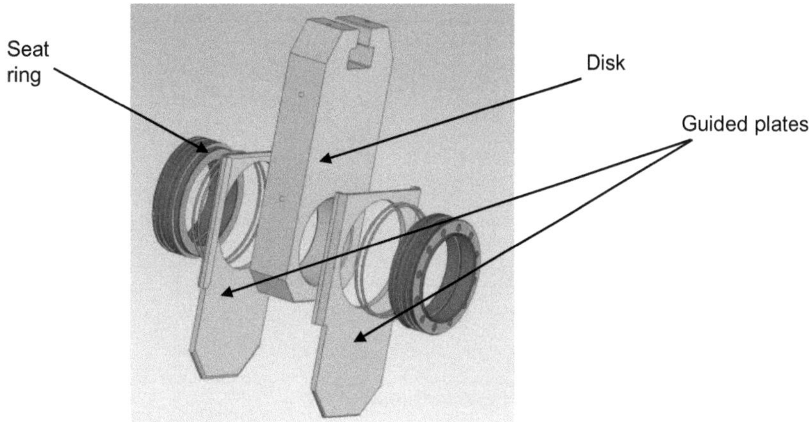

Fig. 4.14 Seats, disk, and disk guiding plates.

Fig. 4.15 Seats of a slab gate valve.

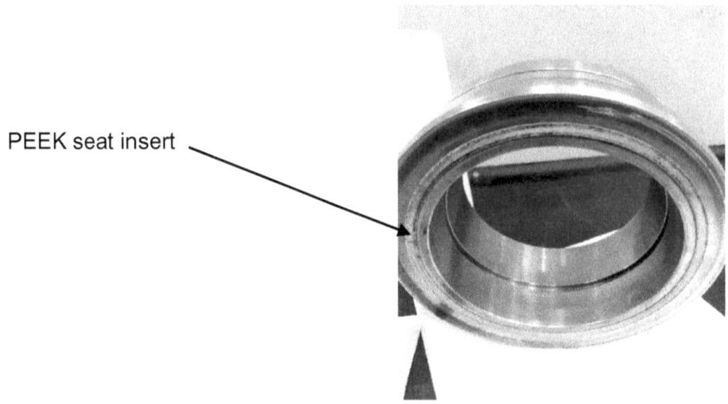

Fig. 4.16 Seat of a slab gate valve with PEEK seat insert.

services. The MtM sealing between the gate valve disk and the seats could be according to leak rate A in ISO 5208, meaning zero leakage. The flat surface between the seats and the disk is considered an advantage; it can provide bubble-tight shut-off or zero leakage capability from the valve. No sealant injection or sealant injection ports are used for subsea valves.

Both the seats and the disk are tungsten carbide coated to prevent galling. The stability of the tungsten carbide on the seat and disk surfaces should be evaluated by an analysis to make sure that, in different load conditions, the tungsten carbide is not removed from these two contact surfaces.

4.2.1.3 Stem

In slab gate valves, the stem can be driven by an actuator or manually by a remote-operated vehicle (ROV) in order to open or close the valve through downward and upward movements. In fact, the stem or shaft transfers the actuator or ROV load to the valve internals. The stem in a gate valve has linear (upward and downward movement) and is not rotated. Nonrising gate valves have threaded connections at the bottom of the stem where the plug rotates and moves upward or downward for cycling of the valve. The threaded connections on the bottom of the stem are located at the pressure boundary of the valve with the possibility of contact with corrosive and erosive internal fluid, thus they can be worn or corroded and cause stem and valve cycling failure. This type of stem design, illustrated in Fig. 4.17, with threads inside the pressure boundary is called "inside screw and yoke" (IS&Y) or nonrising stem. Therefore rising stem gate valve in which the stem threads are on the top of the stem and out of the pressure boundary is the preferred gate valve design, illustrated in Fig. 4.18. A rising stem gate valve is also known as "outside screw and yoke" (OS&Y). The stem of the gate valve should have enough mechanical strength to withstand the operator loads, especially in the case of actuation. Enlarging the stem or upgrading the stem material to a higher mechanical

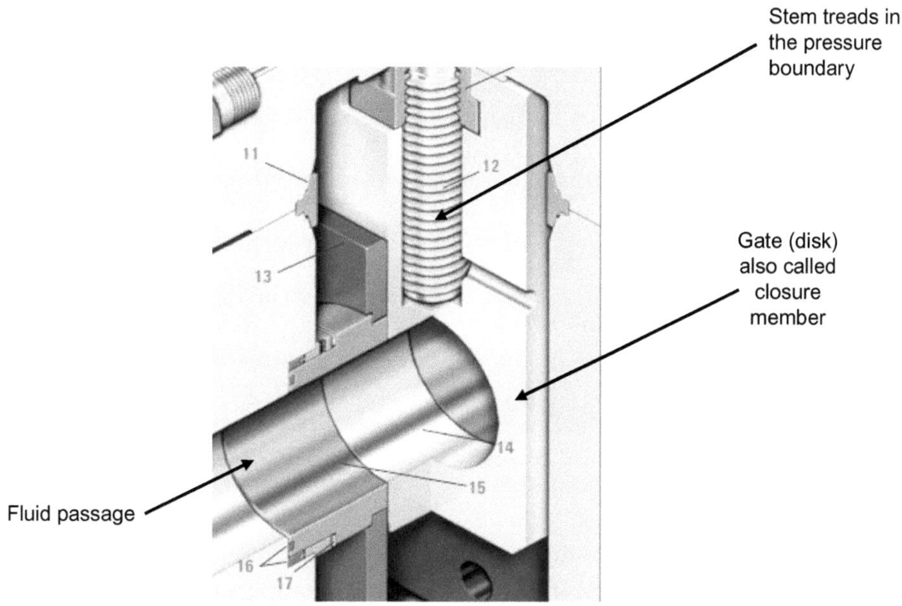

Fig. 4.17 Inside screw and yoke (IS&Y) slab gate valve.

Fig. 4.18 Outside screw and yoke (OS&Y) rising stem slab gate valve.

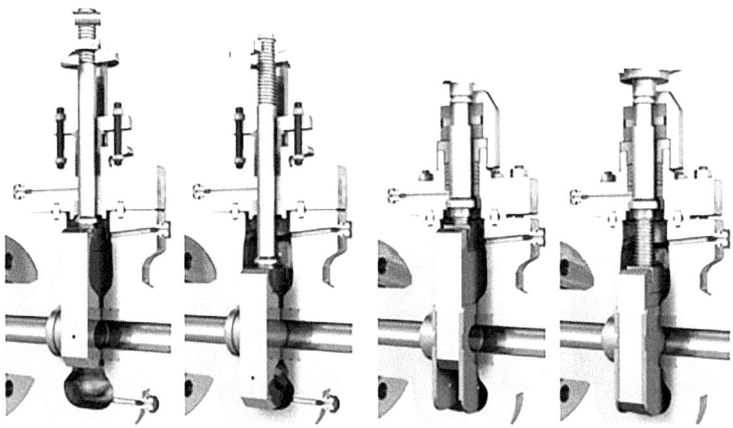

Fig. 4.19 Rising and nonrising stem slab gate valve comparison. *(Courtesy: Norwegian Oil Industry Association.)*

strength material could be two solutions to increase the maximum allowable stem torque (MAST) to withstand the operator force, especially when the valve is actuated.

Fig. 4.19 compares rising and nonrising stem slab gate valves. The two pictures on the left side illustrate a slab gate valve with rising stem and thread connections out of the pressure boundary. The two pictures on the right show threads at the bottom of the stem inside the pressure boundary where the disk moves up through the stem to open the valve without any movement of the stem. In fact, the stem is stationary in the two pictures on the right-hand side.

4.2.1.4 Stem sealing

Stem sealings are used to provide tightness between the stem and the body and bonnet. Two different sets of stem sealing are used for slab gate valves; one seat on the upper part of the stem prevents ingress of seawater to the valve internals such as the stem sealing areas. In general, these seals, called environmental barrier sealing, should consist of two different seals such as an O-ring and a lip seal. Details of sealings are explained in Chapter 13. The other sealing prevents leakage of the internal fluid in the valve to the environment. These sealings are also made up of two different sealing configurations, such as a lip seal and a V-pack. It is not recommended to use metallic sealing or bellows for gate valves, unlike ball valves, due to the high friction between the valve stem and sealing during the upward and downward movement of the valve stem.

Fig. 4.20 illustrates a slab gate valve with the major components explained earlier in this chapter.

A glyd ring (see Fig. 4.21), which is a robust double layer and double acting O-ring, can be used for primary external pressure sealing instead of an O-ring for gate valves where the friction between the stem and sealing is high. This ring is proposed for dynamic

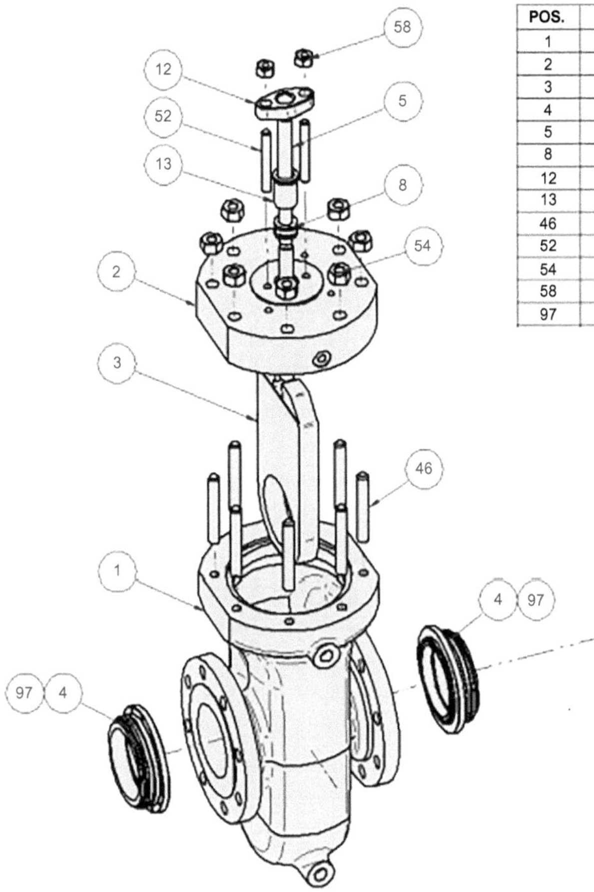

POS.	Part name
1	Body
2	Bonnet
3	Gate
4	Seat
5	Stem
8	Lantern ring
12	Gland flange
13	Gland
46	Stud bolt
52	Stud bolt
54	Nut
58	Nut
97	Lip seal

Fig. 4.20 A slab gate valve including the main parts.

Fig. 4.21 Glyd ring.

applications where low friction and wear are important parameters. A glyd ring, as the outermost external sealing, is placed under the gland and gland flange.

The intention of this paragraph is to provide a discussion about the different stem seals and other components around the stem of a subsea slab gate valve with reference to Fig. 4.22. Item #120 is called a wear ring or "wearing ring" by the valve supplier; it is installed at the bottom section of the stem. Item #120 acts as a lower bearing against the linear motion of the valve stem; it is made of turcite, which is a thermoplastic material. Item #101 is a V-pack sealing made of different layers of Teflon (PTFE), PEEK, and Elgiloy. A V-pack provides the preliminary stem sealing against the internal fluid pressure. Items #8 and 8A are called lantern rings, which are metallic parts made of super duplex in this case; they are located between the different stem seals. Items #103 and 103A are two lip seals that prevent leakage of the internal fluid to the external environment; these lip seals are known as secondary internal stem sealing. Items #104 and 104A are two lip seals that provide stem area sealing against the leakage of external seawater to the valve internals. These two lip seals are known as secondary external stem sealing. Item #120A is a wearing ring, same as item #120; it is made of turcite and acts as the upper bearing around the valve stem. In fact, these two bearings prevent lateral movement of the gate valve stem during linear movement. Item #106 is a glyd ring that acts as the primary external stem sealing to prevent leakage and ingress of seawater to the valve internals. The body and bonnet of the valve are in low alloy steel. Most parts of the stem sealing pockets are highlighted in red (dark gray in print versions), indicating Inconel 625

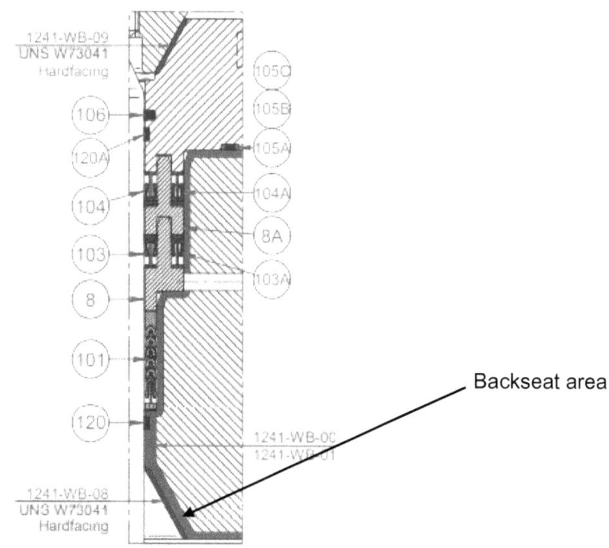

Fig. 4.22 Gate valve stem sealing. *(Courtesy: Advanced Valve Technology.)*

cladding over the body and bonnet to prevent crevice corrosion. The lower part of the stem seal area is the backseat area, which is hard faced with UNS W73041, a stellite 21 material, to prevent galling and friction between the backseat area and the conical part of the stem with which it is in contact.

The aim of this paragraph is to show and explain in detail the different parts of a slab gate valve illustrated in the cross-section drawing in Fig. 4.23. A cross-section drawing or detail drawing shows the different parts of a valve as well as selected materials. Item #1 is the body of the valve, which is the main pressure-containing part. Item #3 is the gate (disk) in open position, which is stopped by the bottom of the valve. The disk has reverse action since the hole for fluid passage is located on the top part of the disk. There is a hole inside the disk to equalize the pressure between the downstream, upstream, and cavity of the valve which leads to less force requirement and a smaller operator selection for the valve. Item #180 is a valve support in plastic material such as high-density polyethylene (HDPE) to prevent wearing between the bottom of the valve (valve body) and the metallic structure. The thickness of this support could be 50 mm, as an example, and it can be connected to the bottom of the valve by cap screws (item #180 A). Item #145 is pup pieces welded to the body of the valve from both ends. The material of the pup pieces should be the same as that of the connected piping. In this case, the valve is in ASTM

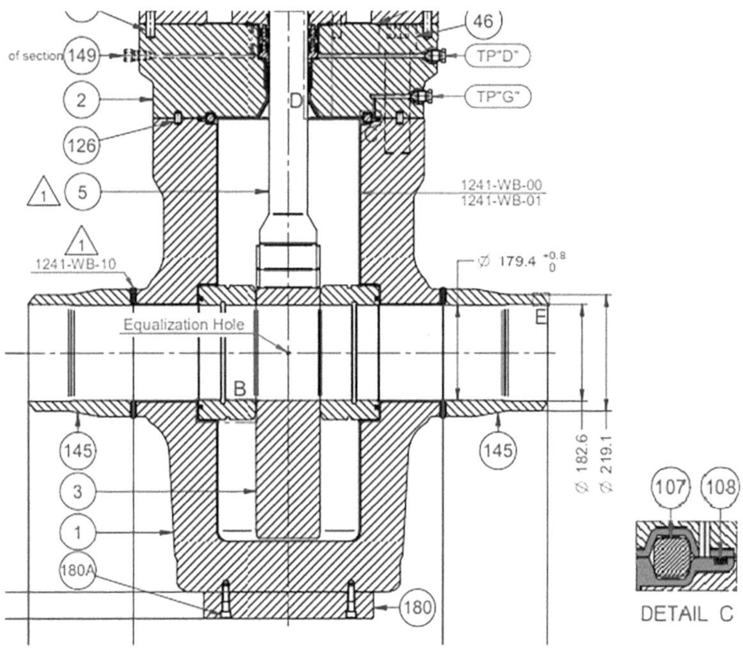

Fig. 4.23 Slab gate valve including Detail C for body and bonnet sealing. *(Courtesy: Advanced Valve Technology.)*

A694 F60, fully cladded in Inconel 625 and welded to 25Cr super duplex piping; the pup piece material is ASTM A182 F55 (forged super duplex). The length of the pup piece(s) should be sufficient to keep the heat generated by welding the piping to the valve in the construction yard away from valves' sealing and internals. It should be noted that the valve sealing could be melted and damaged as a result of the welding heat input. The welding between the valve body in low alloy and the pup piece in duplex would be challenging in this case, due to the fact that the valve thickness is more than 20 mm. The challenge in welding is that the pipe in carbon steel and low alloy steel requires postweld heat treatment (PWHT) after welding to remove the residual stress according to ASME B31.3, process piping code requirements. However, high heat input during the postweld heat treatment could make sigma brittle phase in super duplex. One common technique to solve this issue is to use an Inconel 625 or alloy 59 buttering on the carbon steel and low alloy steel parts and then apply the PWHT. The next step is to use a filler of alloy 625 or 59 to connect the buttered low alloy steel to the super duplex without any PWHT requirement. It should be noted that super duplex is more sensitive to high heat input due to welding compared to 22Cr duplex. In addition, some metallurgists prefer alloy 59 for buttering and as a welding filler instead of alloy 625. Item #5 is the stem of the gate valve, which can be in 25Cr super duplex or upgraded to Inconel 718 or 725 in case of actuation and when the MAST must be increased. The area close to the connection of the stem to the disk is conical, which provides a stop on the backseat when the stem moves upward. The backseat area is located close to the letter "D" in the figure. Item #2 is the bonnet of the valve, which acts as a cover on the body; it has the same material as the body. Item #126 is a dual pin between the body and bonnet to ensure correct alignment between these two components. In fact, there are two small holes in the body and bonnet compatible with the dual pin that enters these two holes. Beside the pins, there are sealings between the body and bonnet, shown in the figure with a letter "C," including one BX metallic gasket based on the API 6A standard (Item #107) providing sealing from both the internal and external sides, as well as an O-ring (Item #108) to provide isolation from the external seawater. One port, called TP "D" is used to test the stem sealing sealability; TP "G" is used to test the sealability of the body and bonnet sealings. The details of these ports are explained in Chapter 14, titled "Testing." Fig. 4.24 illustrates the same valve, installed inside a manifold at the construction yard.

4.2.2 Expanding gate valves

Through conduit expanding gate valves, also called expanding gate valves, are designed, selected, and operated to provide isolation of the piping system or components when the valve is in closed position. Like slab gate valves, expanding gate valves should not be used for flow control (throttling) since the valve internals would be subject to vibration, excessive wearing, and damage. Again, the main parts of expanding gate valves are the body,

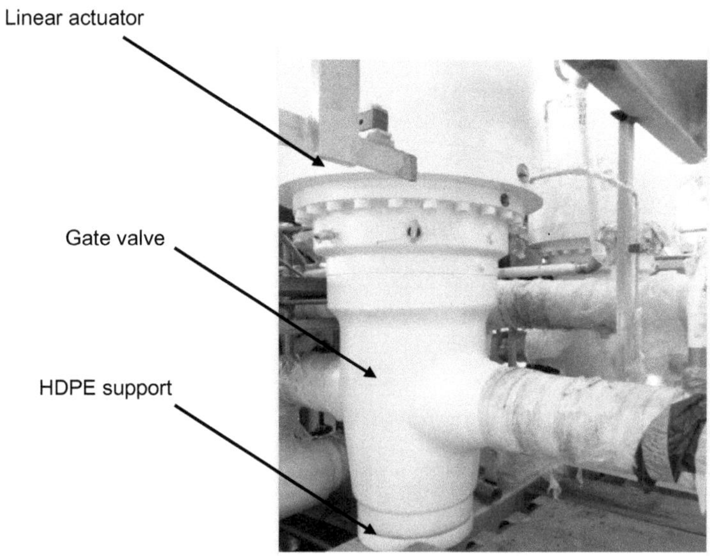

Fig. 4.24 Slab gate valve installed in the manifold at a construction yard.

bonnet, seats, and disk. Fig. 4.25 illustrates an expanding gate valve and its main components. The description of all the components is same as for the slab gate valve components, except for the disk and sealing mechanism. Unlike slab gate valve seat sealing, which is achieved through floating seats and fluid pressure that pushes the disk to the downstream position, an expanding gate valve is a torque sealed valve, which means that the stem force is transmitted to half disks and pushes them strongly toward the seats. For a double expanding gate valve, two half disks are pushed and expanded against both seats strongly in both closed and open positions. However, in a single expanding gate valve, strong sealing between the disk halves and seats is achieved only in closed position. In fact, a double expanding gate valve is a double block and bleed valve as well as a double isolation and bleed type 1 (DIB1). Leakage to the valve cavity is minimal for double expanding valves, and the packing or stem sealing life is extended for this type of valve. Double expanding gate valves are more expensive than both single expanding and slab gate valves. In addition, double expanding gate valves are heavier and more robust compared to slab gate valves. Also, unlike slab gate valves, no sand and only minimum fluid can enter the body cavity in double expanding gate valves.

4.2.2.1 Body and bonnet

The body and bonnet of expanding gate valves are the same as those of slab gate valves; for more information, refer to Section 4.2.1.1. Expanding gate valves have an upper travel stop, as shown in Fig. 4.25, so the backseat is never activated during upward stem movement. Fig. 4.26 illustrates a subsea double expanding gate valve with a pressure relief valve

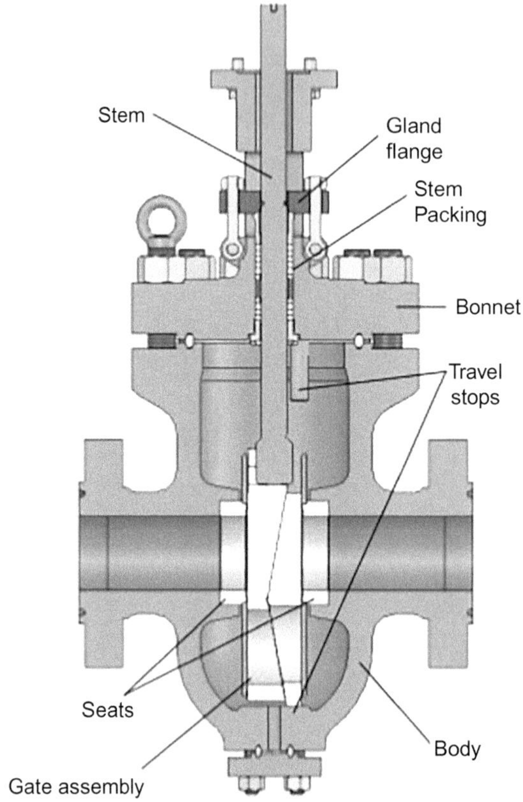

Fig. 4.25 Expanding gate valve.

on the upper body cavity in order to release the cavity pressure. DIB1 valves, such as double expanding in liquid (noncompressible fluid), require a pressure relief valve on the cavity in order to release any overpressurized fluid that has accumulated in the cavity. A relief valve is not required for valves in gas service because gas is compressible. A relief valve, also called a pressure relief valve, is a type of safety valve used to control and maintain the pressure limit in the system. Otherwise, the pressure can increase beyond the acceptable level and cause component or equipment (in this case valve) failure.

4.2.2.2 Gate and seats

Two seats are inserted in the valve body seat pockets, one on each side. The seat and disk arrangement should provide straight flow without interruption suitable for the through conduit design of expanding gate valves. The seats' sealing forces are achieved through the expanding forces of the half disks. The seat surface should be hard faced with tungsten carbide to prevent galling due to MtM contact with the half disks and should be perfectly

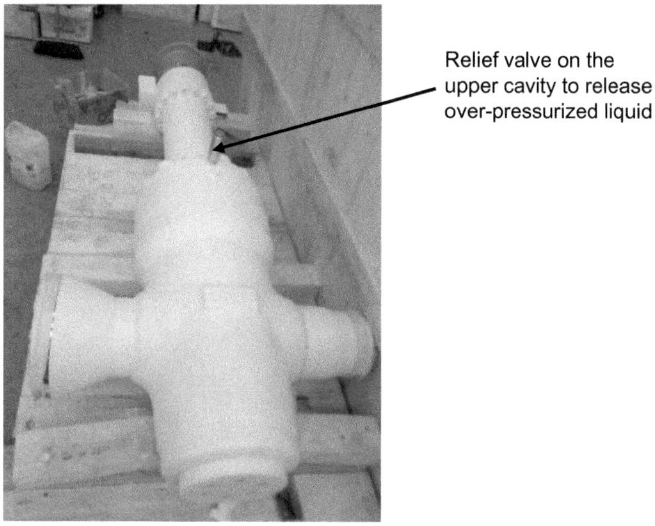

Fig. 4.26 Subsea double expanding gate valve. *(Courtesy: ATV.)*

polished. A seat insert in PEEK material can be placed at the back part of the seat to provide seat and body sealing.

The closure member or disk mechanism for an expanding gate valve is totally different from that of a slab gate valve. The gate or disk in a double expanding gate valve is made of two parts: one female and one male with a sealing or lever between (see Fig. 4.27). One half disk slides against the other one. The stem is connected to the top part of the female section with a T-shape slot connection. A T-shape connection is a robust connection between the stem and disks that provides an antiblowout feature for the stem, as well as flexibility of movement of the stem with the disk sidewise. The contact surfaces between the male and female halves provide lateral expansion of the disks due to stem force application (see Fig. 4.27). The whole disk assembly is placed between two seats with a guided ring or seat skirt, or without a guiding ring. The female section drags the male disk and the whole disk assembly moves up and down between the upper and lower travel limits. At both ends, travel stops limit the movement of the disk assembly and the stem force is transferred to the female part of the disk, creating a wedging effect, as illustrated in Fig. 4.28. A wedging effect, in simple terms, is defined as a situation in which force is applied to the thick end of a wedge on the top and the wedge, in turn, transfers the force to the objects located along both of the slopping sides. Therefore the mechanical force of the stem is transferred to the female disk, which creates disk expansion toward both seats, which provides a double block and bleed as well as a double isolation and bleed feature for the valve.

Fig. 4.29 illustrates a situation when the stem and disk assembly are moving upward. The disk will be stopped at the upper stop limit close to the backseat and the valve closes.

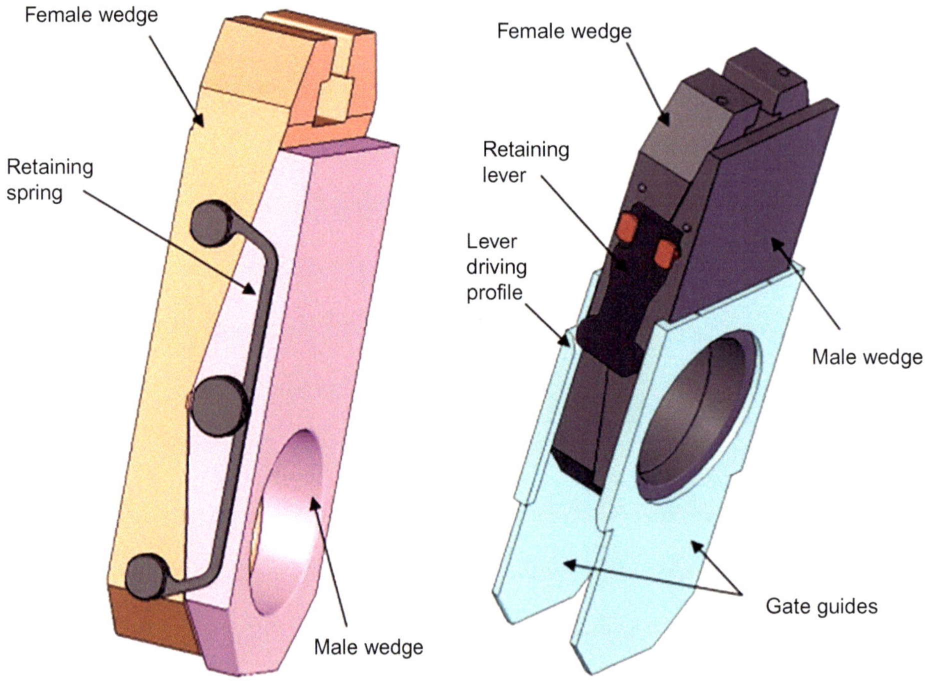

Fig. 4.27 Male and female disk segments. *(Courtesy: Orion Valve.)*

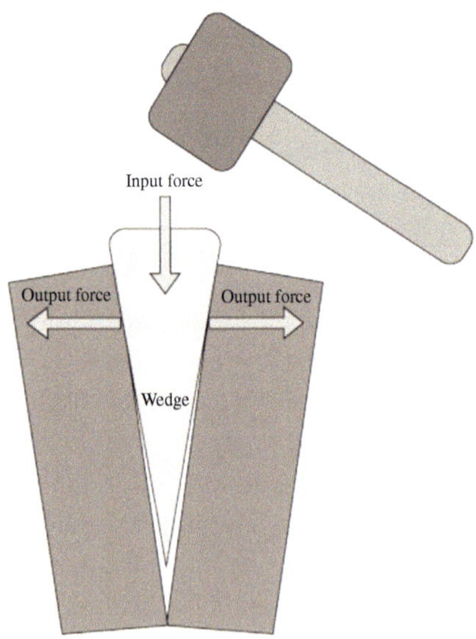

Fig. 4.28 Wedging effect.

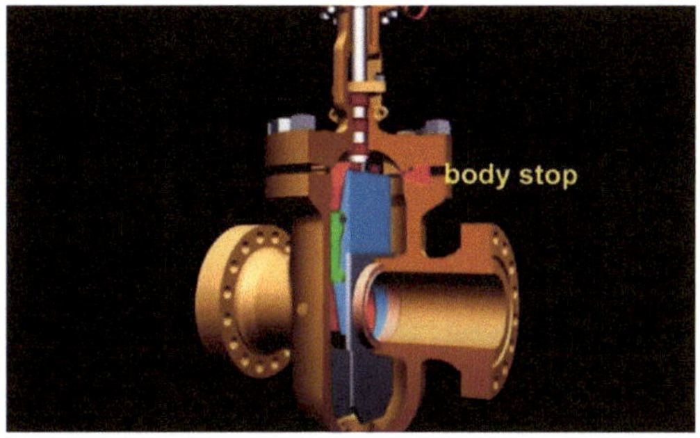

Fig. 4.29 Stem in upward position and the valve is closed.

The stem force transfers to the disk assembly and causes the disk halves to expand toward the seats. After that, the stem and the disk assembly move downward and the valve opens, as illustrated in Fig. 4.30. While the stem is down and the half disks are stopped by the lower body of the gate valve, the stem force is transmitted to the half disks, which expand and push against the seats (see Fig. 4.31). Then the valve stem moves upward and the valve position changes from open to closed position. There is no wedging effect during the stem movement upward from open to closed position (see Fig. 4.32) or stem movement downward from closed to open position. The expansion force on the half disks is reduced significantly and is much less than in the fully open or closed positions. Thus, during the movement between open and closed positions, there is a possibility of fluid ingress to the valve cavity.

4.2.2.3 Stem

Same as a slab gate valve; for more information refer to Section 4.2.1.3.

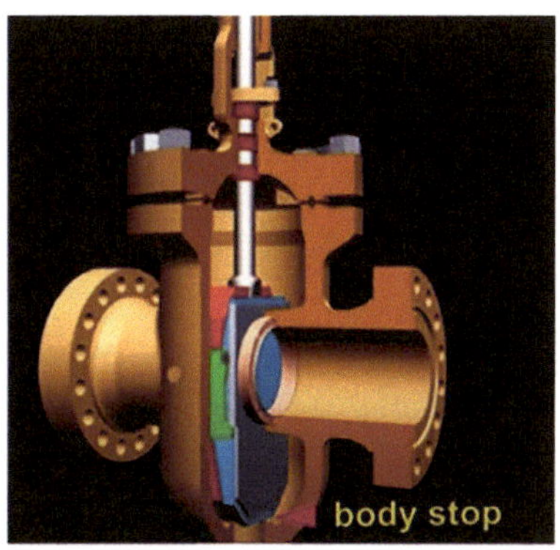

Fig. 4.30 Stem in downward position and the valve is open.

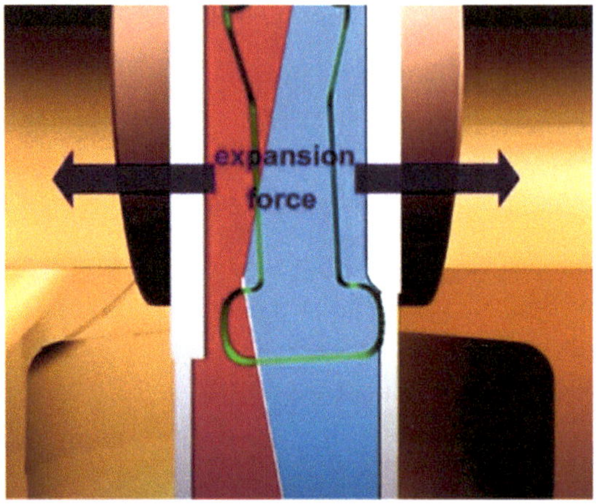

Fig. 4.31 Disk halves are expanded toward the seats in open position.

4.2.2.4 Stem sealing

Same as a slab gate valve; for more information refer to Section 4.2.1.4.

4.3 Through conduit gate valve characteristics

4.3.1 Height

Gate valves have more height than ball valves due to their high body and upward stem and gate movement for rising stem design. The height of a gate valve could be considered a disadvantage, since it can impose a manifold structure with higher height. Usage of a linear actuator on top of a TCG valve could make the height differential even greater compared to an actuated ball valve. One should bear in mind that rack and pinion or scotch and yoke actuators, which are used for ball valve automation, stand horizontally when the valve is connected to a horizontal pipe, whereas linear actuators stand vertically

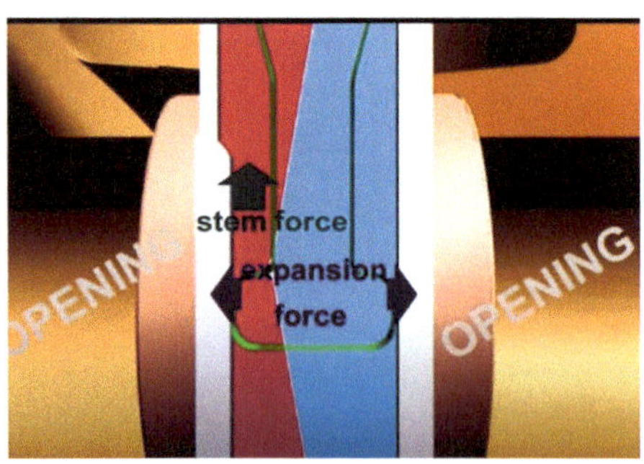

Fig. 4.32 Stem is moving upward from open to closed position and no wedging effect.

Fig. 4.33 High height of an actuated slab gate valve. *(Courtesy: ATV.)*

when the TCG valve is connected to horizontally installed piping. Fig. 4.33 illustrates the height of an actuated slab gate valve.

4.3.2 Speed of operation

Although it has been observed that TCG valves have been made and used in subsea projects for Emergency Shutdown (ESD) or High Integrity Pressure Protection (HIPPS), TCG valves are not quarter turn with a 90-degree rotation of their closure member, so they are not recommended for ESD or HIPPS applications in some of the valve and actuator specifications. Instead, ball valves with a ¼″ turn to cycle between open and closed or closed to open are the most attractive choices for ESD and HIPPS applications in subsea systems.

4.3.3 Weight

In general, TCG valves (both slab and expanding) are heavier and bulkier than ball valves but that is not always the case. There is no standard weight for valves; their weight depends on the valve manufacturer standard. Let's suppose a valve manufacturer produces heavy and bulky ball valves. In that case, a ball valve could be heavier than a TCG valve of the same size and pressure class. Gate valves are generally top entry design with a large and heavy bonnet that could make them bulkier than ball valves, which can be produced in side entry "bonnet-less" design.

4.3.4 Stem and stem seal wear

Stem wear is greater for a gate valve than a ball valve, since the stem moves up and down and it is in contact with all of the stem sealing and packing. However, a ball valve's stem has just a 90-degree rotation, so it has much less contact area between the stem and sealings. Unlike ball valves, it is not proposed to use metallic stem sealing for gate valves as it can wear the stem significantly. Metallic seals such as bellows are more common for large ball valves.

4.3.5 Sensitivity to deep-water depth

This section addresses actuated slab gate valves. Actuated TCG valves generally have a fail-safe closed function on manifolds and Christmas trees. These valves get closed with spring force and upward movement of the stem. The upward movement of the stem must overcome the seawater head pressure on the valve. Therefore there is a requirement to have very large spring, which can increase the size of the actuator accordingly. On the other hand, ball valve actuators, which are explained in Chapter 10, do not require enlargement due to seawater depth and head pressure, as the actuators are installed horizontally and the internals move on the horizontal axis.

4.3.6 Double block and bleed possibility

Subsea small bore ball valves could have a DBB feature, including two block ball valves and a needle valve between in a single body (see Fig. 4.34) for chemical injection lines to be operated by ROV. It is not common to have a single body for double TCG valves.

4.3.7 Sealability between the disk and seats

Gate valves have flat surfaces between the seats and the disk, which facilitates having zero leakage between the seat and disks compared to ball valves. Many subsea gate valves are capable of providing zero leakage (bubble tight) leak rate A as per ISO 5208. The allowable sealing from a gate valve after qualification and 600 cycles could be leak rate B according to ISO 5208. However, the sealing area between the ball and the seats in a ball valve is spherical, so achieving zero leakage is not easy and requires extra lapping between the ball and seats. Lapping is a machining process in which two surfaces are rubbed together with an abrasive between them by hand movement or using a machine. Lapping is done in the valve industry between two components that should provide sealing in order to provide very smooth or low roughness surfaces between them. Fig. 4.35 illustrates the lapping of a ball valve against the seats to improve the sealability of the valve.

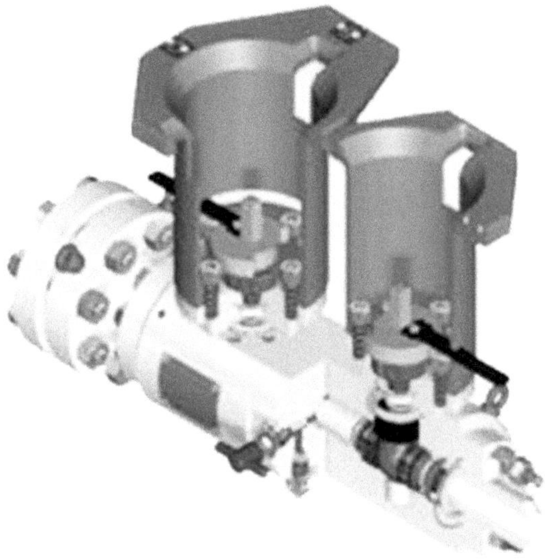

Fig. 4.34 Double block and bleed (two ball valves and one needle between). *(Courtesy: Pacson.)*

4.3.8 Small bore (size) actuated valve selection

Actuated small bore valves in 1″ size and a pressure class of 690 bar, as an example, can be used for chemical injection fluid start/stop purposes. The main question is, which valve is suitable—ball or gate? Four valve suppliers were contacted and all of them confirmed that the valve of choice should be a gate valve. One of the manufacturers confirmed that due to actuation, higher MAST is required and gate valves have more flexibility to enlarge the

Fig. 4.35 Lapping of a ball against two seats.

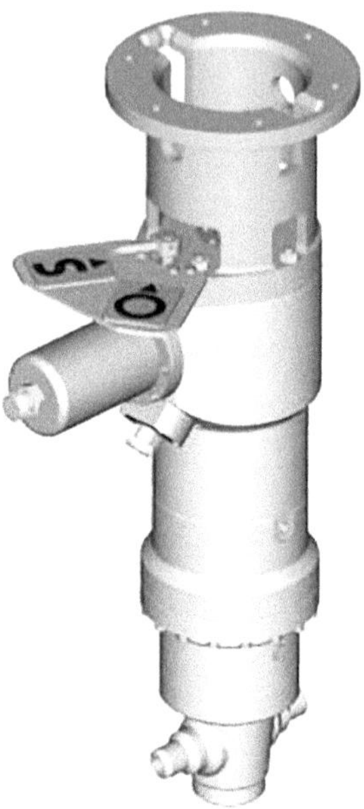

Fig. 4.36 1″ class 690 slab gate valve with hydraulic actuator.

stem through the bonnet due to their top entry design. A 1″ hydraulically operated slab gate valve in pressure class 690 equal to 10,000 psi is illustrated in Fig. 4.36.

4.4 Rotary gate valves

Rotary gate valves, specifically in small sizes of 1″ and less such as ½″, ¾″, 3/8″, and 1″, are designed and used for operation in subsea environments. These valves could be located on chemical injection lines and used instead of needle valves for flow throttling (control) application. There are many advantages associated with rotary gate valves: they are more compact compared to ball and TCG valves, have minimum moving parts and sustain no damage to the valve internals due to overtorquing or quarter turn movement for opening and closing. The rotary gate valves have MtM sealing with reliable operation in particle-containing (dirty) fluid services. Both manual (ROV) (Fig. 4.37) and actuated operation (Fig. 4.38) of rotary gate valves are possible.

There are four holes located at the bottom of the stem on the closure member with a 90-degree distance between them. Two of these holes are blinded at 12 and 6 o'clock, and two are filled in at 9 and 3 o'clock according to Fig. 4.39. Two ports for inlet flow into and outlet flow from the rotary gate valve are located opposite the four holes located on the valve closure member. When two blind holes align with two inlet and outlet flow

Fig. 4.37 ½″ manually operated rotary gate valve. *(Courtesy: Bentley.)*

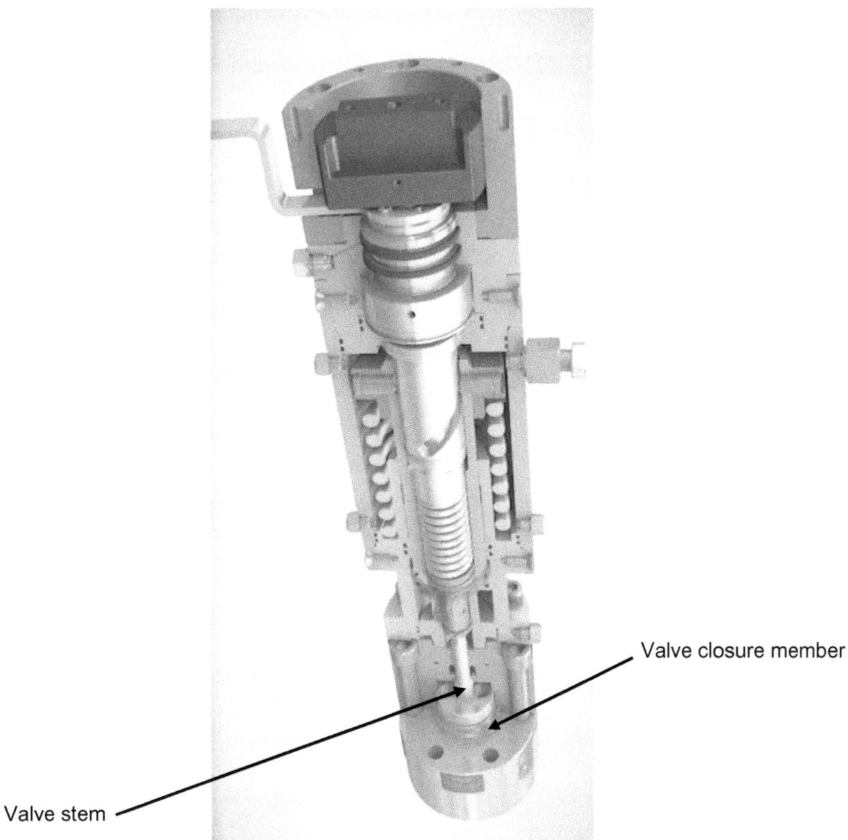

Fig. 4.38 Actuated rotary gate valve in closed position. *(Courtesy: Bentley.)*

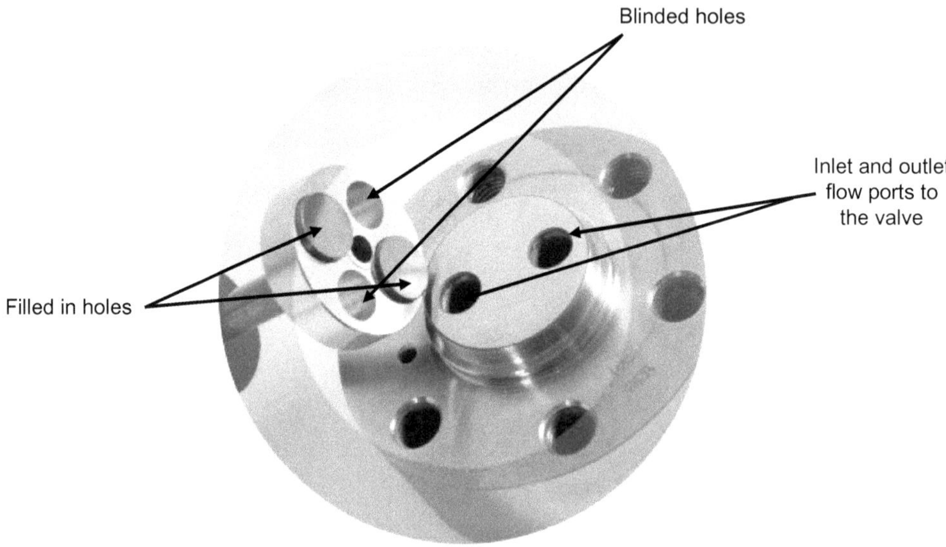

Fig. 4.39 Rotary gate valve closure member and inlet and outlet flow ports. *(Courtesy: Bentley.)*

ports to the valve, the valve is open. Operating the valve with an actuator or manually leads to rotation of the closure member 90 degrees from open to closed position. In fact, two filled-in holes would be aligned to the inlet and outlet flow ports to the valve when the valve is getting closed. The connection of the valve to the piping is perpendicular to the valve stem. Fig. 4.40 illustrates a rotary gate valve in both open and closed positions.

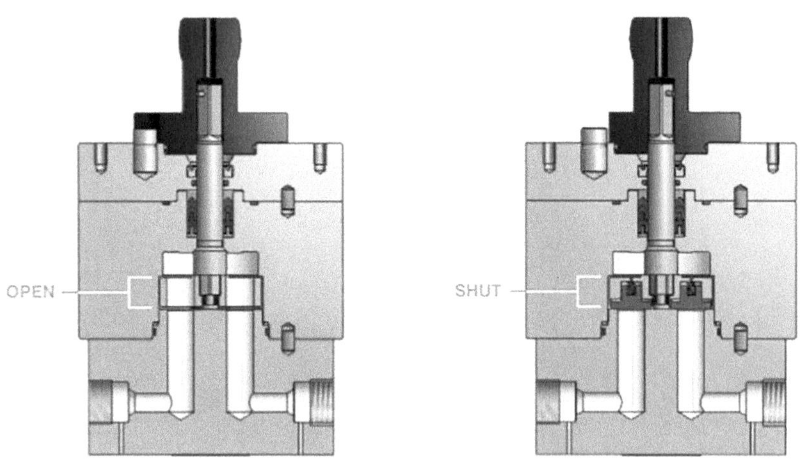

Fig. 4.40 Rotary gate valve in open and closed position. *(Courtesy: Bentley.)*

The valve is supplied with a position indicator so it is possible for the operator to recognize the position of the valve. In addition, the valve is bidirectional, meaning that it can be installed on both flow directions. In fact, the valve can be rotated 180 degrees and be installed.

Rotary gate valves are typically required to be in 25Cr super duplex (material class FF as per API 6A/ISO 10423) or nickel alloy such as Alloy 625 (material class HH as per API 6A/ISO 10423), and in pressure classes of 5000 psi equal to 345 bar, 10,000 psi equal to 690 bar, or 15,000 psi equal to 1035 bar. The standard hydraulic pressure for operation of these valves is 5000 psi equal to 345 bar for hydraulic actuated rotary gate valves. In addition, these valves can be designed and tested for deep-water depths such as 3000 m and a relatively long design life of 30 years.

4.5 Product quality assurance

Achieving a reliable product is dependent on accurate design, manufacturing, testing, and material selection, which are explained in this section.

4.5.1 Design

Gate valves shall be designed accurately and properly taking all the factors into account to ensure safe operation during their design life. Comprehensive methods for designing the valves should take enough safety margins into account against all possible failure modes and possibilities. Some of the main parameters mentioned as follows should be considered in detail for the design of gate valves:

- Design pressure, which includes both internal and external pressure due to water depth;
- Temperature conditions, both minimum and maximum;
- Different applied loads to the valve such as weight, pipeline load, etc. during operation, test, installation, and maintenance. It should be noted that the performance of the valve shall not be affected by the combination of the loads;
- Wearing, galling, corrosion, and erosion values, which are provided by the purchaser;
- Engineering methods and calculations such as finite element analysis (FEA);
- Weight and dimensional data, which can be extracted from the valve general arrangement drawing.

Example 4.1 A 7 1/16″ slab gate valve is used on a Mono-Ethylene Glycol (MEG) line. MEG is not a corrosive or erosive fluid. How many milometers of corrosion allowance and erosion allowance should be considered for this valve?

Answer: MEG is not a corrosive fluid, so the body and bonnet of the gate valve can be selected in carbon steel (e.g., A350 LF2) or low alloy steel (e.g., A694 F60) material. The standard corrosion allowance for piping and valves according to NORSOK L-001, "piping and valve material specification," is 3 mm. However, since the MEG fluid is not corrosive, 1 mm corrosion allowance could be sufficient. There are no particles in MEG service, so there is no need to consider any erosion allowance on the valve.

Example 4.2 A 5 1/8″ gate valve is located on a manifold branch in highly corrosive and erosive fluid service. The body and trim materials should be FF as per API 6A/API 17D requirements. In this case, which material is proposed for the body of the valve, and how many millimeters of erosion and corrosion allowance should be considered?

Answer: Trim FF addresses stainless steel such as 25Cr super duplex or Corrosion Resistant Alloy (CRA) material for sour H_2S-containing fluid. A CRA is defined in API 6A/API 17D as a kind of material that has more than a 50% percentage of alloys such as Inconel 625. Using a 25Cr super duplex body is not recommended, since this material has limited resistance to sour service. Using solid Inconel 625 for the body and bonnet of this valve would make it very expensive. Thus a carbon steel body and bonnet with a 3 mm layer of Inconel 625 is proposed. There is no need to consider corrosion allowance for the body and bonnet, since Inconel 625 as a CRA material is in contact with the fluid. Normally, there is no need to consider corrosion allowance for CRA materials. One-millimeter erosion allowance should be considered on an Inconel 625 overlay to withstand erosion, so the total thickness of the Inconel overlay is three plus one millimeter, equal to four.

Example 4.3 A 5 1/8″ hydraulically operated gate valve is designed for a water depth of 1700 m. The pressure at sea level is equal to 380 bar. What pressure rating shall be considered for the valve?

Answer: The valve is operated at a depth of 1700 m, which is equal to 170 bar water head pressure (see Fig. 4.41). The pressure at 1700 m depth is equal to the pressure at sea level plus the seawater head at depth, which is $380 + 170 = 550$ bar. The minimum API 6A/17D pressure to cover 550 bar is 690 bar, equal to 10,000 psi.

Example 4.4 The weight of a 7 1/16″ gate valve in pressure class 517 bar is given by the valve manufacturer as equal to 1230 kg. What would be the weight of the valve in water?

Answer: The weight of steel is less in water compared to air, and a conversion of 0.875 can be used to obtain the weight of steel in water. Therefore the weight of the valve in water is equal to $1230 \times 0.875 = 1076.25$ kg.

4.5.2 Manufacturing and testing

Very accurate machining of the valve components is one of the factors that guarantees the reliability of subsea valves and the compliance of the manufactured parts with their design. In addition to machining, accurate and correct nondestructive testing (NDT) shall be applied to the machined and welded parts to make sure that there is no defect in the tested materials. Correct assembly and installation of the components are essential

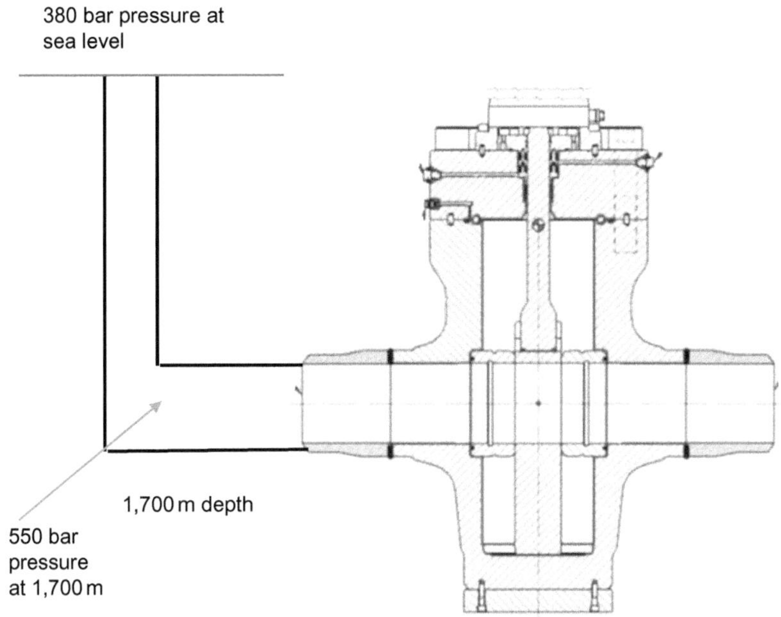

Fig. 4.41 A slab gate valve operated at depth of 1700 m.

key factors to ensure valve reliability. Very strict factory acceptance testing based on standards as well as end user requirements is applied to the products. Finally, dimensional and visual examination is performed to validate the design and quality of the products. Although API 6AV1, qualification in sandy service, is normally applicable for safety valves such as surface safety valves (SSV), boarding shutdown valves (BSDV), and underwater safety valves (USV), end users may ask to apply the sand slurry test as per the above-mentioned standard on gate valves as well.

4.5.3 Materials

Different body/bonnet materials can be selected for valves, such as 25Cr super duplex, Inconel 625, carbon steel or low alloy steel with full or partial Inconel 625 overlay. The stem can be in 25Cr duplex or Inconel 625. Stronger materials could be selected, such as Alloy 718 or 725 to achieve higher MAST. The seat and disk can be in 25Cr super duplex or Inconel 625, and both of these components could be upgraded to a material such as Inconel 725 in case of stem material upgrading. Most of the above-mentioned materials are exotic and highly corrosion resistant to guarantee the reliability of the subsea valve at depth. Full material traceability, normally EN 10204 3.2, is required for pressure-containing parts such as the body, bonnet, and stem. EN 10204 3.1 can be used as the minimum material traceability level for pressure-controlling parts.

Different levels of test certificates and traceability based on the EN 10204 standard are explained in Chapter 10.

4.6 Conclusions

In general, the following characteristics are expected from subsea slab gate valves:

- Through conduit type is defined as a valve characteristic in which, when the valve is in open position, the bore of the valve provides an uninterrupted interior surface across the seat ring and through the valve port with minimum pressure drop. The same characteristic is required and expected from double expanding gate valves. Fig. 4.42 illustrates an expanding gate valve highlighting the seat, bore, and flow path. As illustrated in the picture, the flow can pass the valve bore and seat areas without any interruption.
- Bidirectional means that the valve can be rotated 180 degrees and installed. Bidirectional installation is applicable for both slab and expanding gate valves.
- Rising stem and outside screw and yoke design is applicable for both slab and expanding gate valves.
- Floating gate, with the possibility of tight sealing on the downstream seat, is only applicable for slab gate valves.

In general, the following requirements are expected from subsea expanding gate valves:

- One-piece stem design in forged material and threaded on the upper part to provide an outside screw and yoke (OS&Y) feature for the valve;
- Strong stem sealing design including different sealing components such as a lip seal, chevron V-pack, O-ring or glyd ring;

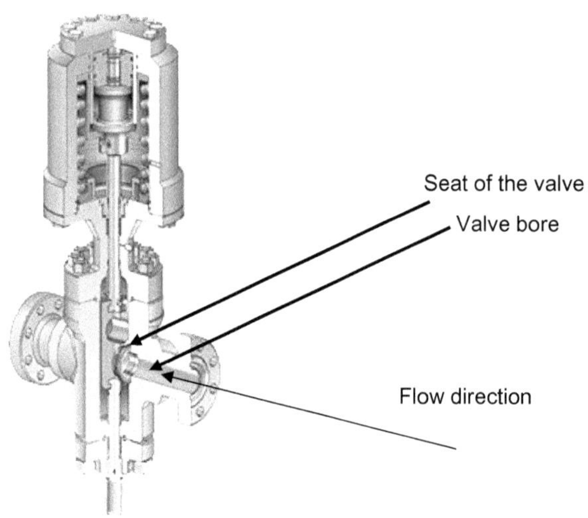

Fig. 4.42 Double expanding gate valve indicating the flow path without any obstruction.

- Expanding gate to provide tight mechanical force-activated sealing on both the upstream and downstream seat to provide both DBB and DIB1 effect as per API 6DSS;
- Heavy and robust body as a pressure-containing part;
- Full bore and through conduit design provider;
- Bidirectional with a preferred pressure or flow direction;
- Perfectly polished and hard-faced sealing.

4.7 Self-assessment questions

This section provides some multiple-choice questions to assess the knowledge of the readers about gate valves in the subsea industry.

1. Which sentence is correct regarding TCG valves?

A. Single expanding gate valves provide DIB1 and DBB features during stem movement.

B. Slab gate valves always provide a DIB2 feature.

C. Double expanding gate valves require a relief valve on the cavity in all service conditions.

D. Slab gate valves do not necessarily provide a DIB feature.

Answer: Single expanding gate valves provide a DIB1 and DBB feature only in closing position and not during opening or other positions of the stem between opening and closing, so option A is not correct. Slab gate valves can provide DIB1, DIB2, or no DIB features depending on the seat arrangement according to API 6DSS and API 6D. Fixed seats in slab gate valves cannot offer DIB1/DIB2/DBB. Standard floating seats in a slab gate valve would provide pressure energized upstream and downstream sealing which is a DIB2 feature. Using two double piston effect seats for a slab gate valve leads to having a DIB1 feature. Thus a DIB2 feature is not always the case for a slab gate valve, so option B is not correct either. Double expanding provides a DIB1 feature and requires a pressure relief on the cavity only when operated in a liquid service, so option C is also wrong. Option D is correct since a slab gate valve with two fixed seats cannot provide a DIB feature. However, a DIB2 feature can be achieved by a standard slab gate valve.

2. Which sentence is wrong?

A. Wedge gate valves are not used in subsea since they are not robust; they are not proposed for sandy service, as subsea production services contain a lot of sand.

B. Rotary gate valves have a quarter turn mechanism like ball valves, unlike other types of gate valves.

C. Slab gate valves are cheaper and more robust compared to double expanding gate valves.

D. The areas between the stem sealing are called a lantern ring.

Answer: All options are correct except for option C. Slab gate valves are cheaper, lighter, and less robust compared to double expanding gate valves.

3. Which sentence is correct regarding the closure member of TCG valves?
   A. The term "double expanding" is extracted from the two half disks used in double expanding gate valves.
   B. Hydraulically actuated double expanding and slab gate valves always have a reverse port closure member.
   C. Gate valves on manifold branches have a reverse port closure member in most cases.
   D. Reverse port closes the valve when the stem is downward.

Answer: Option A is wrong since the name of the valve "double expanding" is taken from the expansion of the half disks at lower and upper disk travel stops due to transferring the stem force to both seats. Option B is not correct either, since hydraulic fail-safe closed actuators have a reverse port, whereas fail-safe open and fail-as-is actuators have a normal port. Option C is correct since most slab gate valves located on the branches of manifolds are fail-safe closed, so they are supplied with reverse port closure. Option D is wrong since reverse port closes the valve with the stem upward and opens the valve when the stem downward.

4. Which sentence is correct regarding ball and TCG valve comparison?
   A. Ball valves have less height compared to gate valves, which could be an advantage in some cases when there is a limitation on height availability.
   B. Gate valves' linear actuator design and sizing can be sensitive to water depth, unlike ball valves with rack and pinion actuators.
   C. Gate valves could provide better sealing compared to ball valves due to the flat surfaces of their contact disk and seats.
   D. All options are correct.

Answer: All the options are correct so option D is the best answer.

5. Which sentence is correct?
   A. Gate valves have higher stem and sealing wear compared to ball valves due to the linear upward and downward movement of the stem.
   B. Gate valves are a good choice of valve selection for HIPPS and emergency shutdown systems compared to ball valves.
   C. Subsea gate valves are supplied with emergency sealant and grease injection ports on both seats and stem sealing.
   D. Rotary gate valves are suitable and economical choices for Christmas tree and manifold branches.

Answer: Option A is correct and stem wear is higher in gate valves with linear stem movement compared to ball valves, which have only a 90-degree rotation of the stem. Option B is not correct since even though gate valves are used for HIPPS application with a fast closing requirement, they are not as fast closing as ball valves or axial valves. Option C is wrong; since subsea valves are maintenance free, no sealant injection is required for emergency maintenance of the seats and stem sealing. Option D is not correct, since rotary gate valves are suitable for small gate valves on the chemical injection line.

Further reading

American Society of Mechanical Engineers (ASME). B31.3, Process Piping, American Society of Mechanical Engineers, New York, 2012.

American Society of Mechanical Engineers (ASME), Carbon, Alloy and Stainless-Steel Pipes. ASME B36.10/19, ASME, New York, NY, 2004.

International Organization of Standardization (ISO) 5208, Industrial Valves—Pressure Testing of the Metallic Valves, fourth ed., 2015. Geneva.

D.R. Mefford, Deepwater Subsea Ball Valves, Cameron, 2014.

Norsok, Piping and Valves, (1999). Norsok L-001. 3rd revision. Lysaker.

Norwegian Oil Industry Association, Valve Technology, 2013. Revision 2.

K. Sotoodeh, Top entry export line valve design considerations, Valve World Mag. 20 (5) (2015) 55–61.

K. Sotoodeh, Pipeline valves technology, material selection, welding, and stress analysis (a case study of a 30 class 1500 pipeline ball valve), in: American Society of Mechanical Engineers (ASME), Journal of Pressure Vessel Technology. Paper No. PVT-18-1043, 2018, https://doi.org/10.1115/1.4040139.

K. Sotoodeh, Valve selection in dirty process service, Valve World Mag. 25 (5) (2018) 58–60.

K. Sotoodeh, The importance of maximum allowable stem torque in valves, SN Appl. Sci. 1 (2019) 433, https://doi.org/10.1007/s42452-019-0445-0. Springer Switzerland.

K. Sotoodeh, Actuator sizing and selection, SN Appl. Sci. 1 (2019) 1207, https://doi.org/10.1007/s42452-019-1248-z. Springer Switzerland.

K. Sotoodeh, Welding a carbon steel body valve to 22Cr duplex pipe: a case study, Valve World Mag. 24 (5) (2019) 38–41.

CHAPTER FIVE

Subsea check valves

Contents

5.1 Introduction

Check valves are one-way or nonreturn valves that open with fluid movement and pressure, and close to prevent backflow of the pressure to upstream equipment such as pumps and compressors. In fact, check valves allow the fluid to flow in only one direction. It can be said that check valves provide some sort of isolation in just one direction. Three types of check valves are common in the subsea sector of the oil and gas industry: swing check, piston check and axial flow or nozzle check valves. Subsea swing check valves are covered by the API 6DSS/ISO 14723 standards, titled "specification for subsea pipeline valves." Additionally, API 6A/ISO 10423 and API 17D/ISO 13628-4 are the applicable standards for swing check valves. With the exception of swing check valves on the pipeline and axial valves installed after subsea pumps and compressors, subsea check valves are normally small bore in a maximum size of 2″.

5.2 Types of check valves

5.2.1 Swing check valves

5.2.1.1 Swing check valve characteristics

Swing check valves contain a disk that swings around a hinge or shaft to open and close the valve. Fig. 5.1 illustrates a swing check valve in an open position with its disk moving

Subsea Valves and Actuators for the Oil and Gas Industry
https://doi.org/10.1016/B978-0-323-90605-0.00009-8

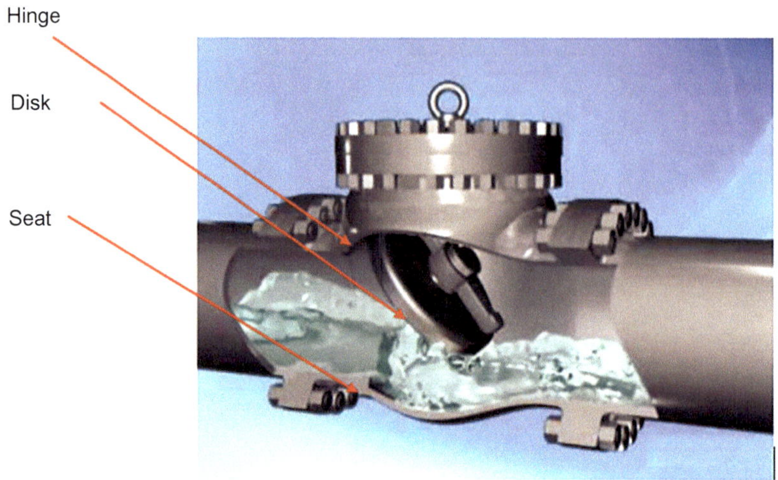

Fig. 5.1 Swing check valve. *(Courtesy: ABV.)*

upward due to fluid pressure in the piping system. The disk or closure member is mounted on the seat and swings around the hinge to close the valve. Swing check valves are widely used in water and wastewater piping systems. These valves are normally cheap and provide very minimal pressure drop or head loss when they are in open position.

Valve head or pressure loss is defined as fluid pressure loss in the valve. Generally speaking, valves create more pressure drop in comparison to piping. If the inlet pressure to a valve is 10 bar and the outlet pressure is 6 bar, it means that a 4 bar pressure drop is created inside the valve. The first important consideration is that a swing check valve is not completely open with 100% fluid passage during operation based on end users' experiences. As illustrated in Fig. 5.1, the disk of the valve swings between 60 and 90 degrees (fully open) during the passage of fluid. The main problem with this valve during operation is that the disk slams against the seat during the closing of the valve. In fact, the long stroke of the disk and the sudden closure of the valve because of the weight of the disk exacerbate the slamming effect in swing check valves. Valve disk slamming stops the fluid movement suddenly and creates water hammering (hydraulic shock). This shock (see Fig. 5.2) is a pressure surge or wave that creates noise and can damage the piping system.

The valve manufacturers decided to find an alternative solution to upgrade swing check valves with a dashpot, shock absorber, or cushioning/damping system, illustrated in Fig. 5.3, to reduce the slamming effect. Although this so-called counterweight solution solves the slamming problem, it adds two new problems. First, the weight of the disk makes the valve close faster, while the dashpot system reduces the speed of closure. Reducing the check valve closing speed leads to less resistance of the valve against backflow. Reduction in the closing speed of the check valve is more problematic when the

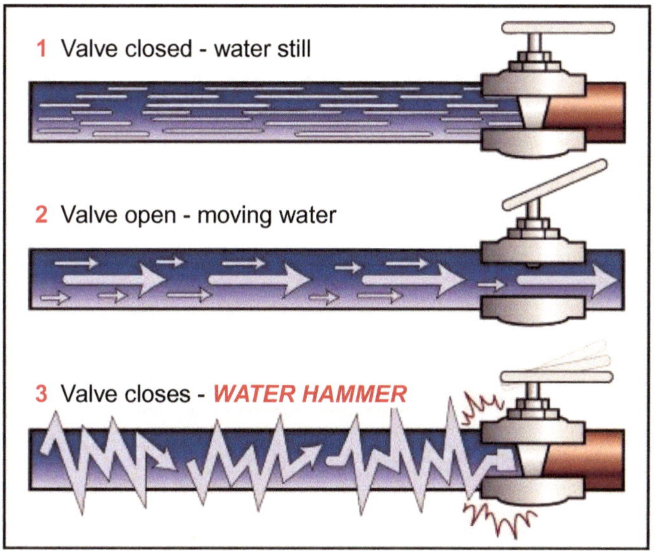

Fig. 5.2 Water hammering in piping systems due to rapid valve closure. *(Courtesy: Livinator.)*

Fig. 5.3 Swing check valve with dashpot. *(Courtesy: ABV.)*

piping system is located downstream after the check valve has transported the fluid to a higher elevation. Second, using a dashpot causes a greater pressure drop inside the valve.

Swing check valves could either be full or reduced bore/port. Subsea swing check valves (Fig. 5.4) with a full bore/port feature are used in subsea pipelines and flowlines since they are piggable. Full bore design has two other advantages: it minimizes pressure drop by providing an unobstructed flow path in the valve and reduces flow turbulence and erosion. Swing check valves have two main functions: one is to prevent backflow and

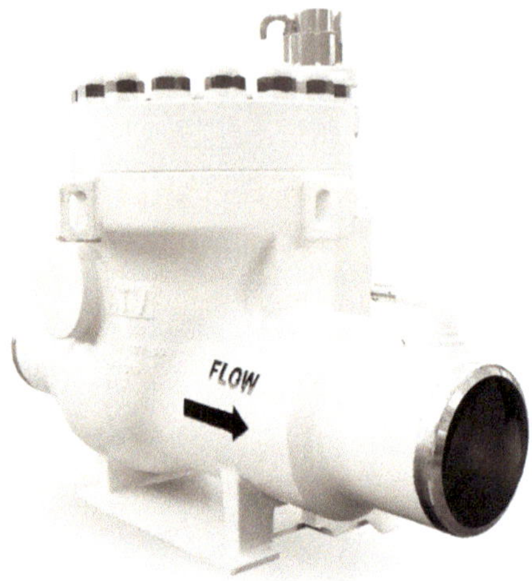

Fig. 5.4 Subsea swing check valve. *(Courtesy: ATV.)*

the other is to be opened by the fluid pressure in the line. Pig refers to a "piping injected gadget" that is shot through the pipeline for different purposes such as cleaning and inspection, as explained in Chapter 1. Since check valves open only in one direction, the flow direction should be marked on the body of the valve as illustrated in Fig. 5.4.

Fig. 5.5 shows a subsea swing check valve installed on a pipeline during pig running. The disk of the valve is up, indicating that the valve is in open position and the pig is passing through the valve. The disk of the valve moves downward and closes the valve

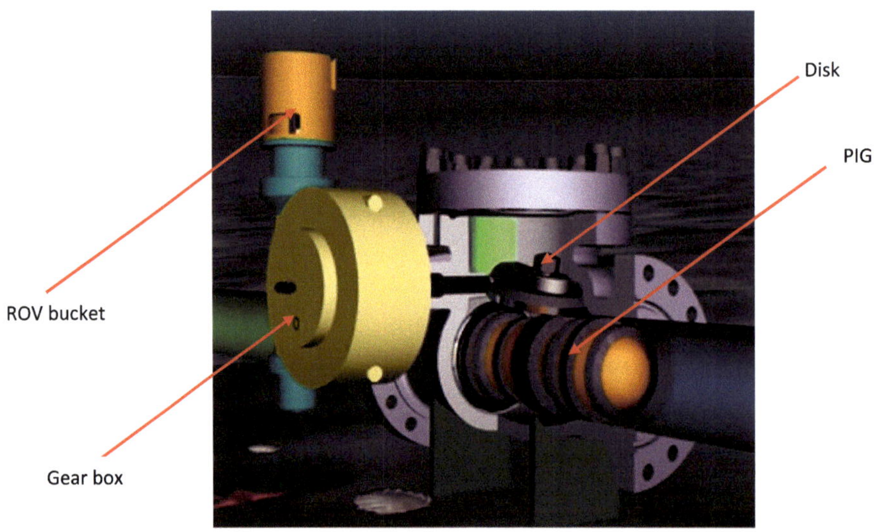

Fig. 5.5 Subsea swing check valve installed on the pipeline in open position during pig running operation. *(Courtesy: Orion.)*

after the pig has completed its passage through the valve, as illustrated in Fig. 5.6. Afterward, an ROV opens the swing check valve by operating the ROV bucket as shown in Fig. 5.7. Now it is possible to run the pig inside the piping from the opposite direction through the fully open valve (see Fig. 5.8). An ROV locks the valve in open position

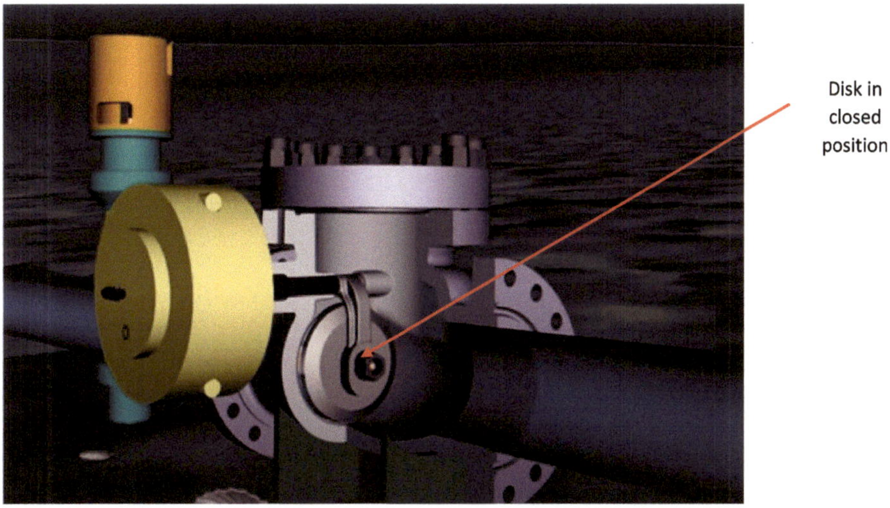

Fig. 5.6 Subsea swing check valve in closed position installed on the pipeline after pig passage. *(Courtesy: Orion.)*

Fig. 5.7 Half-opened ROV-operated swing check valve. *(Courtesy: Orion.)*

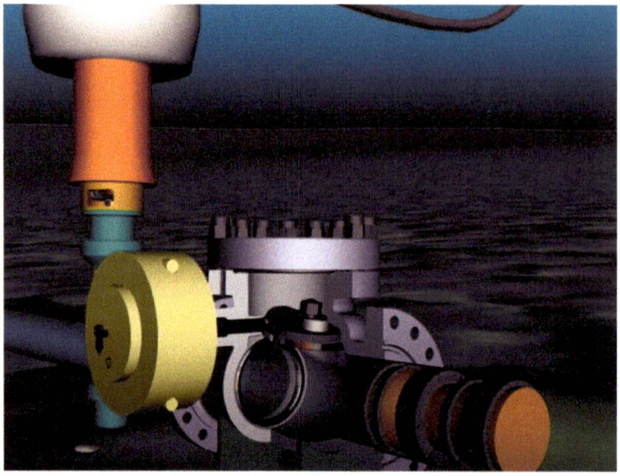

Fig. 5.8 Subsea swing check valve kept open by ROV for reverse pig running. *(Courtesy: Orion.)*

during pig running. After the pig passes, the ROV closes the valve. In fact, adding an ROV to a swing check valve installed on a pipeline facilities pig running from both directions. The important design consideration is to have a local position indicator close to the ROV bucket that is readable by the ROV to show the position of the valve. The position indicator should be made of antifouling material to prevent the growth of bacteria and marine plants. The position indicator should have an Aquasign trademark with ultraviolent protection.

5.2.1.2 Swing check valve components

Fig. 5.9 highlights the different components of a swing check valve. This section aims to provide a general description for each component of the valve.

1. Body/Bonnet (cover)

   The body/bonnet of the valve is the main pressure-containing part, which means that any failure in its function leads to leakage of internal fluid to the seawater. The body and bonnet of the valve should be forged, a process that provides a higher quality than casting. However, a high quality casting process such as investment casting or centrifugal casting could be acceptable instead of forging. ASME Sec. VIII Divisions 01 and 02 are the recognized design standards for optimized design and verification through finite element analysis (FEA). Application of FEA is critical for the body/bonnet since they are the pressure-containing parts. Fig. 5.10 illustrates FEA on the body, bonnet, and bolting of a subsea swing check valve. The body end connection of the check valve in Fig. 5.9 is a flange connection. Butt weld ends with a 100% radiography test on the welds is more common for subsea valves, including swing check valves, due to their higher joint efficiency compared flanged connections. The top entry design of swing check valves provides the benefit of online

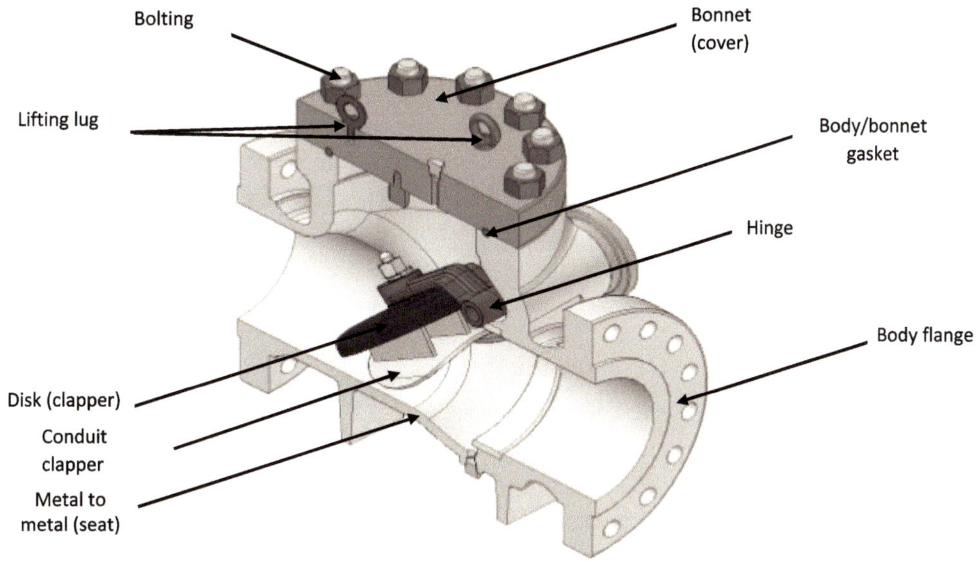

Fig. 5.9 Swing check valve.

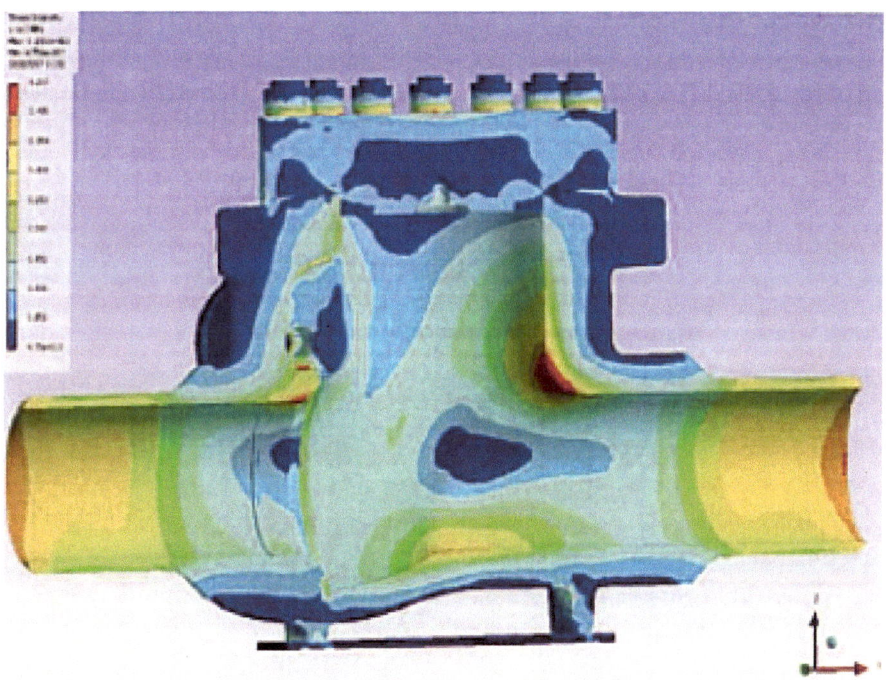

Fig. 5.10 FEA on the body/bonnet and bolting of a subsea swing check valve.

maintenance, meaning that access, inspection, and repair of the valve internals can be achieved by unfastening the body and bonnet bolts and disassembling the bonnet from the valve without any need to remove the valve from the line. However, subsea check valves should be designed for maintenance-free operation during their design life. Sealing between the body and bonnet is achieved through metal-to-metal (MtM) contact. Normally, exotic materials, such as solid 25Cr duplex or Inconel 625, or an overlay of Inconel 625 on carbon steel, are considered for the body and bonnet, so there is no need to add any corrosion allowance to these components. Corrosion allowance is defined as an extra thickness (e.g., 3 mm) normally added to carbon and low alloy steel to mitigate the CO_2 corrosion rate. One or two millimeters of erosion allowance may be added to the body and bonnet, even if they are made of exotic materials, to mitigate the erosion effect.

2. Body/bonnet bolting

Bolting between the body and bonnet is known as pressure-containing bolting and should have high mechanical strength. Different methodologies can be used for the body/bonnet bolting joints, such as:

- Design by analysis method and additional requirements as per ASME Sec. VIII Div. I & II;
- Independent bolt method, design by analysis;
- ASME bolting method for BX joints.

A BX joint is a type of API 6A flange joint that has both raised face and ring joint features.

Design by analysis method is a popular bolt analysis approach in which FEA or an equivalent approach can be used for stress analysis of the bolting and connected joints. Different information and conditions such as the contact surfaces of the bolting and the friction between the bolts and nuts should be considered via design by analysis method.

Low alloy bolting such as B7/L7 bolting and ASTM A194 nuts have high mechanical strength and are widely used for subsea purposes. Coating such as zinc-nickel plate or zinc flake can be used for the coating of the bolts used in subsea environments. Zinc flake coating contains zinc and aluminum and provides mechanical adhesion instead of electrolytically adhesion.

3. Disk

The disk or flapper has a one-piece design and is the main component of the swing check valve that can create fluid pressure drop through the valve, especially when the valve is not in fully open position. The check valve disk or closure member can be moved by fluid pressure, unlike the closure member of ball and expanding gate valves. A computational fluid dynamics (CFD) analysis should be done by the valve supplier for different positions of the disk and different flow rates to make sure that the valve can perform properly without any unacceptable pressure drop. CFD is an engineering tool used to simulate the flow values and conditions of the valve in different situations. Fig. 5.11 illustrates a CFD analysis for a swing check valve to make sure that the pressure drop is acceptable. In fact, a CFD analysis is run in order to optimize different flow parameters such as pressure drop, flow capacity (also called C_v calculations), as well as noise level and fluid dynamic forces. C_v is defined as the flow rate passing through the valve in gallons per minute when the pressure drop between the inlet and outlet of the valve is 1 psi. The disk is normally hard faced with a material such as stellite 6 or 21 or tungsten carbide to minimize the effect of erosion due to particles, the high velocity of the fluid and galling, which can happen when the disk is placed on the seat of the valve during closing action. Galling is defined as a sort of wear caused by the contact between two surfaces. The disk could be designed as a conduit disk or conduit clapper, or a clapper with pig guide facility (see Figs. 5.12 and 5.13) to allow smooth movement and passage of the pig through the swing check valve.

4. Seat

The seat of the swing check valve illustrated in Fig. 5.14 is integral with the body, which provides MtM sealing when it is in contact with the disk. The seat should be hard faced with stellite or tungsten carbide to prevent galling/wear due for the same reasons explained previously in regard to the disk. Seats that are integral with the body are not renewable. Welded, pressed, or threaded seats are renewable. However,

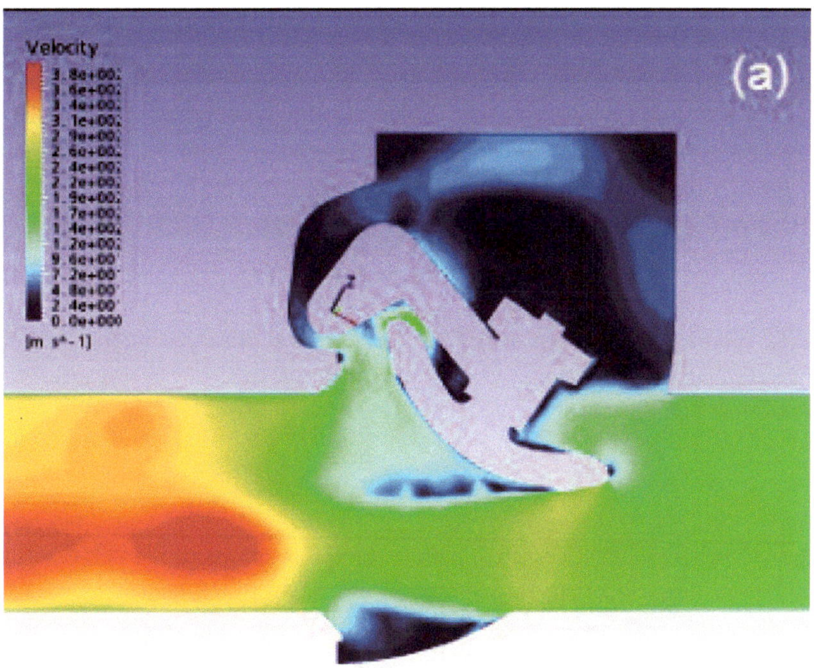

Fig. 5.11 CFD Analysis for a swing check valve. *(Courtesy: ATV.)*

Fig. 5.12 Conduit clapper/disk for easy pig passage. *(Courtesy: Cameron.)*

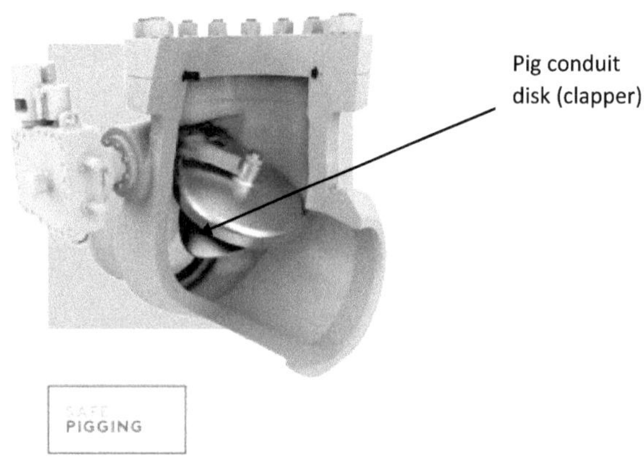

Fig. 5.13 Subsea swing check valve with a conduit clapper/disk for easy pig passage. *(Courtesy: Petrolvalve.)*

threaded connections are not typically allowed for subsea application. There is no need to apply grease for sealing improvement between the seat and disk of a swing check valve.

5. Hinge

A hinge is connected to the body of the valve from one side and to the disk from the other side, which acts as an axis for the disk to move between open and closed positions. Both disk and hinge are suspended from the body of the valve with a hinge pin. The hinge pin, also called a shaft, enters the body of the valve and passes through the hinge. There are two bearings around the hinge pin, one on each end, to prevent lateral movement of the hinge pin and support the shaft for smooth movement. Fig. 5.14 provides more details about the hinge pin/shaft sealing and bearing. Shaft

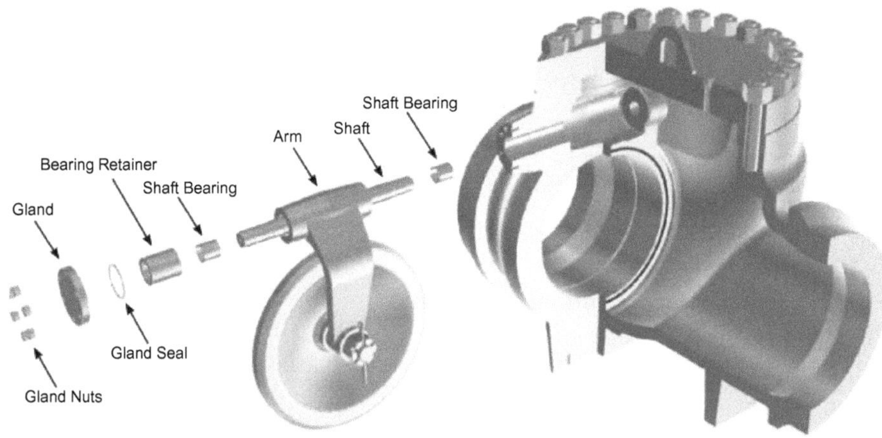

Fig. 5.14 Hinge and hinge pin connections in a swing check valve. *(Courtesy: Cameron.)*

bearings should have Teflon (PTFE) internals to minimize friction and galling with the hinge (shaft).

6. Gasket/sealing

A metallic gasket is located between the body and bonnet to provide sealing. The gasket is normally bidirectional to prevent the leakage of internal fluid to the outside of the valve (seawater) and to prevent the leakage of seawater to the valve internals. These gaskets are typically BX type, defined as per API 6A, which sit completely in the grooves on the body and bonnet and seal both the inner and outer areas as illustrated in Fig. 5.15. If the body of the valve is in carbon steel with an overlay of Inconel 625, then the groove between the body and bonnet should be overlaid with Inconel 625 to mitigate the risk of crevice corrosion.

A separate O-ring in Nitrite Rubber (NBR, Buna N) or Hydrogenated nitrile butadiene rubber HNBR or Viton is normally used to seal the area between the body and bonnet as

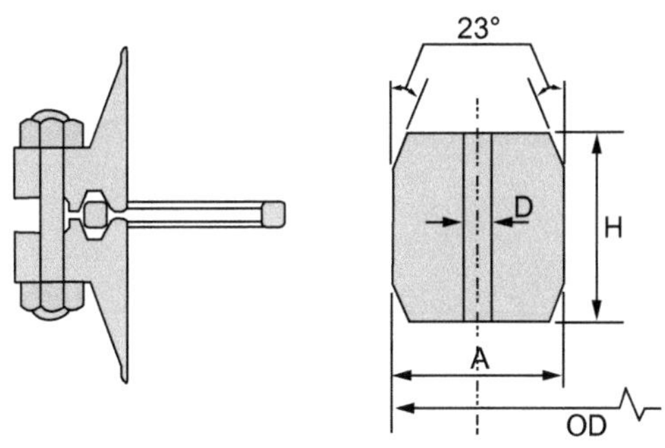

Fig. 5.15 BX ring type joint (RTJ) metallic gasket schematics.

an additional barrier to prevent seawater ingress into the valve internals; this is also known as an "environment seal." Fig. 5.16 illustrates a metallic and soft sealing arrangement between the body and bonnet. NBR is a synthetic rubber copolymer used in valve sealing as an elastomer seal. Elastomers are a class of polymer materials than can be stretched repeatedly to twice its original length with little or no permanent deformation. HNBR is made by hydrogenating NBR, which improves its temperature and ozone resistance significantly. Viton is a brand of fluorocarbon (FKM), another type of elastomer sealing widely used in the oil and gas industry due to its high resistance against chemicals and process services such as hydrocarbon fluid.

This section reviews possible material choices for the swing check valves described in this section. Let's suppose that the crude oil is separated subsea and is free of noncorrosive products like carbon dioxide and hydrogen sulfide. A swing check valve is used on the well-treated oil on an export pipeline. Low alloy steel ASTM A694 F60 could be selected for the valve body and bonnet with an Inconel 625 overlay on the grooves, such as the body and bonnet contact groove where the BX RTG (Ring Type Joint) sits. Low alloy or carbon steel grooves are exposed to crevice corrosion, so it is proposed to apply Inconel 625 on these areas. The metallic gasket between the body and bonnet should be in Inconel 625 material. An O-ring, which is the environmental sealing, can be in Viton, Buna N, or hydrogenated Buna N materials. More information about these soft seals is available in Chapter 13. The disk of the valve is trim, meaning that it is in contact with the fluid service and should be made of the same material as the body or better. In this case, super duplex UNS S32750 is proposed for the disk. 22Cr duplex does not have as much pitting corrosion resistance as 25Cr super duplex, so this material is not as popular as super duplex. The seat ring of the valve should also be in 25Cr super duplex, and the contact between the disk and the seat should be hard faced with tungsten carbide (TC). It is important to note that TC can be used for hard facing of the valve internals subsea, but it can not be exposed to seawater. The valve should have a suitable material for wet storage if it will be submerged in the sea for installation on the sea floor. In that case, Ultimet hard facing is proposed on the disk and seat surfaces as an alternative to TC. Ultimet or Tribaloy are hard-facing materials more suited to valves used in seawater service. Ultimet (UNS R31233), a cobalt alloy, also offers excellent wear resistance properties. Ultimet contains chromium, nickel, molybdenum, iron, and tungsten. Table 5.1

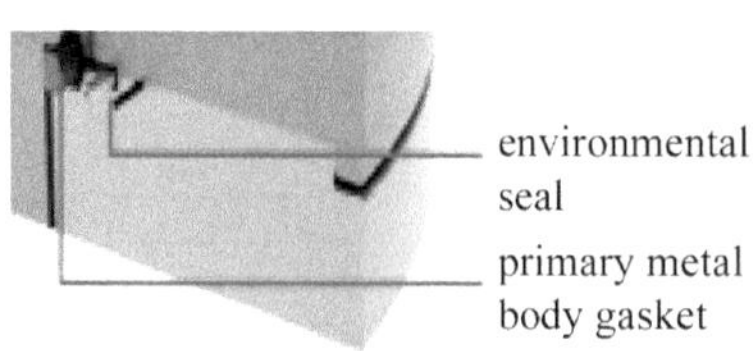

Fig. 5.16 Body/bonnet double sealing.

Table 5.1 Chemical composition percentage of Ultimet.
Nominal chemical composition, weight percent

Co[a]	Cr	Ni	Mo	Fe	W	Mn	Si	N	C
54	26	9	5	3	2	0.8	0.3	0.08	0.06

[a]As balance.

provides the percentage of chemicals in Ultimet: Co stands for Cobalt, Cr stands for chromium, Ni stands for nickel, Mo stands for molybdenum, Fe stands for iron, W stands for tungsten, Mn stands for magnesium, Si stands for silicon, N stands for nitrogen, and C stands for carbon. The hinge, hinge pin, and other internals can also be in 25Cr super duplex.

5.2.1.3 Handling, storage, packing, and preservation

Same as handling, storage, packing, and preservation for axial check valves; refer to Section 5.2.5.3.

5.2.1.4 Installation

The valve should be stored in a proper location away from moisture, humidity, etc. The protective plywood or plastic end caps should be removed and the valve should be inspected carefully before installation to make sure that there is no sign of damage or flaw. Unpacking the valve and removing the protective caps shall be done immediately prior to installation. Flange faces are sensitive parts that should be inspected properly. The lifting of the valves for installation can be done either by sling or by hand, since axial valves in small sizes are not heavy. A lifting sling may be used with a lifting device or crane to lift objects effectively and safely. Slings usually consist of a cable, chain, rope, or webbing, and generally include metal devices such as hooks, shackles, links, or rings. The fit-up of the valve to the pipeline is very important, as any foreign material ingress to the valve from the pipeline can damage the valve and valve internals.

Swing check valves can be installed horizontally without any problem. However, special attention should be paid when the valve is installed vertically. If the swing check valve is installed on a vertical line with a downward fluid direction, the disk of the valve could remain in open position. Swing check valves in sizes as large as 6″ or 8″ can be installed vertically when the fluid is moving upward. The large disk of a check valve in a vertical line with upward-moving fluid could prevent fluid passage due to its heavy weight. The best reference for the direction of installation of valves, including swing check valves, is a document called the "installation, operation and maintenance" (IOM) manual provided by the valve supplier. As an example, the manufacturer of the subsea swing check valve illustrated in Fig. 5.17 recommends installing it either vertically or horizontally without any size limitation.

Fig. 5.17 Subsea swing check valve.

5.2.1.5 Operation

Swing check valves, like other types of check valves, are shipped in closed position. The subsea operation of this valve is limited to a very simple task: fluid pressure is applied from the upstream side and opens the valve; when the upstream fluid pressure stops, the disk returns back to the body or seat and closes the valve. A reverse flow from the downstream side of the valve to upstream could assist in closing the valve. Chattering or vibration of the valve disk due to fluid flow fluctuation between the minimum and maximum amount is a common operational problem for different types of check valves such as swing check valves. As an example, when the flow of fluid in the piping decreases to its minimum or a relatively low amount, it probably cannot keep the swing check valve fully open. This creates significant pressure drop and flow loss inside the valve. Therefore the valve manufacturer should be provided with the minimum flow value inside the valve to design the valve in such a way as to be fully opened with minimum flow. Frequent disk opening and closing causes slamming, which causes water hammering and a higher rate of wear on the disk and seat of the valve as well as the hinge pin and bearing. In addition, the valve manufacturer should ensure that the valve is capable of full opening during normal and maximum flow conditions in the piping system as well as the valve. The other operational problem is leakage from the valve in reverse flow condition. The higher pressure of reverse flow or back pressure assures easier and better sealing of the valve against reverse flow. The valve manufacturer should make sure that the swing check valve can provide sufficient sealing and leakage rate, even during a low level of back pressure from the downstream side such as 100 or 300 psi as examples.

The other important point regarding operation is that the valves should be suitable for operation within the pressure and temperature ranges written on their name plate. Normally, API 6DSS provides the information that is required to appear on the name plate.

5.2.1.6 Maintenance

The valve is designed for subsea operation and should be free of maintenance during its design life—for 25 or 30 years as examples.

5.2.2 Poppet check valve

A poppet check valve (illustrated in Fig. 5.18) is an economical valve that can provide a relatively high flow capacity. Fig. 5.19 illustrates the internals of a poppet check valve. Please note that the internals shown in Fig. 5.19 are not related to the valve shown in Fig. 5.18. The poppet acts like a disk in a swing check valve for controlling the fluid flow. The fluid flow from left to right compresses the spring and pushes the poppet back to make the valve open. A spring is an essential part of most check valves except for swing check valves, and affects the amount of flow passing through the valve. The spring could have different torque values (high or low). Higher spring torques make more resistance against the fluid flow and lower spring torques increase the flow capacity or fluid passage through the valve. Thus spring selection has a direct impact on valve performance. Therefore it is proposed to provide the fluid data as well as the direction of installation of the valve to the valve manufacturer to enable the valve manufacturer to provide the most suitable spring torque. The spring affects the cracking pressure, which is defined as the pressure at which the valve starts opening. In fact, correct spring torque should be selected to provide a desirable cracking pressure to prevent operational problems such as check valve closure member vibration described in the previous section. Inconel X750 is widely used for check valve springs in different sectors of the oil and gas industry due to its high

Fig. 5.18 Poppet check valve. *(Courtesy: Oliver Valve.)*

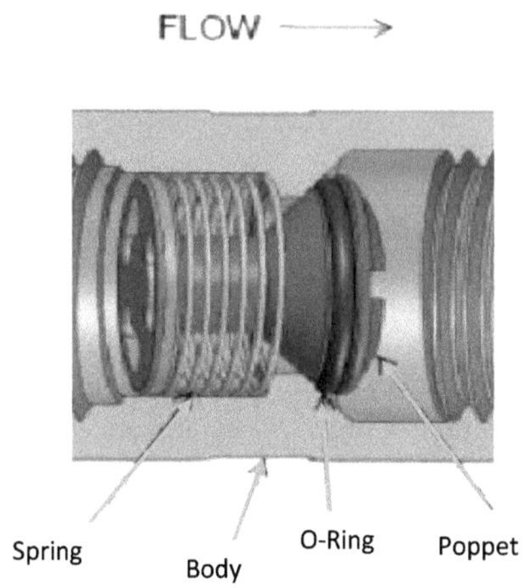

Fig. 5.19 Poppet check valve internals.

mechanical strength and fatigue resistance. However, this material is not suitable for seawater service due to its low pitting corrosion resistance, so an alternative material such as Elgiloy UNS R3003, a cobalt alloy, is proposed in this book according to the industrial experiences of the author. The sealing of the valve is achieved by means of an O-ring in addition to the MtM contact between the poppet and the body. A thermoplastic material such as Teflon (PTFE) or polyether ether ketone (PEEK) can be used for soft sealing instead of an O-ring. The O-ring could be made of NBR, HNBR, or Viton (FKM). The direction of flow should always be marked on all check valves, as illustrated in Fig. 5.18. Check valves shown in Fig. 5.18 can be used either on chemical injection lines, such as those used to inject mono-ethylene glycol (MEG) to avoid hydrate (a solid ice-like formation in the gas service), or on small lines, such as those that supply hydraulic fluid to subsea distribution units (SDUs). These valves are normally in small sizes, such as ½″, ¾″ or a maximum of 2″. However, these valves are available in the market from ½″ to 6″.

5.2.3 Piston check valve

Piston check valves, like poppet check valves, are used in chemical injection or hydraulic lines mainly in small sizes of a maximum of 2″. Fig. 5.20 illustrates a piston check valve for subsea application. This type of valve could have MtM seal achieved between the piston and the seat or body of the valve or it could have a primary soft seal for tight shut-off plus a secondary metal seat. As an example, the piston check valve illustrated in Fig. 5.20 can provide bubble tight shut-off by using an MtM seal plus a secondary polymeric seat ring. The internal arrangement of the piston check valve is similar to that of a poppet check valve, except that the closure member is like a piston.

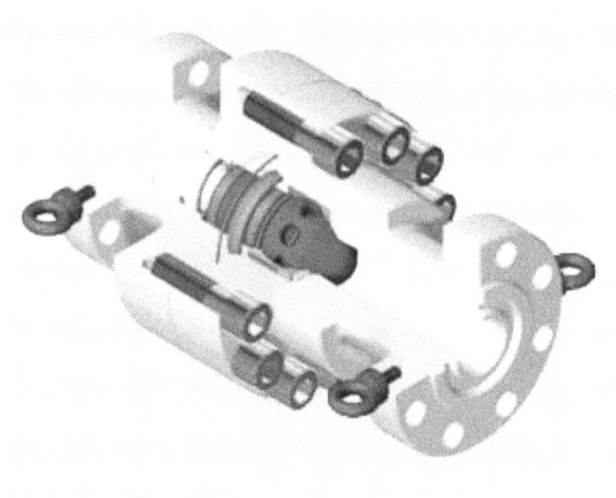

Fig. 5.20 Subsea piston check valve. *(Courtesy: Pacson.)*

5.2.4 Combination check valve

Piston or poppet check valves located on chemical injection lines are used along with a gate valve for chemical injection flow control. The gate valve could be slab or rotary gate and is actuated due to frequency of operation. The check valve is located downstream (after) the gate valve as illustrated in Fig. 5.21. The chemical MEG is injected from the left side of the actuated gate valve, which is normally closed. The valve gets opened by hydraulic supply to the actuator and the production fluid enters the production line after passing the check valve. In fact, the check valve is opened by the fluid flow from left to right and gets closed in the opposite flow direction. When the chemical injection flow is reduced and the injection task has been finished, the check valve is closed to prevent the passing of the production fluid, shown in green (thick gray in print versions) in Fig. 5.21, to the MEG injection line, shown in red (thin gray in print versions). The gate valve is also closed by spring force in the actuator to isolate the chemical injection and production line. The size of the actuated gate valve could be ¾″ or 1″ as illustrated in Fig. 5.22. A rotary gate valve, which can be used for chemical injection as an alternative to a slab

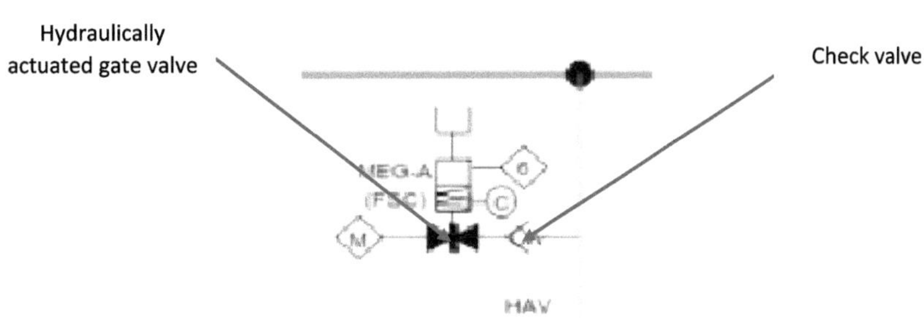

Fig. 5.21 Chemical (MEG) injection line including a check valve and actuated gate valve.

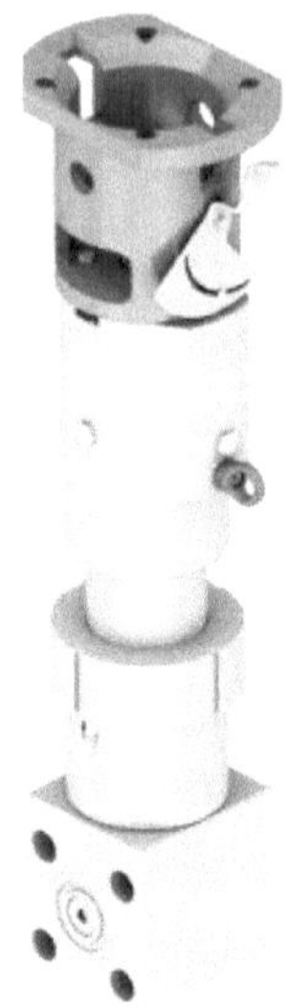

Fig. 5.22 Hydraulically actuated ¾″ slab gate valve. *(Courtesy: Cameron.)*

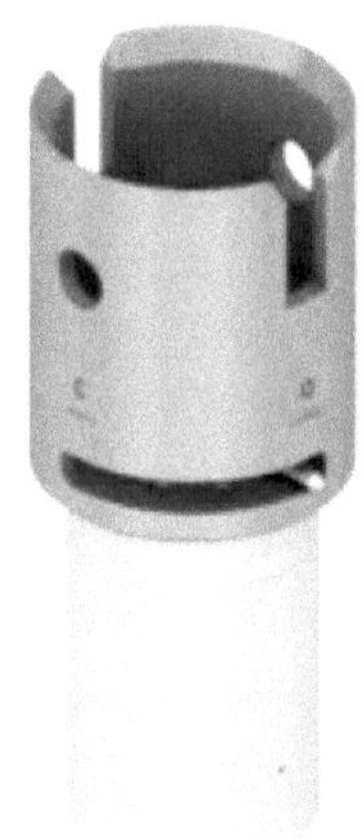

Fig. 5.23 ¾″ Subsea rotary gate valve with low-torque paddle for chemical injection line. *(Courtesy: Cameron.)*

gate valve, is illustrated in Fig. 5.23 with a low-torque paddle for ROV operation. Rotary gate valves have a more compact design than slab gate valves with the possibility of fast, quarter-turn operation. More information about rotary gate valves is available in Chapter 4 of this book. It is possible to combine the gate and check valve together in one valve body or block to save space, weight, and cost. A valve with a check and gate valve combined in one body is called a combination valve.

5.2.5 Axial (nozzle) check valve

5.2.5.1 Introduction and advantages

Axial or nozzle check valves are probably the most expensive, and one of the best technical solutions to prevent the return of the fluid in piping systems without slamming. The name of the valve is probably taken from the internal component of the valve, called a flow nozzle (see Fig. 5.24).

Nonslamming means that although the disk movement is very fast, the disk of the valve sits smoothly on the seats during closing action without any slamming effect, and consequently without water hammering. Two factors provide nonslamming: spring-assisted closing action and a very light disk (see Fig. 5.25). The disk of the valve is so light that it is possible to move it from closed to open position easily by hand. The spring is a key component in this valve; it should be selected properly to provide sufficient torque to facilitate both opening and closing of the valve. The choice of spring depends on different process parameters such as fluid velocity, viscosity, type of fluid, valve direction of installation, etc. As an example, a spring in gas service may need a lower torque value compared to a spring in a service in which liquid is flowing inside the piping system and the valve, in order to be opened easily by gas pressure.

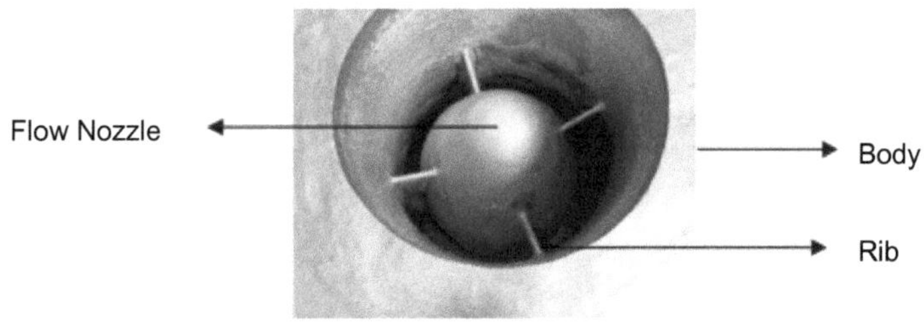

Fig. 5.24 Flow nozzle and internal of an axial flow check valve. *(Courtesy: Valve World.)*

Fig. 5.25 Low mass disk for a nonslam axial check valve moveable by hand. *(Courtesy: Valve World.)*

Axial or nozzle check valves are installed downstream from (after) rotary equipment such as pumps and compressors in the topside sector of the oil and gas industry to prevent the return of fluid from the downstream side toward the pumps and compressors and protect this expensive equipment from reverse flow. Axial valves have several advantages, such as robust design—especially those with a one-piece body that can withstand the loads and vibrations generated by the pumps and compressors; fast opening and closing action due to their low mass disk; very short disk travel distance between full open and closed positions, which provides a good barrier against very fast flow return in closed position; low pressure drop (e.g., 0.1 bar maximum) and relatively good flow capacity due to the venturi effect inside the valve (see Fig. 5.26). Low pressure drop is a key feature of this valve, since it makes the energy loss of the pump and compressor used to increase the pressure of the fluid as minimal as possible.

Venturi effect is the reduction in fluid pressure that results when a fluid flows through a constricted section (or choke) of a pipe. The flow movement shown in Fig. 5.26 is from area 1 (flow inlet zone) to area 2 (venturi area). Area 1, which is the upstream section of the valve, contains the fluid with a higher static pressure than area 2. However, the flow speed is higher in area 2 than area 1 because the cross-section area is less in area 2. As illustrated in the Figure, axial check valves are always reduced bore. The bore area inside the valve in area 2 could be 60% or 70% of the bore in area 1. The differential in bore also provides force from the fluid to move the disk and open the valve. The main advantage of venturi or axial flow is that it is a streamlined flow path without any flow turbulence, which prevents erosion and wearing inside the valve. The robust design of these valves makes them maintenance free, which is an essential requirement for subsea valves. The important design consideration is to apply FEA on the valve components, especially the body as the main pressure-containing part. Additionally, these valves in general can

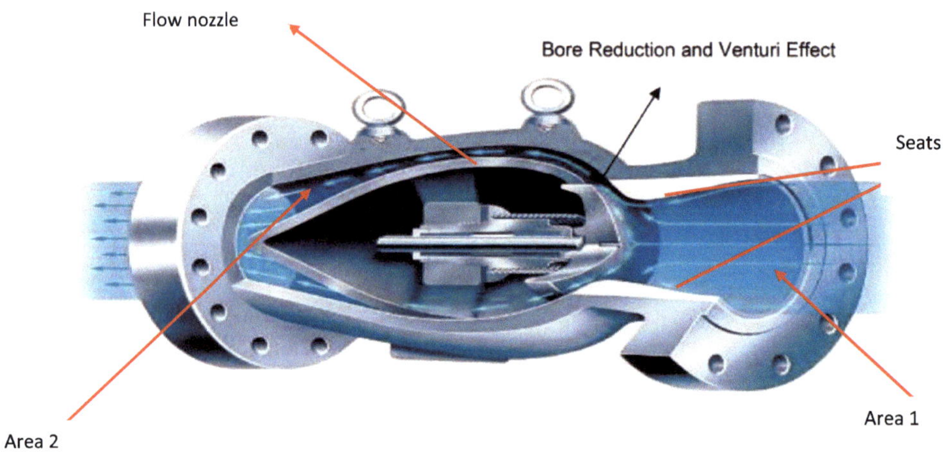

Fig. 5.26 Venturi effect in axial check valves. *(Courtesy: Valve World.)*

provide a tight shut-off feature, i.e. highly reliable sealing with a very low amount of leakage. This type of check valve is proposed for unstable flow and flow types with pulsation. Generally speaking, axial flow check valves can be considered an upgraded design version of swing check valves and other types of check valves. The disk of an axial check valve has spring-assisted movement at the center of the disk, allowing for smooth flow and sudden response time for cycling (opening and closing) without any slamming effect or water hammering. Axial check valve may also be called humble check valves or silent check valves. One of the weak points of axial check valves is that they are not suitable for particle-containing or dirty services. The other disadvantage of this valve is not suitable for pig running.

Fig. 5.27 illustrates a nozzle or axial check valve in both open and closed positions. The image on the left shows the nozzle check valve in closed position. In fact, spring force keeps the disk on the seat in closed position. However, the fluid pressure in the picture on the right side overcomes the spring force and pushes the disk up and off the seat. The spring is compressed in the figure on the right side and the short disk distance from closed to open position is highlighted. As you can see from the valve internals, these valves have a more complex design and manufacturing process compared to other types of check valves.

Fig. 5.28 illustrates a subsea axial check valve. The flow direction shall always be engraved on the body of the check valve according to the relevant standards such as API 6DSS. The valve is welded to the line.

Fig. 5.29 illustrates and compares an axial check valve and a swing check valve during closing. The flow upstream and downstream of the axial check valve is consistent, without any turbulence, whereas the fluid upstream and downstream of the swing check valve is turbulent. The check valve disk provides high slamming force during closing, which could damage the disk and the seat and apply loads to the piping system. The effect of slamming may be exacerbated due to high velocity reverse flow from the downstream

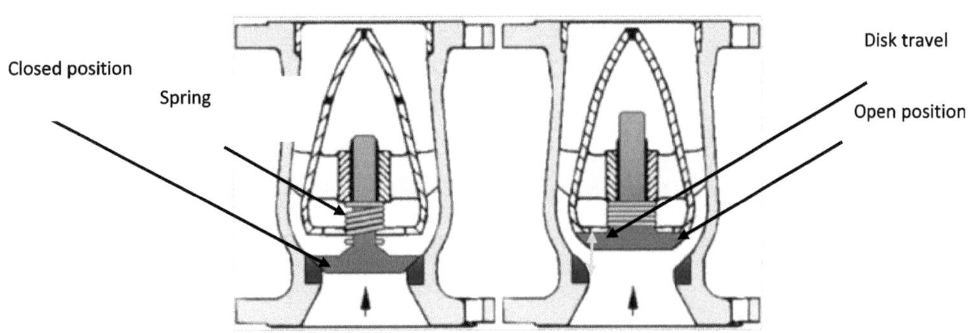

Fig. 5.27 An Axial check valve in closed and open positions. *(Courtesy: Norwegian Oil Industry Association.)*

Fig. 5.28 Subsea axial check valve. *(Courtesy: Mokveld.)*

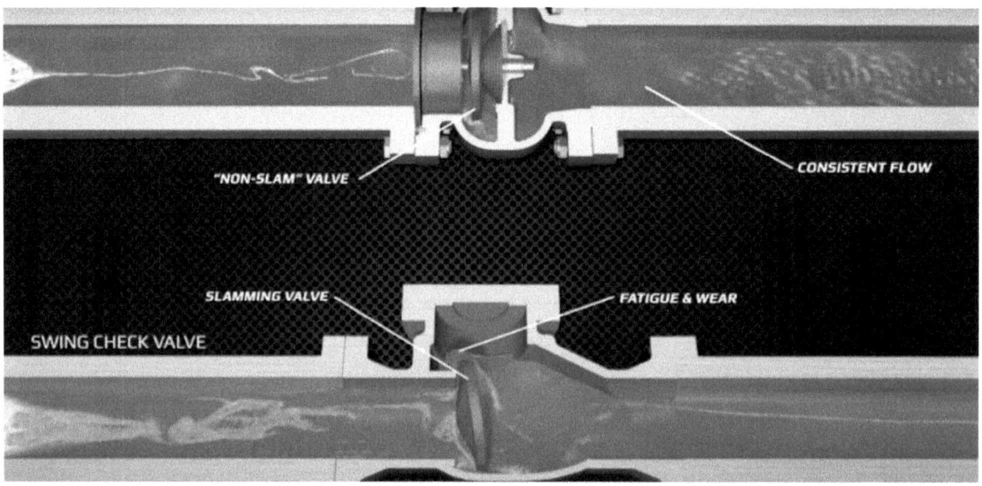

Fig. 5.29 Comparing an axial check valve and swing check valve during closing action.

side. Additionally, high fatigue and wear is associated with the hinge and hinge pin of swing check valves.

5.2.5.2 Case study and lessons learnt from subsea axial check valves used for chemical injection lines

As explained in Section 5.2.4 and shown in Fig. 5.21, axial check valves are selected for chemical injection lines. The axial check valves used for chemical injection lines have a flange connection on one end, according to the API 6A standard, and the other end is welded to the line. The upstream side where the chemical is injected has a flange ending, and an actuated gate valve is also located upstream of the check valve. The check valve is welded to the process line and gas service from the downstream side. These small check valves in the chemical injection line could be in 25Cr super duplex in most cases. The welding of 25Cr super duplex subsea axial check valve to a line in super duplex should be done with extra care to prevent the formation of sigma phase which is brittle. Sigma phase (σ) is a chromium/molybdenum-rich intermetallic phase found in the Fe-Cr-Mo system, which occurs when the material dwells or slow cools through a temperature range of 550–1050°C. This phase can be formed in duplex and super duplex since these materials possess both chromium and molybdenum, and can lead to the formation of a hard and brittle phase. The welding temperature for welding super duplex should be limited to the value given in the welding procedure specification (WPS), typically 150°C, to minimize the risk of sigma phase formation as well as consequences such as cracking. Considering the small size of the welded joints and the limitation on the welding temperature, the heat input would not affect the valve internals. High heat input can melt the valves' soft materials or cause thermal expansion of the valve internals, which can jeopardize the sealability of the valve. In such cases, an integral pup piece should be added downstream of the valve where the welding takes place to protect the valve internals from the welding heat input. The flow direction is from the upstream side (flanged side) to the welded side, as illustrated in Fig. 5.30.

The axial valve design data is summarized as follows:

Valve type: Subsea axial flow or nozzle check valve;

Valve nominal size: 2″;

Valve pressure class: 690 bar equal to 10,000 psi;

Design and operating pressure: 690 bar;

Valve internal diameter: 42.8 mm;

Design temperature: −29°C to +121°C;

Pressure value at minimum and maximum temperature: 690 bar at −29°C and 690 bar at 120°C;

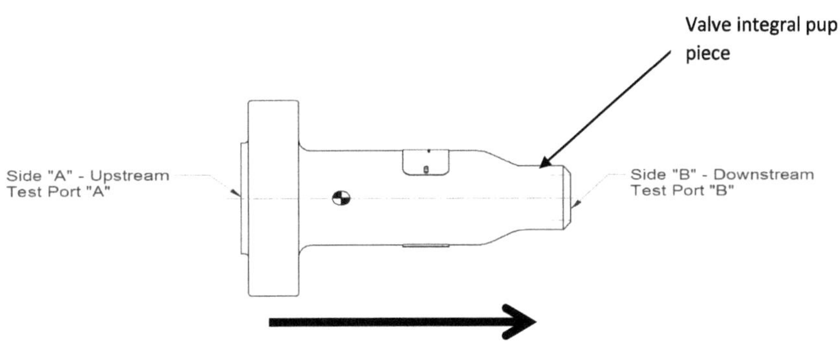

Fig. 5.30 2″ Axial check valve with one end flanged and one end welded with flow direction indicated for a chemical injection line.

Water depth: 2300 m;
Shell or body pressure test: 1035 bar;
Hydrostatic seat test: 863 bar;
Gas body test: 759 bar;
High pressure (HP) gas seat test: 690 bar;
Low pressure (LP) gas seat test: 20 bar;
Test results: pass;
Design Standards: API 6DSS (ISO 14723), API 6A (ISO 10423), API 17D (ISO 13628-4).

Fig. 5.31 illustrates a 2″ subsea check valve and identifies its different components. The body of the valve is the main pressure-containing part and is made of forged super duplex stainless steel as per ASTM A182 F55. For small valves, a super duplex body is more practical and economical compared to carbon steel with an Inconel 625 weld overlay. Other different corrosion resistant alloys such as an Inconel 625 body for this type of valve could be produced by the valve manufacturer. The left side of the body where the flow enters the valve is a flanged connection, according to API 6A 6BX, in 1 13/16″ size. It should be noted that API 6A does not cover a 2″ size, so the closest size is either 1 13/16″ or 2 1/16″. The other end of the valve is welded to the piping through a pup piece integrated with the body of the valve. The ending of the pup piece has butt weld end preparation according to the ASME B16.25 standard. This standard is

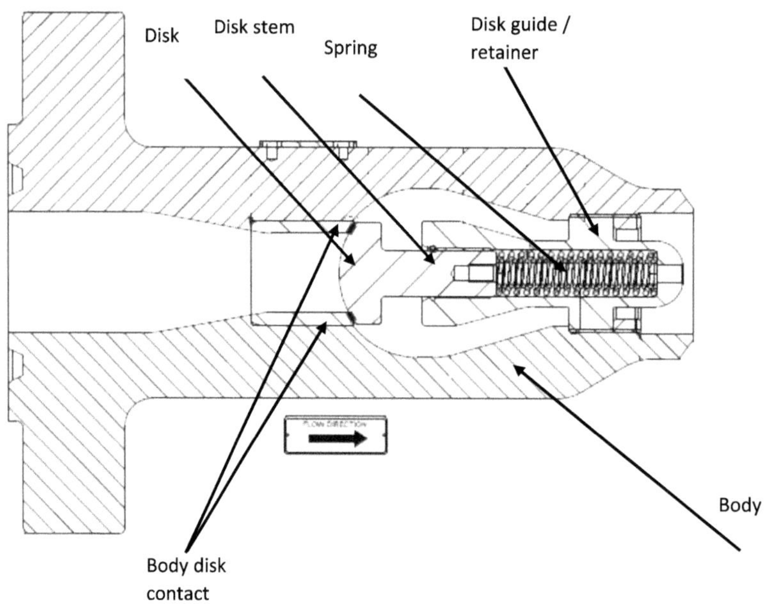

Fig. 5.31 2″ Axial check valve for chemical injection application.

explained in Chapter 2 in more detail. The end-to-end design of this valve is not covered by any subsea valve standards and is designed according to the manufacturer standard. This axial check valve design does not have a separate part as a seat, as illustrated in Fig. 5.31. Seats can be categorized as integral or renewable. The design of the axial valve in Fig. 5.32 has a renewable seat that is inserted into the body of the valve. Renewable seats are considered trim, which refers to the internal part of the valve that has contact with the fluid; it should be made of material at least equal in quality to that of the valve body or better. Renewable seats require additional machining activity and a dye penetration or liquid penetration test as a type of nondestructive test to make sure that there is no defect on the seat after machining. Dye penetrant inspection (DP), also called liquid penetrate inspection (LPI) or penetrant testing (PT), is a widely applied and low-cost inspection method used to check surface-breaking defects in all nonporous materials (metals, plastics, or ceramics). The disk has direct contact with the body of the valve during closure, and the body's contact with the disk during closure acts as the valve's seat. The disk of the valve is a super duplex bar in ASTM A476 S32760. The disk should be designed accurately to provide minimal turbulence and pressure drop across the valve. It is common to apply CFD analysis on the disk in different positions. The body and disk contact surfaces should be hard faced with tungsten carbide to prevent galling and wear due to contact between the two surfaces of the disk and body. The disk is connected from the other side to a disk guide or retainer that facilitates axial movement of the disk back and forth without any vibration or lateral movement. The disk guide or retainer, sometimes called a diffuser, is screwed into

Fig. 5.32 A separate seat after machining for an axial check valve inserted into the body.

the body of the valve. The use of a screwed connection between the body and the disk guide may be considered a weak point, especially in such a high-pressure subsea application. The disk is pushed toward the body for closing by one or two spring coils in Elgiloy material, which is a cobalt alloy. The springs should have sufficient torque to seal the valve even during very low back pressure from the downstream side of the valve. Spring is one of the main components that provide nonslamming in the valve. The spring material should be hard with high mechanical strength and suitable for high numbers of cycles. Inconel X750 is a very common spring material for axial check valves to be used in the topside, onshore sectors of the oil and gas industry. However, Inconel X750 does not have high pitting corrosion resistance in contact with seawater. Subsea valves may have contact with seawater internally during their submersion in the water for installation on the seabed, also called wet storage, so Elgiloy UNS R30003 is proposed alternatively for valves in seawater. It should be noted that the disk is pushed on the body and provides better sealing due to the aid of backflow from the downstream side. The disk of the axial valve in this case is integrated with the disk stem. However, in some cases the disk and stem are two parts that are assembled together (see Fig. 5.33).

The main question is, why does one end of this valve have a flange connection? The reason is that the piping system should be flushed with water to clean the piping inside from particles, welding girth, etc. after fabrication in the yard. Check valves can be disconnected from the injection line through the flange ending as illustrated in Fig. 5.34 and 5.35. The important point is that the axial valves are still connected to the process line, but disconnected from the chemical injection side. The fluid coming out of the axial valve bottom part with the flange connection is water with some particles (see Fig. 5.36). The axial check valve should provide a sealing against the flushing water, since the water

Fig. 5.33 Stem and disk of axial valves in two pieces.

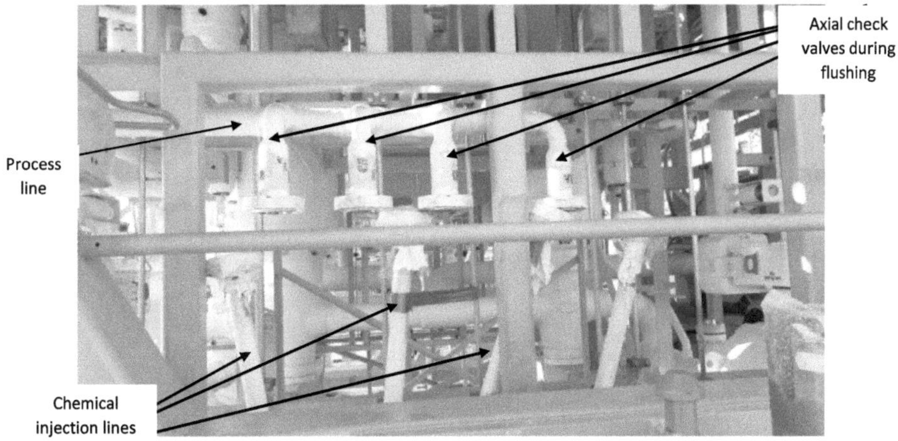

Fig. 5.34 Axial check valves in the yard during flushing-1.

Fig. 5.35 Axial check valves in the yard during flushing-1.

Fig. 5.36 Axial check valves in the yard during flushing-2.

assists with closing the disk of the valve. However, the presence of particles in the water disturbs the sealing of axial check valves.

The case study continued with an inspection of the axial check valve. A borescope inspection was done on the check valve disk and seats, and no damage or scratches were observed as per Fig. 5.37. A borescope is an optical instrument designed to assist with visual inspection of narrow, difficult-to-reach cavities. It consists of a rigid or flexible tube with an eyepiece or display on one end and an objective lens or camera on the other, linked together by an optical or electrical system. After the inspection, water jet and air blowing are injected inside the open valves to clean them and wash out all the debris and particles. After that, the valves are leak tested to make sure that the valve sealing is performing properly.

Certain preventive actions could prevent the flow of debris and particles inside axial flow check valves. Some of these solutions are proposed as follows:

- If the valve flange end is located on the downstream side instead of the upstream side, the valve could be disconnected from the process line during flushing;
- If two ends of the valve are flanged connections, the valve could be disconnected from both the process line and chemical injection line;
- Using a valve design in which it is possible to remove the valve internal during flushing and return it to its place after flushing;
- Protection of the valve internals from the ingress of foreign objects after manufacturing and during storage, transportation, shipment, etc.

5.2.5.3 Handling, storage, packing, and preservation

Figs. 5.38 and 5.39 illustrate blinding the axial check valves with plywood in 5–10 mm after manufacturing the valve to prevent the ingress of any foreign object to the valve inside.

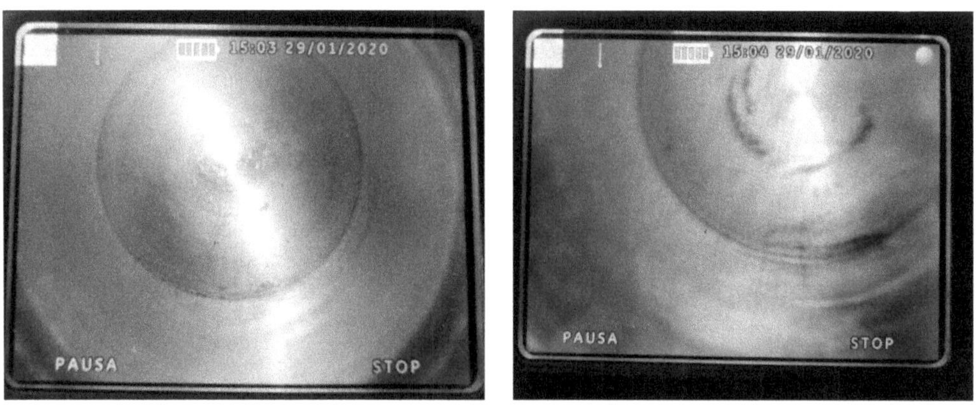

Fig. 5.37 Borescope inspection of axial check valve internals.

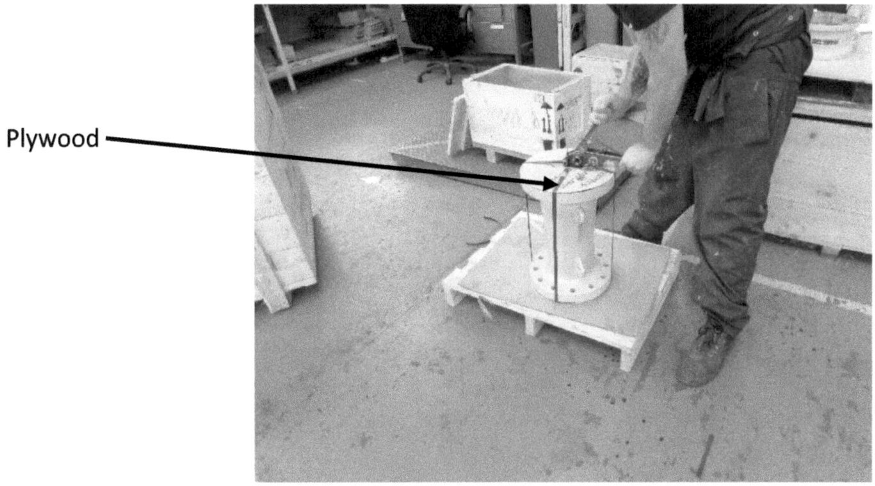

Fig. 5.38 Plywood for blinding the valve ends.

Fig. 5.39 Plywood for blinding the valve ends.

The axial valve in Figs. 5.38 and 5.39 has two end flange connections and can be removed from the line completely during flushing. The type of end protection could be plastic instead of wood or aluminum foil as illustrated in Fig. 5.40. The plywood protections should not be removed from the valves until installation of the valves on the line. Normally, boxes are used for packing the valves as illustrated in Fig. 5.41. For storage, handling, and preservation, the recommendations of the valve manufacturers should be

Fig. 5.40 Subsea axial check valve with an aluminum foil end protection. *(Courtesy: ATV.)*

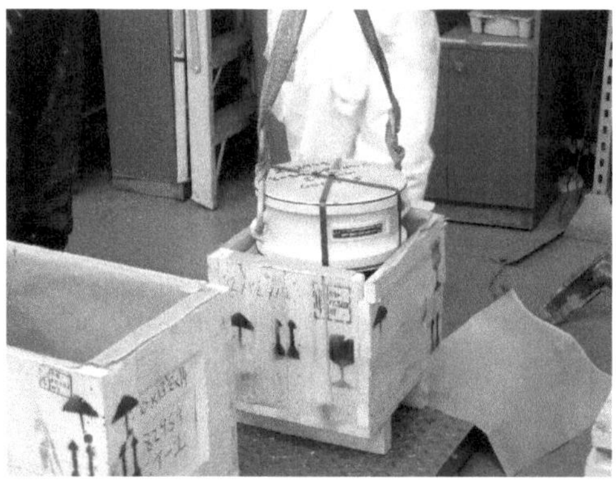

Fig. 5.41 Packing the valves inside a wooden box.

followed. The boxes should provide protection of the valves against snow and rain water. A name plate is fixed by four rivets to the body of the valve, and special care must be taken during handling to prevent the name plates from falling down from the valves and getting lost. The flow direction on the axial flow check valve in Fig. 5.40 is connected to the valve body with rivets, not engraved. In this case, the flow direction plate should be fixed properly to the body to prevent the flow direction template from falling off the valve body. If the valves will be in storage for more than 6 months, an inspection should be done to make sure that the valves are fine and even the end protector plywood or plastic caps can be removed accordingly.

Care must be taken while handling the valve to prevent clashes and possible damage. The surfaces of the flanges, such as the flange faces, are an essential section of the valve that should be protected, and a rubber ring can be placed between the plywood and the flange face to prevent damage to the flange face.

5.2.5.4 Installation

The valve should be stored in a proper location away from moisture, humidity, etc. The end protective plywood or plastic caps should be removed and the valve should be inspected carefully before installation to make sure that there is no sign of damage or flaw on the valve. Unpacking the valve and removing the protective caps should be done immediately prior to installation. Flange faces are sensitive parts that should be inspected properly. Lifting of the valves for installation can be done either by sling or by hand, since axial valves in small sizes are not heavy. A lifting sling may be used with a lifting device or crane to lift objects effectively and safely. Slings usually consist of a cable, chain, rope, or webbing, and generally include metal devices such as hooks, shackles, links, or rings. Fig. 5.42 illustrates the process of lifting an axial check valve with a sling made of chain and hooks.

The fit-up of the valve to the pipeline is very important and any foreign material ingress to the valve from the pipeline can damage the valve and valve internals. Some designs of axial check valves include soft sealing inside the valve that could be easily damaged due to the ingress of particles. The valve should be in closed position when it is installed on the piping. A hydrostatic pressure test is applied to the piping system after installation of the valve at 1.5 times of the design pressure. The pressure can be applied either upstream or downstream from the valve. Axial check valves are suitable for both vertical and horizontal installation. Some limitations may be applied when the valve is installed on vertical piping.

5.2.5.5 Operation

Axial check valves, like other types of check valves, are shipped in closed position. Chattering or vibration is one of the main operational problems of check valves such as axial check valves. The main prevention approach is to select the correct spring torque

Fig. 5.42 Lifting an axial check valve with a sling made of chain and hooks.

to prevent the frequent opening and closing of the valve with flow fluctuation. In addition, the valve should be sized properly, as oversizing the valve can cause chattering and vibration. The subsea operation of this valve is limited to a very simple task: fluid pressure is applied from the upstream side and opens the valve; when the upstream fluid pressure stops, the disk returns to the body or seat and closes the valve. Reverse flow from the downstream side of the valve to the upstream could assist with the closing of the valve.

5.2.5.6 Maintenance

These valves are installed subsea and do not require any maintenance during their design life, which could be 25 or 30 years as examples.

5.3 Important design parameters for subsea check valves

In general, the required design information for a check valve can be split to seven categories, described as follows:

1. General information: This includes the type of check valve (swing, axial, etc.), its size, fluid service, pressure class, temperature rating, bore size and piggability, product specification level (PSL), design standard, design life, and water depth. Chapter 2 of this book provides detailed information about size, pressure class, design standards, temperature class, bore size, and PSL.
2. Design features: This includes body design, which could be a one-piece or two-piece body connection or a body and bonnet connection type; type of connection between pieces (bolted or welded); the sealing direction of the valve, which is always unidirectional for a check valve; seat design, i.e. renewable (welded) or integral; and closure member closing force (spring, weight of the closure member).
3. Situation of the valve in the connected piping: direction of installation (vertical, horizontal), type of connection to the pipe (welded or flanged connection), applicable standard for flange connection (typically API 6A) and, if applicable, whether the valve is welded through an integral pup piece or a separate pup piece welded to the body of the valve (it should be noted that an integral pup piece with the body of the valve is common for small valves of 2″ and less).
4. Valve sealing: Body and bonnet sealing applicable for swing check valves is normally MtM with the addition of an elastomer sealing to prevent environmental leakage to the valve internal and a closure member and a seat or body seat surface sealing, which could be metallic or metallic with an additional soft seal ring.
5. Material: This includes material selection or options for the body/bonnet, disk, seat, bolting, pup piece, gaskets, and sealing, as well as the hard-facing materials. In addition, material traceability level and corrosion and erosion allowance should be considered.

6. Operator information: type of operation (manual or actuated), type of actuator, failure function of the actuator (e.g., fail-safe closed), and hydraulic maximum supply pressure. For a manual valve, the following information is important: ROV class and interface, position indicator type (normally local). For more information about actuators and ROVs, please refer to Chapters 8 and 9.
7. Factory acceptance and qualification test data: acceptable leakage rate from the valve body and the seat during the factory acceptance test (FAT), acceptable leakage rate from the valve seat after the qualification cycles, number of qualification cycles (e.g., 600).

5.4 Axial valves with a stem

A new generation of valves suitable for fluid control and use in high integrity pressure protection systems (HIPPS), called axial control or axial on-off valves, are like axial check valves with a stem, as illustrated in Fig. 5.43. These valves can be used instead of globe valves in various sectors of the oil and gas industry to prevent operational problems such as cavitation in liquid services and wearing associated with globe valves. Globe valves are explained in Chapter 8. Cavitation refers to a phenomenon in which the liquid pressure drops, and vapors or gases are formed and collapse, which exerts high forces on the material, leading to wearing and metal loss.

5.5 Self-assessment questions

This section provides some multiple-choice questions to assess the knowledge of the readers about check valves in the subsea industry.

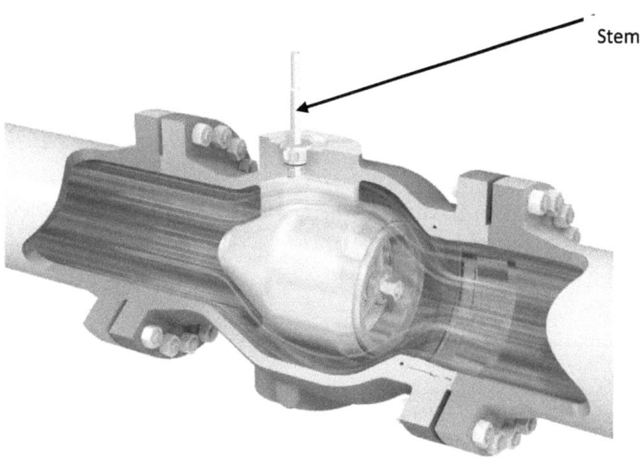

Fig. 5.43 Axial on/off valve. *(Courtesy: Mokveld.)*

1. Which sentences are correct about check valves?
   - **A.** Subsea swing check valves with two pup pieces at both ends have an end-to-end dimension according to either API 6A or API 6DSS depending on the pressure class value.
   - **B.** Subsea axial check valves have a one-piece body design with fast cycling action, but are not suitable for pigging purposes.
   - **C.** Subsea axial check valves are mainly used in subsea applications, either after subsea pumps and compressors or on chemical injection lines.
   - **D.** Subsea swing check valves always produce a slamming effect when closing.

   Answer: Option A is not correct. API 6A and 6DSS provide face-to-face dimension for swing check valves with flange ends. Swing check valves with pup pieces at both ends are not covered by the standard and are made according to the manufacturer standard. Option B is correct since it correctly addresses all the main characteristics of an axial check valve. Additional characteristics of an axial check valve, such as a low mass disk, robust design, venturi flow effect, and low pressure drop, can be added. Option C is also correct since subsea axial check valves are used after subsea pumps and compressors or (mainly) for the injection of chemicals such as MEG and methanol. These two chemicals are injected to the hydrocarbon, mainly gas, in order to prevent hydrate formation. For more information about hydrate, please see Chapter 1. Option D is wrong since adding some extra features like a dashpot prevents swing check valves from slamming.
2. Which sentences are correct about check valves?
   - **A.** Subsea axial check valves do not normally require a position indicator, but swing check valves with ROV override do require a position indicator.
   - **B.** Check valves, unlike ball and gate valves, do not have actuators.
   - **C.** Combination valves, including check valves and gate valves used for chemical injection, provide a more compact and lower weight solution than two separate check and gate valves.
   - **D.** All options are correct.

   Answer: Option A is correct. Subsea axial check valves open and close due to the fluid flow as well as the spring force inside the valve, so they do not normally have an ROV or actuator and do not need to have a position indicator. However, it is possible that the purchaser could ask for an ROV-operated axial check valve with a position indicator. Option B is correct since check valves do not normally have actuators. Option C is correct since a combination valve provides a single body for both a check and gate valve, resulting in a lower weight and more compact solution compared to two single gate and check valves. Option D is the correct answer since all the given choices are correct.
3. Which characteristic of subsea swing check valves makes these valves suitable for pipeline use?
   - **A.** Easier online maintenance possibility due to their top entry feature
   - **B.** Possibility to be upgraded to an ROV-operated version

C. Possibility to have full bore and piggable design
D. Less wearing of the valve internals

Answer: Option A is not correct. Swing check valves do have always top entry design with the possibility of online maintenance. However, subsea valves are maintenance free during their design life. Option B is not correct. Swing check valves can be designed for ROV operation to facilitate pig running from both sides of the valve. Adding an ROV operation feature is not a main reason for selecting subsea swing check valves for pipelines. Option C is correct and swing check valves for pipeline use can have full bore design according to API 6DSS with the possibility of pig running through the valve without any obstruction. Option D is not correct since the internals of swing check valves, especially the hinge and hinge pin, are subject to excessive wearing during valve operation.

4. Which sentences are correct regarding the subsea axial check valves used on chemical injection lines?
A. These valves have one end flanged and one end welded.
B. A pup piece integral with the body can ensure a safe distance between the welding heat and the valves' internals.
C. Axial check valves can be combined with gate valves located on the chemical injection line in a single body to save space, weight, and cost.
D. All options are correct.

Answer: All the options are correct. One end of the valve is a flanged connection to provide the possibility of dismantling the valve from the process line during flushing. The size of the axial check valve for chemical injection is small, so the pup piece is normally integrated with the valve body rather than welded to it. Also, axial check valves can be combined with the upstream gate valve on the chemical injection line to save weight, space, and cost.

5. Which sentence is wrong about axial check valves in general, and in particular those used for chemical injection lines based on the case study in this chapter?
A. Axial check valves are not suitable for particle-containing or so-called dirty services;
B. Three solutions to prevent debris from entering axial check valves are placing end caps or plywood on the valve ends, having both ends flanged, or having the flange end on the downstream side;
C. Axial check valves on chemical lines are normally in small sizes of less than 2″;
D. Welding check valves, especially those in super duplex material, to the piping system in chemical injection lines normally produces a high amount of heat.

Answer: Option A is correct since axial check valves are not suitable for dirty services. Option B is also correct regarding three approaches to preventing ingress of particles to the valve. Option C is correct as well. Option D is wrong since normally the heat of welding axial check valves to the pipe in such a small size and super duplex should not be high.

Further reading

American Petroleum Institute API 615, Valve Selection Guide, first ed., 2010. Washington, DC.

J.V. Ballun, A Methodology for Predicting Check Valve Slam, American Water Works Association (AWWA), 2007.

A. Kruisbrink, The need for dynamic characteristics of check valves, Valve World Mag. 15 (09) (2010) 65–66.

B. Nesbitt, Handbook of Valves and Actuators, 2nd revision, Elsevier, Oxford, 2011.

Norwegian Oil Industry Association, Valve Technology, Norsk Olje & Gass, 2013. 2nd revision.

G. Oxler, Non-return valve and/or check valve for pump system—a new approach, Valve World Mag. 14 (04) (2009) 75–77.

H. Schmitz, What is Water Hammer and How Do I Fix It?, Livintanor, 2018. [online] Available at: https://livinator.com/whats-water-hammer-and-how-do-i-fix-it/. (Accessed 6 June 2020).

P. Smith, R.W. Zappe, Valve Selection Handbook, fifth ed., Elsevier, New York, NY, 2004.

G.A. Provoost, The dynamic characteristic of non-return valves, in: Conference Paper Submitted to 11th Symposium of the Section of Hydraulic Machinery, Equipment and Cavitation. Netherlands, 1982.

K. Sotoodeh, Axial flow nozzle check valves for pumps and compressors protection, Valve World Mag. 20 (1) (2015) 84–87.

K. Sotoodeh, Valve preservation requirements, Valve World Mag. 21 (05) (2016) 53–56.

K. Sotoodeh, Comparing dual plate and swing check valves and the importance of minimum flow for dual plate check valves. Am. J. Ind. Eng. 5 (1) (2018) 31–35, https://doi.org/10.12691/ajie-5-1-5.

Valmetalic, Design and Selection of Check Valves, 2018. [online] Available at: https://www.valmatic.com/Portals/0/pdfs/DesignSelectionCheckValves.pdf. (Accessed 5 June 2020).

CHAPTER SIX

Subsea choke valves

Contents

6.1 Introduction

A choke valve is a type of flow control valve used on Christmas trees and manifolds to reduce the pressure and flow of the fluid coming out of the well. Choke valves allow fluid flow through a very small opening to kill the reservoir pressure while regulating the well pressure. In fact, the valve internals put a restriction on the flow through the valve in order to set and maintain the desired flow rate. Choke valves for flow control are

Subsea Valves and Actuators for the Oil and Gas Industry
https://doi.org/10.1016/B978-0-323-90605-0.00003-7

normally automatic valves that open and close in response to a signal from the valve control panel. Reducing the pressure enables the use of less expensive piping and valves downstream of the choke valve. In addition, choke valves can be used to completely close or open the wellhead flow to the downstream facilities. Choke valves may be used for fluids other than those produced from the well, such as gas and water injection, which are used for advanced reservoir fluid recovery. Thus the materials used to manufacture subsea choke valves should be compatible with the injected chemicals. Choke valves can be called "angle choke valves," since there is a 90-degree flow rotation between their inlet and outlet ports.

The design and advent of subsea choke valves was initiated by transferring topside choke valve technology to the subsea sector. One of the main challenges involved in this technology transfer was to design the valve to withstand the hyperbaric pressure of the seawater head. The hyperbaric effect can be calculated through the ratio of 10:1, meaning that the pressure of 10-m water depth is equal to 1 bar. This means that the installation of these valves at a depth of 2500 m produces 250 bar of extra pressure. For this reason, hyperbaric testing (explained in Chapter 14) is applied to subsea valves, including choke valves, to validate their design in order to prevent the ingress of seawater to the valve internals. Another unique subsea challenge involves the force of sea currents, which makes the installation and maintenance/removal of these valves more difficult. Performing maintenance on such valves poses a huge and costly challenge. The icing and freezing associated with cold ocean regions adds environmental challenges to the design and installation considerations for subsea valves such as choke valves. Choke valve design and operation are considered critical, since choke valves are in contact with thousands of barrels of oil and gas produced from the reservoir, which also contain sand and corrosive compounds. Adding to the challenge, the fluid type and flow rate are different from one production well to the next, and the fluid corrosivity can change over the design life of the plant. In fact, choke valves are subjected to a variety of operation conditions. Flow assurance problems such as wax and hydrate formation add complexity to their design and operation. Therefore, for all these reasons, subsea choke valves can be subject to maintenance and intervention.

Choke valves should be designed robustly with different considerations in mind, such as valve configuration, flow capacity, flow path characteristics, material selection, and ease of maintenance. In addition, it is important to design subsea choke valves to be compact and light, with accurate actuation to provide the reliability, precision, and durability required and expected in the subsea environment. Probably the top three challenging conditions that should be considered in choke valve design are erosion, high flow velocity, and significant pressure drop/fluctuations. Axial control valves are a new generation of valves that can be used for choke application. Flow controlling valves, such as axial and needle valves, are explained in Chapter 8. Choke valves are very important and critical valves, so a separate chapter is dedicated to subsea choke valves. Choke valves are used in different sectors of the oil and gas industry on different units, such as oil or gas production platforms, on surface or subsea wellheads, and on Christmas trees and subsea manifolds.

6.2 Choke valve types

6.2.1 Fixed or positive choke valves

Generally, choke valves are split into two categories: fixed and adjustable. Fixed choke valves do not have any flexibility to be operated under different process conditions; the trim of the position of the plug is set in a fixed position, and the orifice dimension is also fixed. Thus this type of choke valve cannot make any change to the fluid flow or fluid characteristics. Fixed choke valves do not require any operator, such as an ROV or actuator. Fixed bean type trim choke valves, which are explained later in this chapter, are applicable for fixed/positive choke valves. If a positive choke valve is required to change the flow rate, it must be disassembled so that the internals can be changed.

6.2.2 Adjustable choke valves

Adjustable choke valves should have a means of operation, such as an actuator or ROV, to maneuver the valve stem to the required position. This means that they can be adjusted to provide a variety of different flow characteristics and flow rates for the fluid service passing through them. Adjustable choke valves are typically designed for flow regulation and are not proposed for flow isolation purposes.

6.3 Choke valve components

Fig. 6.1 illustrates a typical subsea choke valve with manual operation and highlights the main components. The fluid enters from the bottom of the valve and, after a 90-degree rotation, moves to the right and exits the valve. The body is the main pressure boundary of the valve, including both flange ends connected to inlet and outlet piping. The bonnet or cover is bolted to the body to isolate the body of the valve from seawater. The body and bonnet of the valve could be of different materials, such as low alloy or carbon steel with a full Inconel 625 overlay, duplex and super duplex, or Inconel 625, mainly depending on the service pressure, temperature, fluid service chemical composition, etc. Accurate valve wall thickness calculation as per the American Society of Mechanical Engineering (ASME) boiler and pressure vessel code (BPVC) as well as final element analysis (FEA) according to ASME Section VIII Div. I and II should be applied on a choke valve's body and bonnet—the main pressure-containing parts—to validate the structural integrity of the valve. The body and bonnet are bolted together. These pressure-containing bolts are critical; they can be in low alloy steel materials according to the American Petroleum Institute (API), and may have cathodic protection as well as

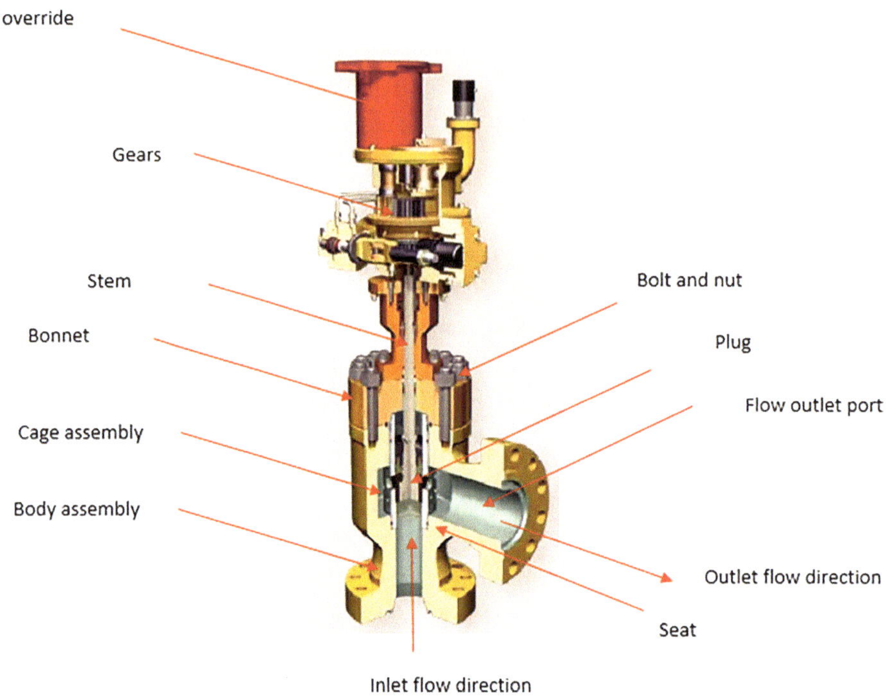

Fig. 6.1 A subsea choke valve with main components.

nickel and zinc coating to prevent external corrosion. The valve in the figure is operated manually by a remotely operated vehicle (ROV), a kind of robot. The rotary movement of the ROV torque tool can be changed to linear movement through gears, which may also be used to increase the produced torque. The produced force is transmitted to the valve stem, which is integrated with the plug of the valve. The plug acts as the main flow control element and/or closure member. The stem moves downward and lowers the plug onto the seats to fully close the valve. The plug works with the cage to control the valve flow rate. The cage is a cylinder (see Fig. 6.2) fitted in the valve outlet port where the plug moves downward. The cage is designed in a way to produce the required fluid characteristic by means of the valve operation, and ensures that the fluid impinges to itself rather than to the valve internals. The stem should be mechanically strong. Different material choices for the stem include super duplex and Inconel 625, 718, and 725. A cage that is integrated with the stem should be made of the same material as the stem. The seat of the valve could be made of the same material as the plug. Tungsten carbide hardfacing should be overlaid on the seat, plug, and cage to minimize the risk of erosion and galling. The valve sealing is metal-to-metal, based on the required leakage rate as per ISO 5208. Choke valves are normally bidirectional to allow injection into the well and to receive and operate in the fluid production from the well in the opposite direction.

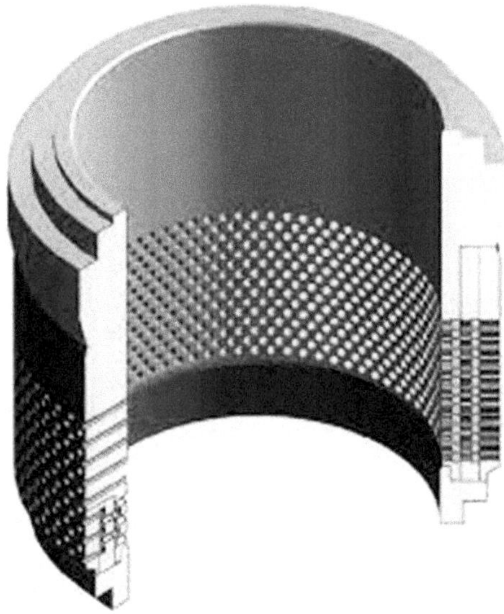

Fig. 6.2 Choke valve cage.

6.4 Choke valve trim design

Different types of trim are available for choke valves as follows:

1. *Needle and seat*: The idea for this type of trim is taken from needle valves. Needle valves are used in small sizes for very accurate flow control and are explained in Chapter 8. Needle and seat trim for flow control purpose is the simplest and least expensive approach, and is arguably the best approach for flow control due to its needle design. The details of needle trim are illustrated in Fig. 6.3.
2. *Multiple flow orifice*: This design uses two disks that seat on each other. Each one contains two holes/circular opening/orifices. Valve manufacturers can produce the disks with other shapes of orifices. The important point is that two flat surfaces should be lapped against each other with a diamond paste, as an example, so that they match exactly and provide sealing. Lapping in the valve industry refers to the process of reducing the roughness of a component with sand or diamond paste to achieve a certain flatness. When the two holes on the lower and upper disks are aligned to each other, the valve is fully open. By rotating one disk on the other, the position of the valve can be changed from fully open to fully closed position. Fig. 6.4 illustrates different shapes of the orifice as well as the percentage of valve opening based on the rotation of the orifices. Fig. 6.5 illustrates three positions of this type of valve: fully closed, throttling, and fully open.

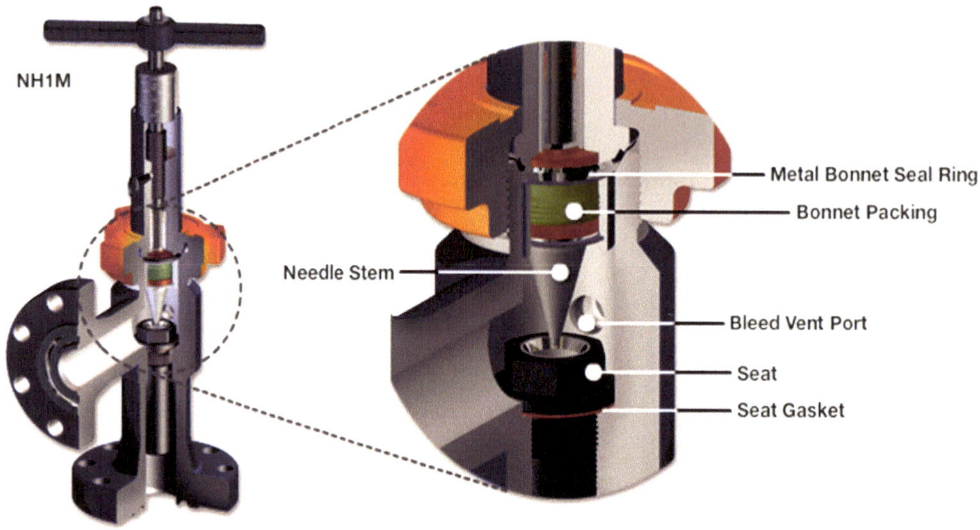

Fig. 6.3 Needle and seat trim for a choke valve. *(Courtesy: Cortec.)*

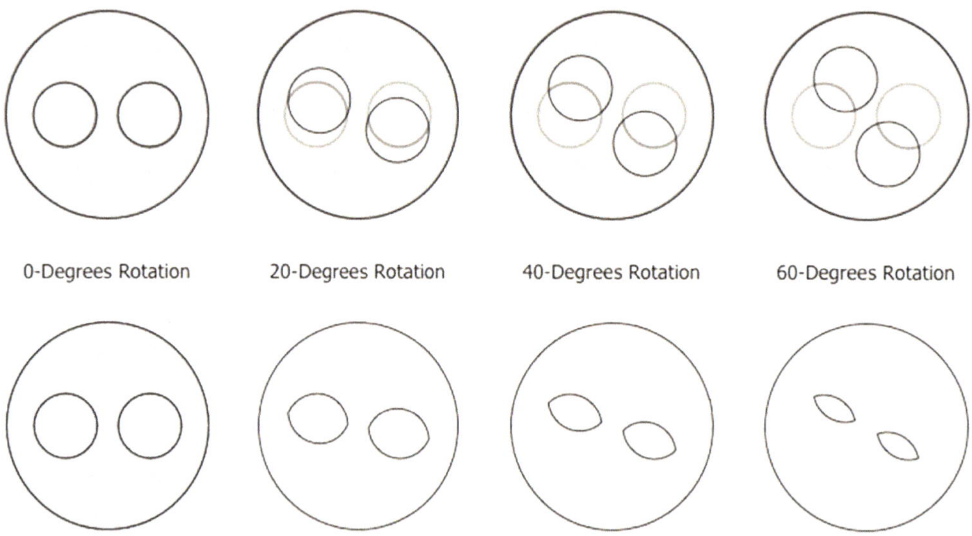

Fig. 6.4 Flow orifice shapes and valve opening percentage based on the rotation of the orifice. *(Courtesy: Piping Designer.)*

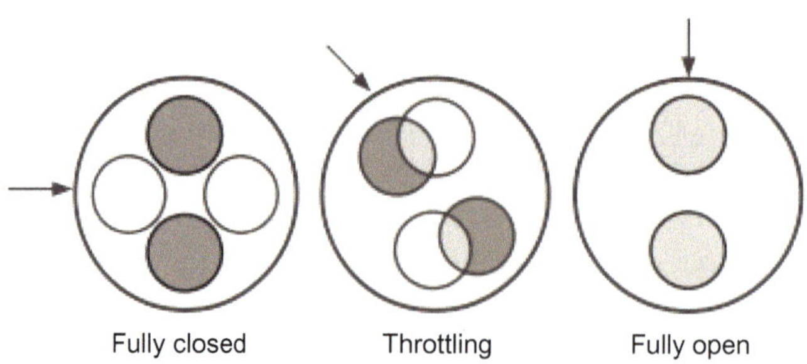

Fig. 6.5 Fully closed, throttling, and fully closed positions in a multiple flow orifice. *(Courtesy: Cameron.)*

Fig. 6.6 Sand erosion on the multiple orifice trim of a choke valve. *(Courtesy: Master Flo.)*

One of the main advantages of multiple flow orifice trim is to prevent the accumulation of wax in the choke valve internals. The other advantage of this type of choke valve internal is extended service life; since the operation action is simple, i.e., achieved through the movement of two disks, it causes less wearing to the valve internals, since the two disks can be made fully in tungsten carbide which is a very hard material. Less wearing leads to less maintenance and downtime which means high cost saving. Quick opening and closing as well as good flow and pressure control are other advantages of this valve. Although multiple flow orifice type trims are at lower risk of sand erosion, Fig. 6.6 illustrates a case of sand erosion in this type of trim.

3. *Fixed bean*: This trim design is applicable for nonadjustable choke valves, which are designed for a specific flow rate without any changing. The fixed bean or bin is a cylinder or rod, typically made from a very hard material. The bean is produced very precisely with a specific bore or flow passage area. Different materials can be selected for the bean, such as stellite 6 or 21, hard nickel alloys like Inconel 718 or 725, or tungsten carbide. The size or diameter of the bean is variable and could be from 1/64″ equal to 1.5 mm up to 250/64″ equal to 100 mm. Bean trim choke valves are typically used for pressure classes of 5000, 10,000, and 15,000 psi and could be in a size range between 1″ and 7 1/16″. Fig. 6.7 illustrates a fixed bean choke valve.
4. *Plug and cage*: The combination of a plug and cage can be used to control the flow as illustrated in Fig. 6.8. Fig. 6.9 illustrates a choke valve with plug and cage trim.

   The ports or holes in the cage are sized to provide the required flow characteristic and flow capacity for the choke valve. The bottom part of the plug tip, as well as the inner part of the cage, should be hardfaced to prevent erosion. A solid wear sleeve could also be provided to further reduce the risk of erosion. Fig. 6.10 illustrates a plug and cage with two additional wear sleeves: a retaining sleeve and an external sleeve. The external wear sleeve could cause the flow to be impinged into itself and the upward area of the cage to prevent the effect of throttling fluid energy that could otherwise destroy the valve internals. The retaining sleeve protects the valve body against

Fig. 6.7 Fixed bean choke valve. *(Courtesy: JVC Flow Control.)*

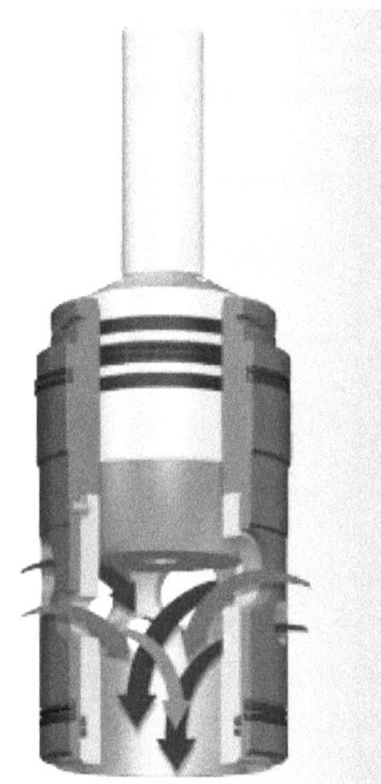

Fig. 6.8 Plug and cage trim for a choke valve. *(Courtesy: Cameron.)*

particles and could be considered less important compared to the external sleeve. The combination of an accurately designed plug and cage with an external sleeve and multistage trim most likely provides the best protection against erosion and cavitation.

The cage should have thick walls and be made of hard material to provide better protection against solids and erosion. A guide ring to guide the plug and reduce side loading and vibration is an additional feature that could be added to the trim of the valve. Fig. 6.11 illustrates the flow passage through the cage and plug of a valve in fully open position. Fig. 6.12 illustrates a choke valve with cage and plug trim in fully closed position.

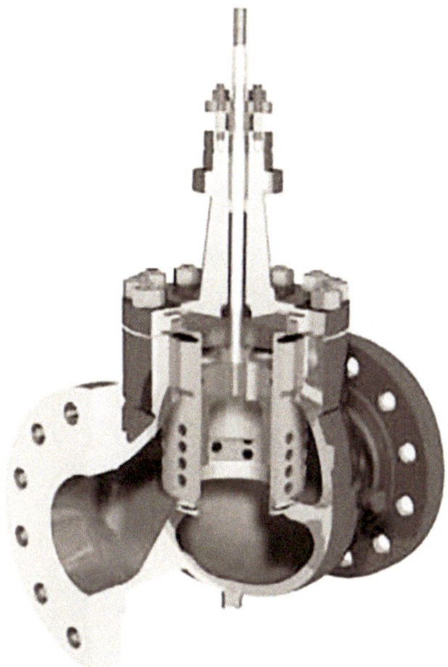

Fig. 6.9 Plug and cage choke valve.

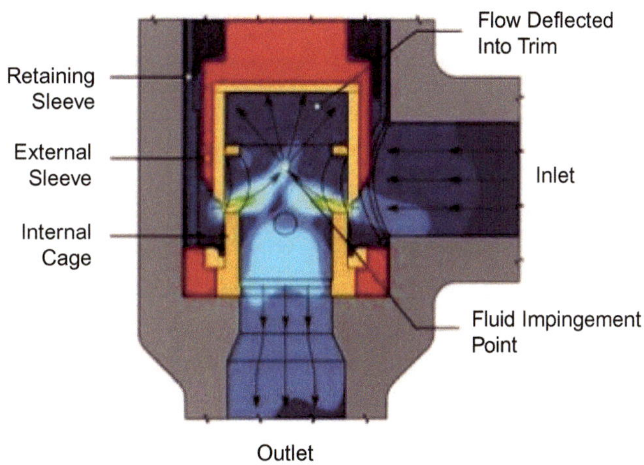

Fig. 6.10 Plug and cage design with additional wear sleeves. *(Courtesy: Master Flow.)*

The advantage of plug and cage trim is its effectiveness in high-flow capacity and multiphase flow. The plug can seat on the cage and provide very good flow sealing. Adding an elastomer sealing can provide tight shut-off, meaning zero leakage for the valve. Multistage plug and cage design is an alternative design used to drop the pressure in two stages rather than one, in order to reduce the impact of the

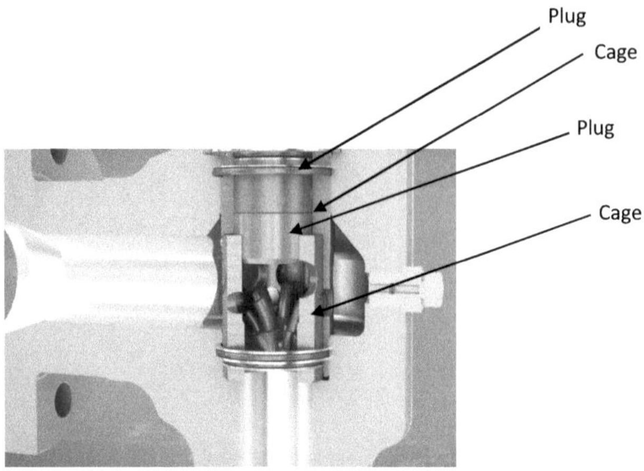

Fig. 6.11 Choke valve with plug and cage in fully open position.

Fig. 6.12 Choke valve with plug and cage in fully closed position.

pressure drop. Less pressure drop causes less risk of cavitation and lower fluid velocity, which reduces the risk of wearing substantially. As an example, the fluid velocity in a multistage plug and seat could be 75% less than in a similar, single-stage trim type. However, sand erosion still poses a risk for plug and cage trim choke valves, as illustrated in Fig. 6.13.

6.5 Pressure, temperature, and material classes

Subsea choke valves are typically designed according to the requirements of API 17D and/or ISO 13628-4. Typical pressure classes addressed by this standard are 5000, 10,000, and 15,000 psi. The important point is that choke valves have two different pressure classes on the endings; one end of the choke valve is connected to the tree or Christmas tree, and has a pressure rating equal to or higher than the pressure rating of the

Fig. 6.13 Sand erosion in the plug and cage of a choke valve. *(Courtesy: International Journal of Research and Technology.)*

tree. The other end of the choke valve has a lower pressure rating due to pressure reduction in the choke. Practically speaking, the whole choke valve should be designed equal to the tree pressure rating as the minimum pressure rating. The temperature rating of the choke valve should be in accordance with API 17D and API 6A or the relevant ISO standards. The temperature classes are explained in Chapter 2. The minimum temperature rating for a choke valve actuator should be between 2°C and 66°C. Material classes for choke valves, like other types of valves, follow the API 6A and 17D standards and are designated as AA, BB, CC, DD, EE, FF, or HH. The first three material classes are used for nonsour services, which are not very common. It should be noted that even if the reservoir fluid is nonsour, the fluid corrosivity could change to sour after a while due to the injection of seawater to the reservoir and the activity of bacteria, so most field materials are designed based on sour service requirements. Material classes DD, EE, FF, and HH are those used for sour services. More information about material classes is explained in Chapter 2.

6.6 Operational problems

Some of the common operational problems associated with choke valves are summarized in this section.

6.6.1 Erosion

Erosion in choke valves is the most severe problem. Solids and particles in the production fluid are the main cause of choke valve failure. Frequent and fast contact of sand with different valve components can be intensified due to high fluid velocity and cause severe erosion. In fact, the impingement of sands in the valve internals could be exacerbated by the effect of high fluid velocity and cause jet impingement or high-speed jet fluid flow. It should be noted that sand erosion is more severe with higher impact at a lower opening percentage of the valve. Abrasion and wearing of the choke valve

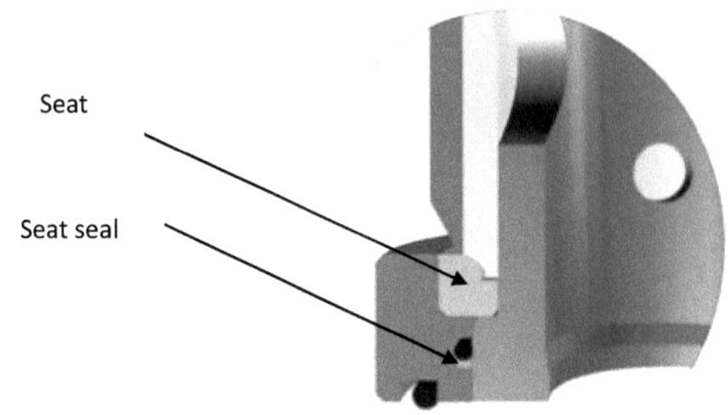

Fig. 6.14 Special seat design in a subsea choke valve. *(Courtesy: Master Flow.)*

internals are not limited to particles; other factors like the velocity and turbulence of the fluid can jeopardize the choke valve restriction and cause the well to fill up with fluids and fail. It is important to know that gas services in general are more erosive than liquid services. Critical locations where erosion can occur are the trim components, such as the plug and cage, due to their exposure to high pressure drop and fluid velocity. Although a strong hardfacing alloy such as tungsten carbide is applied to the valve internals, excessive sand production from the well can damage even the hardfacing alloy. Reducing the fluid velocity is another way to reduce erosion. The other erosion prevention measure taken by some valve manufacturers is to use wearing sleeves. Life cycle endurance testing, which is explained in detail in Chapter 14, is essential for choke valves to evaluate their long-term wear resistance. Fig. 6.13 illustrates sand erosion in both the plug and cage of a choke valve.

The problem of erosion is not limited to a choke valves' trim. Other parts of a choke valve, such as the body and seat, are at risk of erosion. As explained before, one sleeve can be installed inside the valve to protect the valve body from particles. The seat of the valve should be designed away from the throttling area and high fluid velocity zone. Fig. 6.14 illustrates a choke valve seat design with an elastomeric sealing between the seat and the flow nozzle port to prevent or minimize the contact of the flow with the seat during operation.

6.6.2 Cavitation

The other main cause of choke valve failure is pressure drop due to the 90-degree rotation of the fluid inside the valve, as well as the position of the plug to reduce the pressure, which can cause cavitation and irregular pitting on the valve internals. Cavitation is the formation and sudden collapse of vapor bubbles in the flowing liquid. Using two or three chokes in a series to reduce the pressure gradually in different stages is a good approach to preventing cavitation. The other outcomes of cavitation are noise and vibration, which can cause choke valve failure. Cavitation is more severe when the valve is less open due to higher pressure drop. Special attention should be given to the piping installed after the choke valve due to the possibility of high fluid velocity and erosion. The piping downstream or after the choke valve could have a lower temperature compared to the upstream piping, as pressure reduction can lead to temperature reduction. Sand erosion can be modeled using computational flow dynamics (CFD) to validate the design of the choke valve and to predict the effect and severity of sand erosion on the selected valve design. CFD can simulate different opening conditions as well as flow patterns and characteristics inside the valve.

6.6.3 Corrosion

The fluid produced from the reservoir is untreated and contains a large number of impurities and corrosive products. Moreover, seawater could be injected into the reservoir, which changes the nature of the produced oil and gas and introduces new types of corrosion, such as crevice and pitting. The presence of seawater inside the choke valve can cause yet another type of corrosion—galvanic corrosion—which occurs due to contact between two different metals with different electrical potential. Specifically, the less Noble metal acts like an anode and is corroded as a result of contact with the more Noble metal. Also, erosion can make the material vulnerable and intensify the corrosion effect. Combined corrosion and erosion is known as erosion-corrosion. Given all these risk factors, choke valves should be designed to be operated in severe service conditions. The injection of corrosion inhibitor could prevent or minimize the effect of corrosion on the valves. Fig. 6.15 illustrates erosion-corrosion on the body of a choke valve installed on a Christmas tree.

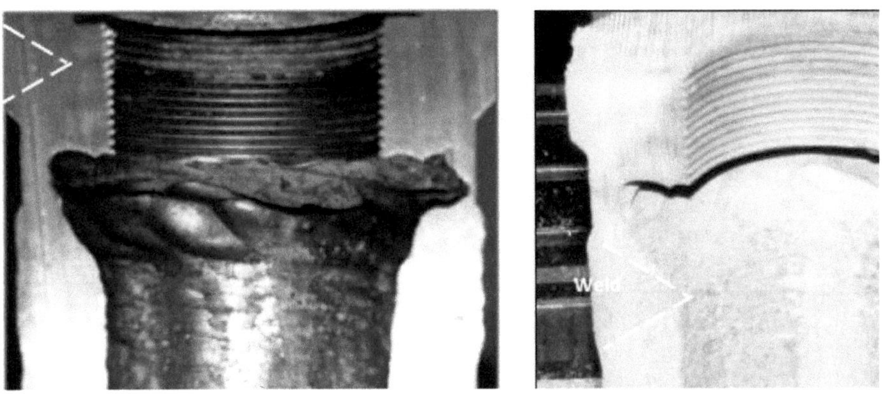

Fig. 6.15 Erosion-corrosion in the body of a choke valve installed on a Christmas tree. *(Courtesy: National Association of Corrosion Engineers.)*

6.6.4 Icing and freezing

Seas and oceans in cold regions of the world pose additional environmental challenges for subsea equipment, including valves such as choke valves, due to icing. Ice can apply a load to the valves and cause valve failure. Icing is also a concern for valves used topside in freezing and very cold environments.

6.6.5 Pipe vibration

Vibration and loads from the connected piping can increase wearing in the body and internals of choke valves. Vibration in the valve can be due to the effect of the fluid flow; this is called flow-induced vibration. Of course, the design integrity and reliability of choke valves, like other types of subsea components and facilities, are essential. However, maintaining the design integrity and desirable flow pattern and capacity in the choke valve is difficult due to constant process fluid alternation in the valve, including significant pressure drops and noise in the fluid. The noise produces vibration in the valve, which can become more critical when it matches the specific acoustic resonance in the pipe. Fig. 6.16 illustrates the effect of high flow velocity and flow-induced vibration on the trim of a choke valve.

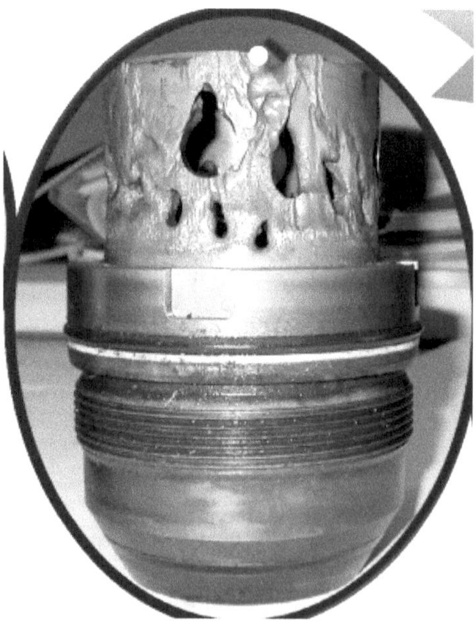

Fig. 6.16 Effect of high fluid velocity and flow-induced vibration on choke valve trim. *(Courtesy: Kentintrol.)*

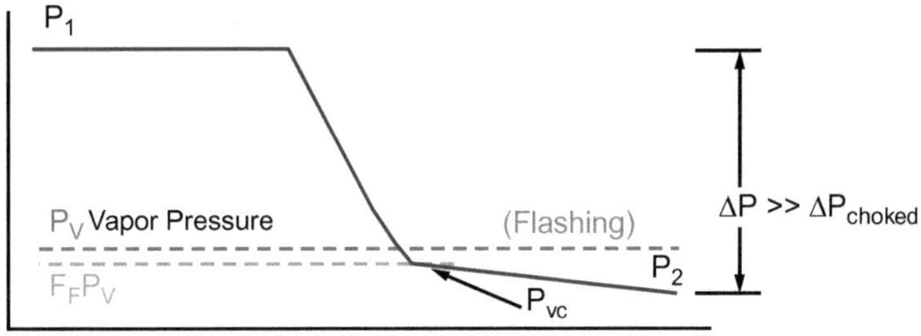

Fig. 6.17 Flashing chart.

6.6.6 Flashing

The main types of fluid passing through choke valves in the subsea industry are oil and gas. Combinations of oil and gas can be considered a liquid; each liquid type has a vapor pressure, and if the liquid pressure falls below the vapor pressure, bubbles can be generated. The generation of these bubbles is known as flashing. The bubbles can cause severe erosion in the valve and downstream piping. If they collapse and burst, they can produce a high amount of load and cavitation, which were explained before as two essential types of choke valve operational problems. Fig. 6.17 illustrates a pressure chart in which the operating pressure of the fluid moving to the choke valve is P_1; the pressure is reduced to P_2 in the choke valve, which is below the vapor pressure value equal to P_V. The flashing occurs at P_V which is named vapor pressure and shown in green line (dark gray in print versions).

6.7 Valve maintenance

Choke valves are used for critical applications, so they may require maintenance or intervention during the design life of the plant. When it comes to maintenance, three categories of subsea check valves are defined. The selection of one of these three options or categories depends on the water depth and location of the field. The three categories are fixed or nonretrievable choke, diver-retrievable, and deep water retrievable with remote-operated vehicle (ROV). ROVs are explained in detail in Chapter 9. The intervention and maintenance categories are also applicable for choke valve actuators.

6.8 Choke valve retrieval

The two types of choke valves in terms of retrievability are fixed (nonretrievable) and retrievable choke valves. Retrievable choke valves are also divided into two types: retrieved by diver or ROV.

6.8.1 Fixed (nonretrievable) choke valves

Fixed (nonretrievable) choke valves are not subject to any maintenance or intervention during the design life of the plant and could be installed at any depth. This type of valve is often installed on a choke module, which can be retrievable. Fixed choke valves have been made for over 50 years in the subsea sector of the oil and gas industry. Fig. 6.18 illustrates a fixed choke valve with the following main characteristics:

1. The size range is between 2″ and 8″. Different parameters affect valve sizing, such as flow capacity and velocity.
2. Standard pressure classes according to API 17D, such as 5000 and 10,000 psi and a maximum 15,000 psi;
3. Suitable for operating temperatures from −46°C to 121°C;
4. May be in API 6A/API 17D material classes like FF and HH, such as Inconel 625;
5. Suitable for different flow characteristics, such as linear and equal percentage (explained later), with the possibility of having different flow coefficients or C_V values. The C_V value refers to the amount of flow passing through the valve in gallons/minute when the pressure drop created by the valve is equal to 1 psi and the

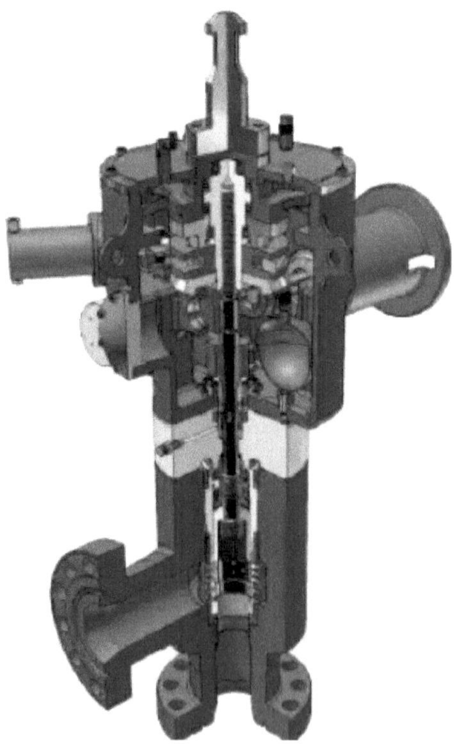

Fig. 6.18 Fixed choke valve.

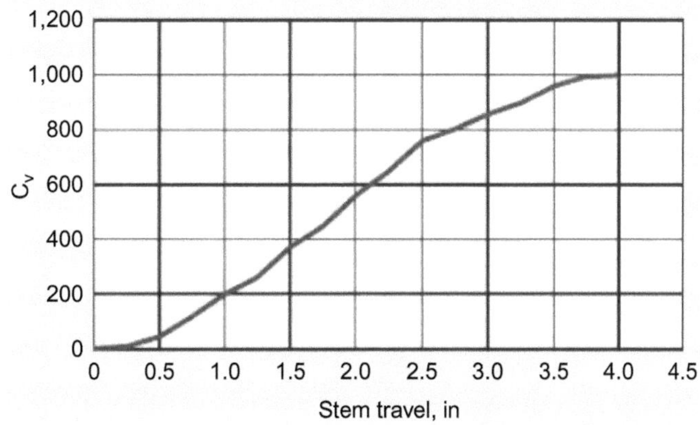

Fig. 6.19 Flow capacity of a nonretrievable choke valve based on stem travel.

temperature is 60°F. Fig. 6.19 is a chart showing the relationship between stem travel in inches and the C_V of the valve. Stem travel indicates the distance that the stem travels upward off the seat. A C_V of 1000 corresponds to full opening of the valve, when the stem is farthest from the seat. The flow curve of choke valves, like that of other types of valves, differs from one valve supplier to another, and also depends on the type of trim.

The linear flow rate provided by choke valves is an advantage; it means that the amount of flow passing through the valve increases linearly with the stem travel and valve opening or closing percentage. Equal percentage is a flow characteristic in which the amount of flow passing through the valve is relatively small compared to the initial opening percentage. As an example, opening just 40% of the valve leads to passing approximately 12% of the flow. Fig. 6.20 illustrates both linear and equal percentage flow curves. The flow curve of a choke valve could be linear or equal

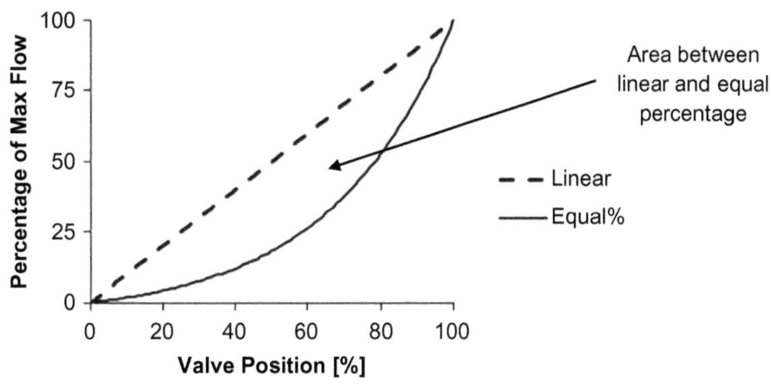

Fig. 6.20 Linear and equal percentage flow curves.

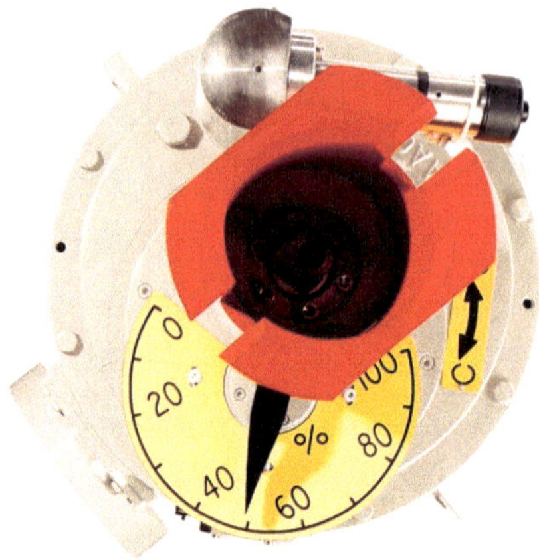

Fig. 6.21 Choke valve position indicator (half open). *(Courtesy: Kentintrol.)*

percentage, or correspond to the area between linear and equal percentage as shown in the figure. The flow characteristic of the valve is dependent on the valve internals or trim, especially the plug and cage of the valve. In general, the flow capacity of choke valves is a very essential parameter that is determined according to the requirements of ISA 75.01.01, "Control valve sizing equation" and ISA 75.02, "Control valve capacity test procedure." Some commercial software has been developed by choke valve manufacturers in order to size choke valves and determine their flow capacity and flow coefficient or C_V value.

6. Compatible with single or multiphase flow production, water and gas injected into the reservoir and chemicals;
7. Actuated with a hydraulic actuator;
8. Possibility of having a remote position indicator through transducers and electrical signals. Fig. 6.21 illustrates an ROV bucket and the position indicator of a subsea choke valve in 50% or half-open position. The position indicator is normally located at the same level as the ROV base.
9. Design life could be up to 25 years without considering the valve internals wearing.

6.8.2 Retrievable choke valves

6.8.2.1 Diver-retrievable choke valves

Diver-retrievable valves are installed in shallow water depths of a maximum of 200 m. The valve internals and actuator can be retrieved, replaced, and assembled by a diver. The visual position indicator should be connected to the valve to provide visibility of the valve position to the diver. This type of operation for choke valves is not popular, since diver operation on such a high-pressure and critical flow application is not

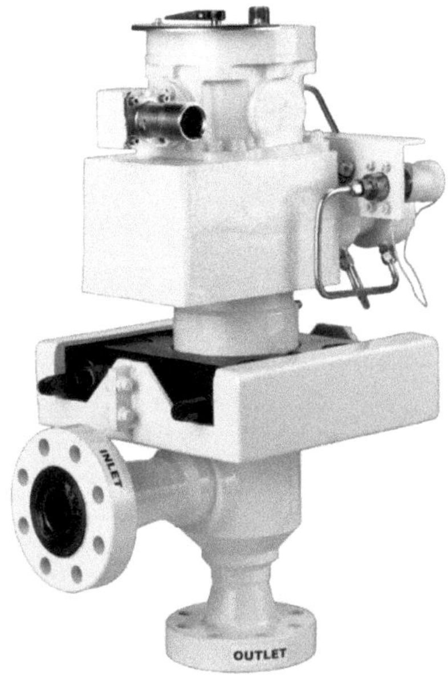

Fig. 6.22 Diver-operated subsea choke valve. *(Courtesy: Mater Flow.)*

considered a health, safety, and environmentally (HSE) friendly solution. Fig. 6.22 illustrates a diver-operated subsea choke valve.

6.8.2.2 Deep water retrievable via ROV

These types of choke valves are installed and operated at water depths deeper than 200 m where it is difficult or impossible for a diver to reach. Intervention tools, called remote-operated tools (ROT) or running tools are used for the intervention. ROV-operated retrievable choke valves can be installed on Christmas trees and should be designed in a way to prevent ROV impact and damage due to dropping objects on the choke valves during ROV intervention. The ROV impact load could be 1.5–3 tones with a speed of 0.5–1.5 m/s. The choke valve internals as well as the top part of the choke, including the actuator, can be disassembled (retrieved) from and installed back into the valve according to the illustration in Fig. 6.23. The clamp style or lockdown mechanism can be released from the choke body, so that the actuator, trim, some seals and the top part, which includes a clamp or lockdown mechanism, can be separated from the rest of the valve. Retrievable choke valves have a more compact and light design compared to fixed choke valves, because of the need for intervention and transportation of some parts of the former to the surface for maintenance.

Fig. 6.24 illustrates a ROT or running tool on a choke valve. There are some pins located on both the retrievable part of the valve and the running tool in order to align

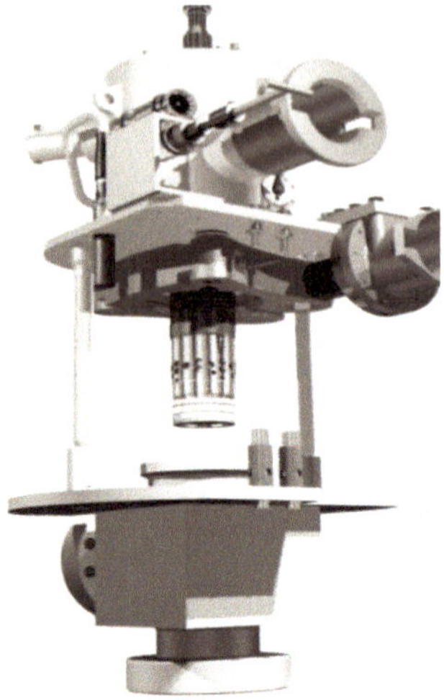

Fig. 6.23 Retrievable choke valve. *(Courtesy: Kentintrol.)*

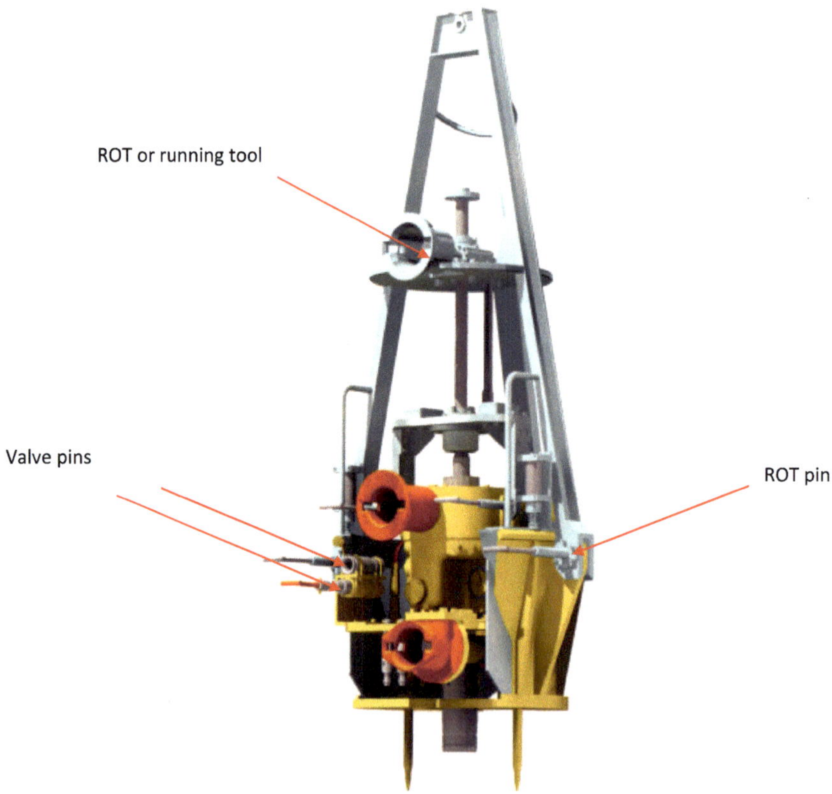

Fig. 6.24 Running tool or ROT installed on a subsea choke valve. *(Courtesy: Kentintrol & Subsea 7.)*

the tool properly with the valve for removal and assembly operations. The pins should be made of high-strength materials such as super duplex. Two female pins are shown on the valve in the picture where the male pins belonging to the ROT enter. In addition, transportation structures may be used for transporting the top part of the choke valve. Transportation structures should have high mechanical strength to withstand the loads in the sea and, more importantly, the weight of the top section of the choke valve during transportation. Therefore when a choke valve is installed on a Christmas tree and requires replacement or intervention due to high wearing or any other operational problem, sufficient space and clearance should be considered around the Christmas tree for the implementation of this type of operation, which is done by ROT.

6.9 Operation and actuation

The performance of actuated choke valves is largely dependent on their actuator. Actuators for choke valves should be lightweight, suitable for the required design life of the plant, and able to precisely open and close the valve according to the required flow opening percentage. Different actuator and manually operated options are available for choke valves, and are explained in this section.

6.9.1 Stepping actuators (standard design)

Stepping actuators produce linear force and motion, so they are suitable for subsea choke valves. The actuation of choke valves with this type of actuator takes place in different number of steps. Subsea stepping actuators work with hydraulic fluid and force. This type of actuator works by the supply of hydraulic oil which is supplied to the actuator after passing through the control system including the solenoid valves. The hydraulic force or power works in cooperation with the gears. The drive nut, also called the stem nut, provides the linear movement of the stem in the choke valve. This type of actuator is quite common for fixed, nonretrievable choke valve actuators. The actuator may contain one hydraulic cylinder as illustrated in Fig. 6.25, or two hydraulic cylinders as shown in Fig. 6.26. The actuator is typically equipped with ROV override to be operable in the event of actuator malfunction or a problem in the control system. There are two important considerations for the actuator; first, to apply a pressure compensation system to prevent the actuator from imploding due to seawater head pressure, and second, to install a position indicator in antifouling materials to prevent marine growth and provide visibility of the valve position for the ROV or diver during operation. Stepping actuators can be standardized to work with a hydraulic pressure of 345 bar equal to 5000 psi in the sea. Electrohydraulic stepping actuators, in which an electrical motor supplies the hydraulic fluid to the actuator, are also available in the industry. In fact, electrohydraulic actuators are considered a type of hydraulic actuator that works with an electrical motor to provide movement and supply the hydraulic fluid.

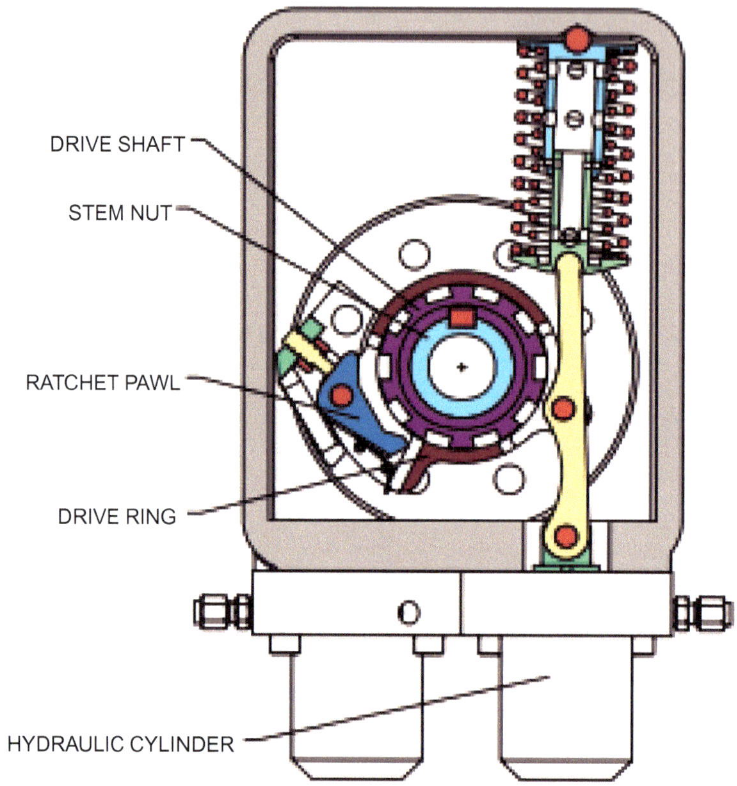

Fig. 6.25 Stepping actuator with one cylinder. *(Courtesy: Master Flow.)*

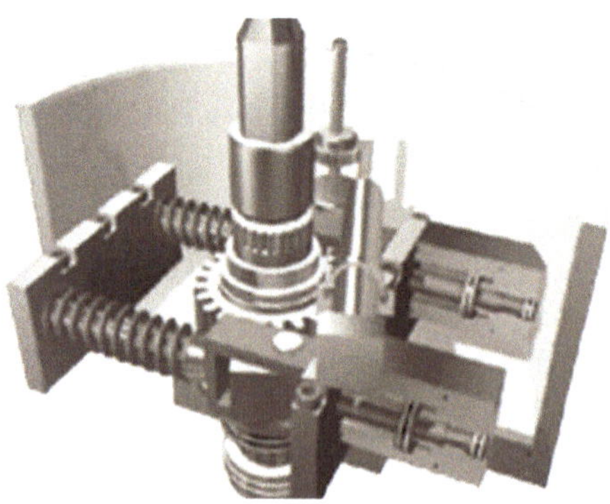

Fig. 6.26 Stepping actuator with two cylinders. *(Courtesy: Schlumberger.)*

The following requirements, according to API 17D, apply to hydraulic actuators for choke valves:

- The standard hydraulic supply pressure to the actuators should be 3000 or 5000 psi or according to the valve data sheet.
- The opening force or torque should not exceed 90% of the rated working pressure. As an example, if the working pressure of the hydraulic supply to the actuator is 345 bar, a fail-safe closed actuator should be opened with a maximum hydraulic pressure of $0.9 \times 345 = 310.5$ bar.
- The other parameters that are important and effective in actuator design are temperature rating and water depth. The depth of water could affect the spring torque in the actuator as well as the piston and cylinder sizing.
- The actuator should be sized in a way to prevent damage to the choke valve and its internals.
- The effect of seawater depth and head on the linear motion of the actuator should be considered.
- Hydraulic actuators require a pressure compensation system.
- The manufacturer of the actuator should provide the required data for the actuator, such as the operating torque input; maximum torque capacity and damage torque, which can break the valve stem; the type and size of the ROV override; material class; temperature rating; the number of ROV turns necessary for closing and opening the actuator and the swept volume of the oil or the amount of oil displacement inside the actuator.
- Actuators are typically equipped with ROV override so that they can be operated manually.
- The factory acceptance tests for actuator are explained in Chapter 14.

6.9.2 Stepping actuators for retrievable operation

This type of stepping actuator contains some modifications compared to the standard one to be compatible with retrievable operation.

6.9.3 Electrical actuator

Electrical actuators provide some advantages over hydraulic types, such as environmentally friendly design, fast and reliable operation, reduced complexity, and lower cost. Traditional hydraulic or electrohydraulic choke valves are mostly equipped with stepping actuators as well as one or two hydraulic cylinders. The choke valve may undergo several steps in order to cycle from open to closed, or closed to open positions. Adding an electrical motor on top of the choke valve could facilitate the operation of the valve more precisely without any need for stepping. One of the positive points in using electrical motors is the possibility of standardizing them to different water depths, unlike hydraulic actuators. Using hydraulic actuators in different depths leads to design modification in the piston, cylinder, and spring of the actuator. Electrical actuators typically contain an electrical motor, electrical connectors, gears, and an ROV or diver interface.

6.9.4 Manual gears

Typical subsea valve gear boxes with a bucket for ROV override can be selected for manually operated choke valves. A worm gear box, shown in Fig. 6.27, is used for producing linear motion applicable to the choke and gate valve stem. A worm type gear box contains a wheel and a worm, highlighted in the figure. The ROV or diver force is transmitted to the wheel and the rotation of the wheel is transferred to the worm, which is connected to the valve stem. The worm is moved linearly and makes the stem move linearly as well. Fig. 6.28 illustrates a subsea worm gear, also called a worm wheel gear. The gear has a closed pressure compensation system to equalize the pressure of the gear box to the seawater head to prevent it from imploding. The gear box is operated through a

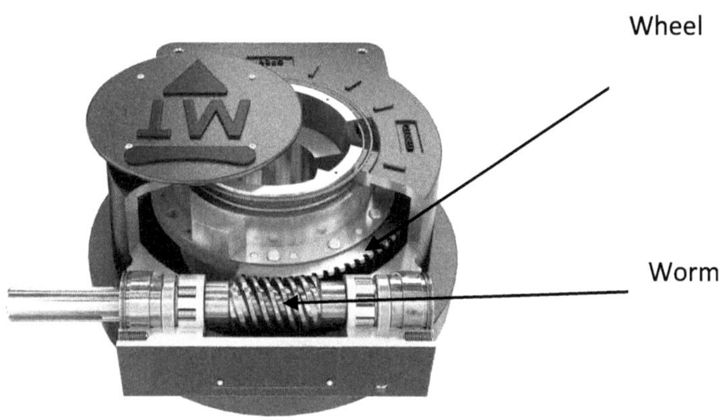

Fig. 6.27 Worm gear. *(Courtesy: Schlumberger.)*

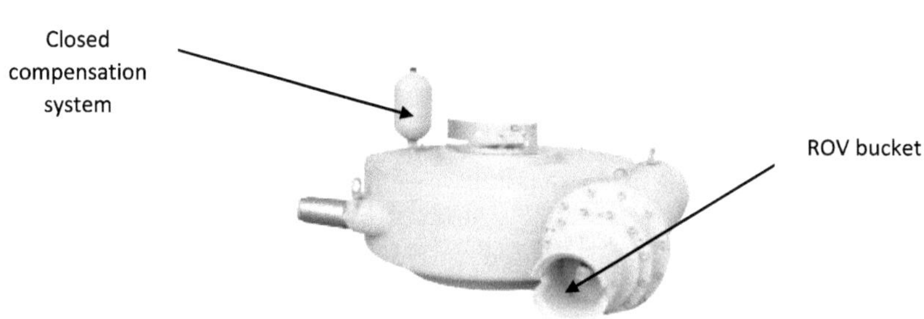

Fig. 6.28 Subsea worm gear. *(Courtesy: Rotork.)*

bucket designed for an ROV torque tool or diver operation. The bucket is designed according to API 17H/ISO 13628-8, which is explained in more detail in Chapter 9.

6.10 Testing

Qualification testing and factory acceptance testing (FAT) are applicable for choke valves, like other subsea valves. Sand slurry testing, according to the requirement of API 6AV1, is also applicable to this valve to validate its design and operation in sand slurry fluid. The sand slurry test is fully explained in Chapter 7.

6.11 Summary

The valve and actuator purchaser should define the following requirements as the minimum for subsea choke valves:

- Size and pressure rating, for example 5 1/8″ and pressure class 15,000;
- Governing standards, such as API 17D;
- Valve end connections, which could be flanged connections on both ends as an example. Both the size of the valve endings and their pressure classes should be clarified;
- Minimum and maximum temperature at which the valve is expected to operate;
- Design water depth;
- Type of operator and the required source for operation; as an example, a hydraulic actuator and 207 bar hydraulic pressure;
- Type of choke (fixed or adjustable);
- PSL level; PSL 3G is quite normal, which adds the requirements of a high-pressure gas test to the valve;
- Retrievable type (diver, ROV, none);
- Type of operation (manual with ROV or diver or both, actuator);
- Type of actuator (e.g., electrical).

Some of the choke valve requirements and expectations during operation could be good controllability, reliability, and safety; high erosion and corrosion resistance of the valve trim, which leads to a long trim life; good shutdown capability with no leakage if the choke valve is required to shut down the fluid; and low cost, including both capital and operational costs.

Further reading

American Petroleum Institute (API) 6A, Specification for Wellhead and Tree Equipment, twenty-first ed., API, Washington, DC, 2018.

American Petroleum Institute (API) 17D, Design and Operation of Subsea Production Systems, Subsea Wellhead and Tree Equipment, second ed., API, Washington, DC, 2011.

Cameron, A Schlumberger Company, Multiple Orifice Valve Chokes, 2020, [online] Available at: https://www.slb.com/-/media/files/sur/brochure/mov-chokes-br. (Accessed 14 September 2020).

International Organization of Standardization (ISO) 13628-4, Petroleum and Natural Gas Industries—Design and Operation of Subsea Production Systems—Part 4: Subsea Wellhead and Tree Equipment, second ed., International Organization of Standardization, Geneva, Switzerland, 2010.

Kentintrol, Subsea Chokes and Control Valves, 2010.

Mcilvanie Company, Choke Valve Decision Guide, Scope and Action Plan, 2017, [online] Available at: https://home.mcilvainecompany.com/images/Choke_Valve_Decision_Guide_4-5-17.pdf. (Accessed 13 September 2020).

G. Navas, I.C. Grigorescu, Erosion-Corrosion Failures in Wellhead Choke, National Association of Corrosion Engineers (NACE), 2011. Conference paper. Paper # 11247 Houston-Texas.

N.E. Ligterink, Flow Induced Vibration from Chokes in Subsea Gas Production Systems, Society of Petroleum Engineers (SPE), 2012. Conference paper. Houston, Texas.

A. Rubio, A.H. Zeid, J. Meyer, Electrically Actuated Subsea Chokes May be Next Major Leap in Subsea Innovation, 2017, Offshore [online] Available at: https://www.offshore-mag.com/drilling-completion/article/16755968/electrically-actuated-subsea-chokes-may-be-next-major-leap-in-subsea-innovation. (Accessed 15 September 2020).

S. Tattersall, Choke valve technology in subsea environment, SAGE J. (2016), https://doi.org/10.1177/0020294016640556.

CHAPTER SEVEN

Safety valves

Contents

7.1 Introduction

A surface safety valve (SSV) is an automatic wellhead valve assembly that closes upon loss of power supply as per the API 6A definition. An SSV is actuated by a surface safety actuator. An underwater safety valve (USV) is defined as an automatic valve

Subsea Valves and Actuators for the Oil and Gas Industry
https://doi.org/10.1016/B978-0-323-90605-0.00005-0

assembly installed at an underwater wellhead location that closes upon loss of power. A USV is actuated by an underwater safety actuator. The actuator goes to closed position automatically when power is lost and opens when power supply is available. A boarding shutdown valve (BSDV) is defined as an automatic valve assembly installed between an underwater production system and a surface facility that closes upon loss of power supply. The actuator for a BSDV valve is called a boarding shutdown actuator. Fig. 7.1 provides a schematic for these three kinds of safety valves. Most SSVs and USVs are gate valves, since they are located on Christmas trees. BSDVs can be either gate or ball valves.

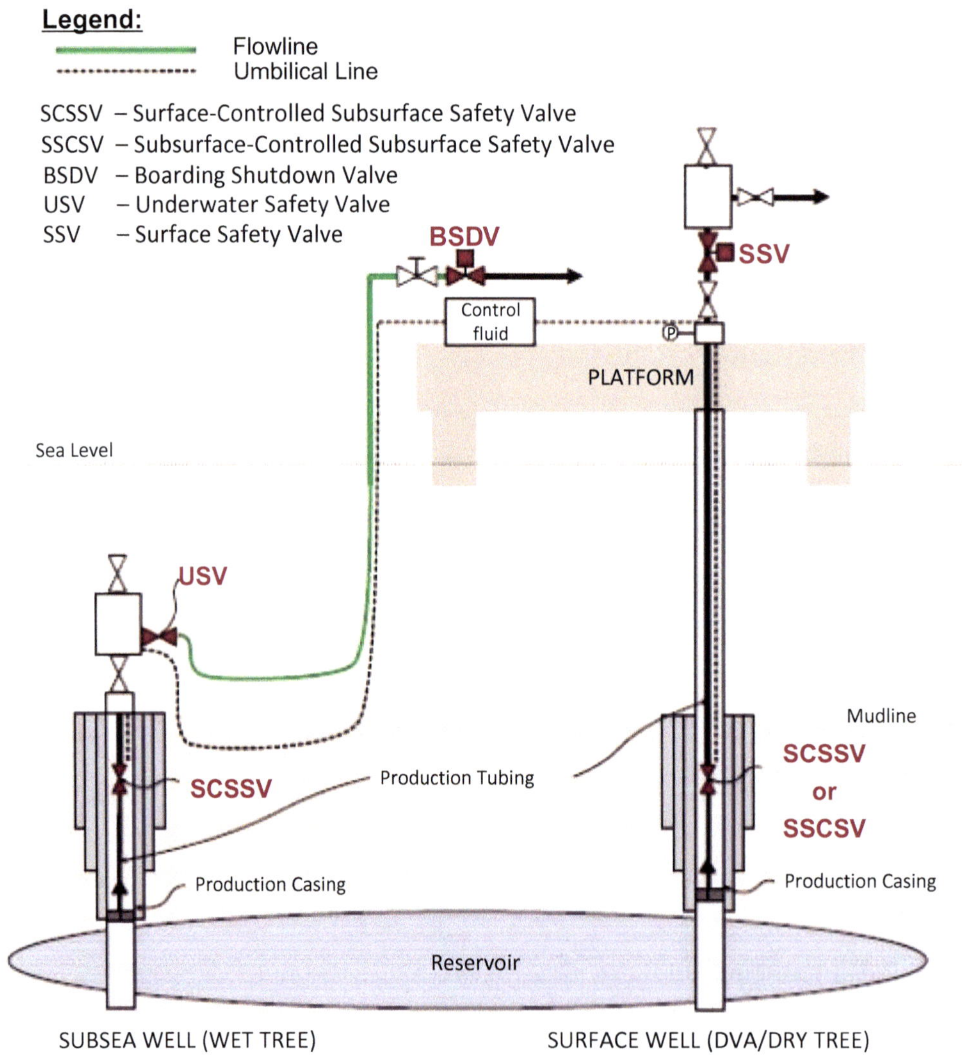

Fig. 7.1 Schematic of safety valves covered by API 6AV1. *(Source: US Department of Energy.)*

7.2 Surface safety valves

An SSV (see Figs. 7.2 and 7.3) is a hydraulically actuated through conduit gate (TCG) valve with a fail-safe close (spring return) function installed on a wellhead, platform, or ship. A TCG valve has a reverse port that closes the valve when the stem moves upward. The SSV is used to quickly shut down the well and protect upstream production units such as the choke and manifold in the event of well problems such as failure, overpressure, or leakage. This valve has an emergency shutdown (ESD) function and is operated remotely by an ESD system. An ESD system is designed to prevent or minimize the negative consequences of emergency or hazardous situations and events such as leakage, uncontrolled flooding of the hydrocarbon, leakage of the fluid inside the system to the environment, breakage, fire, etc. An ESD valve, which can be abbreviated as ESV or ESDV, is an actuated valve designed to stop the flow and close the line in a very quick manner. The shutting down action of an ESDV provides protection against any harm to people, the environment, and the ecosystem. Any undesirable condition, such as overpressure, can be detected by sensors and reported to the control room. The control room can send a signal to the ESDV to close as quickly as possible. Blow down valves (BDV) also have a high safety function like ESDVs, but they should be opened in the event of any undesirable condition in the line.

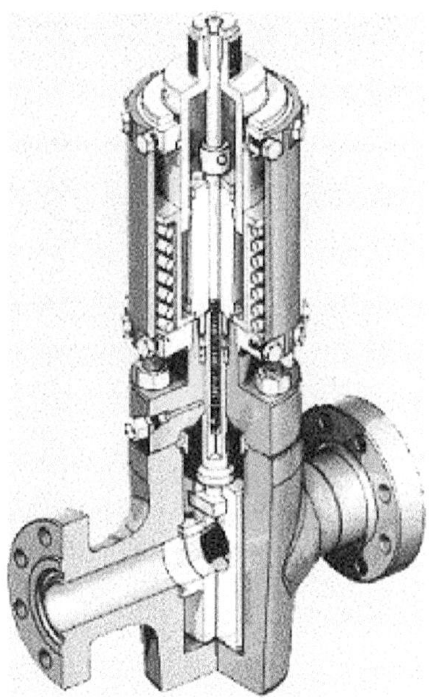

Fig. 7.2 Surface safety gate valve highlighting the internals.

Fig. 7.3 Surface safety gate valve. *(Courtesy: Schlumberger.)*

Actuators for ESD valves should meet the requirements of International Electrotechnical Commission (IEC) standards 61508 and 61511. IEC 61508 is an international standard that includes methods for how to design, apply, and maintain automatic safety protection systems. This basic functional safety standard is titled "Functional safety of electrical/electronic/programmable electronic safety-related systems." IEC 61511 is also an international technical standard that sets out practices for the engineering of the systems to ensure safety implementation for industrial processes. In fact, IEC 61511 was developed as a process sector of IEC 61508 with the title "Safety instrumented system for the process industry." Fig. 7.4 illustrates the difference between these two IEC standards.

SSVs, also called safety shutdown valves, are used for the production of oil and gas from wells or for testing the wells. The main purpose of well testing is to get a fluid sample from the formation or reservoir to understand the characteristics of the oil reserve and the properties of the hydrocarbon, such as viscosity, pressure, gas/oil ratio (GOR), etc. Therefore SSVs are used in high-pressure settings, from 5000 to 15,000 psi as an example, for systems with a high flow rate and in corrosive fluid services. The hydrocarbon produced from the reservoir is untreated and can contain high concentrations of carbon dioxide (CO_2) and hydrogen sulfide (H_2S). In Fig. 7.5, an SSV (item #10) is located on the vertical line exactly after the master valve (item #12). Item #11 is a fail-safe closed hydraulically operated actuator belonging to an SSV. Item #9 is a tee to make a branch from the main production line on the Christmas tree. Item #8 is a wing valve located

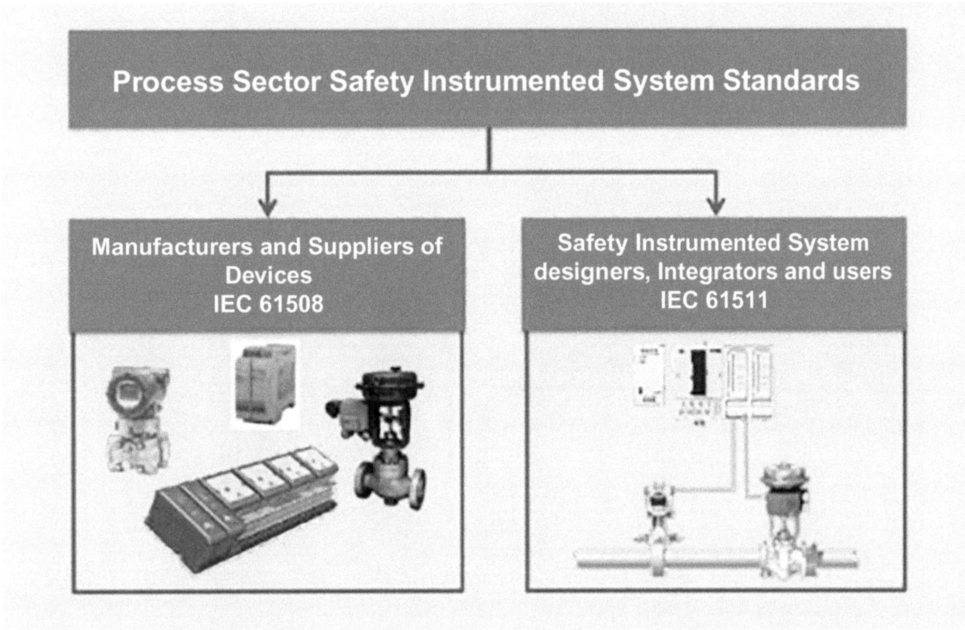

Fig. 7.4 Comparison between IEC 61508 and IEC 61510 standards. *(Courtesy: Instrumentation and Control Engineering.)*

close to item #9, a choke valve. The wing valve is normally open and is closed when the production line should be isolated from the well. Normally, a kill wing valve is installed on the opposite side of the wing valve for treatment or control and monitoring of the well conditions. It should be noted that an SSV is not necessarily installed on a vertical line as illustrated in Fig. 7.5. Both the SSV and master valve can be installed horizontally on a horizontal tree.

7.3 Underwater safety valves

A USV is an automatic valve assembly, installed at an underwater wellhead location, that closes on loss of power supply. When the term USV is used in an International Standards like API 6A or API 17D, it is understood to include a USV valve and a USV actuator. A USV is the same as the SSV installed on the wellhead in Fig. 7.5, but it is installed on a subsea tree instead of a land-based tree. A USV can be also installed on a horizontal line where it is installed on a vertical tree. The following USV requirements are taken from 30 CFR250, subsection H, sections 250.833 and 834, addressing oil and gas production safety systems:

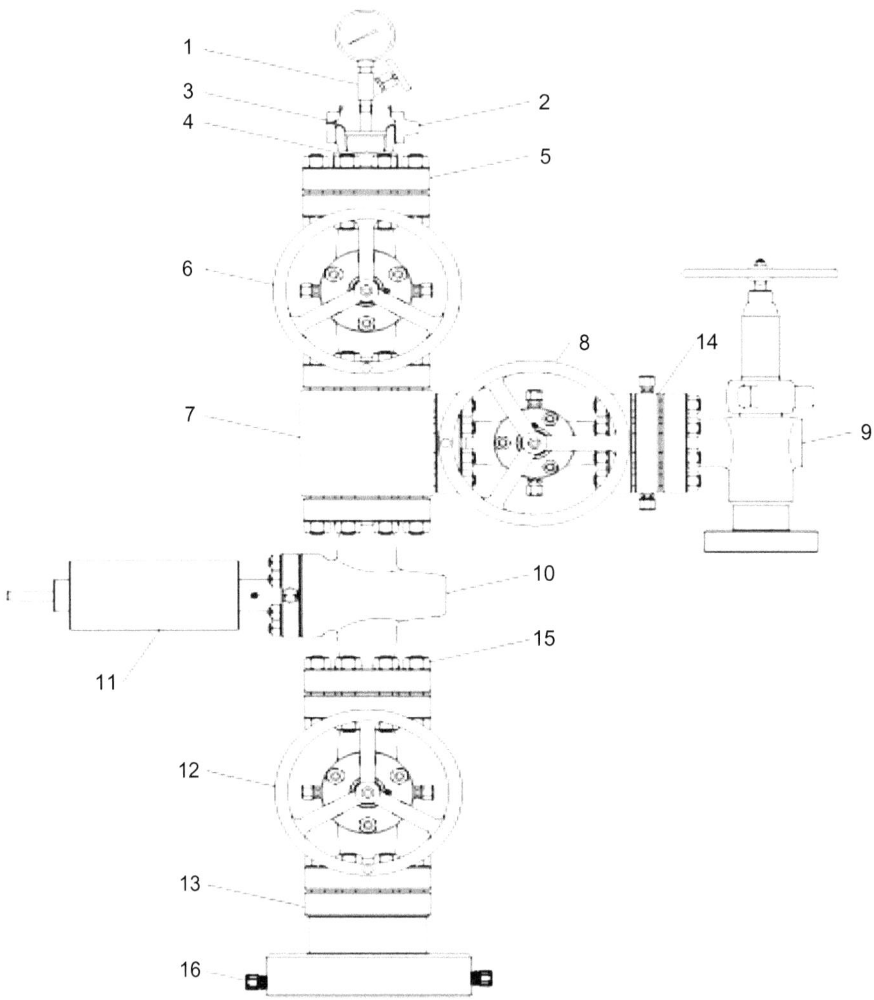

Fig. 7.5 Surface Christmas tree. *(Courtesy: API 6A.)*

- All USVs or valves that act as USVs, which could be a wing or master valve, should be designed and qualified according to API 6A and API 6AV1 standards.
- The operator must install, maintain, inspect, repair, and test USVs in accordance with the requirements of 30 CFR250, subsection H, as well as API 6AV2, "Installation, maintenance, and repair of surface safety valves and underwater safety valves offshore."
- The Christmas tree may contain two or more valves as USVs. In this case, the primary USV, called USV1, must be located upstream of the choke valve. If USV1 fails to operate properly or shows a leakage rate greater than the allowable leakage rate,

the switch from USV1 to USV2 (secondary USV) should take place. When two or more USVs are used, they may be placed in a single valve body.
- USVs may have nonstandard bore and face-to-face dimensions.

7.4 Boarding shutdown valve

Offshore boarding valves, also called boarding shutdown valves (BSDVs), are installed on subsea pipes and used to isolate production platforms and floating, production, storage, and offloading (FPSO) units from the riser side to protect the platform and personnel. Definition of riser is provided in Chapter 1. These valves may also be called riser isolation valves or riser shutdown valves. The barrier concept involves the use of a BSDV required by 30 CFR250, subpart H, titled "Oil and gas production safety systems." This requirement ensures a maximum level of safety for production facilities and for the people who work topside. The BSDV is known as the most critical component in the subsea sector of the oil and gas industry, and requires very strict design and testing criteria. The following requirements for BSDVs are taken from 30CFR 250 subsection H, sections 250.835 and 836:
- Ensure that the internal design pressure of the pipeline, riser, and BDSV is fully rated for maximum source pressure and compatible with the design requirements set forth in 30CFR 250, subpart H, unless the authorization company sets a new requirement.
- BSDVs shall be designed according to the requirements of API 6A and API 6AV1, which will be explained in the next section. Additionally, BSDVs shall be fire rated for 30 min and pressure rated for the maximum allowable operating pressure. API 6FA, which is a fire test standard, can be used for the fire safety test of the valves.
- A single BSDV can be located on the platform to protect multiple subsea flowlines, pipelines, and wells that are tied into a single pipeline riser.
- BSDVs may have nonstandard bore and face-to-face dimensions.
- The location of the BSDV should be within 10 ft of the first point of access to the boarding pipeline riser (e.g., within 10 ft of the edge of the platform if the valve is horizontal, or within 10 ft above the first accessible working deck).
- The operator shall install, inspect, maintain, repair, and test all new BSDVs as well as other BSDVs that are removed from service for manufacturing and service in accordance with API STD 6AV2, "Installation, maintenance, and repair of surface safety valves and underwater safety valves offshore." BSDVs shall be tested during operation on a monthly basis and not exceeding every 6 weeks. The valve should provide zero leakage. If a BSDV cannot comply with the zero-leakage requirement, it should be repaired or replaced.
- In any emergency situation that impacts the facilities' performances such as a storm, hurricane, ice, earthquake, etc., all subsea wells including all safety valves should be closed.

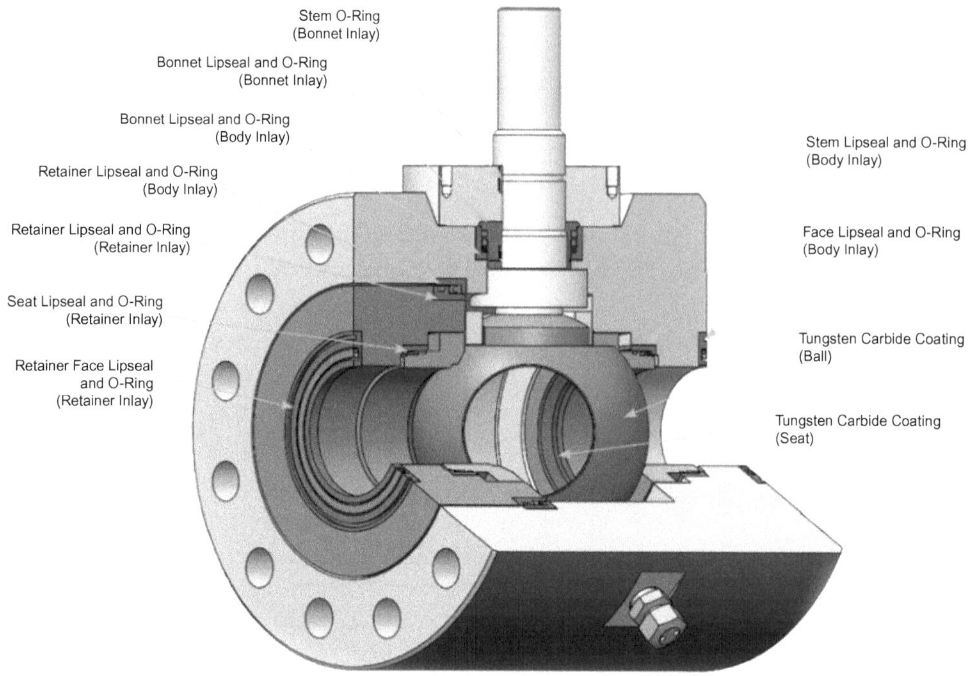

Fig. 7.6 BDSV, ball type. *(Courtesy: CORTEC.)*

Fig. 7.6 illustrates a 5 1/8″ ball valve in a pressure class of 15,000 psi with a fail-safe close actuator used as a BSDV and made based on API 6A, API 6AV1, and API 6FA fire safety test requirements. The BSDV should have metal-to-metal seat contact between the ball and seats as illustrated in Fig. 7.6. The red parts (dark gray in print versions) in the figure indicate Inconel 625 overlay and the blue (dark gray in print versions) indicates tungsten carbide hard facing.

7.5 SSV, USV, and BSDV requirements

7.5.1 Design

Safety valves shall be designed to operate, without damage to the safety valve or safety valve actuator, when the valve is actuated open or closed, pressurized or depressurized, under any internal valve body pressure within its pressure rating, and for USVs, under any external pressure up to sea pressure at the maximum depth rating.

7.5.2 Product safety level

The PSL indicates the different levels of requirement for quality control, qualification, testing, and documentation of materials. In addition, PSL affects the testing duration and requirements for subsea products such as valves. PSL is defined for five product specification levels as per API 6A and API 17D. More information about PSL is provided in Chapter 2. According to API 6A, 2018 edition, PSL1 is not applicable for safety valves. PSL2, PSL3, PSL3G, and PSL4 are product safety levels for safety valves. In practice, end users ask for PSL3G for safety valves in many cases. The important point is that API 6A specifies at minimum PSL2 for SSVs and USVs, and at minimum PSL3 for BDSVs. This means that higher product quality in general is expected from BDSVs as per the API 6A standard.

7.5.3 Actuation

Actuators for safety valves can either be hydraulic or electrical, single acting or double acting. More information about actuators is given in Chapter 10. A standard hydraulic pressure of 345 bar equal to 5000 psi for hydraulic actuators is very common. The temperature rating of an actuator should be between a minimum and maximum value. The maximum temperature range of an actuator should be at least 65°C. The minimum temperature range could be equal to the minimum value of the temperature classes provided in API 6A, or could be based on a value given in the project specification. In addition, actuator force should be sufficient to fully open and fully close the valve even in the most severe opening and closing conditions. A specific feature of SSVs and BDSVs is a heat-sensitive lock-open device to keep the valve in open position when the actuator is out of service for maintenance or not installed yet. A heat-sensitive lock-open device is defined as a device installed on an SSV actuator to maintain the SSV in fully open position until exposed to sufficient heat to cause the device to release and allow the SSV to close. The application of a heat-sensitive lock-open device is only applicable to SSVs and BDSVs. When included, the lock-open device shall keep the safety valve in fully open position at an atmospheric pressure up to 65°C when the valve body is pressurized to its rated working pressure and the actuator supply pressure is bled to atmospheric condition. Other types of openers may be used on the safety valves, such as a mechanical opener or hydraulic opener, to keep the valve in fully open position without an actuator.

7.5.4 Sand slurry testing

Sand slurry testing, according to both API 6A and API 6AV1, shall apply to three categories of safety valves: SSVs, USVs, and BSDVs during valve qualification. The details of the sand slurry test are defined in API 6AV1, "Specification for verification test of wellhead surface safety valves and underwater safety valves for offshore service." Sand slurry qualification is called PR2F in the API 6A standard. Three service classes are defined for safety valves in both API 6A and API 6AV1: standard service class I, and sandy service classes II and III. The sand slurry requirements are only applicable for sandy service class II and III safety valves.

The test descriptions and stages based on API 6AV1; 2008 edition are provided as follows:

The official test shall be applied on a 2 1/16″ flanged end with a 5000-psi rated working pressure. Successful completion of the official test qualifies all the sizes and pressure classes of SSV/USV/BSDVs of the same basic design and material for a qualification level of PR2F. Any change in design and material, especially those that affect the sealing mechanism, requires a new qualification.

After installation of the valves to the test section or bench, place the valve in closed position. Water at the working pressure of the valve and nitrogen at a pressure of 2000 psi should be applied to both seats of the valve in different stages. No leakage is allowed from the seat during the tests. After that, sand slurry should be circulated through the valve when the valve is in fully open position for 50 h. After 50 h, the same seat tests that were done in the beginning with nitrogen and water should be repeated, and again no leakage is allowed from the valve. The stabilization period during the leakage test at all stages should be 3 min. Again, sand slurry is circulated through the valve and the valve is moved from open to closed position and from closed to open position 500 times. The differential pressure across the valve is increased to approximately 400 psi. Again, the leakage test is applied to the valve and no leakage is allowed from the seats. Diagram 7.1 summarizes the slurry tests as per API 6AV1.

The following points should be considered during the sand slurry test for safety valves:

- Validation of an SSV or USV or BSDV with sand slurry would validate the other two types of the safety valves with the same design, type, and bore sealing mechanism.
- Actuated safety valves should be tested with the assembled actuator on the top.
- Scaling is applicable to this test, meaning that the validation of a safety valve in a product family validates other valves in the same product family. Valves with the same design configuration, functional operation, and design stress level are placed inside one product family. Testing and validation of a product family validates other product families that are one size larger or one size smaller, as well as products of the same pressure rating or less.

7.6 Subsurface safety valve

7.6.1 Introduction

Subsurface safety valves (SSSVs) are used in almost all subsea (offshore) wells and some onshore wells. An SSSV is a kind of safety valve installed in the production tubing down the well; it gets activated to stop uncontrolled well flow. The activation of an SSSV is normally done in combination with a surface safety valve (SSV). As illustrated in Fig. 7.1, there are two types of SSSV: Surface controlled subsurface safety valves (SCSSV) and Subsurface controlled subsurface safety valve (SSCSV). An SCSSV is a type of SSSV controlled from the surface and installed by wireline or as an integral part of the tubing.

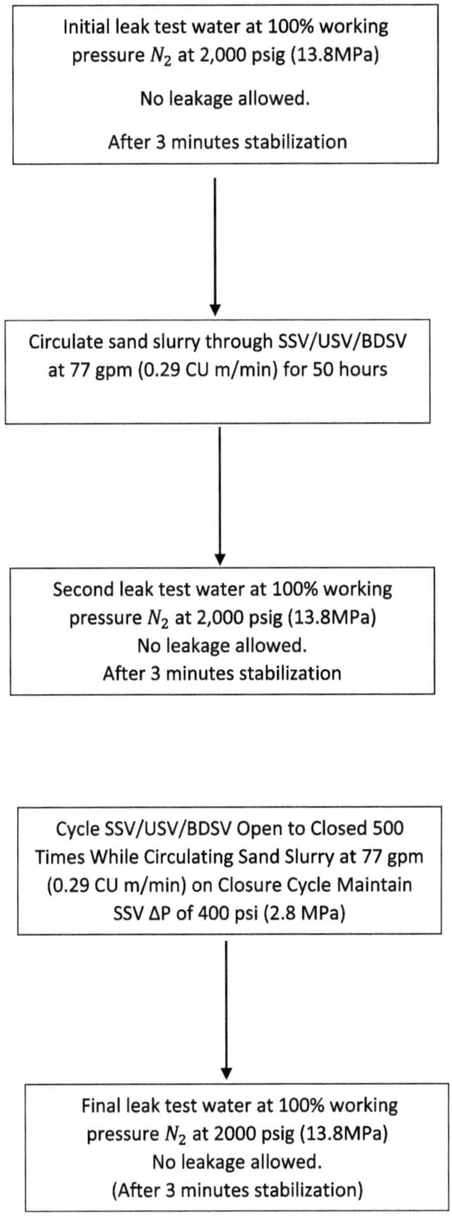

Diagram 7.1 PR2F sandy service SSV/USV/BSDV qualification program according to API 6AV1.

An SSCSV is opened and closed by the flow characteristic of the well, and acts as an excess flow valve or excess flow check valve. An SCSSV is normally controlled by hydraulic fluid via a control panel located on the surface. An SSCSV is designed to be closed automatically when a predetermined flow condition occurs in the well; the valve is opened and closed by pressure differential/flow velocity across the valve.

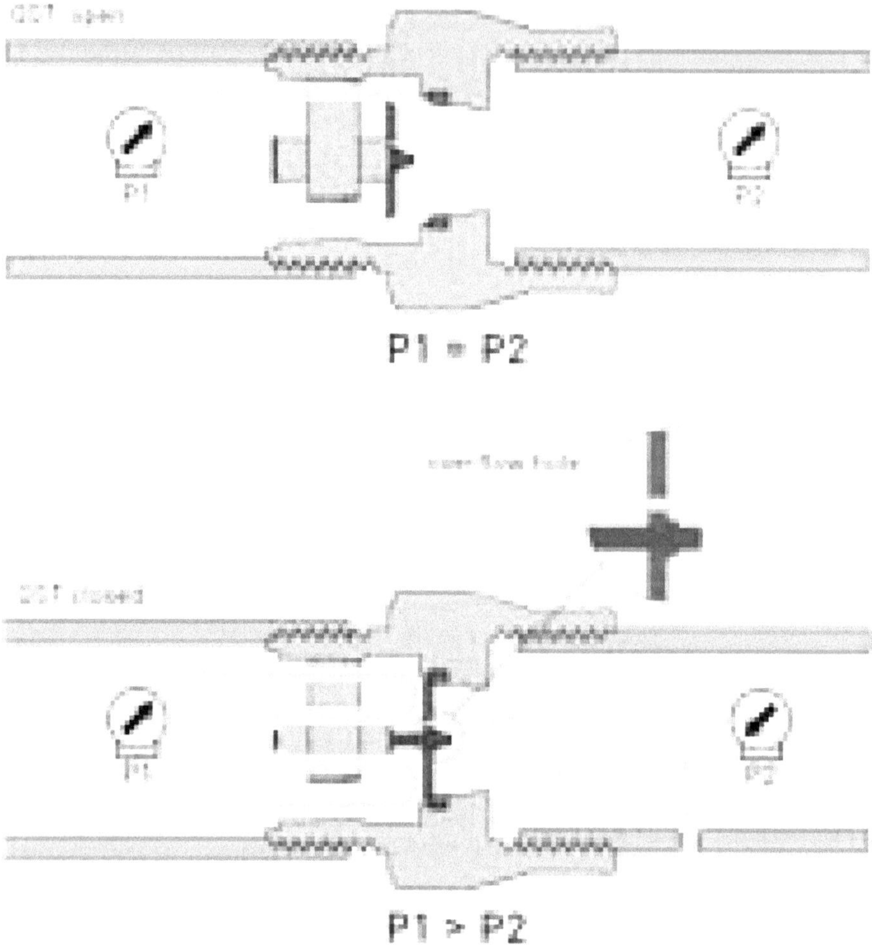

Fig. 7.7 Excess flow (check) valve mechanism.

The mechanism of an excess flow valve is shown in Fig. 7.7. The illustration at the top of Fig. 7.7 shows the valve in open position. The downstream and upstream fluid pressures are equal or almost equal, so the flow can move freely from upstream to downstream. However, a leakage in the downstream side, as illustrated in the lower part of the figure, reduces the pressure on the downstream side and an abnormal pressure differential is created between upstream and downstream. In this case, the high flow on the upstream side has a strong tendency to move to the downstream side, so this high amount of flow closes the valve. SSSVs may be installed in the well to prevent uncontrolled flow in the casing-tubing annulus; these SSSVs can be called annulus safety valves (ASVs). An ASV valve has the following characteristics:

- Provides a flow barrier in the annulus with fail-safe closing;
- Is normally controlled from the surface;
- Is most likely operated separately from the tubing SSSV;
- Allows semisubmergible pump installation;
- Takes the reservoir or well pressure from the downstream side and gas or water injection pressure from the upstream.

Most of the safety valves have an emergency shutdown function and are connected to an ESD system. The purpose of installing SSSVs is to protect humans and the environment, as well as production assets from blowout, human error in operation during the production of hydrocarbon, downhole equipment failure, fire or leakage. Downhole equipment are the facilities which are installed inside the well such as pumps. A blowout is defined as an uncontrolled release of crude oil and/or natural gas from an oil or gas well after failure of the pressure control system. Different reasons, such as change of pressure in the reservoir, can cause blowout. Nowadays, a blowout preventor is installed in the wells to prevent blowout occurrence.

Subsurface safety valves have evolved from their original 1940's design as a simple device; now they are complex devices that are integral with a well completion system. Downhole valves or SSSVs are installed in almost all wells. The requirements for these valves are specified in API 14A, called "Subsurface safety valve equipment." The installation of these valves is according to API recommended practice 14B, called "Design, installation, operation, test, and redress of subsurface safety valves systems." The safety valve could be tested with hydraulic pressure. It should be noted that intervention such as maintenance and repair of a subsea well is very difficult and costly—it can exceed many millions of dollars, not to mention loss of production, damage to the environment, and harm to people, so designing and selecting a reliable safety valve is a very essential task.

7.6.2 Basic closure mechanisms

Two basic closure mechanisms exist for SSSVs: one is a ball valve design and the other is a flapper or check valve assembly. The ball valve closure mechanism, illustrated in Fig. 7.8, contains a closure member in the shape of ball with a hole inside. When this hole is aligned with the production tubing, the flow inside the tubing passes through the valve.

Ball functionality can be weak in sandy services or in oil composition that has a tendency to generate wax and paraffin, so the other closure mechanism, called a flapper or check valve design (see Figs. 7.9–7.11), is more popular in the oil and gas industry. A flapper valve mechanism is less susceptible to malfunction than the ball type assembly. Debris or solids in the flow stream, as well as paraffin and scale, are less likely to prevent movement and functioning of a flapper check valve than a ball valve. In addition, a ball valve can be damaged more easily in the event of a dropped wireline tool or other equipment. A flapper type disk can be flat, curved, or contoured. Two types of curved and

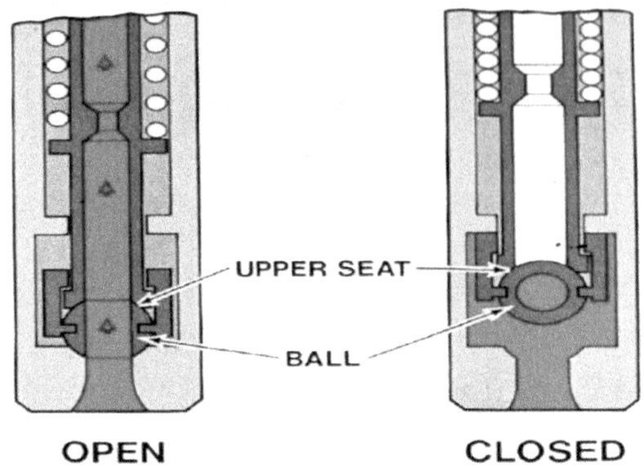

Fig. 7.8 Subsurface safety valve with ball type closure.

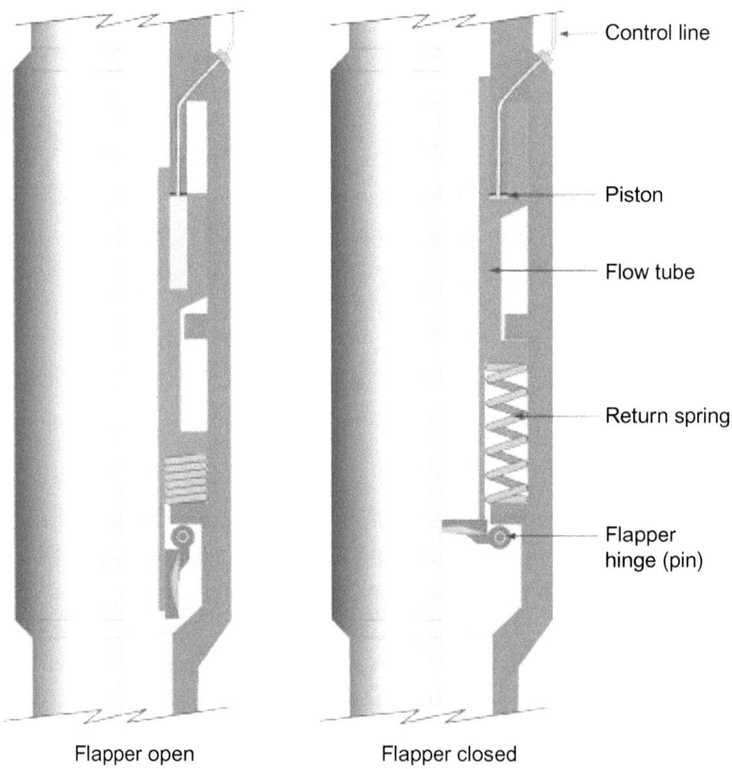

Fig. 7.9 Subsurface safety valve with flat flapper type closure.

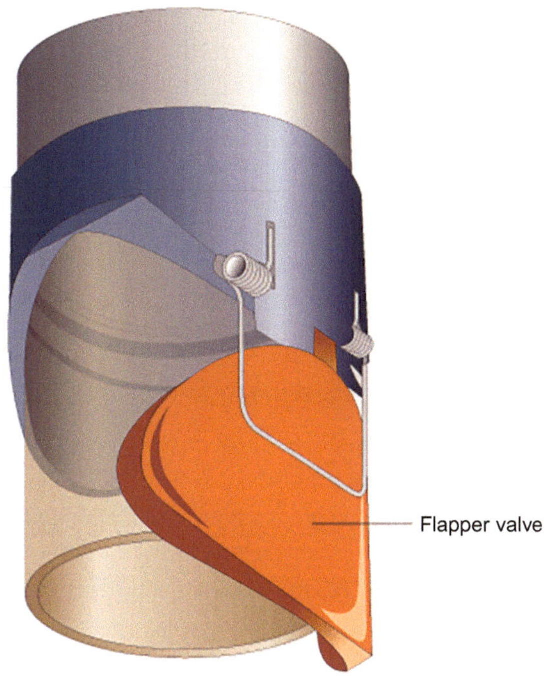

Fig. 7.10 Curved flapper or check valve on an SSSV in open position. *(Courtesy: Schlumberger.)*

Fig. 7.11 Countered flapper for an SSSV. *(Courtesy: Halliburton.)*

contoured flappers offer some advantages, such as providing very accurate and centralized closure of the valve and enlarging the flow area in the tubing. Fig. 7.11 illustrates a contoured flapper with excellent metal-to-metal sealing. The establishment of a high flow area through the safety valve is very important to minimize well flow pressure loss and well flow production reduction. Therefore API 14A proposes applying computational flow dynamics (CFD) as a part of the design verification to predict and evaluate safety valve performance under different conditions, like opening or slam closure, to ensure acceptable flow assurance conditions. Fig. 7.12 illustrates the result of CFD on a safety valve in different angles between the flapper and the vertical plane.

7.6.3 Retrieval tubing vs wireline (slickline)

Retrieval tubing is a type of subsurface safety valve that is run and retrieved as a part of production tubing. More information about the production tubing in the well is given in Chapter 1. Wireline retrievable safety valve is a type of safety valve in which the principal components can be run and retrieved by a wireline. Wireline is related to running a cable inside the well for logging or getting information from different formation layers inside the well.

The wireline retrievable is operated through a control line. A control line is located on the top of the piston and spring illustrated in Fig. 7.15, and when the control fluid is injected to this line as per Fig. 7.13, right-hand side illustration, the fluid enters the piston area and pushes the production tubing down. The control line is generally connected at the surface and also to an emergency shutdown system that automatically shuts down the well when an unsafe condition is detected in the well. There is normally a control panel with pressure gauges and a monitoring system at the surface host facilities. Therefore the spring gets compressed, the tubing pushes the flapper downward and makes the valve open and flow from the well moves inside the tubing. On the opposite side, when the fluid flow and pressure to the control line stop, the tube moves upward, the spring is expanded, and the flapper closes the valve and tubing bore. The SSSV flapper shown in Fig. 7.9 has a retrievable tubing mechanism, meaning that it works as an integral part of the production tubing string. The main benefit of this system is that the valve does not make any restriction against production in open position, as the valve internal diameter is equal to the tube internal diameter. The full diameter bore provided by this type of valve also permits access to the lower part of the wellbore with tools and instruments for flow control, monitoring, and other services. In summary, a retrievable tubing safety valve is installed to provide a flow barrier in the production tubing string; it contains three different main parts: housing, a closure element, and a control mechanism.

A wireline retrievable (slickline) safety valve (Figs. 7.13 and 7.14) is installed on a landing nipple. The nipple has some ports to provide communication between the

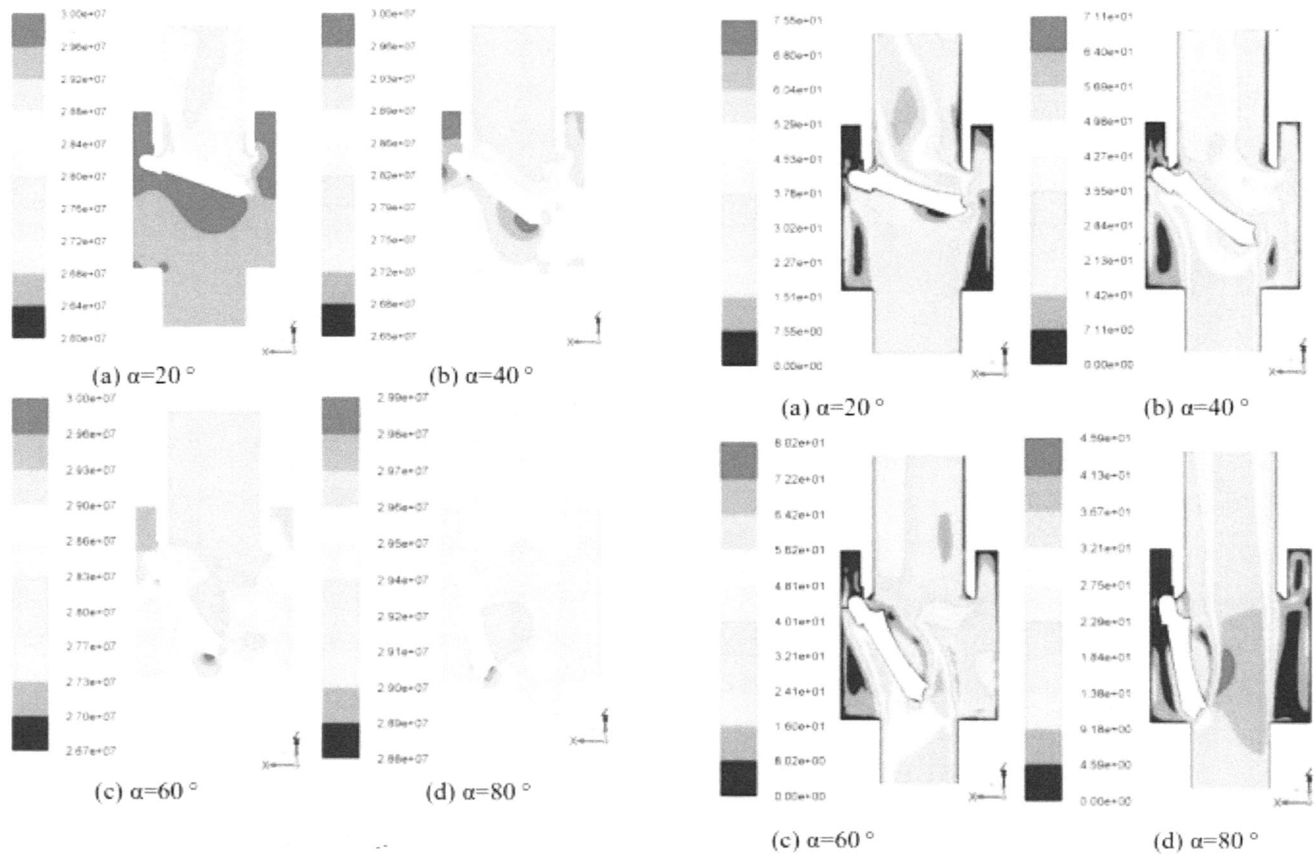

Fig. 7.12 CFD analysis on an SSSV at different angles of flapper with vertical bore. *(Courtesy: International journal of heat and technology.)*

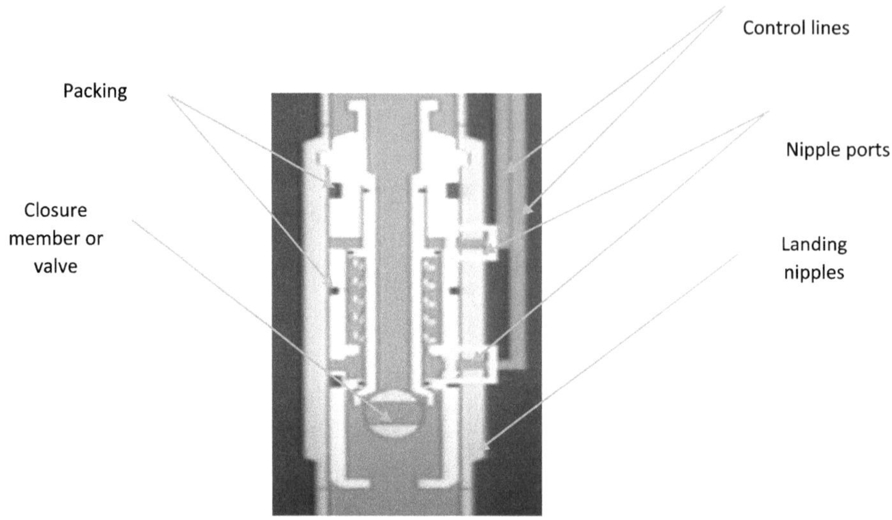

Fig. 7.13 Wireline retrievable safety valve.

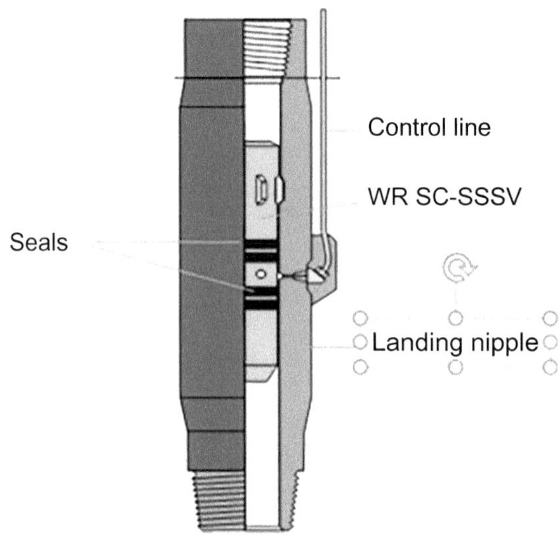

Fig. 7.14 Wireline retrievable safety valve.

control lines and the valve or valve closure member. Sets of packing are located between the safety valve and landing nipple. The disadvantage of a wireline retrievable safety valve is the smaller bore of the valve compared to the line. A wireline retrievable safety valve, like a tubing type valve, is installed to provide a flow barrier in the production tubing string and contains three main elements: housing, a closure element, and a control mechanism.

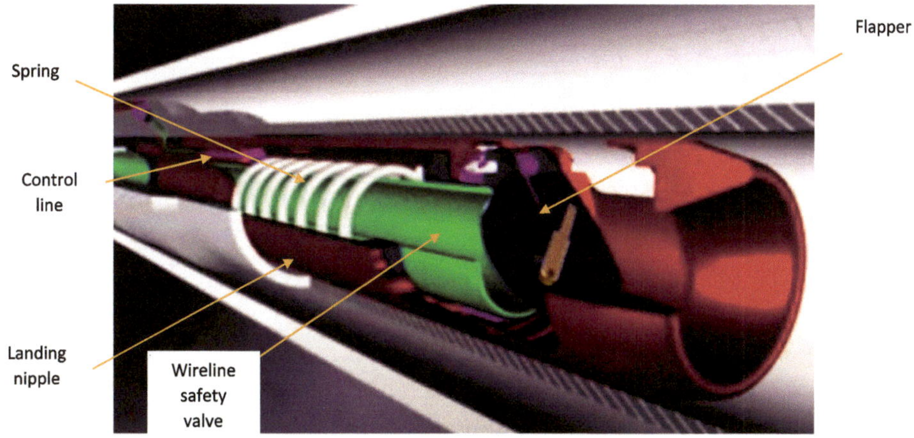

Fig. 7.15 Wireline retrievable safety valve during opening controlled from surface.

Fig. 7.15 illustrates a wireline retrievable SCSSV during opening of the flapper while hydraulic fluid is injected to the valve from a surface control line.

7.6.4 Self-equalizing capability

One of the main characteristics of an SSSV is self-equalizing capability. When an SSSV is closed, high differential pressure is applied across the valve closure member or element. Opening the valve in this condition can be difficult, and in some cases impossible, because a relatively large loading condition from the well is applied to the small surface of the valve. One of the fatal failures like human loss of life can happen during the opening of the valve. The failure occurs because of the extreme reservoir pressure accumulated on the bottom side of the valve. This problem can be more serious when high hydraulic pressure is supplied from the surface to open the valve. Having an SSSV without sufficient self-equalizing capability may deform or damage the valve during opening. In addition, a high amount of wearing and cavitation can occur on the valve closure member. Thus pressure equalizing could be necessary before opening the valve. One of the weak points of pressure equalization can be the provision of a leak path in the valve. Leak path creation in the valve means possibility of leakage can be occurred. The other weak point of self-equalizing capability is that it increases the erosion rate and accumulation of debris in the valve annulus. Thus SSSVs with self-equalizing capability should be made of erosion-resistant materials. The main advantage of equalizing is to have equal pressure upstream and downstream of the valve, which can return the well to production more quickly, reliably, and smoothly. Fig. 7.16 illustrates a pressure equalization mechanism for an SSSV. The mechanism shows a flow path, highlighted in yellow, from the downside of the flapper all the way to the top and passing the ball and spring to the tube.

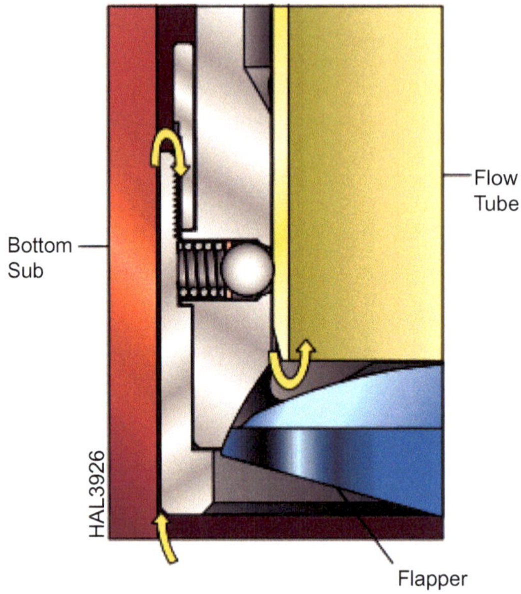

Fig. 7.16 Pressure equalization mechanism. *(Courtesy: Halliburton.)*

7.6.5 Hydraulic control system

A hydraulic control system is only applicable to SCSSs. Fluids that are used as hydraulic fluid are categorized into three main groups, summarized as water-based hydraulic fluids, synthetic oils, and mineral oils. SCSSVs are operated by hydraulic fluid that is controlled and supplied from the surface. The following important parameters should be taken into account during the selection of hydraulic control fluid:

1. *Cleanliness*: Hydraulic fluid should be clean and free from particles and solids. There are different applicable standards, such as SAE-AS-4059 and NAS-1638. The SAE standard is titled "Aerospace fluid power-cleanliness classification for hydraulic fluids." Hydraulic fluid used for hydraulic actuators and other applications should be clean without particle contamination. Different cleanliness classes for hydraulic fluid are defined in this standard. As an example, hydraulic fluid should maintain a cleanliness level of NAS-1638 class 7 or better, which provides hydraulic fluid with a filtering level of a maximum of 40-μm particle-containing fluids. Small amounts of debris and particles in the hydraulic system can lead to malfunction of the valve. Proper education and training of personnel is important in regard to this matter.
2. *Response time*: Response time is defined as the duration from the time the hydraulic fluid is released from the surface until the valve is opened, or more importantly, the duration from when the control system on the surface stops the hydraulic supply until closure of the valve. Many parameters affect the response time, such as control line

size and length, hydraulic fluid velocity and viscosity, well temperature, etc. As an example, cold temperature can increase the viscosity of the fluid, so it might be a good idea to select low-viscosity fluid for lower temperature environments.

3. *Temperature and pressure effects*: High temperature could disturb the functionality of the oil-based hydraulic system. In addition, the type of fluid should be thermally stable for high-temperature conditions. In general, safety valve control fluid should function properly within a wide range of pressure and temperature. Thus the control fluid must be compatible with all expected operating and design conditions.
4. *Fluid viscosity*: Hydraulic viscosity is an important factor, since it affects pressure loss in the control system as well as the system or valve response time. Very viscos hydraulic fluid is not allowed in deeper wells, since it slows down the response time and movement of the valve.
5. *Type of hydraulic fluid*: Hydraulic fluid can be water based or oil based. Water-based fluid is more environmentally friendly and may contain more than 50% water with some percentage of glycol. In addition, water-based fluids have lower viscosity and probably better movement and minimization of the valve response time. Thus water-based fluids may be more suitable for safety valves installed at deeper depths. However, they can cause downhole corrosion; thus any cross contamination of water-based hydraulic fluids with oil-based fluid should be prevented. The type of hydraulic fluid to be used should be given to the valve manufacturer in order to select the valve type and its material accordingly.
6. *Biological growth*: Both water- and oil-based hydraulic fluid may deteriorate due to biological and marine growth in the well. Therefore some inhibitors should be added to the hydraulic fluid to prevent the effect of bacteria and marine growth in the well on the hydraulic fluid.
7. *Long-term fluid stability*: It may last long time between hydraulic injection to the well till operation of the valve in subsea system. To provide a good hydraulic fluid stability, antioxidants or emulsifiers may be added to the hydraulic fluid.

7.6.6 Functional requirements

The valve purchaser shall define in detail the functional requirements of the SSSV through data sheets and specifications, as examples, that include the following data:

- Fluid or environmental conditions such as different corrosion possibilities and requirements such as sour service; ISO 15156 requirements as well as required erosion allowance due to particles and erosion;
- Different applied loads on the valve, possible cycling conditions, and the amount of opening and closing of the valve during its design life;
- Applicable regulations and industrial standards such as API 14A/API 14B;
- Design life (e.g., 20 years);

- Design and operation conditions, such as pressure and temperature; standard pressure ratings could be 5000, 10,000, and 15,000 psi as examples.

In addition, different classifications of SSSVs should be specified as follows:

- Tubing vs annulus safety valves;
- Tubing vs wireline;
- Subsurface vs surface-controlled system;
- Hydraulically, electrically, or mechanically controlled valve;
- Nonequalizing vs equalizing type valve;
- Ball vs flapper closure member.

7.6.7 Material selection

Material selection is largely dependent on the process fluid conditions. Production fluid, in general, is erosive and corrosive and has the potential to create scale and organic solids that can be coupled with very high pressure and high temperatures. Thus extra care must be taken to select suitable materials for the process conditions and design life. The wetted parts of the valve could be coated with a Teflon-based product applied to the surface to protect the metal from deposition and scaling. In addition, Polytetrafluoroethylene (PTFE) or Teflon can provide some degree of lubrication and reduce friction. An example of material selection for different components of an SSSV is given as follows:

- Some flow-wetted parts, such as the nipple, in 9Cr-1Mo, 13Cr martensitic stainless steel or AISI 4130 materials;
- Flapper in Inconel 718/Inconel 925;
- Flapper seat in Inconel 718/Inconel 925;
- Flapper pins in Elgiloy UNS R30003;
- Flapper spring in Elgiloy UNS R30003.

Additional material considerations, including requirements of ISO 156156 for sour service and hydrogen sulfide-containing environments, as well as applying scale-resistant materials to the valve wetted parts. All soft sealing materials, such as Teflon or Viton, or other rubber seals should be compatible with the hydraulic fluid and should be suitable for high temperature.

Elgiloy is a hard and corrosion-resistant cobalt alloy widely used for the spring of subsea valves instead of Inconel X750. Inconel X750 has high mechanical and fatigue resistance, but it can be corroded in seawater service. All of these material characteristics and applications are explained in Chapter 12. In general, all of these materials have high mechanical strength due to the high-pressure application in the wellbore.

The subsea sector of the oil and gas industry is moving toward deeper areas and harsher operating environments, which brings more challenges, especially in designing SSSVs. Material selection is critical for increasing valve endurance against corrosion degradation. In addition to more corrosion- and erosion-resistant material selection and tighter manufacturing tolerances, more restrictive tests are required to improve the reliability of safety valves in deeper water areas. The valve supplier should provide the following information for the traceable metallic components, such as their chemical composition percentage and limits, heat treatment conditions, melt practices, and mechanical properties such as tensile and yield strength, elongation, and hardness. The Charpy impact test may be required in some cases.

When it comes to nonmetallic materials, the valve supplier should provide specifications for their handling, packing, and storage, and should include the following information:

- Material type;
- Mechanical properties such as tensile, elongation, specific gravity, hardness, etc.
- Cure date, batch number, and material identification.
- Applicable testing standards.

7.6.8 Design verification

Secondary tools, per the API 14A definition, are tools and other equipment to be used with SSSVs to perform a secondary function such as permanent lock open or temporary lock open, etc. In general, each SSSV and secondary tool should undergo a design review, design calculation, and physical test to be reviewed and approved as a part of product qualification. Different factors, such as material properties, minimum and maximum operating and design pressure and temperature should be taken into account for the design review, calculation, and test. Design calculation and analysis could be performed in a software such as Ansys as a form of Finite element analysis (FEA). As it mentioned earlier in this chapter, CFD should be performed to ensure the flow assurance compatibility of the designed valve. If the valve should be used for a temperature greater than 177°C and rating working pressure (RWP) greater than 15,000 psi, this condition is called a high pressure-high temperature application. Historically, API 14A has defined four classes of SSSV: standard service, sandy service, stress cracking, and mass loss corrosion service. Now there are six validation grades, cross-referenced in Table 7.1 with the historical classes in the older revisions of API 16A. There is no clear definition about validation grades, but each validation grade could be referred to a type of service. As an example, if SSSV is used in a sandy service, it can be known as validation grade V4-2.

The validation grades hierarchy from high to low are V1-H, V1, V2, V3, V4-2, and V4-1. Table 7.2, extracted from API 14A, indicates the design validation grade coverage.

Design validation for safety valves may be performed via scaling, which means that a valve of a specific size, type, and model may be validated in design if the same product family by the same supplier has been designed and validated before. Scaling has some limitations as per API 14A, which are listed as follows:

Table 7.1 Cross-reference of service classes to validation grades.

Validation grade	Historical class of service (API 14A)
V4-1	Class 1—Standard service
V4-2	Class 2—Sandy service
V3	None—New
V2	None—New
V1	None—New
V1-H	None—New
NA	Class 3—Stress cracking service
	Class 4—Mass loss corrosion service

Courtesy: API 14A, 2015.

Table 7.2 Design validation grade coverage.

Validation grade	Grades covered
V1-H	V1-H, V1, V2, V3,V4-2,V4-1
V1	V1, V2, V3, V4-2, V4-1
V2	V2, V3, V4-2, V4-1
V3	V3, V4-1
V4-2	V4-2, V4-1
V4-1	V4-1

Courtesy: API 14A.

- Scaling shall not be used to validate products with higher rated working pressure (RWP) or a higher temperature design than the base case;
- The RWP shall not be reduced by more than 50% of the base design;
- The scaled valve should have the same design as that of the design validated valve;
- A valve with an equalizing characteristic cannot be scaled from a valve with a non-equalized feature.

7.6.9 Testing, storage, and preservation

After production and assembly of the SSSVs, the following checking and tests are proposed to be performed by the author based on industrial experiences:

- Cycling the valve 4–5 times by switching between pressurizing and depressurizing the valve with hydraulic fluid. The opening and closing time, as well as the pressure that leads to opening the valve, should be according to the valve data sheet or specification.
- Pressurizing the control line and checking that the valve moves to complete open position with hydraulic pressure in the control line.

- Testing the valve from below where the valve should provide sealing against the upstream flow.
- The pressure test shall not be done with water rather than hydraulic due to possible corrosion due to the water, and the possibility of getting different results by performing the tests.
- The safety valve should be stored in horizontal rather than vertical position to prevent flattening of the sealings.
- The safety valves should be fitted with end caps to protect the valve internals from ingress of any external object, dust, or particles, and the tubing connections for the control line should be plugged.

It should be noted that the API 14A standard covers many more tests beyond those mentioned in this chapter; these include functional tests, a temperature cycle test, a differential opening test, a self-equalizing test, a gas and liquid leakage test, operating life tests, a combined operation load test, a dynamic seal system test, as well as posttest activities such as a nondestructive test (NDT). Table 7.3 provides a general description of some of these tests.

Table 7.3 Test purposes of SSSV (API 14A).

Annex identification	Annex title	General description of content	Purpose
E	Alternative requirements for closure mechanism minimal leakage	Provides alternative leakage acceptance criteria for the functional test	Provides more stringent leak rate acceptance criteria
I	Extended sand endurance testing	Enhanced sand endurance testing	Evaluates the ability of the valve design to dose and seal in sandy conditions
J	Combined loads operational test	Validation of closed end rated performance envelope limits	Confirms the ability of the SSSV to operate at the limits of the performance envelope
K	Gas slam closure testing	Testing requirements for high-rate slam closures	Evaluates closure of SSSV in increased flow rate gas wells
L	Dynamic seal system test	Testing requirements for primary dynamic seal systems at intermediate positions at static conditions	Evaluates gas sealing integrity of the dynamic seal system

7.6.10 Installation/setting depth

Correct installation of subsurface safety valves to make sure that the well production flow is shut in during an emergency condition is very important. The setting depth of an SSSV installation is an important parameter that depends on factors such as the possibility of earthquake, hydrate and wax formation and scale formation. Also, the outside and inside diameter of the SSSV can influence casing and tubing configuration at the installed depth. Five different depth categories for installation of SSSVs are introduced as follows:

- Shallow depth, which is considered to be located between the surface or seabed and 300 m below the seabed or surface;
- Medium depth, which can be considered between 300 and 600 m below the surface;
- Deep depth, which is deeper than 600 m;
- Packer seat, which means that the SSSV is set at the depth of packer; production packers are an integral part of the completion process of an oil or a gas well. These are devices that seal the tubing to casing annulus and force the produced fluids from the wellbore into the completion tubing.
- Below packer: the SSSV is set below the production packer. A production packer is a standard component of the completion hardware of oil or gas wells used to provide a seal between the outside of the production tubing and the inside of the casing, liner, or wellbore wall.

The recommended minimum setting depth is below the cratering depth (or 50 ft below the deepest pile penetration depth on offshore structures or 100 ft below ground level on land locations, if deeper).

Some of the factors that should be considered for SSSV installation depth are listed as follows:

- To provide protection from abnormal undesirable events and conditions such as earthquake,
- To reduce tubing production loss;
- To minimize wax and hydrate formation inside the valve;
- To enable downhole jobs in the well without removing the valve;
- To place the valve as close to the reservoir as possible to provide downhole shut-in and safety means if there is a catastrophic event during well testing.

For tubing safety valves, deeper installation of the valve provides better protection of the well. However, the installation and maintenance of the valves becomes more challenging and time consuming. Additionally, the temperature at deeper depths is higher, which could affect the sealing materials such as polymeric sealing. Also, the response time of the actuator in case of control from the surface can be increased due to the valve installation at deeper depths in the well.

7.6.11 Operation manual

The supplier/manufacturer of the safety valve shall provide the following operational data as a minimum in the valve operation manual:

- The full operation data of the valve used for validation of design through testing and calculations;
- All parameters of the control system;
- All data required to keep the valves in fully open or closed position;
- All special tools needed for the installation and operation of the valve;
- All required training for the installation and operation of the valve.

7.7 Safety relief valves

7.7.1 Introduction

A relief valve is an excessive pressure relief device actuated by inlet static pressure having a gradual lift generally proportional to the increase in pressure over opening pressure. A pressure relief valve (PRV) is used to control or limit the pressure in a system. Conventional types of PRVs are equipped with spring housing suitable for closing the valve. PRVs are mainly used for discharging the extra pressure of liquid services. Fig. 7.17 illustrates a PRV including the main components.

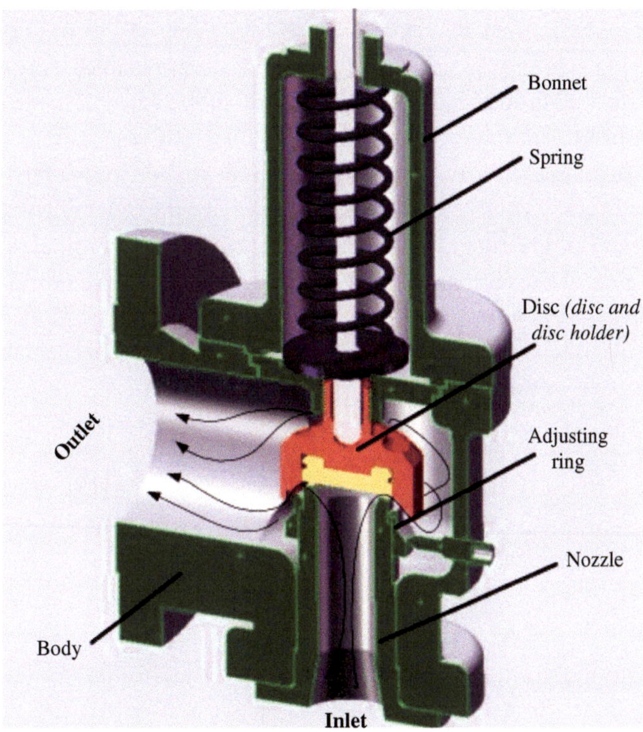

Fig. 7.17 Pressure relief valve including the main components.

This paragraph explains how the safety valve in Fig. 7.17 works. Fluid with excess pressure enters the inlet port of the check valve, also called the nozzle, overcomes the spring force, compresses the spring, pushes the disk up, and opens the valve. Afterward, the flow exits the valve through an outlet, so the spring is expanded and pushes the disk down toward the seat to close the valve. PRVs are used on the subsea valves' body/bonnet or on the compensation system. The compensation system is explained in Chapter 11.

The design of PRVs used for subsea applications and their compensation system is not necessarily the same as the valve illustrated in Fig. 7.17. An inline relief valve is shown in Fig. 7.18 which is has different concept of design and operation compared to the PRV illustrated in Fig. 7.17. Fluid with excessive pressure moves from the left side to the right, pushes the disk back, opens the valve, and drains from the safety bleed hole.

Three applications of this valve are reviewed in this part.

7.7.2 Cavity release of DIB1 valves (applicable for liquids)

Double isolation and bleed type 1 (DIB1) valves, as per the API6DSS definition, have two double piston effect seats, in the case of a ball valve or slab gate valve, or a double expanding gate valve. In fact, a ball valve or a slab gate valve requires to have two double piston effect seats to be a DIB1. However, a standard double expanding gate valve is DIB1 as per API 6DSS definition. A double piston effect seat provides sealing in both directions: one from the line toward the cavity and one from the cavity toward the line. Thus if fluid is trapped in the valve cavity, it cannot be released to the line. The trapped fluid could be either liquid or gas. Liquid fluid is not compressible, so accumulation and overpressurizing of the liquid could damage the body of the valve. Therefore the proposal is to have a pressure relief valve on the cavity of DIB1 valves, such as expanding gate valves, in order to release the overpressurized liquid in the cavity. Fig. 7.19 illustrates a PRV on the cavity of a double expanding gate valve. Fig. 7.20 illustrates a 3″ double expanding gate valve in pressure class of 2500 equal to 420 bar nominal pressure after the pressure test as a part of the factory acceptance test, installed on the liquid service (water)

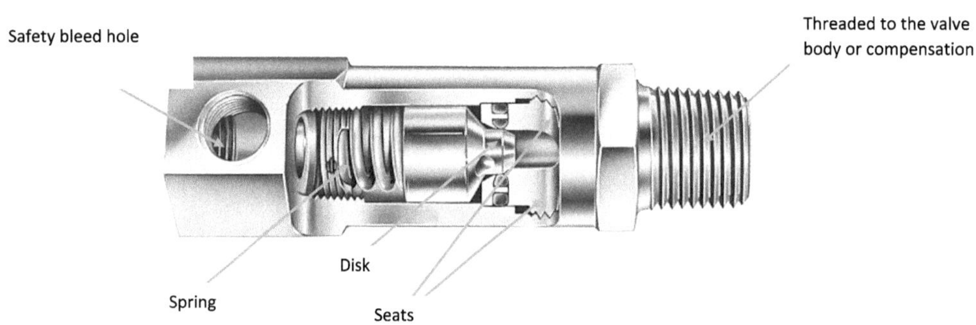

Fig. 7.18 Inline pressure relief valve.

Fig. 7.19 PRV on the cavity of a double expanding gate valve.

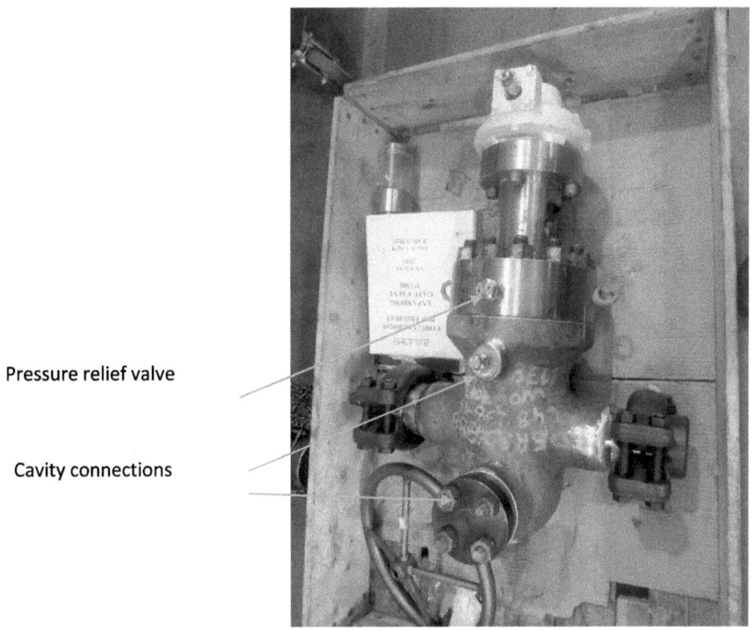

Fig. 7.20 3″ class 2500 expanding gate valve.

piping connected to and under a topside separator. The valve has a safety relief valve as well as a body cavity vent and drain connections.

7.7.3 Pressure relief on the body/bonnet of the valve

All actuators should be designed to prevent pressure buildup inside the actuator casing due to leakage from the valve. Thus a provision for relief of pressure buildup could be included on the valve bonnet to release the leakage from valve through the stem area to the external environment rather than to the actuator. Fig. 7.21 illustrates a pressure relief valve on the bonnet of a subsea slab gate valve to drain the possible leakage from the stem area to the environment and prevent actuator pressure buildup. Fig. 7.22 illustrates a safety relief valve installed on the bonnet of a slab gate valve, and a relief port. The relief port is located between the primary external stem sealing, which is a Glyd ring or O-ring, and a secondary external stem sealing, which is a lip seal. Any internal leakage from the secondary internal stem seal, a lip seal would pass the other lip seal with a secondary external stem sealing function, and would release to the relief port and drain to the water through the relief valve.

Fig. 7.21 PRV on the bonnet of a subsea slab gate valve.

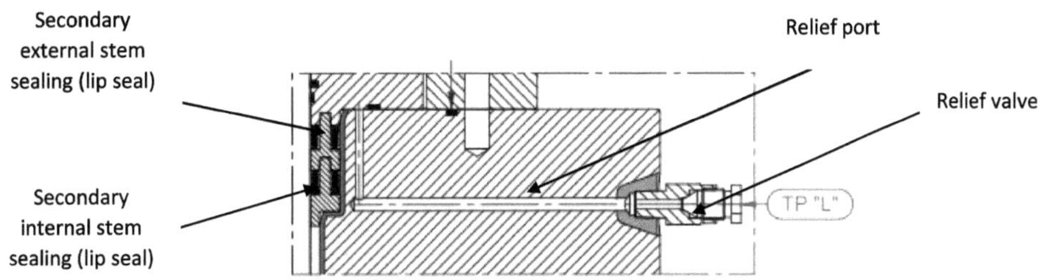

Fig. 7.22 Relief valve installation on the bonnet of a subsea slab gate valve.

7.7.4 Relief valves installed on pressure compensation

Subsea actuators, unlike valves, are not designed thick enough since the actuator shells are not pressurized by internal fluid. Since actuators cannot be designed with thicker walls due to cost and feasibility of design, an appropriate compensation system must be provided. Subsea actuators at depth can be imploded by the water head pressure if they are not properly pressure compensated. The pressure in the compensation system is in equilibrium with the seawater pressure at the installed depth. It is important to know that compensation systems are not limited to actuators; other components, such as the ROV bucket, gear box, and stem extension of the valve, could be pressure compensated. Fig. 7.23 illustrates a closed pressure compensation system installed on the linear actuator of a TCG valve.

The closed compensation system illustrated in Fig. 7.24 is supplied with two inlet and outlet pressure relief valves. Two pressure relief valves should be installed on a closed

Fig. 7.23 Closed compensation system for a gate valve linear actuator.

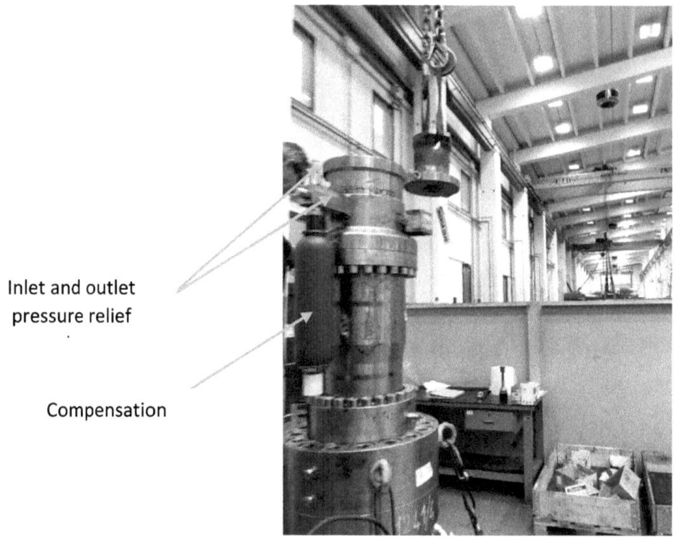

Fig. 7.24 Inlet and outlet pressure relief valves on the pressure compensation system.

pressure compensation system in order to keep the internal pressure equal to the seawater head pressure. One of the relief valve's functions is to release the internal pressure of the cavity to the seawater if the cavity pressure exceeds the seawater pressure by 5 or 7 bar as an example; this type of valve is called an outlet relief valve. The other pressure relief valve could release the seawater pressure to the internal capsule if the seawater pressure exceeds the internal compensation pressure by 5 or 7 bar. This relief valve can be called an inlet relief valve. One of the main considerations for pressure relief and safety valves is to calculate the set pressure.

The set pressure, which is measured in pounds per square inch (PSIG), is the pressure at which a safety or pressure relief valve opens. The set pressure of the valve must not exceed the maximum allowable working pressure (MAWP) of the boiler or other vessel. Assuming a relief pressure of 6 bar as an average for the safety valves, it is possible to calculate the set pressure through Eq. (7.1):

PRV set pressure calculation

Set pressure = relieving pressure − over pressure − atmospheric pressure

Over pressure = 10% set pressure

Atmospheric pressure = 1 bar

$$\rightarrow \text{Set pressure} = \text{relieving pressure} - 0.1\,\text{set pressure} - 1\,\text{bar} \rightarrow 1.1\,\text{set pressure}$$
$$= \text{relieving pressure} - 1 \rightarrow \text{Setpressure} = \frac{\text{relieving pressure} - 1}{1.1} = 5/1.1 = 4.55\,\text{bar} \quad (7.1)$$

The other important parameter that should be considered is the minimum discharge area for the PRV; this can be calculated according to Eq. (7.2).

Minimum required discharge area for PRV

$$A = \frac{15.9Q\sqrt{G}}{K_W K_V \sqrt{\Delta P}} \tag{7.2}$$

where:

A: minimum required effective discharge area of PSV (mm^2);
G: specific gravity of the liquid flowing (either seawater or compensation oil);
Q: required relieving capacity at flowing temperature (mm^3/s);
ΔP: Relieving pressure (kPa);
K_V: Flow correction factor due to the viscosity of the fluid at flowing conditions;
K_W: Capacity correction factor due to back pressure;
K_V can be calculated according to Reynolds number according to Chart 7.1.

The Reynolds number is the ratio of the fluid inertial force to its viscous force, which is calculated according to Formula (7.3). It is possible to determine whether the fluid is laminar or turbulent from the Reynolds number. A Reynolds number greater than 4000 is turbulent, and less than 2300 is laminar. Any Reynolds number between 2300 and 4000 is categorized as transient flow.

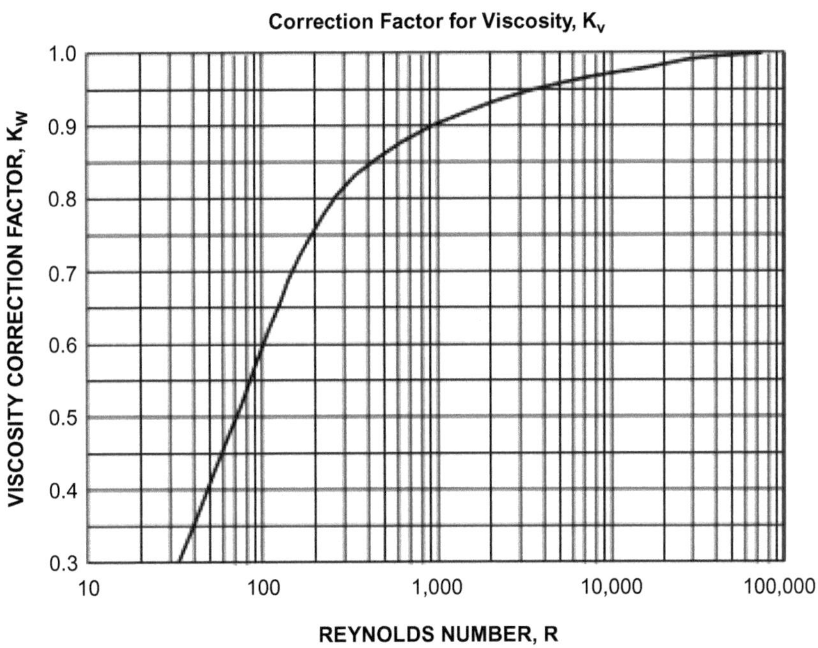

Chart 7.1 Flow correction factor based on Reynolds number.

Reynolds number calculation

$$R = \frac{\text{inertial force}}{\text{viscous force}} = \frac{\rho v d}{\mu} \tag{7.3}$$

where:

R: Reynolds number (unitless);
ρ: Fluid density (kg/m^3);
V: Velocity (m/s);
D: Pipe or valve diameter (m);
μ: Viscosity of the fluid (Pa s).

7.8 Subsea isolation valves

The function of a subsea isolation valve (SSIV) is similar to that of a BSDV: to protect the platform and personnel from unintended release of hydrocarbon. Like a BSDV, an SSIV is an actuated valve with a fail-safe closed mechanism that is connected to an emergency shutdown (ESD) system. The actuator should have a safety integrity level requirement (SIL) level of 3, as an example, according to IEC 61508 and IEC 61511 requirements. Unlike a BSDV, which is installed on the surface, an SSIV is installed subsea as a stand-alone component connected to the pipeline or inside a structure such as a pipeline end manifold (PLEM) (see Figs. 7.25 and 7.26). An SSIV is normally a ball valve with a hydraulic or electrical retrievable actuator. An actuator retrievable tool (ART) is used to dismantle and remove the actuator from the valve. The SSIV could be DIB2 with one double piston effect seat, or DIB1 with two double piston effect seats to provide more reliable isolation.

Fig. 7.27 illustrates a schematic for an SSIV and its installation structure. Fig. 7.27 shows more detailed information about the SSIV connection to the flowlines, the umbilical used for actuation as well as the foundation. The distance from the SSIV to the

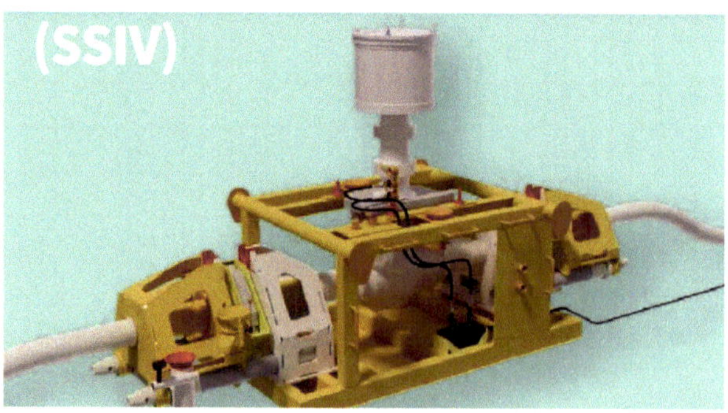

Fig. 7.25 SSIV installed in a PLEM structure in *yellow*. *(Courtesy: NOV.)*

Fig. 7.26 SSIV inside the structure template in *yellow* color before installation. *(Courtesy: iStudio technologies.)*

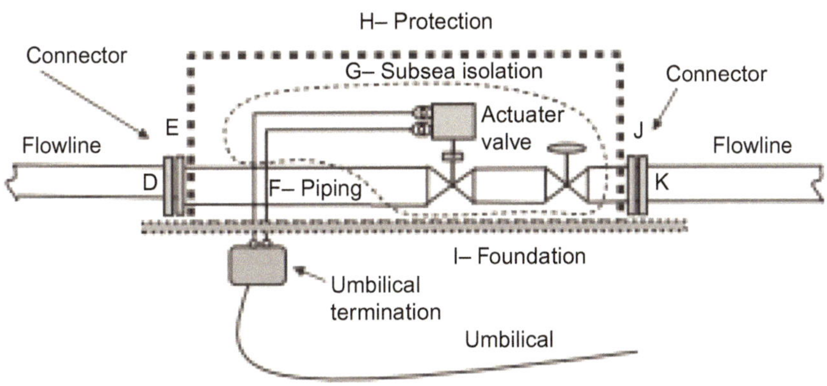

Fig. 7.27 Schematic of an SSIV in a subsea production system. *(Courtesy: Society for Underwater Technology.)*

platform should be defined accurately to protect the platform from fire and jet fire effects. A jet or spray fire is a turbulent diffusion flame resulting from the combustion of a fuel continuously released with some significant momentum in a particular direction or directions. Jet fires can arise from releases of gaseous, flashing liquid (two phase) and pure liquid inventories.

The main components of a SSIV system are listed as follows:

- On-off bare stem ball valve
- Hydraulic actuator with spring return and fail-safe close function (e.g., rack and pinion actuator)

- Base support and protection cover
- Piping and connected flanges
- Control umbilical
- Control system
- J tube and its top flange for umbilical hang-up

7.9 Self-assessment questions

1. Which sentence is correct about safety valves?
   A. Safety valves are used both underwater and on land, and they are always according to API 6A and API 17D requirements.
   B. Pressure relief valves are always used on the body cavity of a double isolation and bleed type 1.
   C. Subsurface safety valves can either be operated or controlled from the surface through a hydraulic control line, or from the subsurface, as an example, through pressure and velocity in the well.
   D. One pressure relief valve is sufficient to be installed on a compensation system to provide pressure balance with seawater head.

Answer: Option A is not completely correct. Safety valves can be installed in the water or on the land. However, the governing standards for these valves are not limited to API 6A and 17D. API 14A is applicable for subsurface safety valves. Other standards, such as API 521 or API 527, are applicable for pressure safety valves. Option B is not completely correct either, since a pressure relief valve is installed only on the body cavity of double isolation and bleed type 1 (DIB1) valves that handle liquid services. There is no requirement for a pressure relief valve on DIB1 valves in gas service, since gas is compressible. Option C is correct, since SSSVs could either be controlled from the surface (these are called SCSSVs) or from the subsurface (these are called SSCSVs). Option D is incorrect, since the compensation system installed on the actuator spring chamber normally has two relief valves for providing pressure balance.

2. Which sentence is correct about safety valves?
   A. Underwater safety valves (USVs) and surface safety valves (SSVs) are always installed on a vertical line on the Christmas tree.
   B. USVs and SSVs are normally closed and are opened in case of emergency.
   C. Boarding shutdown valves (BSDV) and surface safety valves (SSV) are installed above the sea level.
   D. A ball type closure for subsurface safety valves is more reliable than a flapper type.

Answer: Option A is not correct, since USVs and SSVs can be installed on a horizontal line as well as a vertical line on Christmas trees. Option B is wrong, since USVs and SSVs are normally open, and are closed by the actuator in case of emergency. Option C is correct, since BSDVs and SSVs are installed above the sea level. Option D is wrong, since a ball type closure for an SSSV is less reliable than a flapper type.

3. Which sentence about additional tests on SSSVs is wrong?
   - **A.** The purpose of a sand slurry test is to evaluate operation and sealing of the valve in sandy services.
   - **B.** Gas slam closure testing confirms the smooth closure of SSSVs when the flow rate of gas in the wells is high.
   - **C.** The combined loads operational test confirms the ability of the SSSV to perform and operate during a combination of loads.
   - **D.** The static seal system test evaluates the sealability of the static seals in the gas service.

Answer: All options are correct except for option D. The correct test is a dynamic seal test to evaluate the performance of dynamic seals in the gas service. Dynamic seals are defined as seals that are placed between at least one moving part of the valve. Dynamic seals are more exposed to wearing and leakage compared to static seals, since they are between at least one moving part. On the other hand, static seals are between nonmoving or rotating components.

4. Which sentence is wrong about hydraulic fluid for control of SCSSVs?
   - **A.** The cleanliness of the hydraulic fluid is important, since debris could disturb the function of the valve.
   - **B.** Many parameters affect the response time of the valve opening, such as control line size, hydraulic velocity, hydraulic viscosity, well temperature, etc.
   - **C.** Marine growth in the well cannot affect the hydraulic supply to the valve installed in the well.
   - **D.** The type of hydraulic fluid is not selected by the valve manufacturer.

Answer: Option A is correct, since debris in the hydraulic fluid can disturb the functionality of SSSVs. Option B is correct, since all the mentioned parameters affect the response time of the valve opening. Option C is wrong, since marine growth does affect the quality of the hydraulic fluid supplied to the well and the valve. Option D is correct since the end user or purchaser should identify the type of hydraulic fluid for operation of the valves in the specification or valve data sheets.

5. Which sentence is wrong about material selection for SSSVs?
   - **A.** High mechanical strength materials such as AISI 4130, Inconel 718, and Inconel 925 are common for the wetted parts and closure members of SSSVs.
   - **B.** Elgiloy is a corrosion- and erosion-resistant spring material.
   - **C.** Materials for sour, hydrogen sulfide services should be compatible with the ISO 15156 standard.
   - **D.** Soft materials such as Viton, Buna N, etc. are not applicable for SSSVs.

Answer: Options A, B, and C are correct. Option D is not correct since soft sealing is applicable for SSSVs.

6. Which sentences are correct about the actuators of SSVs/BSDVs/USVs?
   A. Hydraulic actuators are the only type used for these valves.
   B. The actuators for these valves are spring return with fail-safe closed and open functions.
   C. There is no standard temperature class for the actuators, but the hydraulic supply pressure of 5000 psi equal to 345 bar can be considered a standard pressure.
   D. Lock open devices can be used to keep the valve open during actuator maintenance.

Answer: Actuators for safety valves can be hydraulic or electrical, so option A is not correct. Actuators for safety valves only have a fail-safe closed function, so option B is not correct. There is no standard temperature class for actuators and 5000 psi equal to 345 bar hydraulic pressure could be considered a standard pressure class for the actuators. Therefore option C is correct. Option D is also correct, since lock open devices keep the valves open while the actuator is dismantled from the valve.

7. Which sentence is not correct about subsea isolation valves?
   A. SSIVs are connected to an emergency shutdown system.
   B. SSIVs are always installed in a structure like a PLEM.
   C. SSIVs could be double isolation and bleed type 1 or 2 to provide better isolation.
   D. SSIVs' main function is to protect the host platform or ship, facilities, and personnel.

Answer: Option A is correct. SSIVs are connected to an ESD system, so the requirements of safety integrity level and safety analysis report preparation based on IEC 61508 and IEC 61511 are applicable to these valves. Option B is not completely correct, since an SSIV could be installed on the pipeline as a stand-alone component without any need to have a structure. Option C is correct and SSIVs could have nonstandard sealing like one seat double piston or both seats double piston effect to provide better sealing. In addition, the actuators of SSIVs are normally retrievable. Option D is correct since the main function of SSIVs is to provide a safety barrier for the protection of the platform or ship, host facility, and personnel.

Further reading

American Petroleum Institute (API) 6A, Specification for Wellhead and Tree Equipment, twenty-first ed., API, Washington, DC, 2018.

American Petroleum Institute (API) 6AV2, Installation, Maintenance, and Repair of Surface Safety Valves and Underwater Safety Valves Offshore, first ed., API, Washington, DC, 2014.

American Petroleum Institute (API) 14A, Specification for Subsurface Safety Valve Equipment, twelfth ed., API, Washington, DC, 2015.

American Petroleum Institute (API) 17D, Design and Operation of Subsea Production Systems, Subsea Wellhead and Tree Equipment, second ed., API, Washington, DC, 2011.

Crosby Valve Inc (Ed.), Pressure Relief Valve Engineering Handbook, first ed., Crosby Valve Inc, Wrentham, MA, 1997.

Drilling for Gas, [online] Available from:https://drillingforgas.com/completion/design/safety-valves.html, 2020. Accessed 26 August 2020.

International Electrotechnical Commission (IEC), IEC 61508 Functional Safety of Electrical/Electronic/ Programmable Electronic Safety-Related Systems Part 1: General Requirements, second ed., (2010) Geneva Switzerland.

International Electrotechnical Commission (IEC), IEC 61508, Functional Safety of Electrical/Electronic/ Programmable Electronic Safety-Related Systems, Part 4: Definitions and Abbreviations, second ed., (2010)Geneva, Switzerland.

International Electrotechnical Commission (IEC), IEC 61508, Functional Safety of Electrical/Electronic/ Programmable Electronic Safety-Related Systems, Part 2: Requirements for Electrical/Electronic/Programmable Electronic Safety-Related Systems, second ed., (2010) Geneva, Switzerland.

International Electrotechnical Commission (IEC), IEC 61508, Functional Safety of Electrical/Electronic/ Programmable Electronic Safety-Related Systems, Part 6: Guidelines on the Application of IEC 6158-2 and IEC 6158-3, second ed., (2010) Geneva, Switzerland.

International Organization of Standardization (ISO) 10423, Petroleum and Natural Gas Industries- Drilling and Production Equipment-Wellhead and Christmas Tree Equipment, fourth ed., (2009) Geneva, Switzerland.

Legal Information Institute, Code of Federal Regulations (CFR). Title 30. Mineral resources. Chapter II. Bureau of Safety and Environment Enforcement, Department of Interior, Subchapter B. Offshore, Part 250 Oil and Gas and Sulphur Operations in the Outer Continental Shelf. Subpart H. Oil and Gas Production Safety Systems, (2012).

B. Nesbitt, Handbook of Valves and Actuators: Valves Manual International, first ed., Elsevier, Oxford, 2007.

V.N Singh, S. Khandelia, SCSSV Training Manual, Well Services Mumbai, 2003.

United States Department of the Interior Minerals Management Service, Notice to Lessees and Operators of Federal Oil, Gas and Sulphur Leases in the Outer Continental Shelf, Gulf of Mexico Region. NTL No. 2009-G36, (2010).

CHAPTER EIGHT

Flow control valves

Contents

8.1 Introduction

Some valves are suitable for flow control purpose, also called regulation or throttling of the fluid. A flow control valve adjusts the amount of flow passage through the valve and changes the flow rate inside a valve; this has an impact on other fluid process parameters such as temperature, pressure, and level. Globe valves have been widely used in different sectors of the oil and gas industry excluding subsea. Globe valves are covered mainly by American Petroleum Standard (API 602), titled "Gate, globe, and check valves for sizes DN 100 (4″ size) for the petroleum and natural gas industries." Other standards include BS 1873 "Specification for steel globe and globe stop and check valve (flanged and butt-welding ends) for petroleum, petrochemical and allied industries" as well as ISO 15761 "Steel gate, globe and check valves for sizes DN 100 and smaller, for the petroleum and natural gas industries." Although some books and standards such as the API 615 valve selection guide may allow the usage of globe valves for fluid isolation, it is not common to use globe valves solely for that purpose in the oil and gas industry. Alternatively, ball and other types of gate valves such as wedge, slab, and expanding gate are suitable options for fluid isolation or stoppage.

Subsea Valves and Actuators for the Oil and Gas Industry
https://doi.org/10.1016/B978-0-323-90605-0.00014-1

Fig. 8.1 A T-pattern globe valve 4″ class 300 during the test.

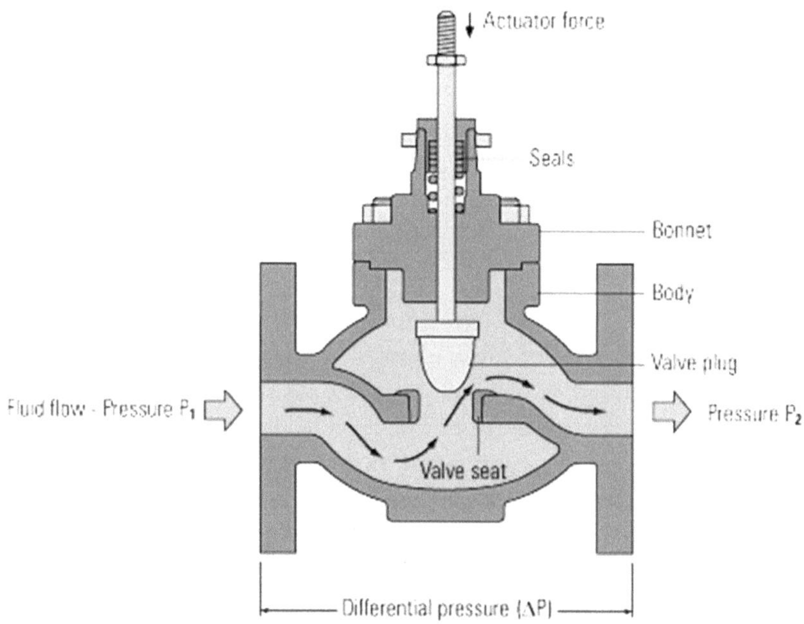

Fig. 8.2 A tee pattern globe valve. *(Courtesy: Valve World Magazine/Elsevier.)*

T-pattern (or tee pattern) globe valves (illustrated in Figs. 8.1 and 8.2) are very common in the oil and gas industry. The flow enters the valve illustrated in Fig. 8.2 from the left side and reaches the middle part of the valve where the plug (disk) and seats are located. The disk is off the seat so the fluid makes two 90-degree turns and leaves the valve from the outlet port highlighted with pressure P_2. These two 90-degree rotations of the fluid result in substantial pressure drops inside the valve. There is a narrow area below the plug where the velocity of the fluid is very high and the pressure is very low. The pressure in this narrow area can drop below the vapor pressure of the liquid which causes some bubbles to be formed. These bubbles of gas which are separated from the liquid in the narrow area due to the high pressure drop can recover the pressure and collapse firmly in a way that creates pressure waves. This is called cavitation and can damage different parts of the globe valve such as its seats and plug. Cavitation can be considered a type of corrosion; it can lead to metal loss or create pits on the valve trims (internals). Fig. 8.3 illustrates the effect of cavitation on a globe valve plug.

The American Petroleum Institute (API) released the first edition of its new globe valve design standard (API 623) in 2013 to control and avoid operation problems in globe valves such as cavitation, vibration, and leakage. API standard 623 is called "Steel globe valves—flanged and buttweld ends, bolted bonnets." This standard covers the following features of globe valves: bolted bonnet, outside screw and yoke, rotating rising and

Fig. 8.3 Cavitation on the plug of a globe valve. *(Courtesy: Valve World Magazine.)*

nonrising stems, rising and nonrising handwheels, Y-pattern type globe valves from 2″ to 24″ and pressure classes of 150 (pressure nominal (PN) = 20 bar), 300 (PN = 50 bar), 600 (PN = 100 bar), 900 (PN = 150 bar), 1500 (PN = 250 bar), and 2500 (PN = 420 bar). API 623 standard specifies hard facing on the seat, plug, and guided disk, especially for high-pressure classes. The stem diameter in API 623 follows the principles of API 600 "Cast steel gate valves standard" with different values. The stem diameter values in API 623 are larger than other globe valve standards like BS 1873 to avoid stem and plug separation (breaking). API 602 and ISO 15761 cover globe valves for sizes up to and including 4″. However, it should be noted that globe valves are not used for subsea applications due to the risk of cavitation. Cavitation is one of the most important operation problems and challenges for subsea valves used for flow regulation such as choke valve. Y-pattern globe valves produce much less pressure drop compared to T-type globe valves, and are thus much less exposed to cavitation risk (see Fig. 8.4). As an alternative to globe valves, needle valves and axial control valves are used for flow regulation purposes. Choke valves are also used for production fluid control and flow control purposes which are discussed in Chapter 6.

8.2 Needle valves

8.2.1 Introduction

The name of the needle valve is taken from the valve plug, also called a closure member, plunger, or stem tip, which is in the shape of a needle. The remarkable feature of this valve is the long and tapered needle-like point at the end of the stem. Needle valves are typically used to control flow in small piping sizes such as between ¼″ and 2″ and for small amount of flow rate regulation consequently. There are many similarities between needle and globe valves, but the main difference between these two valves is

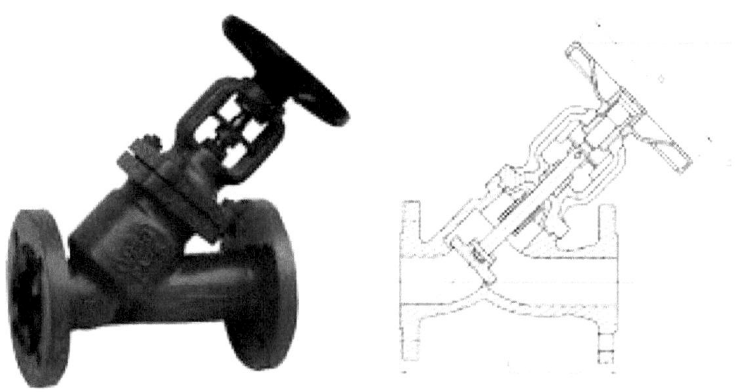

Fig. 8.4 Y-type globe valve. *(Courtesy: Rako Valve.)*

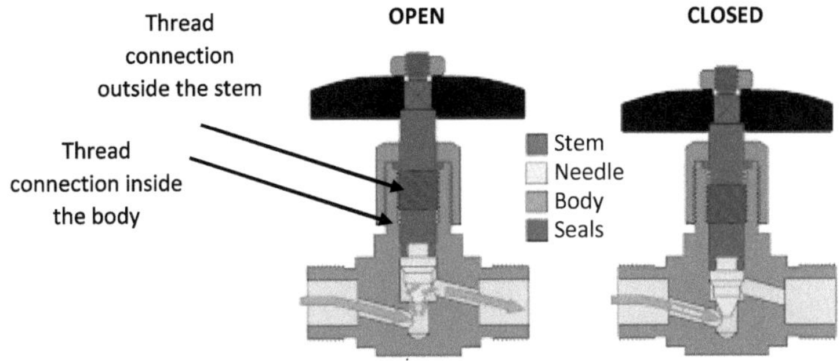

Fig. 8.5 Needle valve in open and closed positions. *(Courtesy: Assured Automation.)*

that a needle valve is normally used in small sizes and has a higher accuracy of flow regulation than a globe valve. Fig. 8.5 illustrates a needle valve in open and closed positions. The needle moves upward and off the seat so the flow in green color (dark gray in print versions) enters the valve from the left side, passes the plug, and exits the valve from the right side in open position. The needle is on the seat in closed position and blocks the fluid when closed so the fluid flow in green color (dark gray in print versions) gets stuck below the needle in closed position. A needle valve has the capability to regulate and isolate the flow from the opposite direction where the fluid may enter the valve from the connection on the right and move out from the connection on the left. The first important point is that needle and stem are normally a one-piece component. Additionally, rotating the bar in the black color on the top of the valve makes the stem move up through the thread connections on the outside of the stem and inside the body of the valve. Subsea needle valves have different applications such as chemical injection and hydraulic connection and distribution, especially to the Christmas tree, or connection to instrumentation such as pressure gauges. Subsea needle valves can be used on the hydraulic control line which is used to deliver hydraulic fluid to the subsea safety valve for actuation and to open the valve. Detailed information about safety valves is provided in Chapter 7.

8.2.2 Design features

Some of the important design features of subsea needle valves are listed as follows:

Antiblowout stem: The stem is normally designed for a maximum amount of torsion or torque from the operator. A backseat feature could be provided as an antiblowout safety provider. Backseat acts as a stop against blowing out of the valve stem upward. The stem of the valve should have an area which has larger diameter than the normal diameter of the stem which could sit on backseat stoppage area and prevents extra movement of the stem outwards. More information about backseat is available for slab gate valves in Chapter 4.

Robust stem design: The stem of needle valves is designed in such a way as to withstand the maximum Remotely operated vehicle (ROV) input torque without any damage. ROV stands for remote-operated vehicle which is a robot used for different purposes such as operating the valves. In case of having a ROV bucket, it should be designed based on API 17H/ISO 13628-8 requirements which are explained in detail in Chapter 9. The other important design feature is to keep the weakest part of the stem out of the pressure envelope of the valve to avoid stem leakage in case of stem breakage from the weakest area and minimize other negative consequences that could occur after stem breakage.

Stem packing/sealing: Stem packing and sealing prevents leakage from the inside of the valve to the seawater atmosphere. The needle valve could have a packing gland design with positive force to compress the stem sealing in place and provide tight sealing. Fig. 8.6 illustrates the load (shown in letter F) coming from the gland flange and the gland on the packing vertically which is transferred into horizontal force and stem sealing. The type of packing material affects the axial sealing loads. As an example, graphite packing can expand less axially compared to Teflon packing. However, graphite has advantages such as high corrosion and fire resistance compared to PTFE (polytetrafluoroethylene) or Teflon, and is thus suitable for a very wide range of operating temperatures. The essential parameter is the torque applied to the gland flange bolt. The other important factor is the friction of packing with stem which increases the torque for the operation of the valve.

The sealing capability of the compression packing is defined as the ability to expand laterally against the stem and stuffing box wall when stressed by tightening the gland. The stress applied laterally by the stem sealing due to applied axial stress from the gland depends on Poisson's ratio of material (μ), which is expressed using Eq. (8.1). Poisson's ratio is defined as the ratio of transverse strain to longitudinal strain, as shown in Eq. (8.2).

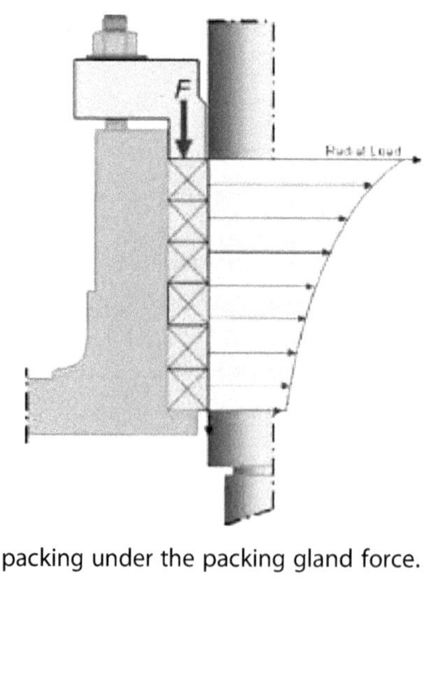

Fig. 8.6 Sealing of the stem packing under the packing gland force.

Relationship between packing axial and lateral forces

$$\delta_1 = \delta_a \left(\frac{1-\mu}{\mu} \right) \tag{8.1}$$

where:

δ_1 = lateral stress;

δ_a = axial stress;

μ = Poisson's ratio (ratio of lateral expansion to axial compression of an elastic solid compressed between two faces).

Calculation of Poisson's ratio

$$\mu = \frac{\text{Lateral strain}}{\text{Longitudinal strain}} \tag{8.2}$$

Nonrotating tip: A nonrotating tip (also called a stem tip) has an advantage for high cycle applications to extend the life of the valve by preventing galling between the seat and the plug. This feature provides straight contact between the seat and plug during closing action without rotating. Fig. 8.7 illustrates a nonrotating stem tip or plug for a needle valve. This feature normally comes with a nonrotating stem requirement. Nonrotating stem requirement means that the stem as well as connected integral plug has just linear movement without any rotation.

Metal-to-metal sealing: The sealing between the needle or plunger and the seats is metal-to-metal.

Port design: Normally needle valves are full bore, e.g., 25 mm bore for a 1″ valve or 19 mm bore for a ¾″ valve.

Environment sealing capability: A subsea needle valve that is subject to a hyperbaric test has an environmental sealing to prevent the ingress of seawater to the inside of the valve. Needle valves, like other subsea valves, may be installed in very deep depths such as 3 km.

Body/bonnet connection: If the valve has a body/bonnet design, the body and bonnet can either be bolted together or threaded.

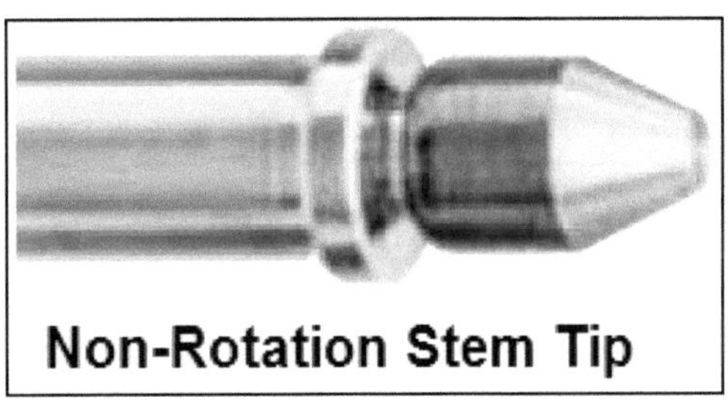

Fig. 8.7 Nonrotating stem tip (plug). *(Courtesy: Hoke needle valve.)*

Operator: A different type of operator is possible for needle valves such as tee bar, cross or X bar, as well as ROV bucket. Different examples of needle valves with ROV bucket, cross or X bar are shown in the next section.

Design life: Subsea needle valves are suitable for maintenance-free operation during their design life which could be 20, 25, or 30 years.

Low operating torque: The operating torque for needle valves is normally low in both flow conditions of under and above the seat and plug.

Material class: Different material classes are available, especially for sour service based on ISO 15156 such as trim class FF and HH. The definition of each trim class is given in Chapter 2. Sour service and sour corrosion associated with hydrogen sulfide is explained in Chapter 12.

Traceability of material: Subsea needle valve materials should have full traceability as per EN 10204 3-1 or 3-2 requirements. EN 10204 3-1 requires that the mill test certificate for the valves components be validated by an inspector authorized by the manufacturer but independent from the manufacturing department. EN 10204 3-2, which can be applicable for the pressure-containing parts of the valves such as body, bonnet, and stem, indicates that the test certificate issued by the mill is validated by an inspector nominated by the manufacturer but independent from the manufacturing department, by an inspector nominated by the purchaser, or by an inspector from a regulatory organization.

8.2.3 Review of different needle valve designs

8.2.3.1 General review

Fig. 8.8 illustrates a subsea needle valve in a typical size range between 3/8″ and 1″ supplied with a ROV bucket.

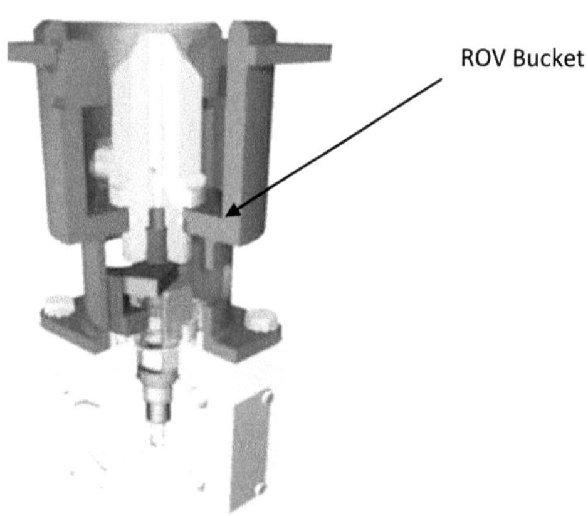

Fig. 8.8 Subsea needle valve with ROV bucket. *(Courtesy: Pacson.)*

Fig. 8.9 Subsea needle valve with X bar.

Fig. 8.9 illustrates a subsea needle valve equipped with an X-shaped blue bar (dark gray in print versions) at the top suitable for operation of both diver and ROV. More information about ROV is available in detail in Chapter 9. This type of needle valve is manufactured from ¼″ to 2″ with a pressure class of 10,000 psi equal to 690 bar. The operating temperate range of this product is from −29°C to 121°C. The valve is qualified according to a pressure-temperature (PR2) test and a hyperbaric test.

A needle valve is typically designed based on API 6A/API 17D. It has a standard pressure rating of up to a maximum 20,000 psi and a typical temperature range between different temperature classes in API 6A/API 17D standards from −29°C to 121°C. Needle valves with temperature resistance as high as 177°C are also available on the market for such as high-temperature application. The machining of the needle plugs should be very accurate since even a small variation in machining the plug or needle head could have a high impact on the flow rate as well as the valve flow characteristics. In addition, accurate machining of the needle head can guarantee the accurate sitting and sealing of the needle on the seats. These valves are typically qualified for pressure-temperature (PR2), hyperbaric as well as endurance tests with 600 cycles.

Needle valves can be made in the form of double block and bleed (DBB) valves that include three needle valves in one body to save space and weight compared to a condition where these three valves are located on three different bodies. Such DBB valves provide a double isolation feature with the possibility of draining or bleeding the fluid trapped between two valves. DBB valves, compared to single isolation valves, have higher safety because they provide double barriers for blocking the fluid; if one valve fails, the second could provide isolation against leakage of the valve to the pressure boundary outside the valve. Fig. 8.10 illustrates a DBB needle valve that includes two 2″ needle valves in a pressure class of 10,000 psi operated by a ROV through a low torque paddle according to ISO 13628-8 type A. There is a small size needle valve such as ½″ located on the body of the valve between the two needles to drain the accumulated pressure between the two

Fig. 8.10 DBB needle valve 2″ class 10,000 psi. *(Courtesy: Galperti Engineering.)*

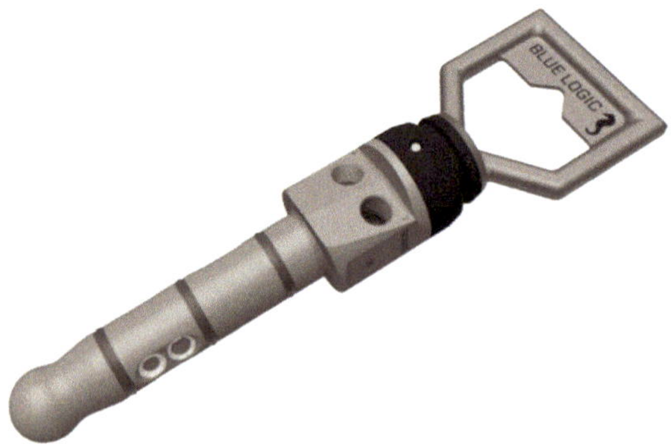

Fig. 8.11 Hot stab. *(Courtesy: Blue Logic.)*

2″ needle valves. In this case, the bleeder of the DBB is connected to a hot stab. A hot stab (see Fig. 8.11) is a subsea hydraulic coupling device that transmits hydraulic fluid from a topside hydraulic power unit to energize subsea equipment. Essentially, it is a hydraulic quick-acting coupling designed for deep-sea application. Two major types of stabs are used: live stabs and plug stabs. Most likely, the valve illustrated in Fig. 8.10 is used for supplying hydraulic oil to the Christmas tree for the operation of hydraulically operated components.

DBB selection, usage, and isolation philosophy is very common for high-pressure class and/or hazardous services. The face-to-face or end-to-end dimensions of DBB valves is not covered by standards such as API 6DSS and API 6A. The face-to-face or end-to-end dimension is determined by the valve manufacturer for DBB valves. The compact design of DBB valves is very important, especially in the offshore sector of the oil and gas industry where space is limited.

Fig. 8.12 illustrates a DBB valve with three needle valves in one body with a side entry design that provides access to the valve inside sidewise by unbolting the blind flange connections at both ends. The operator interface of these needle valves is ROV bucket rather than low torque paddle as shown in Fig. 8.10. A DBB valve could be made of two ball valves with a needle valve between as a bleeder in one body. This type of valve provides only double isolation without any possibility of fluid regulation.

Fig. 8.13 illustrates another type of DBB needle valve that includes three needles in one body: two are used for isolation and one for bleeding or draining. A pressure-temperature qualification test with 200 cycles has been performed on this valve and the valve has passed the test successfully. The valves have metal-to-metal sealing between the needle (plug) and the seats. This valve is called a light duty needle valve combination with a small size of ¼″ with a 6 mm bore and a pressure class of 690 bar equal to 10,000 psi and very low operating torque values. Its temperature range is between 2°C and 121°C.

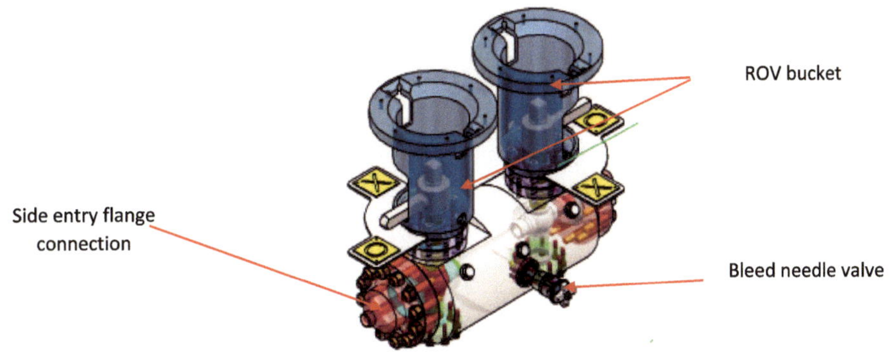

Fig. 8.12 DBB side entry needle valve with ROV bucket interface. *(Courtesy: Galperti Engineering.)*

Fig. 8.13 Double block and bleed needle valve. *(Courtesy: Oliver Valvetek.)*

Multiple PTFE (Teflon) seals are used for stem packing and it has a nonrotating stem tip during closing action when the stem tip sits on the seats to prevent galling and friction. The valve has a product quality level (PSL) of 3 and all the pressure-controlling and pressure-containing parts have an EN 10204 3-1 certificate.

Fig. 8.14 illustrates another type of DBB needle valve for more serious applications in larger sizes such as ¾″ (19 mm bore) and 1″ (25 mm bore) with a pressure class of 15,000 psi suitable for installation at a depth of around 3 km. The normal operating torque of this valve is 136 N m and the damage torque that breaks the stem is 542 N m. The valve operator is a cross or X bar that can be opened or closed by a ROV with three turns.

Installation of DBB needle valves can be performed on a mono-flange as illustrated in Figs. 8.15 and 8.16. A mono-flange combines three valves as well as their functions in one

Fig. 8.14 Double block and bleed needle valve. *(Courtesy: Oliver Valvetek.)*

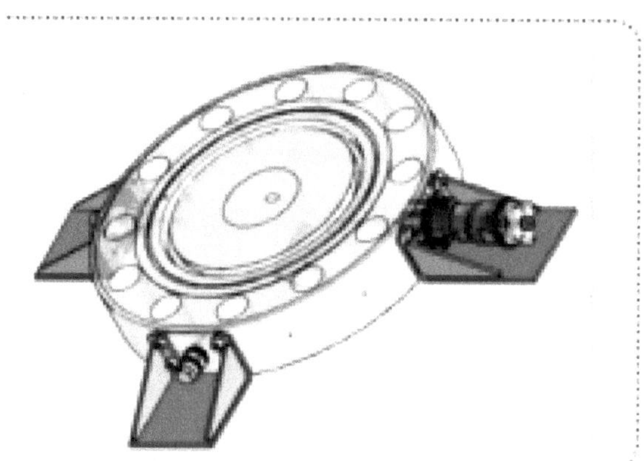

Fig. 8.15 Double block and bleed needle valve on a mono-flange. *(Courtesy: Galperti.)*

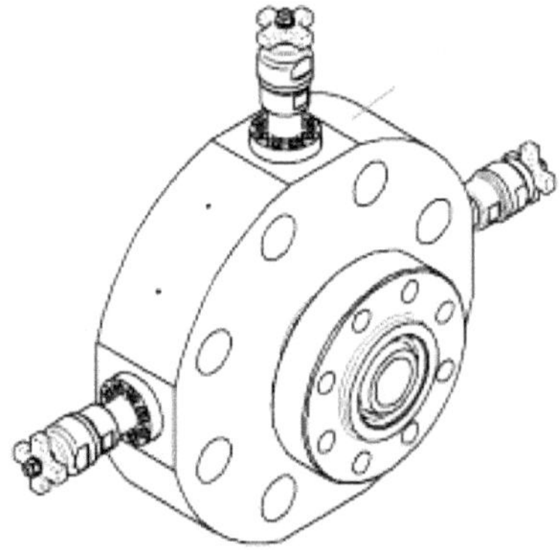

Fig. 8.16 Double block and bleed needle valve on a mono-flange. *(Courtesy: Galperti.)*

compact body. The main advantage of a mono-flange is not limited to its compact design. It is important to have a very short distance between the process fluid and the connected instrument to provide high speed control. Additionally, in case of vibration generation, longer distance could increase the impact of the vibration. Mono-flange DBB needle valves offer other advantages such as less weight, lower cost, and fewer leakage points. It should be noted that single needle valves can be installed on a mono-flange.

8.2.3.2 Fire test for needle valves

Needle valves that are used for surface wellheads should be fire tested according to API 6FA or another equivalent standard such as ISO 10497. API 6FA is a more common standard for the fire safety of valves made based on the API 6A standard. Valves in the oil and gas industry are at risk of fire so these valves should be designed in such a way as to perform a certain amount of sealing in case of fire and continue to operate properly. In fact, fire and explosion are the main risks and areas of concern in different sectors of the oil and gas industry such as pipeline, refineries, and petrochemical plants. The safest way to deal with fire is to prevent it. However, fire prevention strategies are not always successful and fires can happen. Therefore valves should be designed and tested to be fire safe. A fire test is carried out in order to evaluate the resistance of the valve against fire under conditions defined in the international standards. The complete test can be monitored and logged by computers. Fig. 8.17 illustrates a fire test on a valve.

This section provides a short description of a valve fire test based on API RP 6FA. The "RP" included in the name of the standard stands for recommended practice.

The valve should be tested close to the water with the stem in a horizontal position. The temperature of the flame (fire) should be between 761°C and 980°C; the burning period is 30 min. Two thermocouples are installed on the body of the valve to monitor the temperature. A maximum of seven performances are required from the valve during the test. Most of the valve performances are in relation to the valve seat and external

Fig. 8.17 Fire test on a valve.

leakages. However, valve operability and pressure release from the self-relieving seats are included among the tests. These tests are highlighted as follows:

(1) Through seat leakage (high test pressure)—during the burn period. The maximum leakage from the seat during 30 min of burning should not be greater than 400 mL/in. of the valve per minute.

(2) Body external leakage (high test pressure)—during the burn and cool-down period. The maximum external leakage should not be more than 100 mL/in. of valve size per minute during the 30 min burning period plus cooling down to 100°C.

(3) Through seat leakage (low test pressure)—after cool down. This test is applicable for valves in pressure class 600 and lower. The leakage from the seat during 5 min should not be more than 40 mL/in. of valve size per minute.

(4) Body external leakage (low test pressure)—after cool down (valve in closed position). This test is applicable for valves in pressure class 600 and lower. The maximum external leakage from the valve during 5 min of testing should not be greater than 20 mL/in. of valve size per minute. It should be noted that the high- and low-pressure values are given in the API RP 6F standards.

(5) Operation of test valve after fire test: The valve should be capable of being unseated from its closed position against a high test pressure differential and moved to an open position.

(6) Body external leakage in open position: The body external leakage of the valve in open position should not be more than 200 mL/in. of valve size per minute.

(7) Pressure relief provision: The self-relieving seats should be tested to ensure that the seats release the accumulated pressure in the cavity to the line.

Note: When it comes to the operation of the test from closed to open position, it is possible to apply thermal analysis on the valves' seats and closure members to see if the expansion of these valves' internals could obstruct the opening of the valve.

When it comes to fire safe design consideration for valves, Atmospheric Explosible (ATEX) should be discussed. ATEX is the name commonly given to the two European Directives for controlling explosive atmospheres. Equipment used in potentially explosive atmospheres such as offshore platforms should have a certificate of conformance to ATEX and equipment directive 2014/34/EU. The objective of the ATEX guidelines is to clarify specific matters and procedures concerning equipment and protective systems intended for use in potentially explosive atmospheres. That valves, especially those with an actuator, are subject to ATEX means that the design of the valve and actuator should not cause any weak point for fire to occur. It should be noted that ATEX is not applicable for subsea valves in general. The reason why fire test and ATEX have been discussed is their applications for the surface valves on the wellhead and Christmas trees.

8.2.3.3 Detailed review of a needle valve with a part list

This section provides a detailed review of needle valve design including its different components and general features. Fig. 8.18 illustrates a subsea needle valve. Item #1 is the body of the valve which is known as the main pressure-containing part; its failure to function leads to the leakage of internal fluid to the sea. The body of the valve could be in 25Cr super duplex as an example. Other exotic materials are available for the valve body upon the request of the purchaser. Item #2 is called the bonnet or cover which appears to be integrated with the body. The bonnet, like the body, is a pressure-containing part and its failure also leads to the leakage of internal fluid to the sea. The connection of body and bonnet is not clear in the picture; bonnet and body could be integral or bolted to allow for valve dismantling for maintenance; this is not the case normally in subsea operation since the valves should be maintenance free. The bonnet of the valve should be made of the same material as the body which is 25Cr super duplex in this case. Item #3 is the bonnet top which is bolted to the bonnet through item #15 which is a bolt. The bolt could be low alloy steel with nickel-zinc plated type coating according to API 20E or 25Cr super duplex as per API 20F standard. Item #4 is the needle or plug which is the main pressure-controlling component. The movement of the needle upward increases the flow passage through the valve, and the downward movement of the needle reduces the flow passage of the valve. The material of the needle can be 25Cr super duplex as it is in this case, the same as the body, with tungsten carbide coating on the sealing surface with a seat to prevent galling. Item #5 is a stem that can be integral with the needle. The stem is a

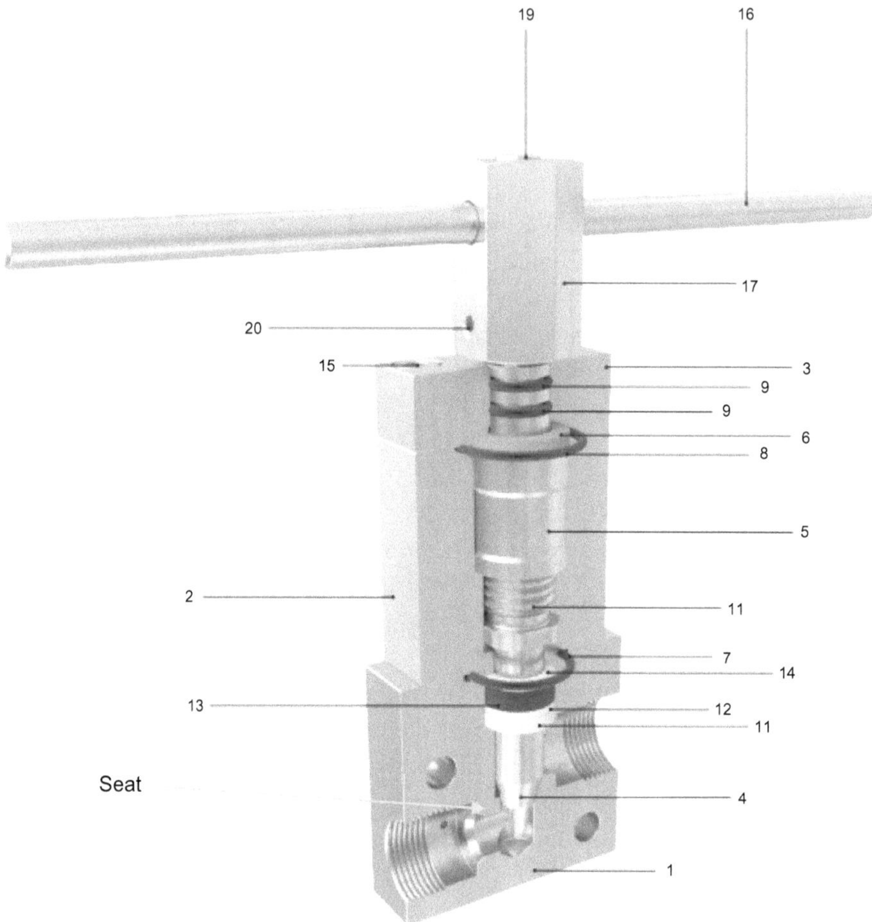

Fig. 8.18 Subsea needle valve designed by Parker. *(Courtesy: Parker.)*

pressure-containing part of the valve that transfers the operator force to the valve internals to open and close the valve. The stem material should have high mechanical strength; 25Cr super duplex is a strong stem in this case. The force required for the operation of a needle valve is not high, so 25Cr super duplex is normally strong enough to withstand the transmitted load from the operator for valve operation. However, materials with higher strength such as Inconel 718 and 725 are available. Item #60 is a thrust washer or bushing that can provide lubrication for the stem and prevent the stem movement due to axial load. Item #7 is a sealing; it could be in Viton, Buna N, or hydrogenated Buna N, and its main function is to seal the body and bonnet contact area. Different seals mentioned in previous sentence are covered in Chapter 13. Item #8 is another sealing on the top part of the bonnet that seals the area between the bonnet top and the bonnet which has same material as the other sealing (Item #8). One of the main important components that is not highlighted in Fig. 8.18 is the seat of the valve which is in 25Cr super duplex with a tungsten carbide hard facing to prevent galling and friction between the disk and seat. The valve has metal-to-metal seat sealing with a provision of tight bubble shut-off. Tight bubble shut-off could be interpreted as leak rate A as per ISO 5208 meaning zero leakage. A nonrotating needle is selected for this valve as in other designs to prevent galling and friction between the needle and the seats. The valve is designed for a depth of 3 km with the possibility of having a ROV bucket instead of a bar handle.

Item #9 indicates two O-rings that are used to seal the stem and prevent the internal fluid from leaking into the seawater through the stem area. Item # 11 illustrates some threads on both the stem and bonnet of the valves facilitating the movement of the stem upward in the bonnet as the stem of the valve is nonrising and nonrotating. Items #12 and #13 are a sealing that could be in Polyether ether ketone (PEEK) material. Item #16 is the handlebar of the valve which can be rotated by an operator for valve operation. Again, 25Cr super duplex can be selected for the T-bar material. Item #17 is the handle hub in 25Cr super duplex which connects the handlebar force to the stem.

8.3 Axial control valves

8.3.1 Introduction

The design of axial control valves is similar to that of axial check valves with an additional stem component. Axial check valves are installed inline; they open with fluid flow (see Fig. 8.19) and close with spring force without any need for an operator. Therefore axial

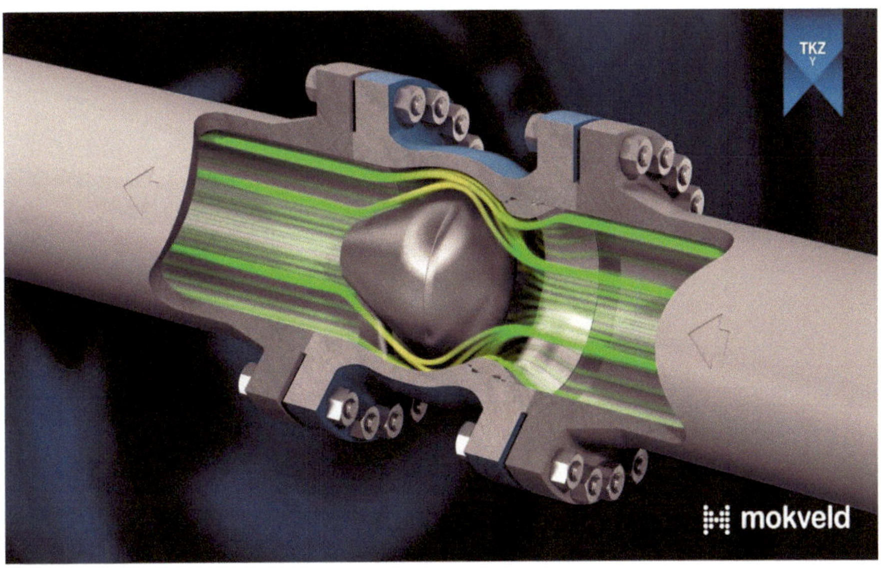

Fig. 8.19 Axial check valve in open position. *(Courtesy: Mokveld.)*

check valves do not have any stem to provide operator access for valve operation. More information about axial check valves is available in Chapter 5. Axial valves are specifically addressed in API 6DSS standard.

However, axial control valves do require a stem to transfer the operator force to the valve internal. Fig. 8.20 illustrates a flow control axial valve.

Fig. 8.21 illustrates an axial control valve in closed condition. The disk of the valve is located on the seats and provides sealing. The stem has moved downward and pushed the valve internals including the seat forward, so the disk is placed on the seats to close the valve. The obturator in this type of valve has a cylindrical shape that moves back and forth

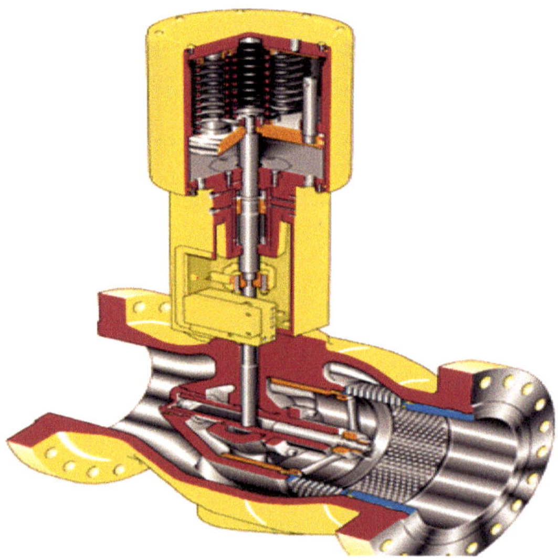

Fig. 8.20 Axial control valve.

Fig. 8.21 Axial control valve in closed position. *(Courtesy: Mokveld.)*

parallel to the fluid flow direction. A typical ending of an axial control valve can either be flanged like the one in Figs. 8.20 and 8.21 or welded for subsea applications.

The opposite condition is when the stem moves upward so the valve internals are pulled back and the disk moves away from the seats, opens the valve, and allows the fluid to pass through the valve (see Fig. 8.22).

The important point is that an axial control valve does not necessarily stay in a full open or closed position. Other intermediate positions between open and closed can be adjusted by the operator to keep the valve in flow regulation mode. Axial valves come in two categories depending on the function and application of the valve. An axial on/off valve is only used for stop and go of the fluid so it has just two modes: "on" and "off." This type of valve does not have the possibility of flow control. The other type is an axial control valve which is used for flow regulation.

8.3.2 Main design features

Some of the main design features of axial valves are listed as follows:

- The one-piece body construction and bonnet-less design of this valve reduces leakage from body of the valve. The weight of the valve body and the whole valve in total is much lighter than the weight of a globe valve of the same size and pressure class;
- Triple steam sealing prevents any leakage from the valve internal to the seawater and from the seawater to the valve internal. The stem sealings are not in contact with the process fluid and the sealings are only in contact and friction with the stem seal during a short period of time during valve stroke;

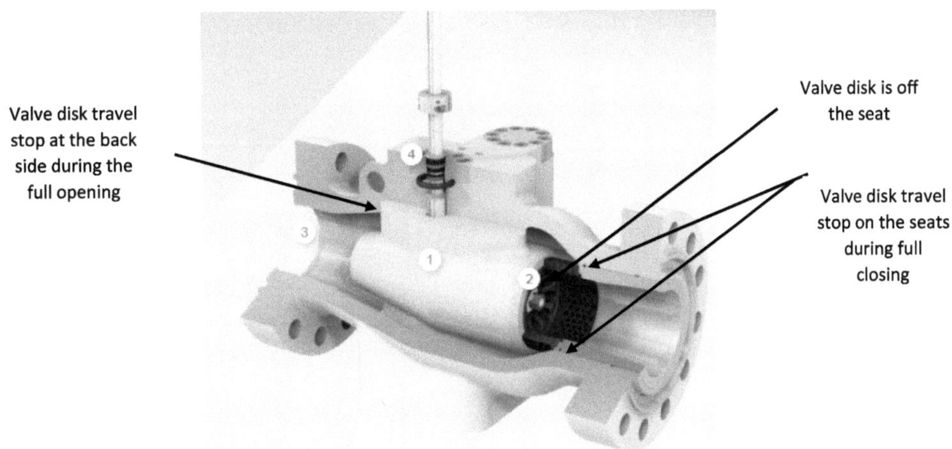

Fig. 8.22 Axial control valve in open position. *(Courtesy: Mokveld.)*

- Qualification tests of the valve at subsea depth through several cycles such as pressure-temperature, endurance, and hyperbaric testing. The number of cycles could be 600 in total; this includes 200 cycles for each category of the qualification test;
- Venturi effect inside the valve to provide very smooth flow with a very small amount of pressure drop that significantly minimizes the risk of wearing and erosion for the valve internals; however, it should be noted that valves used for flow regulation do cause a pressure drop, so having a pressure drop is not a concern for an axial control valve unlike axial check valves. However, body shape and flow nozzle shape should be optimized in a CAD software like SolidWorks.
- Axial valves according to API 6DSS "Subsea pipeline valves" standard are reduced bore. The venturi shape flow path inside the valve is narrower than the inlet and outlet flow paths;
- Less flow turbulence, less noise and vibration are associated with operation of this valve;
- The travel stop for the disk is located inside the body of the valve as illustrated in Fig. 8.22 and not on the actuator. The actuator is just used to control the flow of fluid by opening and closing of the valve.
- Actuators installed on axial valves have linear movement with vertical orientation when the valve is installed on a horizontal line. Therefore installing the actuator on top of the valve can increase the height of the actuated valve considerably (see Fig. 8.23). It should be noted that a supplier could provide a rotary actuator standing horizontally for axial valves but this is not common. The advantage of installing an actuator horizontally for axial valves is that it saves space on the vertical plain.
- The valve normally provides metal-to-metal sealing with tungsten carbide hard facing on the disk and seat surfaces; a secondary soft sealing could be added for axial valves used in the nonsubsea sector of the oil and gas industry.
- Tight bubble shut-off meaning zero leakage can be expected from this valve even though it has metal-to-metal sealing;

Fig. 8.23 Axial valve height increase due to installation of the actuator.

- An axial valve has a flow obturator on the path of the fluid flow so this valve is not piggable;
- It is recommended to contact the valve supplier for usage of this valve in dirty (particle-containing fluids). However, most reputable axial valve suppliers allow the use of axial valves in dirty services;
- Nonslam operation means that the disk of the valve contacts the seat slowly without any slamming effect;
- The valve can be installed on both vertical and horizontal lines;
- Choice of valve actuation could be electrical or hydraulic;
- The valve has a preferred flow direction that should be marked on the body of the valve (see Figs. 8.24 and 8.25);

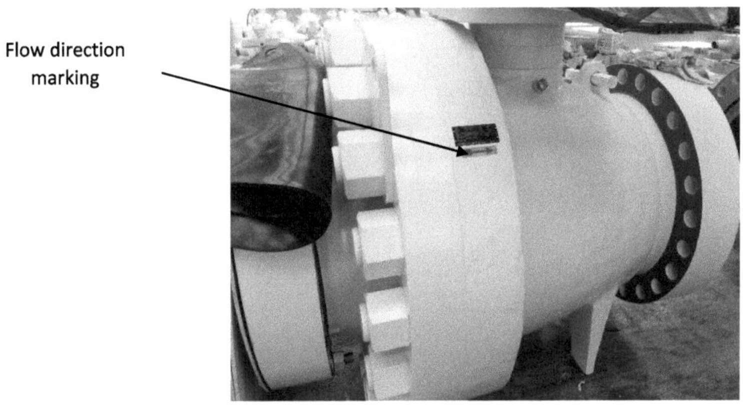

Fig. 8.24 Flow direction marking on the body of an axial valve.

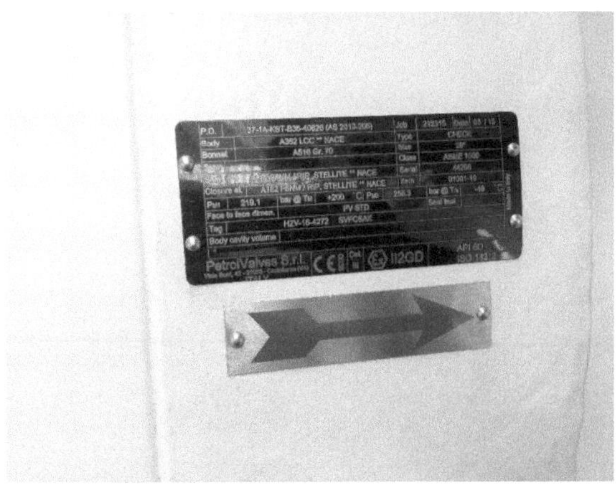

Fig. 8.25 Flow direction marking on the body an axial valve under the name plate or tag plate.

- Axial control valves can be used as choke valves for reducing the pressure of produced flow from the well; these are called "choke axial flow valves";
- Axial isolation valves are a good choice for a high integrity piping (pressure) protection system (HIPPS) due to their very fast closing possibility. Fig. 8.26 illustrates an axial valve with a hydraulic actuator for HIPPS application. HIPPSs are explained in detail in Chapter 15. The fast closing or opening of the valve is due to the low mass of the disk as well as its short travel distance. The other important feature is the pressure balance disk which can be opened and closed without the effect of upstream and downstream pressure on actuation force which leads to a low torque requirement for operation the valve;
- Pressure balance can be achieved since there are some holes on the disk of axial control valves (see Fig. 8.27) that allow fluid passage so that the pressure upstream and downstream from the disk is almost equalized. This makes the required operation force or torque for the valve very low. It should be noted that the fluid passing through the holes does not leak to the downstream piping; it accumulates inside the axial nozzle as illustrated in Fig. 8.28. The holes inside the disk minimize the noise level inside the valve as well;
- The design of axial valves provides fewer moving components without the need for maintenance which is ideal for subsea applications;
- Different types of trims (cage) are available to change the flow characteristic of the fluid. Fig. 8.29 illustrates the different flow characteristic of the valves: quick opening,

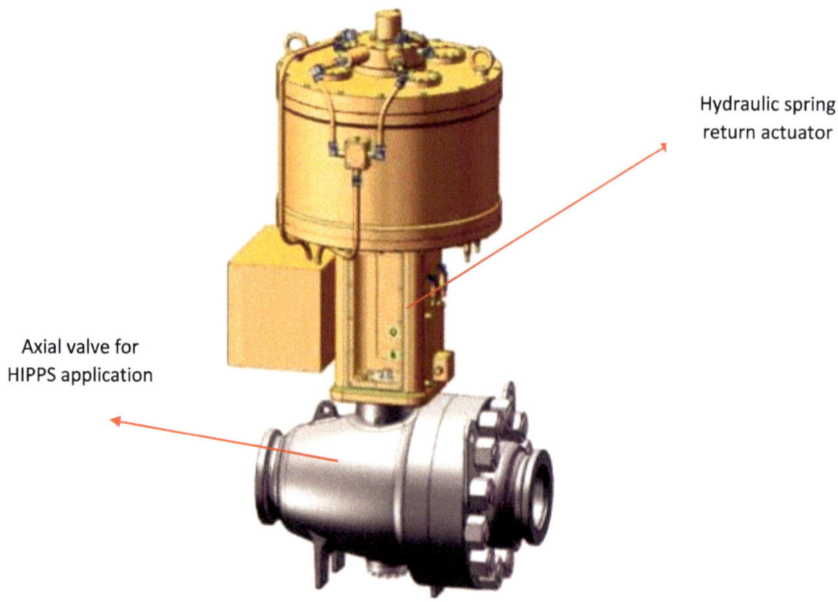

Fig. 8.26 Axial valve with hydraulic spring return actuator for HIPPS application.

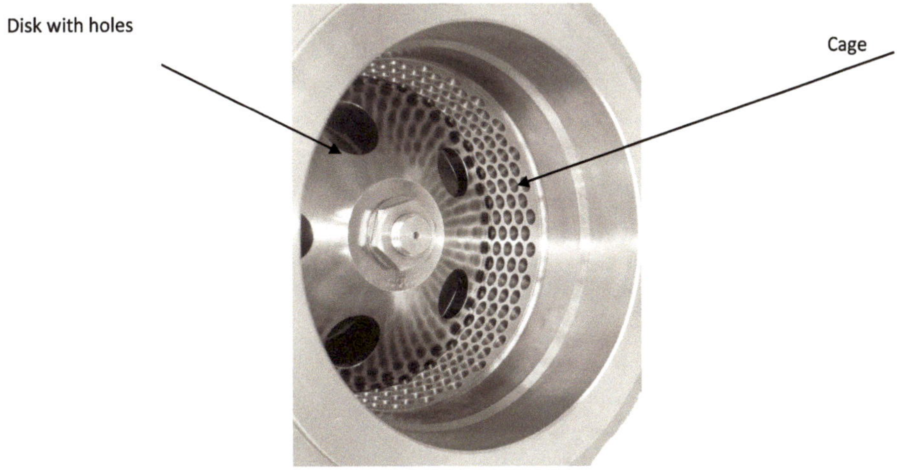

Fig. 8.27 Axial flow disk, cage, and seat.

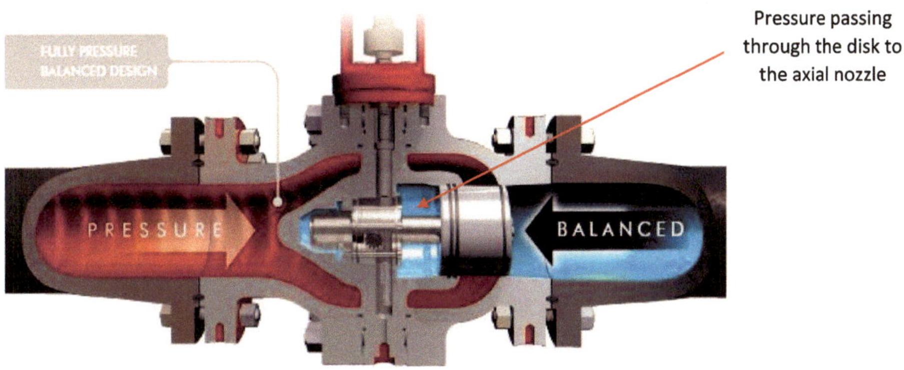

Fig. 8.28 Pressure balance design of axial valves.

linear, and equal percentage. Quick opening means that a large amount of flow is passing through the valve as the valve begins to open. In fact, a small opening of the valve especially in the beginning can lead to a high amount of flow. Linear flow means that the change in the flow is equal to the opening percentage. As an example, opening a valve 20% leads to passing 20% of the fluid through the valve. Equal percentage means that the valve releases a relatively low amount of fluid in the beginning of opening stage. As an example, a valve that is getting open 30% to just start releasing the flow has equal percentage flow characteristic.

A detail of the flow marking which is located under the name plate of the valve is shown in Fig. 8.25.

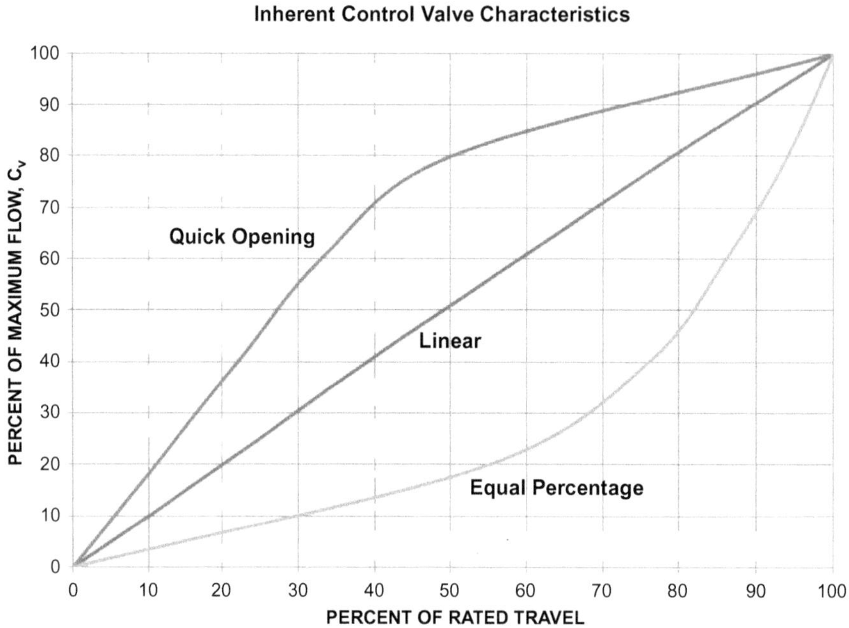

Fig. 8.29 Flow characteristics of a valve.

8.3.3 Detailed design review of an axial control valve

Fig. 8.30 illustrates an axial control or on/off valve schematic including the main components extracted from the API 6DSS standard. Three types of ending are illustrated in this figure: flanged ends with a raised face, a welded end connection, and flanged ends with a ring type joint (RTJ) from top to bottom. The end connection on the top showing a flanged end connection with a raised face is not common for subsea application; it is normally used for low-pressure applications such as Class 150 (PN = 20 bar) and 300 (PN = 50 bar) which are not applicable in subsea. Both RTJ flanges and welded end connections are used for subsea applications. A welded end connection subjected to a 100% nondestructive test such as radiography guarantees 100% joint efficiency between the valve and the connected piping without any possibility of leakage. However, zero leakage is not expected from a flange connection. Valves with welded end connections are supplied with a pup piece that could be an integral part of the valve body or welded to the valve body from both sides. The length of the pup piece is very important as it must provide sufficient distance between the valve internals, including both soft and metallic material, and the pipe welded to the valve in the construction yard. This distance is especially important for soft materials since they can be melted by the welding heat at the yard. The face-to-face or end-to-end dimensions of an axial control valve are given in

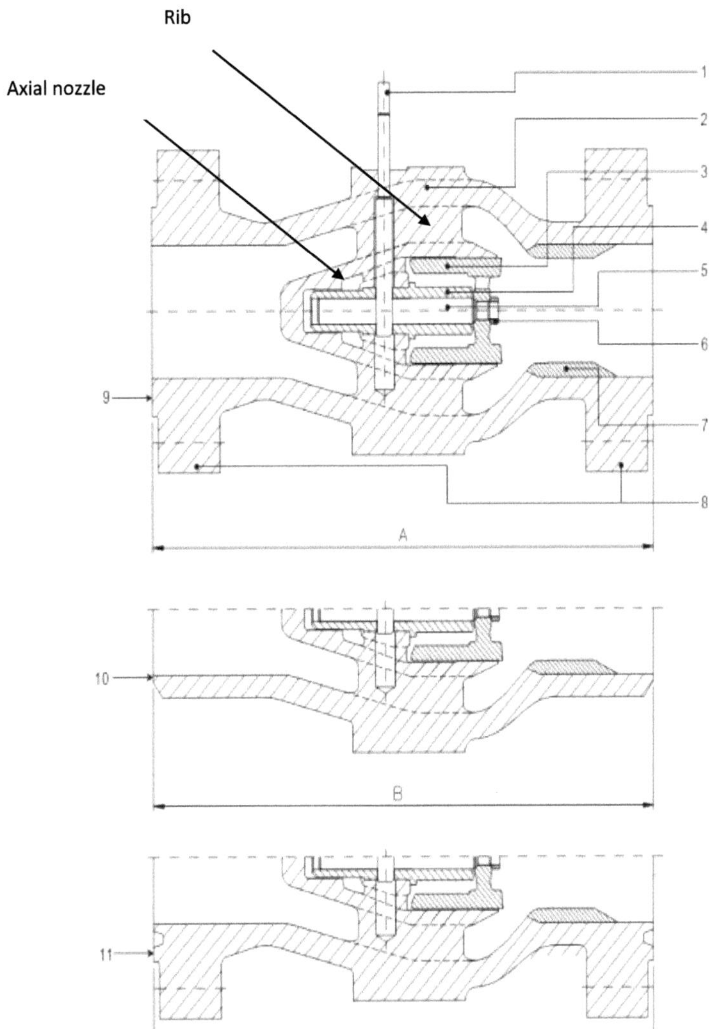

Fig. 8.30 Axial control valves. *(Courtesy: API 6DSS.)*

Table C5 of the standard and Table 8.1 for pressure class 1500 equal to 250 bar and pressure class 2500 equal to 420 bar.

Face-to-face dimension is applicable for flanged end connections with a raised face. However, "end-to-end" is the correct expression for valves with butt weld ends prepared according to ASME B16.25 and ASME B31.8 and flanged ends with an RTJ face. It should be noted that end-to-end for butt weld end axial valves could be based on manufacturer standards as an alternative. The other point is that even though Table 8.1 provides the end-to-end and face-to-face dimensions of an axial on-off valve, the

Table 8.1 Face-to-face and end-to-end dimensions of axial control/on-off valves.

NPS	DN	Raised face *A*	Welding end *B*	Ring joint *C*	Raised face *A*	Welding end *B*	Ring joint *C*
			Class 1500			Class 2500	
2	50	14.50 (368)	–	14.63 (371)	17.75 (451)	–	17.88 (454)
2½	65	–	–	–	–	–	–
3	80	18.50(470)	15.00 (381)	18.63 (473)	22.75 (578)	16.55 (420)	23.00 (584)
4	100	21.50 (546)	18.00 (457)	21.63 (549)	26.50 (673)	18.00 (457)	26.88 (683)
6	150	27.75(705)	24.00 (610)	28.00 (711)	36.00 (914)	24.00 (610)	36.50 (927)
8	200	32.75 (832)	29.00 (737)	33.13 (841)	40.25 (1022)	29.00 (737)	40.88 (1038)
10	250	39.00 (991)	33.00(838)	39.38 (1000)	50.00 (1270)	33.00 (838)	50.88 (1292)
12	300	44.50 (1130)	38.00 (965)	45.13 (1146)	56.00 (1422)	38.00 (965)	56.88 (1445)
14	350	49.50 (1257)	40.50 (1029)	50.25 (1276)	62.75 (1594)	40.50 (1029)	–
16	400	54.50 (1384)	44.50 (1130)	55.38 (1407)	70.00 (1778)	44.50 (1130)	–
18	450	60.12 (1527)	48.00 (1219)	61.00 (1549)	–	–	–
20	500	65.50 (1664)	52.00 (1321)	66.38 (1686)	–	–	–
22	550	–	–	–	–	–	–
24	600	76.26 (1937)	61.00 (1549)	77.36 (1965)	–	–	–
26	650	–	–	–	–	–	–
28	700	–	–	–	–	–	–
30	750	–	–	–	–	–	–
32	800	–	–	–	–	–	–
34	850	–	–	–	–	–	–
36	900	–	–	–	–	–	–
38	950	–	–	–	–	–	–
40	1000	–	–	–	–	–	–
42	1050	–	–	–	–	–	–
48	1200	–	–	–	–	–	–

Dimensions in inches (mm).
Courtesy: API 6DSS.

information in the table can be used for axial control valves. Item #1 is the stem of the valve which provides an interface between the operator and the valve internals. The stem should be made in a high strength material such as 25Cr super duplex, Inconel 718, or Inconel 725. Item #2 is the body of the valve which is the main pressure-containing part. Different materials such as low alloy steel A694 F11 or F22 with an Inconel 625 overlay on the sealing areas (partial overlay) or a full overlay is an option especially for large sizes as it is a more economical and cost-effective choice compared to exotic, corrosion-resistant alloys such as 25Cr super duplex. Alternative materials could be 25Cr super duplex or Inconel 625 body materials. Many parameters affect the material selection for the body, especially corrosion and erosion. Items #3 and 4 are the piston and piston rod guide, respectively. The linear movement of the stem is transferred to the piston rod (item #5). The piston rod can move back and forth by the up and down movement of the stem.

The piston rod is connected to the piston rod guide and the piston which move back and forth with the piston rod movement. Both piston rod guide and piston control the movement of the rod axially without any lateral movement. These three components, piston, piston rod guide, and piston rod, are considered trim and are in contact with the fluid. Corrosion-resistant alloys (CRAs) such as 25Cr or Inconel 625 are proposed to be selected for the piston, piston rod, and piston guide rod. An additional item called the cage, which could be in different shapes with different hole patterns inside (see Figs. 8.31 and 8.32), can be installed between the seat and piston rod arrangement to achieve the required flow characteristic and flow capacity for flow control. Item #6 is called the piston nut which connects the piston rod and piston guide rod to the piston. The nut should be made from the same material as the other valve internals. Item #7 shows a seat ring which provides a stop against the obturator for full closure of the valve. The seat material should be like the other internal materials such as 25Cr super duplex or Inconel 625 with tungsten carbide or stellite hard facing to prevent galling and friction with the valve closure member during closing. Item #9 is the raised face of the flange end connection which is not used in subsea. Item #10 shows a weld end and item #11 is an RTJ ending of a flange face. Two other parts highlighted in the figure are the axial nozzle and the rib. The axial nozzle is located in the center of the body and the piston rod is located inside of that. The connection of the axial nozzle to the body is done through the ribs (see Fig. 8.33).

The cage mechanism depends on the pattern and the number of holes, and the same mechanism is used for the control valves. Control valve refers to a globe type valve with a

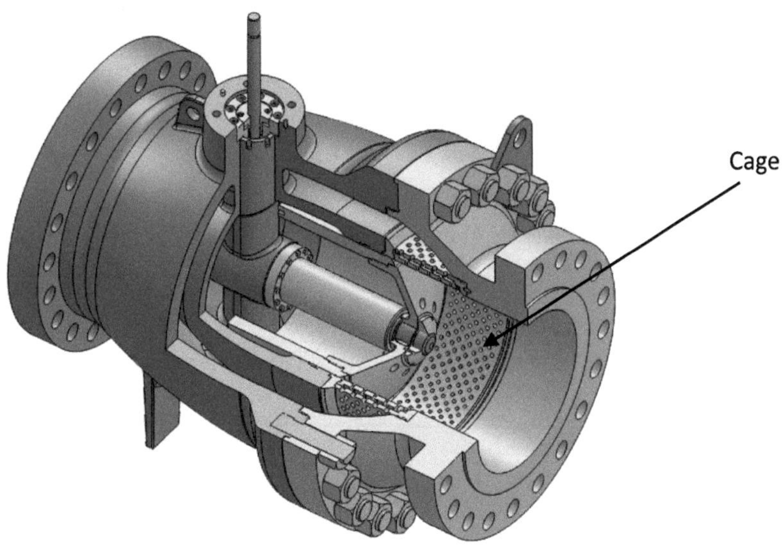

Fig. 8.31 Axial control valve with cage highlighted. *(Courtesy: FG Valvole.)*

Fig. 8.32 Different cages for an axial valve.

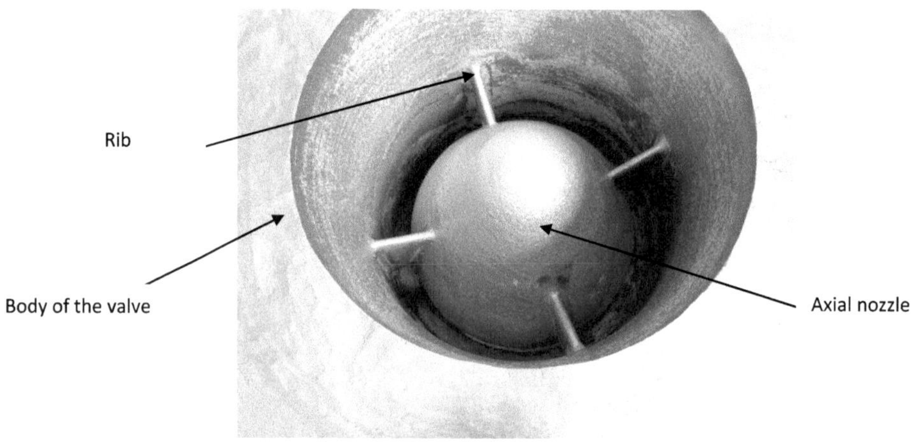

Fig. 8.33 Axial nozzle and ribs inside the axial valve.

diaphragm actuator. The control mechanism of the control valve is dependent on the type of cage. Different types of control valve cages are illustrated in Fig. 8.34.

8.3.4 Control valve philosophy

As mentioned before, control valves are used to regulate flow, which can lead to changes of pressure, temperature, or fluid level in the system. For example, variation in the fluid flow can change the pressure or adjust the two other above-mentioned parameters to the

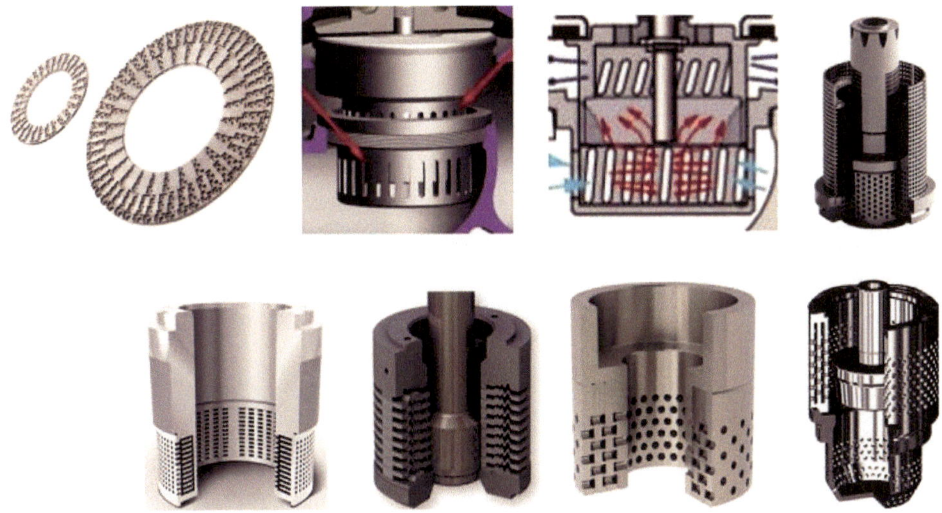

Fig. 8.34 Control valve cages.

required levels. Thus we can say that control valves control the process variables to be as close as possible to the desired set point. In general, control valves control different variables such as pressure, temperature, and liquid level by changing the flow rate.

Control valves are always supplied with an actuator that receives a signal and makes the valve move accordingly. Like other types of valves, many aspects should be taken into account in the design of control valves such as material selection, fluid mechanics, actuation, etc. The performance of the control valve is largely dependent on the actuator. The control valve is the final element of the control loop, which is the combination of different blocks working together to satisfy the process needs. If there are deviations in a process variable such as pressure, different elements in the control loop work together to achieve the required or set pressure. To achieve the required process variable, different sensors are installed on the piping or equipment to measure process parameters such as pressure, temperature, and level. After capturing the process data, the information is transmitted to the control room through transmitters. The control room analyzes the measured parameter and compares it to the set point. Then the control panel sends a signal to the control valve to open or close more in order to meet the process requirement. Thus the control valve is known as the last element in the control process. Fig. 8.35 shows a control loop that includes all the required elements:

8.3.5 More pictures of axial valves

(See Figs. 8.36–8.42.)

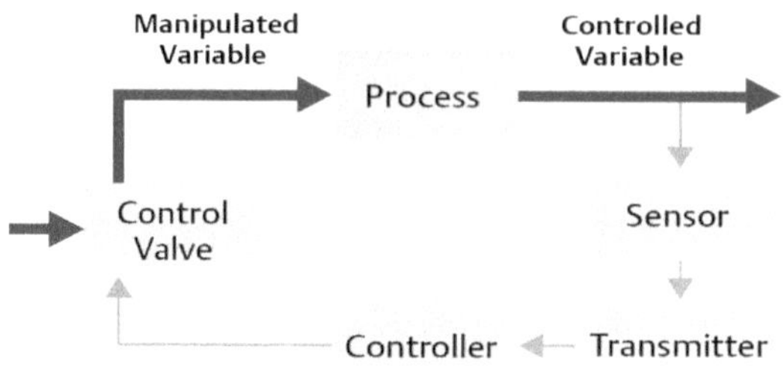

Fig. 8.35 A control loop. *(Courtesy: Springer.)*

Fig. 8.36 28″ Class 1500 axial control valve with a hydraulic actuator during the pressure and factory acceptance test.

Fig. 8.37 Axial valve after assembly and test.

Fig. 8.38 Axial valve: Before packing, the control panel is wrapped with aluminum foil for preservation.

Fig. 8.39 Drilling a bolt hole in the body flange of an axial valve.

Fig. 8.40 Inside the axial valve body after machining.

Fig. 8.41 Packing an axial valve.

Fig. 8.42 Packing an axial valve.

8.4 Self-assessment questions

1. Why are needle valves a good choice for flow regulation in the subsea sector of the oil and gas industry?
   - **A.** Needle valves are not exposed to cavitation like globe valves.
   - **B.** The taper-shaped plug or closure member gives needle valves the capability of very precise regulation.
   - **C.** These valves as a form of DBB can be installed on a mono-flange to save space, weight, and cost.
   - **D.** High torque or force is not required for operation of these valves.

Answer: Needle valves, like globe valves, are exposed to cavitation so option A is not correct. However, these valves are subject to high numbers of cycles during the qualification program for subsea application. The valve should not show any sign of damage due to cavitation or any other cause after a high number of cycles. Option B is correct since the taper shape of the plug gives needle valves the capability of very precise regulation. Needle valves are used as a type of double block and bleed (DBB) on a mono-flange with advantages such as weight, space, and cost saving. However, having DBB needle valves on the mono-flange does not add any value to the flow regulation capability of the valve. Option D is correct, as needle valves do not require high torque or force for operation, but this feature does not improve the flow regulation capability of the valve. Therefore the correct answer is option B.

2. Which sentence is not correct about the fire test for needle valves?
   - **A.** Subsea needle valves do not need a fire test according to the relevant international standards.
   - **B.** Needle valves installed on the surface wellhead are required to be fire tested as per API 6FA.
   - **C.** The fire test based on API 6FA is limited to leakage measurement from the valve.
   - **D.** The best way to handle fire is to prevent it. But since it is not always possible to prevent fire, valves in the nonsubsea sector of the oil and gas industry should be designed and tested to withstand fire to some extent.

Answer: Options A, B, and D are correct. The fire test based on API 6FA includes seven test stages and not limited to a pressure test. In case of fire, the valve should be operable, especially from closed to open position. Therefore Option C is not correct.

3. Which sentence about axial control valves is not correct?
   - **A.** Axial control valves have very short operating time due to their low mass disk and the short distance between the disk and seats.
   - **B.** Axial control valves have good flow capacity and a smooth flow characteristic due to the venturi effect inside the valve.

C. Axial control valves are not susceptible to cavitation.

D. Axial control valves are only used for the subsea sector of the oil and gas industry.

Answer: Axial control valves, like axial check valves, have a very short operation time due to their low mass disk and short disk travel distance. Therefore option A is correct. Axial valve design provides a venturi effect for the fluid flowing inside the valve which leads to a very smooth flow characteristic. Thus option B is also correct. Smooth flow and low turbulence flow make the risk of cavitation in this type of valve almost zero, so option C is also correct. Option D is not correct, as axial valves are used topside on the platforms for flow regulation or fast opening blow-down applications.

4. Which sentences are correct about axial valves?

A. Axial valves are only suitable for installation on a horizontal line.

B. Axial valves can be installed on a pipeline that is subject to pig running for cleaning the pipe;

C. Axial valves provide very smooth flow without turbulence inside the valve due to the venturi effect.

D. Axial valves are a new generation of valves to be used instead of globe valves to minimize cavitation and provide better fluid flow.

Answer: Axial valves can be installed on both horizontal and vertical lines so option A is not correct. Axial valves have an obturator or closure member in front of the fluid flow path so they cannot be installed on a pipeline that should be pigged. Therefore option B is not correct. Axial valves provide very smooth flow without turbulence so option C is correct. Option D is also correct since axial valve manufacturing and usage in the piping system in oil and gas industry is relatively new; axial valves are a good replacement for globe valves to reduce cavitation risk and provide smooth flow. Thus both options C and D are correct.

5. What could be the main reason that makes axial valves suitable for subsea applications?

A. Axial valves can be installed on both vertical and horizontal piping.

B. Axial valves have a one-piece body design that makes them robust.

C. Axial valve design involves the least number of moving parts and provides maintenance-free operation.

D. The usage of the stem guide bar assures axial movement of the disk for opening and closing of the valve.

Answer: The fact that axial valves can be installed on both vertical and horizontal lines does not give any value added to this valve for subsea operation so option A is not correct. The robust design of the valve due to its one-piece design could reduce the leakage possibility from the body of the valve but again this is not a good reason to select this type of valve in subsea. Therefore option B is not correct either. Option C is correct since maintenance-free operation is very important in subsea. Option D is very important for the safe operation and sealing of the valve but again it is not a main reason to select axial valves for subsea. Thus option D is not correct.

6. Which sentence is correct about axial control valves?
   - **A.** The face-to-face and end-to-end dimensions of the axial valves can be taken from API 6DSS/API 6A in all applicable pressure classes for subsea.
   - **B.** Different cages with different hole patterns could be used for axial control valves to achieve different flow characteristics and flow capacity.
   - **C.** The only additional component in an axial control valve compared to an axial check valve is a stem.
   - **D.** The pressure balance disk does not affect the operation torque of axial valves.

Answer: It is possible to find the end-to-end dimensions of axial valves in pressure class 1500 and 2500 in API 6DSS. However, the pressure classes in API 6A could be 5000, 10,000, or 15,000 psi which are not covered in API 6DSS. In addition, axial valves' end-to-end dimensions are not covered in API 6A so option A is not correct. Option B is correct since different types of cages can be used inside axial valves to change the flow characteristic and flow capacity. Option C is not correct since axial control valves have a stem and cage while axial check valves do not. Option D is not correct since having equal pressure on both side of the disk reduces the operating torque for valve operation.

7. Which sentence is incorrect about axial valve actuators?
   - **A.** A linear actuator is normally installed on axial valves which increases the height of the valve plus actuator.
   - **B.** A hydraulic spring return actuator could be a correct choice for an axial valve.
   - **C.** The force produced by the actuator is transmitted to the piston and the disk through the stem for opening and closing.
   - **D.** Single acting linear actuators are rarely used for axial valves.

Answer: Options A, B, and C are correct. However, single acting linear actuators which are equivalent to spring return actuator type (explained in Chapter 10) are the correct choice for axial valves. Thus Option D is not correct.

8. Which sentence is wrong about control philosophy?
   - **A.** A control valve is the last element of the control loop.
   - **B.** A control loop includes a sensor, transmitter, control room, and control valve.
   - **C.** A control valve is used only to control the flow as a process parameter.
   - **D.** A control valve normally has an actuator.

Answer: Option C is not correct since a control valve has the responsibility to adjust all the process parameters such as fluid, pressure, temperature, and level. A control valve is the last element of the control loop receiving a signal from the control room. In addition, a control valve comes with an actuator to control the position of the valve closure member.

9. Which sentence is correct about valve flow characteristics?
   - **A.** Opening a valve 30% leads to passing 30% of the total flow capacity of the valve. This valve has a linear flow characteristic.
   - **B.** A valve is opened 50% while the fluid passage from the valve is just 20% of the total flow. This valve has a flow characteristic of quick opening.

C. If we need to have small flow passage in the beginning of the opening of the valve, we need a linear flow characteristic trim.

D. A valve releases 25% of the total flow capacity at 27% opening. This valve has a quick opening characteristic.

Answer: Option A is correct and the percentage of the valve opening is equal to the percentage of the flow capacity so the characteristic is linear. Option B explains an equal percentage flow characteristic since the amount of flow at half opening is relatively low. Therefore option B is not correct. Option C is wrong because having a small flow at the beginning of the valve opening is equal percentage and not linear. Option D is very similar and close to linear flow characteristic so it does not address quick opening characteristic.

10. Which parameter is not considered an advantage of an axial valve over a globe valve?

A. Less risk of cavitation.

B. Less possibility of leakage due to one piece, bonnet-less design.

C. Shorter face-to-face/end-to-end dimensions.

D. More smooth flow inside the valve.

Answer: Options A, B, and D are correct. Option C is incorrect because an axial valve does not necessarily have a shorter face-to-face compared to a globe valve.

11. Which sentence is wrong about needle valves' design feature?

A. The nonrotated tip or needle provides less galling between the needle and the seats.

B. Needle design provides very good fluid control rangeability.

C. Subsea needle valves could be supplied with a ROV bucket or different shapes of bars.

D. Gland and gland flange force on the packing does not provide enough force for sealing the packing rings.

Answer: Options A, B, and C are correct. Option D is not correct since gland and gland flange force provides lateral force to be changed to axial force for stem sealing by packing.

12. Which choice contains a correct statement about the material selection for axial control valves?

A. The body of ASTM A694 F22 could be selected for the body of the valve without any Inconel 625 overlay for noncorrosive fluid.

B. Super duplex is a good material for the stem of axial control valves due to its high mechanical strength.

C. ASTM A694 F22 cannot be used for the body of the valve due to the risk of corrosion in seawater such as pitting.

D. Super duplex is strong enough for the stem of the valve so there is no need to select any other strong material for the stem.

Answer: Option A is not correct since an Inconel 625 partial overlay is required on the body of a valve made of low alloy F22 grade to prevent crevice corrosion. Partial Inconel

625 overlay should be done on the sealing surfaces and grooves where fluid can be trapped under the sealing and cause crevice corrosion. Option B is correct; super duplex has relatively high mechanical strength including both yield and tensile strength. Option C is not correct since low alloy F22 for the body of the valve would be subject to coating as well as cathodic protection to prevent external corrosion due to the seawater. Option D is wrong since super duplex cannot provide enough strength in case of high actuator force so a stronger material such as Inconel 718 and 725 is proposed as an alternative to super duplex in case of high torque from the actuator.

Further reading

American Petroleum Institute (API) 6DSS, Specification for Subsea Pipeline Valves, third ed., API, Washington, DC, 2017.

American Petroleum Institute (API) 20E, Alloy and Carbon Steel Bolting for Use in the Petroleum and Natural Gas Industries, first ed., API, Washington, DC, 2012.

American Petroleum Institute (API) 20F, Corrosion Resistant Bolting for Use in the Petroleum and Natural Gas Industries, first ed., API, Washington, DC, 2015.

American Petroleum Institute (API) 602, Gate, Globe and Check Valves for Sizes DN100 (NPS 4) and Smaller for the Petroleum and Natural Gas Industries, tenth ed., API, Washington, DC, 2016.

American Petroleum Institute (API) 615, Valve Selection Guide, first ed., API, Washington, DC, 2010.

American Petroleum Institute (API) 623, Steel Globe Valves -Flanged and Buttwelding Ends, Bolted Bonnets, first ed., API, Washington, DC, 2013.

American Petroleum Institute (API) 6FA, Specification for Fire Test for Valves, third ed., API, Washington, DC, 2011.

American Society of Mechanical Engineers (ASME), Buttwelding Ends. ASME B16.25, ASME, New York, NY, 2017.

American Society of Mechanical Engineers (ASME), Gas Transmission and Distribution Piping System. ASME B31.8, ASME, New York, NY, 2018.

Fisher & Emerson Automation Solutions, Control Valve Handbook, fifth ed., Fisher & Emerson Automation Solutions, New York, 2017.

International Organization of Standardization (ISO) 10497, Testing of Valves- Fire Type-Testing Requirement, third ed., 2010. Geneva, Switzerland.

CHAPTER NINE

Remote-operated vehicle (ROV) valve operation

Contents

9.1 Remotely operated vehicles

9.1.1 Introduction and history

A Remotely operated vehicle (ROV) is a free-swimming submersible craft used for the operation of subsea valves and actuators. These vehicles can be defined as state-of-the-art subsea (underwater) robots. ROVs are designed to perform underwater tasks without any need to put a diver in the subsea. A ROV can explore the subsea world while the operator or pilot remains in a comfortable condition on the surface. Without the development and usage of ROVs, the ability of humans to explore underwater would be limited and in some cases impossible. The first ROV was designed and built in 1953 by French diver Dimitri Rebikoff. The US navy started using ROVs in the 1960s for the recovery and maintenance of underwater facilities. "ROV" is a common name and umbrella term for

Subsea Valves and Actuators for the Oil and Gas Industry
https://doi.org/10.1016/B978-0-323-90605-0.00011-6

autonomous underwater vehicle (AUV), unmanned underwater vehicle (UUV), and unmanned untethered underwater vehicle (UUUV). AUV is equipment used in water with the ability to position itself without interference from surface control. UUV/UUUV is equipment used in water with the ability to position itself without a cable to the surface.

The arm of the robot is called a "manipulator"; it is connected to the ROV frame and carries a torque tool for the operation of the valve. There are normally two or three fingers at the end of the arm that can handle an object such as a torque tool for operation. A ROV normally has two arms or manipulators for the performance of tasks such as operation or intervention. Intervention refers to different tasks that are done subsea such as maintenance, repair, inspection, retrieval, and maintenance. Fig. 9.1 illustrates a ROV that operates the valves with one arm through a ROV bucket in orange color. ROVs are typically connected to the surface control station by umbilical cables. Power, control, and video signals are transmitted through the umbilical and an operator from the surface control station controls the ROV. Usually the source of power for ROV movement is electrical or electrical and hydraulic.

Fig. 9.2 illustrates in detail how a ROV arm is put in the torque tool inside the ROV bucket for operation. The type of subsea tasks that can be performed by ROV is limited to general operation support such as the operation of valves and the installation and replacement of small and light objects. Alternatively, Remote-Operated Tools (ROTs) are used for the installation and replacement of heavy objects subsea. As an example, an actuator retrieval tool (ART) may be considered a type of ROT that can remove an actuator from a valve subsea and send it to surface for maintenance or transfer it to another valve for operation. In general, intervention tasks could be done by ROV or

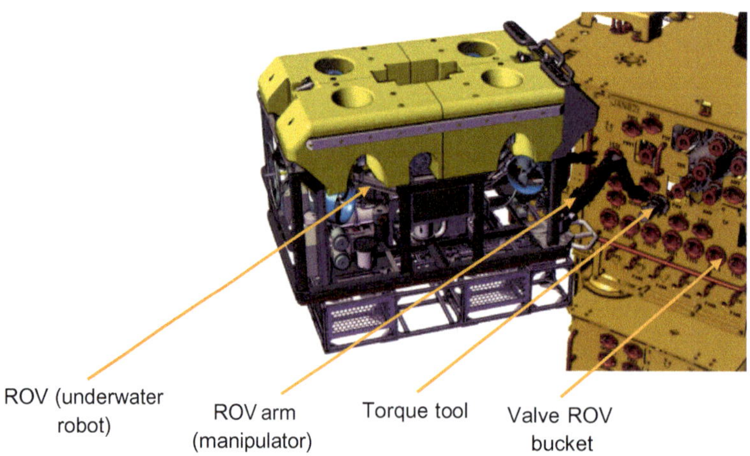

Fig. 9.1 Valve operation by ROV.

Fig. 9.2 ROV operation underwater on a bucket in class 4.

by ROT or by a combination of both. Normally, if both ROV and ROT are required for intervention, the ROV acts as a backup for the ROT.

Therefore subsea valves can be operated manually by a remote-operated vehicle (ROV) or actuated. An actuator is a mechanical device installed on the top of a valve for the automatic opening and closing of the valve. Valves that are not opened or closed frequently can be operated manually with a ROV and/or by a diver. Subsea valves installed in shallow water depths such as 80 m could be operated by both diver and ROV.

9.1.2 ROV key components

Pilot Control is not a component of a ROV. Normally a control room is installed on the surface. Surface control provides an interface for an operator or pilot to control the ROV and get feedback from it via its camera view.

The key components of a ROV system are highlighted in Fig. 9.3; they are explained as follows:

Frame: This is a structure that can protect the ROV internals. Components such as a camera, lights, and a tether are installed on the frame. Most ROVs are built in rectangular frames that are easy to manufacture and modify. However, some ROVs can be built with highly specialized frames to reduce drag force, enhance appearance, and prepare the ROV for special tasks.

Tether: Almost all ROVs have tethers in order to get electrical power from the surface and send signals to the surface. Drawing electrical power from the surface enables the pilot to control the ROV at depth, permits the ROV to send signals to the surface, and enables

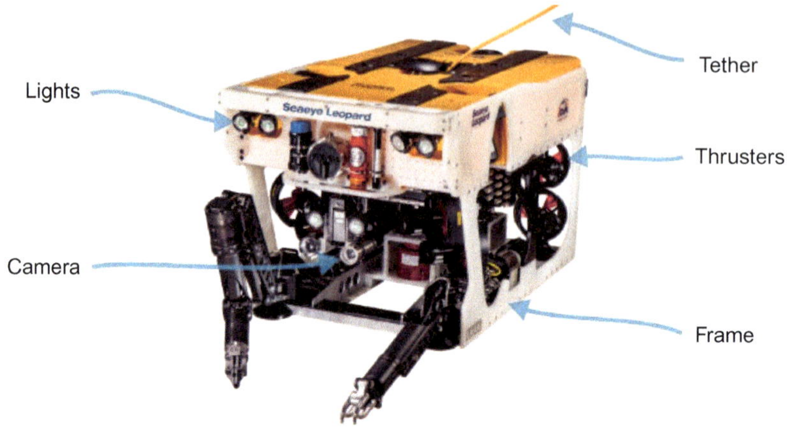

Fig. 9.3 Key components of a ROV. *(Courtesy: Blue Robotics.)*

the operator or pilot to see the ROV through its camera. Alternatively, in rare cases, optical or acoustic modems on the surface enable wireless operation. The design of the tether should prevent mechanical damage such as bending or water head pressure.

Lights: There is no sunlight underwater at times, so lights are required to provide illumination for the operation of ROVs.

Camera: A camera provides images from the underwater path and allows the surface pilot or operator to monitor the operation of the ROV. A minimum of two views from underwater should be provided by the camera.

Thrusters: Thrusters are electrically or hydraulically powered propellers used to maneuver the ROV.

The minimum connection points for additional equipment to be connected to ROVs are listed as follows, based on the Norsok U-102 standard for ROVs:

- Electrical connection suitable for 12, 24, and/or 250 V direct current (DC) and 5 A electricity. Alternatively, the electrical connection could be 110 and/or 220 V alternative current with 1 A current.
- Signal connection in 1 quad or 2 data channels through optical fiber. Fiber optics, or optical fiber, refers to the medium and the technology associated with the transmission of information as light pulses along a glass or plastic strand or fiber. Fiber optics is used in long-distance and high-performance data networking.
- One video connection for additional cameras that shall provide a view of at least two sides at the same time.
- Transponders: The observation ROV shall be equipped with a mini transponder, preferably continuously charged from the ROV and, if relevant, compatible with the system on board the platform/vessel for navigational purposes and emergency recovery. Transponders are devices that emit signals in response upon receiving signals.

9.1.3 ROV standards, dimensions, and weight

There are three standards that address ROVs: the American Petroleum Institute (API 17H), the International Organization for Standardization (ISO 13628-8), and the Norwegian Petroleum Standard (Norsok U-102). There is no standard size for ROVs. Annex A of API 17H/ISO 13628-8 describes the typical dimensions of different ROV types, including length, width, and height as well as weight values. Table 9.1, extracted from API 17H annex A, provides ROV weight and dimensional data.

9.1.4 ROV work classes

Five work classes are defined for ROVs according to the Norsok U-102 standard:

1. *Class 1—Pure observation*: The capability of these ROVs is limited to video observation. These ROVs in general are small in size and fitted with lights, cameras, and thrusters. These vehicles cannot perform any underwater task or intervention.
2. *Class 2—Observation with payload option*: ROVs in this work class have the capability of carrying some additional components such as a cathodic protection measurement system, additional sensors, and video cameras.
3. *Class 3—Work class vehicles*: Typically, these ROVs are large enough to carry additional sensors and manipulators. ROVs in work class 3 are larger and heavier than those in work class 1 and 2. Work class 3 ROVs are divided to three categories based

Table 9.1 Different types of ROV with weight and dimensional values.

	Length		Width		Height		Mass	
ROV	**mm**	**(in.)**	**mm**	**(in.)**	**mm**	**(in.)**	**kg × 10^3**	**(klb)**
Challenger	2475	(97.44)	1400	(55.12)	1325	(52.17)	1.71	(3.76)
Hysub 60	2510	(98.82)	1420	(55.91)	1480	(58.27)	1.77	(3.89)
Pioneer	1620	(63.78)	1620	(63.78)	1620	(63.78)	1.34	(2.96)
Super Scorpio	2500	(98.43)	1500	(59.06)	1700	(66.93)	2.25	(4.95)
Triton	2440	(96.06)	1420	(55.91)	1320	(51.97)	2.27	(5.00)
Trojan	2200	(86.61)	1600	(62.99)	1600	(62.99)	1.84	(4.05)
MRV	2300	(90.55)	1500	(59.06)	1560	(61.42)	1.90	(4.18)
SCV	2500	(98.43)	1500	(59.06)	1700	(66.93)	2.81	(6.18)
Examiner	2000	(78.74)	1950	(76.77)	1800	(70.87)	1.73	(3.82)
Diablo	2466	(97.09)	1500	(59.06)	1750	(68.90)	1.94	(4.27)
Triton XL	2440	(96.06)	1480	(58.27)	1620	(63.78)	3.06	(6.74)
Hydra magnum	2400	(94.49)	1400	(55.12)	1700	(66.93)	2.03	(4.47)
Hydra millennium	3000	(118.11)	1524	(60.00)	1700	(66.93)	2.22	(4.90)

In order to take advantage of ROVs of opportunity, interfaces should be designed with a mass of 2500–3000 kg (5510–6612 lbs).

Note: This list is not exclusive and is indicative of the different sizes of vehicle available.

Courtesy: ISO 13628-8.

on power: class 3A are vehicles with less than 100 horse power (HP), class 3B have a range of 100–150 HP, and class 3C have more than 150 HP.

4. *Class 4—Seabed working ROVs*: These types of ROVs are capable of performing underwater tasks such as pipeline and cable trenching and other remote underwater tasks on the seabed. This class of ROVs is much larger and heavier than work class 3 types.
5. *Class 5—Prototype or development vehicles*: Any ROV that does not belong to the four above-mentioned classes are considered to belong to this category. Autonomous underwater vehicles (AUV) are considered to be class 5 or a new class.
6. *Class 6—Autonomous underwater vehicle (AUVs)*: This is a robot which can travel underwater without any operator requirement. AUV can be named unmanned undersea vehicle (UUV).

Fig. 9.4 illustrates a 3 tonne Triton XL inside a frame capable of performing a wide range of underwater tasks and interventions. This ROV is designed for very deep water depths and harsh environmental conditions and can support the following tasks:

- Deep water and ultra-deep water construction work and installation;
- Subsea cable burial and maintenance;
- Deep-water maintenance and recovery;
- Remote tool deployment;
- Platform inspection and maintenance;
- Drilling task support and well completion activities;

Fig. 9.4 Triton XL.

The length, width, and height of this tool are 3226, 1803, and between 2000 and 2148 mm, respectively. It has been rated for 2000-m water depth and a payload of 200 kg. Payload is defined as additional weight that can be safely added to the ROV over and above its empty weight.

9.1.5 ROV personnel and control system

The personnel working with a ROV should have technical training and education about such areas as hydraulic, electric, and electronic remotely operated systems and robotics. Personnel working with ROVs can be split into the different categories of superintendent, supervisor, and pilot trainee.

The control station should be capable of monitoring the following parameters of the ROV:

- Electric voltage and current;
- Hydraulic pressure;
- Duration of ROV movement;
- Number of ROV turns for opening and closing a valve, or operation of the valve in general;
- Reading the depth of the ROV;
- Alarm for any water entering the ROV;
- Alarm for failure to function or move.

Additionally, the equipment and components in the control room should be designed to withstand shocks. The control room is a workplace for the people who are working with the ROV, so the noise level there should be as minimal as possible and conform to rules and regulations such as Occupational safety and health administration (OSHA). The atmosphere inside the control room should be dry without any humidity. From an architectural point of view, the position of the lights inside the control room is important and should minimize reflection on the screens and video monitors to protect the eyesight of the operator. The control room should have air conditioning, but the air conditioning outlet should not directly face the operators in the control room.

9.2 ROV bucket

The ROV torque tool enters the ROV bucket (see Figs. 9.5 and 9.6), also called a receptacle, which is connected to the top of the operator. The engagement of the ROV torque tool in the receptacle is done through two slots (see Figs. 9.5 and 9.6) that are machined and sized accurately according to ISO 13628-8 or API 17H CL.4 to guarantee very precious fitting. Metallic ROV buckets are in carbon steel and plastic ROV buckets are in Devlon. Devlon V-API is a thermoplastic soft material with a proprietary formula developed by James Walker; it includes additives that allow the material to perform at

Fig. 9.5 Metallic ROV bucket including slots in carbon steel.

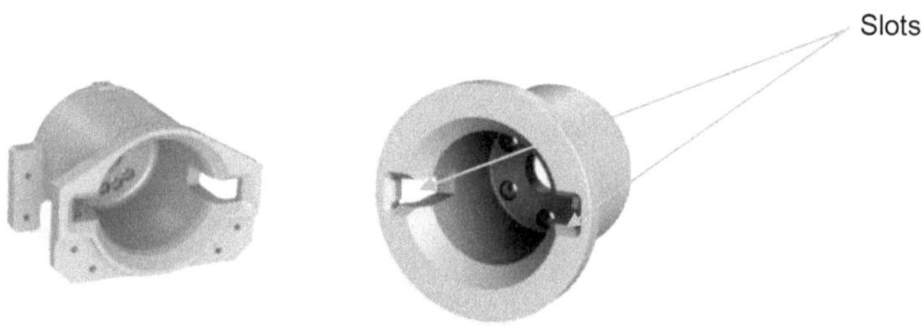

Fig. 9.6 Nonmetallic ROV bucket including slots in Devlon.

higher pressures and temperatures than the majority of soft valve seat materials. The ROV bucket could be supplied with lifting points for the installation of lifting lugs for safe transportation (see Fig. 9.7).

The ROV bucket is located between the torque tool and the nut or drive and the valve gear box. A valve gear box is used for increasing torque. For this reason, the operation type of the valve in Table 9.2 is given as ROV plus gear. The force is transmitted from the torque tool to the nut or override and to the gear box afterward if there is a need to increase the torque value or change the direction of the torque. The torque or load will be transferred to the valve stem and to the valve internals such as the closure member and the seats. A stem nut (see Fig. 9.8) is located around the valve stem of gate valves to change the rotary motion of the ROV bucket to linear movement.

Fig. 9.9 illustrates a ROV-operated manual ball valve in a small size such as ¾″. Item # 1 in the figure is the ROV bucket, which could be class 2 for a ¾″ ball valve in pressure class 10,000. More information will be provided for ROV class selection later in this chapter. The torque tool sits inside the ROV bucket (item #1) and the torque is transferred from the "nut" or "key" or "ROV bucket shaft override" (item #2; see Fig. 9.10) to the valve stem (item #3). There is no gear box required in this design since there is no

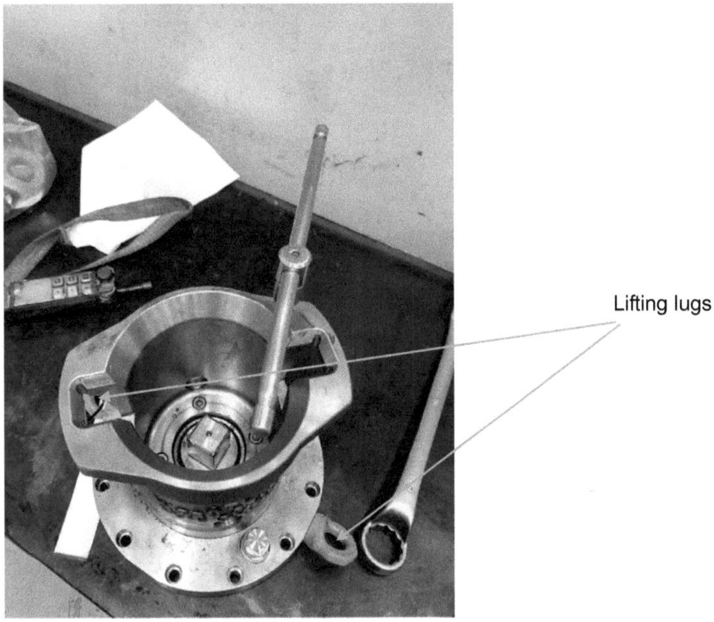

Fig. 9.7 ROV bucket with lifting lugs.

Table 9.2 ROV operation turns for cycling a 7 1/16″ manual slab gate valve in pressure class 10,000 psi.

Type of valve	Size	Pressure class	Operator type
Slab gate valve	7 1/16″	10,000 psi	ROV + gear
Total number of turns to open and close the valve	**Minimum disengagement turns**	**Maximum disengagement turns**	
32	0.1	0.5	

need to increase the input torque or change the rotary torque or force to linear movement for a quarter turn ball valve. Item #4 is a part of position indicator which will be reviewed and discussed later in this chapter.

The ROV override moves clockwise to close the actuator and the valve. After a certain amount of rotation to close the valve completely, the ROV should be rotated to the opposite side to a certain degree, called disengage turns, to disengage the ROV torque tool from the receptacle. The interface of a ROV bucket to a valve for a manually operated valve could be direct (see Fig. 9.9) or through a flange connection (Fig. 9.11). Direct connection of the ROV to the valve is called direct drive, meaning that there is no gear mechanism. On the other hand, the ROV bucket could be used in connection with the

Fig. 9.8 Stem nut.

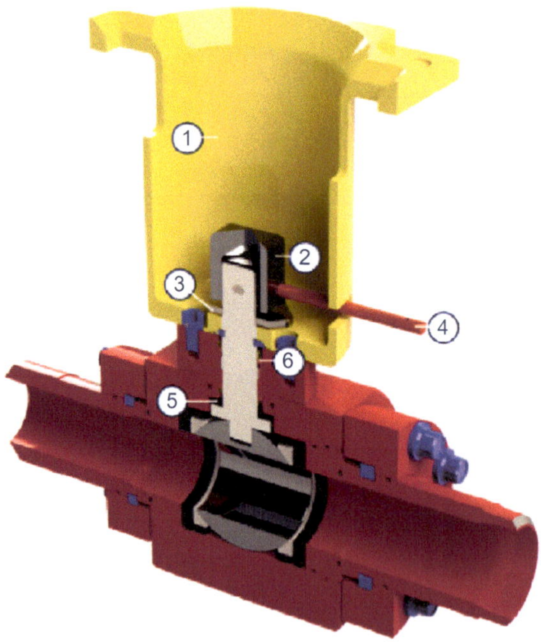

Fig. 9.9 ROV direct drive-operated ball valve. *(Courtesy: Combet Valves.)*

Fig. 9.10 Nut/key/override shaft.

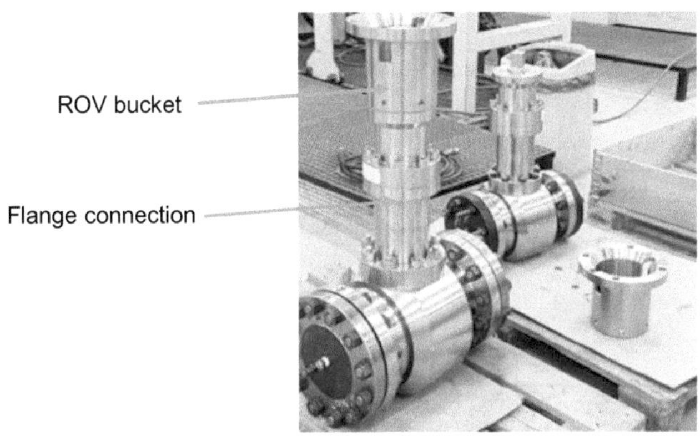

Fig. 9.11 ROV connection to the valve through flange connection and gear box.

Fig. 9.12 ROV bucket with gear box and pressure compensation.

gear box to transmit torque to the valve. In the first case, the valve is operated via direct drive ROV and in the second case, the valve is operated via ROV plus gear.

The gear box of the valve is normally pressure compensated to prevent implosion due to the seawater head pressure. Figs. 9.12 and 9.13 illustrate a ROV bucket connected to a gear box that is pressure compensated through a closed compensation system. Compensation systems are explained in Chapter 11. Fig. 9.14 illustrates a typical ROV interface supplied by a valve manufacturer. The interface of the valve in small sizes such as a ½″ ball valve in pressure class 690 bar equal to 10,000 psi, or needle valves and other low torque applications for ROV operation, is called a rotary low torque paddle interface. Fig. 9.15 illustrates a low torque paddle for a double needle valve with a bleed between connected to a hot stab. The other important consideration associated with ROV bucket design is the antifouling requirement. Any object submerged in water can be fouled by marine growth and bacteria; it is the designer's responsibility to ensure that the subsea component is clear and readable during its design life. ROV interfaces such as ROV buckets should be made in antifouling materials, paint, and coating.

Figs. 9.16 and 9.17 illustrate a gear box for an 18″ ball valve. The type of gear illustrated is a spur gear, also called a straight cut gear, which is one of the simplest types of gears. A spur gear is a typical kind of cylindrical gear that is parallel to the shaft. Fig. 9.18 illustrates two spur gears that rotate together; the rotation is transferred to a shaft that is connected to one of the gears and rotated 90 degrees. There is a connection to the gear box for oil or grease injection for the purpose of lubrication. Lubrication of the gear box is very important to prevent galling and wearing of the gear box teeth due to contact and

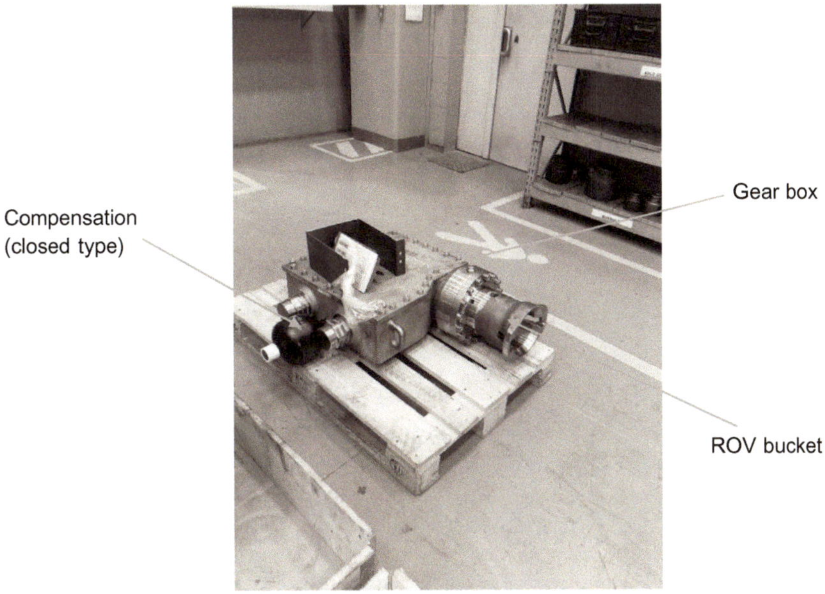

Fig. 9.13 ROV bucket with gear box and pressure compensation.

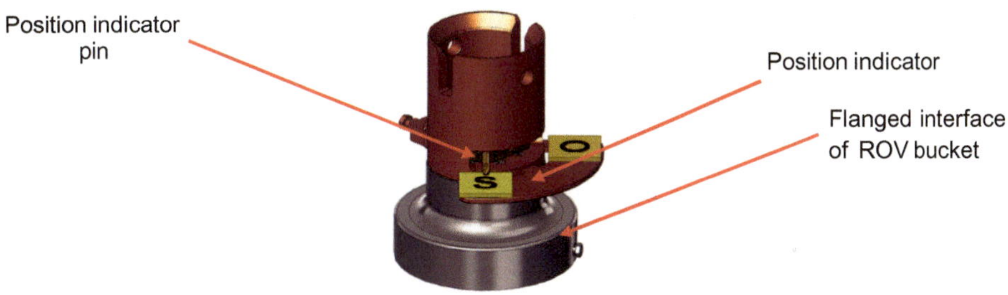

Fig. 9.14 Typical ROV connection supplied by Petrol Valve.

Fig. 9.15 Low torque paddle ROV interface for a double block and bleed including two needles and a bleed between. *(Courtesy: Galperti Engineering.)*

Fig. 9.16 Spur gear box for an 18″ ROV-operated ball valve.

movement together. Some end users may ask for 100% filling of the grease or oil inside the gear box. The gear box is normally pressure compensated; two relief valves can be connected to the gear box to adjust the compensation pressure to keep the gear box pressure equal to the seawater head. Two nuts installed on the gear box act as travel stops to

Fig. 9.17 Spur gear box for an 18″ ROV-operated ball valve.

Fig. 9.18 Spur gear connections and movement together.

prevent overtravelling the ball valve closure member. Overtravelling of the ball means movement of the ball beyond 90 degree rotation. The whole gear box is placed in a casing as illustrated in Fig. 9.19.

Fig. 9.20 illustrates a disassembled gear box plus a ROV bucket and compensation capsule.

Fig. 9.19 Gear box casing.

Fig. 9.20 Spur gear parts disassembled from the valve with ROV bucket and compensation capsule.

Fig. 9.21 illustrates gears on the top of a 7 1/16″ slab gate valve in a pressure class of 690 bar equal to 10,000 psi. The gear is removed from the top of the gate valve after its qualification tests during disassembly and inspection (see Figs. 9.22 and 9.23) and checked visually to make sure it is not damaged or worn due to frequent cycling.

Fig. 9.21 Gears on the top of a slab gate valve.

Fig. 9.22 Gear disassembled and removed from the top of the valve.

Fig. 9.23 Gear disassembled and removed from the top of the valve.

9.3 Position indicator

A position indicator is located at the bottom of the ROV bucket; it should be readable during valve operation. The indicator shows two letters: "S" for shut and "O" for open. The small pin showing the letter "S" means that the valve is in closed position. The position indicator should be made of antifouling material to prevent marine growth and should always be readable. An Aquasign trademark is proposed for the position indicator with ultraviolet (UV) stability based on industrial experiences. For a ball valve with quarter turn movement of the ball from open to closed position, the position indicator would move from "O" to "S" during valve closure. As an example, the position indicator on the ball valve in Fig. 9.24 shows that the valve is in open position. The position indicator can be connected to the valve stem through a high-strength screw in some valve designs.

Fig. 9.25 illustrates a position indicator located inside a housing and cover. The cover is the part with the letters "O" and "X" indicating the position of the valve. The main part of the position indicator is located inside the housing and includes components such as a piece of pipe, sealing, screws, and an indicator bar. The position indicator sealings should be dual, one as a backup for the other, since they are dynamic sealings, i.e., sealings used for moving components.

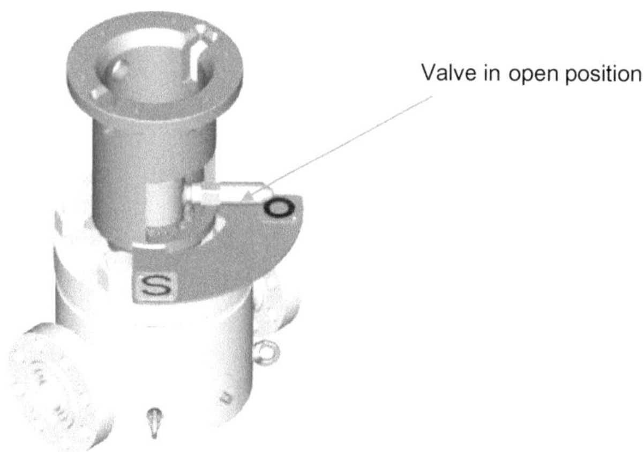

Fig. 9.24 Position indicator showing the valve in open position.

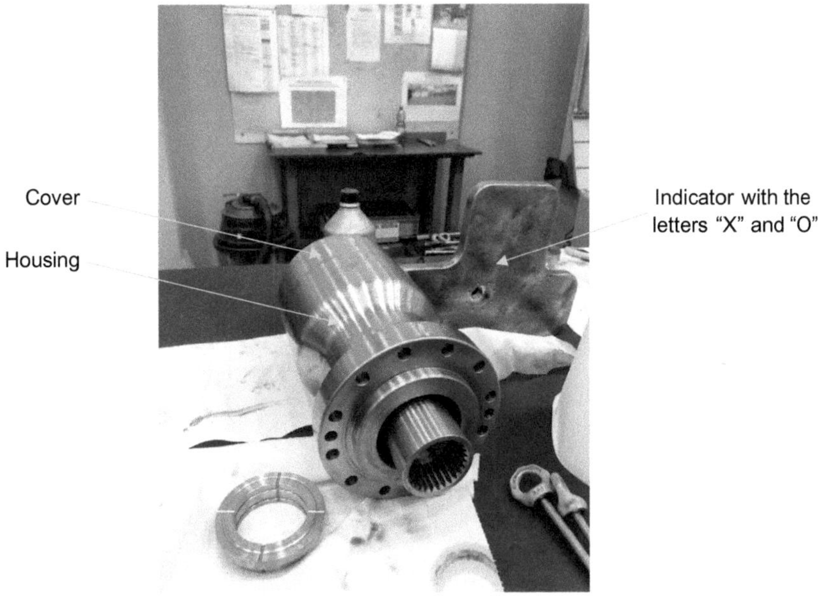

Fig. 9.25 Position indicator.

9.4 ROV turns and operation

There is always a minimum and maximum number of opening/closures turns and disengage turns. As an example, Table 9.2 provides the number of turns required for ROV operation as well as the minimum and maximum number of disengage turns for a 7 1/16″ slab gate valve produced based on API 6A in pressure class 10,000 psi equal to 690 bar.

The number of ROV turns is given in its installation, operation, and maintenance (IOM) manual and general arrangement (GA) drawings. The number of ROV turns is measured during the factory acceptance test (FAT).

ROV-operated subsea valves are also called manually operated valves since they are operated manually without the use of actuators. Manually operated valves in other sectors of the oil and gas industry such as topside are operated by a human operator by moving the valve lever or handwheel. Subsea valves can also be operated via handwheel or lever by a human diver. Fig. 9.26 illustrates a small ball valve less than 1″ in size with lever operation for subsea application suitable for diver operation. More information and examples are available in Section 9.8.

Fig. 9.26 Lever-operated subsea ball valve. *(Courtesy: Oliver Valve.)*

9.5 ROV override

Usage of a ROV bucket is not limited to manually operated valves. Using a ROV to override an actuator is common for actuated valves. ROV override allows the operation of the valves independently from the actuators. ROV override as a backup for valve actuation is beneficial, especially in the case of actuator failure to function. Fig. 9.27 illustrates a rack and pinion spring return (fail-safe closed) actuator with ROV override. The actuator contains two spring housings and two pistons and cylinders, which is called symmetrical design. This design provides a more robust actuator for high numbers of cycles (opening and closing) and faster operating speed due to its double numbers of springs housing and double volume of hydraulic oil. A fail-safe closed actuator closes via spring force and opens via hydraulic force or pressure. If the actuator gets out of order, ROV override would be the only tool for operating the actuator. Rotating the ROV counterclockwise opens the actuator (instead of hydraulic pressure) without compressing the spring. The ROV override on the actuator should be operated and rotated clockwise (look at the blue rotation symbol in Fig. 9.27) to close the valve and actuator. The blue clockwise rotation angle is shown beside the spring housing with a blue arrow on the springs indicating the expansion of the springs. This means that the clockwise rotation of the ROV for closing the actuator and the valve is the same as that of the spring expansion inside the actuator for closing the valve. In fact, operating and rotating the ROV bucket clockwise closes the valves and actuators in the same way the spring expansion action does.

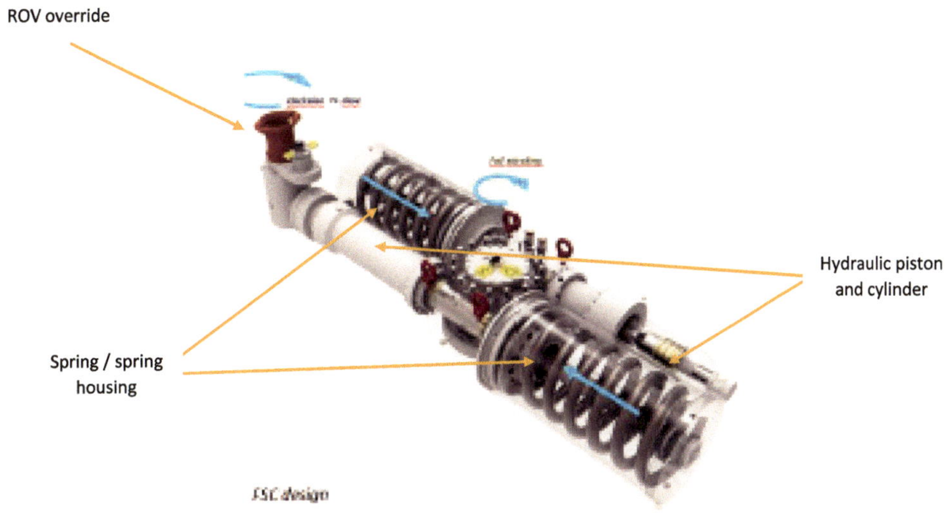

Fig. 9.27 Symmetrical spring return rack and pinion subsea actuator. *(Courtesy: Galperti Engineering.)*

Table 9.3 indicates the same information mentioned in Table 9.2, including total number of ROV torque tool turn for opening and closing of the valve, minimum and maximum disengagement turns for the same valve type, and operated with a hydraulic linear spring return actuator and ROV override:

When it comes to valve and actuators, ROV bucket installation is not limited to actuators and gear boxes installed on valves. Some actuators can be designed and operated as retrievable. Coupling between the spool piece and operator could be guaranteed by a clamp system. A clamp system provides alignment between a valve and actuator and acts as a clutch for removing the actuator from the valve and cycling the actuator independently from the valve. The clamp arms can be blocked and unblocked by a ROV tool through the ROV bucket located on the clamping system (see Fig. 9.28).

Table 9.3 ROV operation turns for cycling a 7 1/16″ actuated slab gate valve in pressure class 10,000 psi.

Type of valve	Size	Pressure class	Operator type
Slab gate valve	7 1/16″	10,000 psi	Actuator + ROV override
Total number of turns to open and close the valve	**Minimum disengagement turns**	**Maximum disengagement turns**	
283	2.54	3	

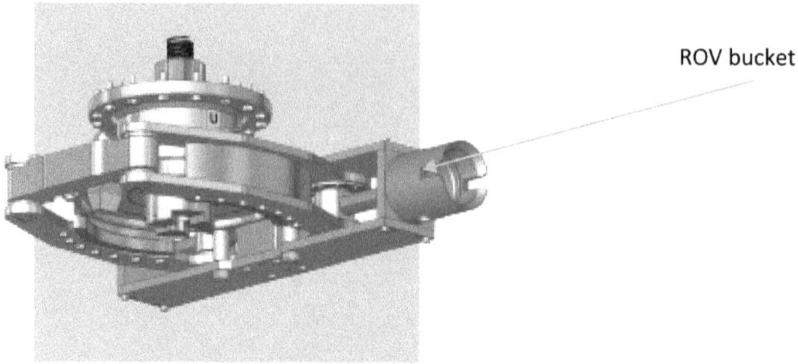

Fig. 9.28 Clamp system for actuator retrieval. *(Courtesy: Petrol Valve.)*

9.6 ROV bucket design considerations

The different parameters important for the design and selection of a ROV bucket are listed as follows:

1. *ROV type interface and class*: ROV interface is selected based on API 17H or ISO13628-8, or part 8 of ISO 13628 standard, titled "Remotely Operated Vehicle (ROV) on subsea production systems." Seven classes of ROV are defined in API 17H/ISO 13628-8, and each class has a maximum design torque capacity that is defined in Table 9.4. The higher the class number, the greater the torque input capacity.

   ROV bucket classes 2, 3, and 4 are common for subsea valves. Class 3 is less popular than the other two classes for the ROV bucket. In addition, a low torque paddle could be used for smaller valves such as 0.5″ ball valves in 10,000 psi. The ROV class selection for manual valves depends on the valve size and pressure class. ROV override class, which is used for actuators, can also be dependent on the hydraulic pressure supplied to the actuator as well as the valve size and pressure class. Higher ROV

Table 9.4 ROV classes and maximum torque capacity as per API 17H/ISO 13628-8.

Class	Maximum ROV bucket design or input torque N m (lbf; ft)
1	67 (50)
2	271 (200)
3	1355 (1000)
4	2711 (2000)
5	6779 (5000)
6	13,558 (10,000)
7	33,895 (25,000)

Table 9.5 ROV bucket dimensions.

Dimension	Class 1	2	3	4	5	6	7
A square	17.50 (0.687)	17.50 (0.687)	28.60 (1125)	38.10 (1.50)	50.80 (2.00)	66.67 (2625)	88.90 (3.50)
B	154.0 (6.06)	154.0 (6.06)	154.0 (6.06)	154.0 (6.06)	190.5 (7.50)	243.0 (9.56)	243.0 (9.56)
C min	41.0 (1.62)	41.0 (1.62)	41.0(1.62)	41.0 (1.62)	63.5(2.50)	89.0 (3.50)	89.0 (3.50)
D	38.0 (1.50)	38.0 (1.50)	38.0 (1.50)	38.0 (1.50)	57.0 (2.25)	82.25 (3.25)	82.25 (3.25)
E	32.0 (1.25)	32.0 (1.25)	32.0 (1.25)	32.0 (1.25)	38.0 (1.50)	44.5 (1.75)	44.5 (1.75)
F	82.5 (3.25)	82.5 (3.25)	82.5 (3.25)	82.5 (3.25)	127.0 (5.00)	178.0 (7.00)	178.0 (7.00)
G min.	140.0 (5.51)	140.0(5.51)	140.0 (5.51)	140.0 (5.51)	140.0 (5.51)	222.0 (8.75)	435.0 (17.13)
G max.	146.0 (5.75)	140.0 (5.75)	146.0 (5.75)	146.0 (5.75)	146.0 (5.75)	228.0 (9.00)	441.0 (17.38)
H	181.0 (7.12)	181.0 (7.12)	181.0 (7.12)	181.0 (7.12)	206.0 (8.12)	–	–
J	12.7 (0.50)	12.7 (0.50)	12.7 (0.50)	12.7 (0.50)	–	–	–
K	168.5 (6.63)	168.5 (6.63)	168.5 (6.63)	168.5 (6.63)	–	–	–
M	25.4 (1.00)	25.4 (1.00)	25.4 (1.00)	25.4 (1.00)	–	–	–
N	194.0 (7.63)	194.0 (7.63)	194.0 (7.63)	194.0 (7.63)	–	–	–

As an alternative to dimension *A*, end effector shapes as found in Annex D for the appropriate torque range may be used.
All dimension tolerances are as follows:

$0{,}x \pm 0.5\ (0.020)$

$0{,}xx \pm 0.25\ (0.010)$

$C:\ \begin{matrix}+1.27\\0\end{matrix}\left(\begin{matrix}+0.05\\0\end{matrix}\right)$

Note 1: Chamfer on the end of the end effector profile 45′ × 1.65 (0.06)max.
Note 2: Clearance behind antirotator slots [*E* × F × 50.8 (2)] is to allow for licking feature option provided by some tools.
Dimensions in millimeters (in.).

bucket classes are normally selected for larger and higher pressure valve classes. The following examples give some ideas and guidelines for ROV bucket class selection. The minimum size of valve that can be operated with ROV is ½″ according to API 17H-ISO 13628-8. Dimensions of ROV buckets in both millimeters and inches are given in Table 9.5 based on the dimensional letters provided in Fig. 9.29 according to API 17H/ISO 13628-8.

The important consideration is that all classes (1, 2, 3, and 4) have the same end-to-end length for the ROV bucket (parameter K), but the dimension of the nut in the middle of the ROV bucket (parameter A) is different for these classes.

Example 9.1 Table 9.6 provides ROV bucket classes for some of the manual subsea valves based on industrial experience:

2. *ROV bucket maximum rated torque capacity*: Maximum torque capacity for each ROV bucket class is given in the standard according to Table 9.4. As an example, a 2″ ball valve in ROV bucket class 4 could have a minimum operational torque equal to 1350 N m and a maximum operational input torque equal to 2711 N m as per API 17H/ISO 13628-8 standard, and a damage torque of 20% above the maximum input torque equal to approximately 3250 N m.

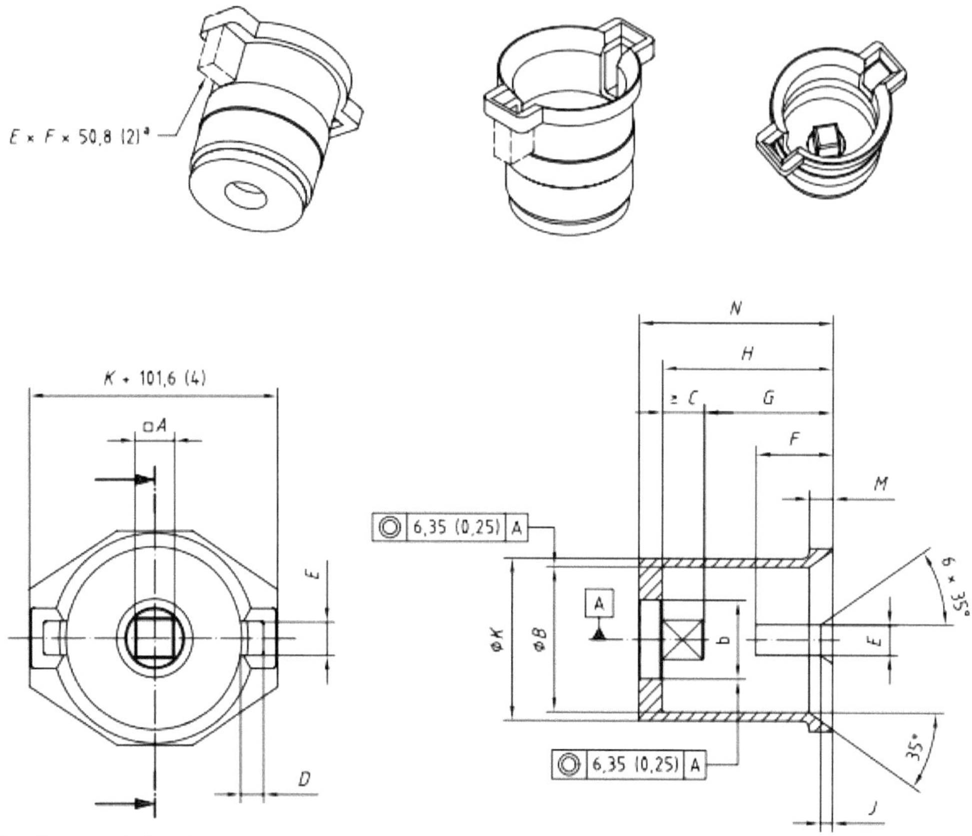

a Clearance both ends.

Fig. 9.29 Rotary ROV bucket.

Table 9.6 ROV bucket class selection for manual valves.

Manual valve	ROV bucket class
22″ ball valve 7500 psi	4
16″ ball valve 7500 psi	4
10″ ball valve 7500 psi	4
7 1/16″ slab gate valve 7500 psi	4
1 13/16″ slab gate valve 7500 psi	3
1″ slab gate valve 15,000 psi	3
1″ slab gate valve 7500 psi	2
¾″ ball valve 10,000 psi	2
½″ ball valve 10,000 psi	Low torque paddle

3. *Material*: Metallic material such as carbon steel or soft material such as Devlon can be selected for the ROV bucket.
4. *Coating*: Orange colored coating is quite common for ROV buckets which could be according to Norsok M-501 standard system 7C.
5. *Number of turns for opening and closing*: This information is given by the valve supplier in the General arrangement (GA drawings) and the Installation, operation, and maintenance (IOM) manual for valves and actuators.
6. *Speed of torque tool*: This is normally provided by the purchaser or end user in the specification of valves and actuators. However, the valve supplier could provide the data about the maximum speed of the torque tool that the ROV bucket can withstand—in simple terms its "maximum speed limit." More information about the speed of the torque tool inside the ROV bucket is given in Section 9.8.
7. *ROV accessibility*: There are two types of ROV access to the valve bucket: one is from the top and the other is from the side. When it comes to valves in the manifolds, large (above 2″) ball and gate valves are normally installed on the horizontal line with the ROV bucket on the top so the access is from the top. However, smaller valves can be installed on the vertical line with the stem on the horizontal and providing side access for ROV operation. Fig. 9.30 illustrates a 10″ ball valve installed on the horizontal line with the stem and ROV extension standing vertically with an access from the top on the manifold roof. Valves with a horizontal orientation and horizontal stems also have side access for the ROV for valve operation.

Fig. 9.30 10″ ball valve with ROV access from the top of the manifold.

Fig. 9.31 10″ ball valve with ROV access from the top of the manifold.

Fig. 9.31 illustrates small bore ½″ ball valves with a low torque paddle ROV interface in orange color installed on the vertical line, with horizontal stem orientation and low torque paddle installation so the ROV access is from the side of the manifold. Fig. 9.32 shows low torque paddles for these valves from the outside of the manifold. Small valves with low torque paddles on the top interface with the steel structure through a square plate at the ROV receptacle.

Fig. 9.32 Side view of low torque paddles.

8. *ROV tool operation closing and opening movement*: Normally the ROV torque tool moves clockwise to close and counterclockwise to open. ROV tools normally have rotary motion that can be converted to linear motion in a slab gate valve with linear stem movement. The transfer of the torque tool's rotary motion to the valve stem is done through a key or nut, or a ROV bucket shaft as explained before.

9.7 ROV bucket extension

A ROV bucket may be extended to reach to the level of the manifold roof to provide access to the ROV for safe operation. The requirements for an extension of the ROV bucket, including the length of the extension, should be provided by the valve purchaser to the valve supplier or manufacturer. The position indicator should be extended along with the ROV bucket in order to be readable by the ROV. As an example, Fig. 9.33 illustrates 8″ and 10″ lines on one manifold. 7 1/16″ slab gate valves are located on the branches of piping with a nominal pipe size (NPS) of 8″. Gate valves occupy a lot of space vertically when they are installed on the horizontal line. In addition,

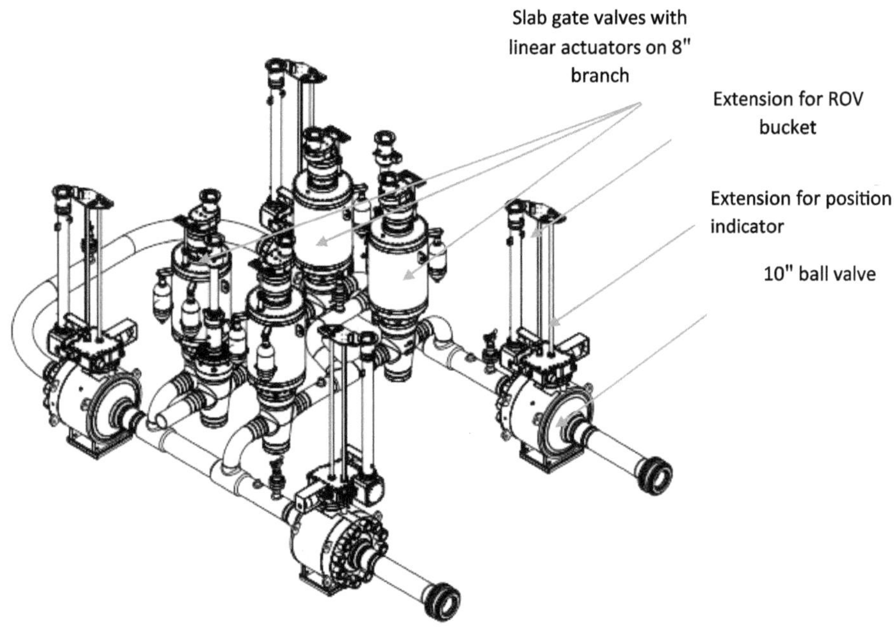

Fig. 9.33 ROV bucket extension for ball valves on a manifold.

linear actuators, which stand vertically when the valve is on the horizontal line, increase the height of the actuated slab gate valves. Thus the height of the manifold and the manifold roof is affected by the height of linear actuated gate valves. 10″ ball valves located on a 10″ line are shorter than slab gate valves, so they need an extension for the ROV bucket and position indicator as illustrated in Fig. 9.33. The key question is, what would happen if the ball valve ROV bucket were not extended? In that case, the ROV bucket of the ball valve would be inside the manifold and thus inaccessible for the ROV. The ROV is normally large and cannot pass inside the manifold, so it operates the valves from outside the manifold envelope.

The following types of valves are used inside a manifold:

1. 20″ ball valve in pressure class 7500 equal to 517 bar with ROV type operation; the height of this valve with a ROV on top is 2.7 m.
2. 20″ ball valve in pressure class 7500 equal to 517 bar with a hydraulic double acting actuator; the height of this valve including the actuator installed on the top is 3.7 m.
3. 16″ ball valve in pressure class 7500 equal to 517 bar with a hydraulic single return actuator; the height of this valve plus the actuator installed on the top is 2.4 m.
4. 10″ ball valve in pressure class 7500 equal to 517 bar with ROV type operation; the height of this valve plus the ROV on top is 1.7 m.
5. 8″ slab gate valve in pressure class 7500 equal to 517 bar with ROV type operation; the height of the valve is 2 m.

Question: In this case, which valves require an extension for the ROV bucket and position indicator and what should the height of the extensions be?

Answer: The highest valve is the 20″ ball valve with a double acting actuator at 3.7 m, and its height governs the height of the manifold. Thus the manifold height should be 3.7 m and the other valves would require an extension to reach that height. Thus the required extensions are as follows:

1. The 20″ ball valve with ROV requires a 1-m extension (3.7 − 2.7 = 1 m).
2. The 16″ ball valve with a spring return actuator requires a 1.3 m extension on the ROV override bucket extension (3.7 − 2.4 = 1.3 m).
3. The 10″ ball valve with manual or ROV operation requires a 2 m extension (3.7 − 1.7 = 2 m).
4. The 8″ slab gate valve with ROV operation requires a 1.7 m extension (3.7 − 2 = 1.7 m).

The conclusion is that slab gate valves do not necessarily have a governing effect on the height of the manifold. In addition to the type of valve, the size, pressure class, and type of the actuator affect the height of the valve and the manifold. The other key point is that the position indicator extension should be at the level of the ROV bucket base or a little bit higher.

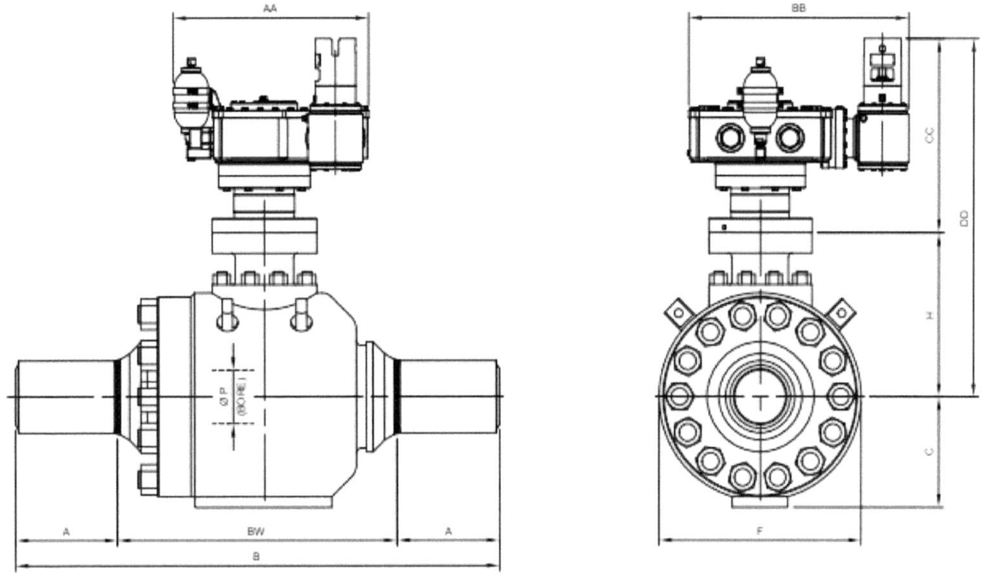

Fig. 9.34 ROV-operated ball valve GA drawing.

The question is, how to get the valve height from a GA drawing. Fig. 9.34 illustrates a front view (right side) and side view (left side) of a ROV-operated ball valve.

The height of the valve can be obtained from the picture on the right by finding the sum of parameter C (the distance from the bottom of the valve to the center of the valve bore), parameter H (the distance from the center of the valve bore to the valve top flange), and parameter CC (the distance from the top flange of the valve to the top of the ROV bucket). Parameter DD, which is the distance from the center of the valve to the top of the ROV bucket, is an important dimension for calculating the height of the valve. Fig. 9.35 illustrates a GA drawing of a through conduit slab gate valve with a linear actuator on the top of the valve. The height of the actuated valve can be calculated from the picture on the right by adding parameter C, which is the distance from the bottom of the valve to the center of the valve bore, to parameter CC, which is the distance from the center of the valve bore to the top of the ROV override bucket.

Fig. 9.36 illustrates a self-supporting ROV bucket and position indicator extension. Fig. 9.37 illustrates an extension of a ROV bucket for a manually operated subsea valve. The material of the extension could be the same as the material of the stem and should have good resistance to seawater. Materials such as 25Cr super duplex and Inconel 725 are suitable.

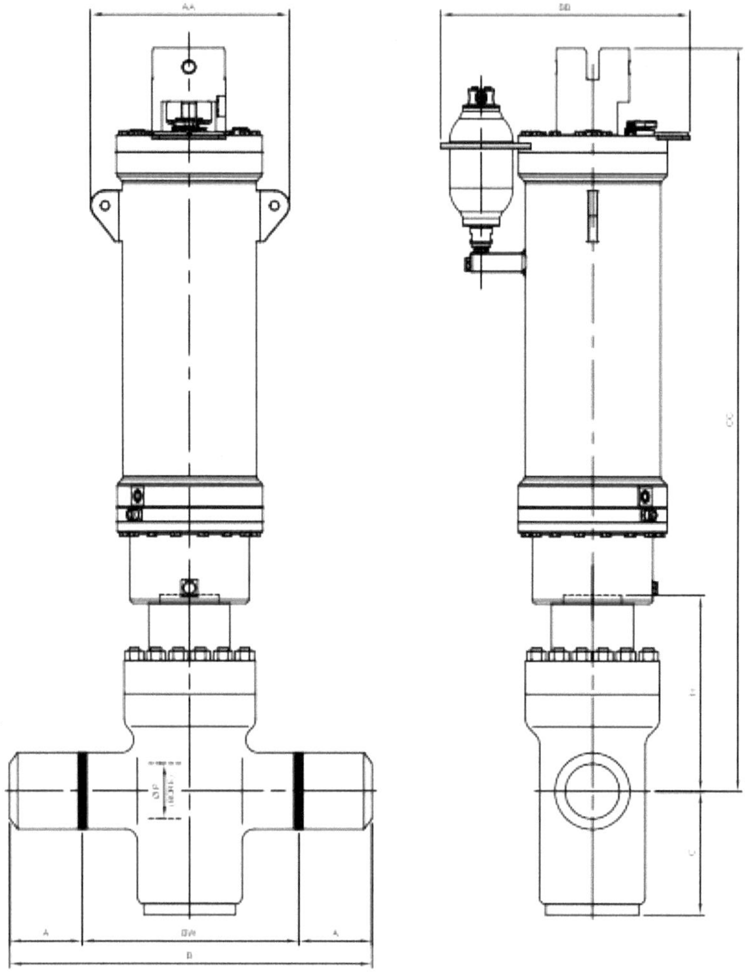

Fig. 9.35 Linear actuator-operated slab gate valve GA drawing.

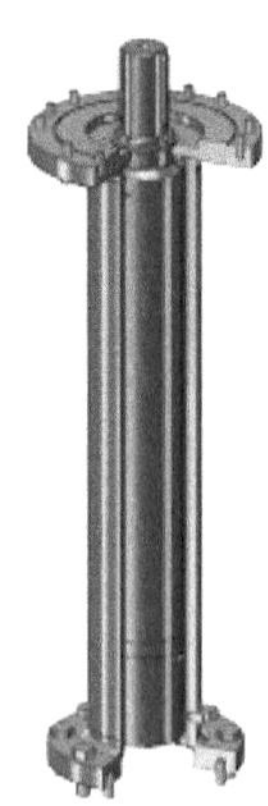

Fig. 9.36 Self-supporting ROV bucket and position indicator extension. *(Courtesy: Petrol Valve.)*

Fig. 9.37 ROV bucket extension.

9.8 Torque tools

The following torque tool requirements are stated in Norsok U-007:

1. Torque tools should be tested in advance for the required depth before being used for subsea operation. The testing of the torque tool should be done in a wet environment to replicate subsea conditions. The proposal in the standard is to use a tested and field-proven torque tool.
2. Torque tools should be calibrated before subsea operation to make sure they produce the correct amount of torque for the valve operation. If a torque tool provides too high an amount of torque for the valve operation, this could damage some of the valve components such as the stem. Facilities that put limitations on the torque shall not be used.
3. Torque tools that are used for torque ranges up to 2700 N m with ROV bucket interfaces 1, 2, 3, and 4 are considered medium torque values. Torque values above 2700 N m up to 13,500 N m are considered high torque values. Fig. 9.38 illustrates a torque tool suitable for classes 1, 2, 3, and 4 (medium torque values) as per the API 17D requirement. ROV torque should be created by the torque tool alone without any support from the surface facilities.
4. ROV tool handles should be easily grabbed and held by the ROV manipulator. The torque tool should not be out of control of the ROV during the operation.
5. The ROV torque tool should have the same shape, dimensions, and tolerances as the ROV bucket as per the relevant standard (typically API 17H/ISO 13628-8).
6. The packing of the ROV tools is important and should be robust and suitable for the offshore environment. The inclusion of part lists, drawings, and an operation manual for the torque tool is required.

Fig. 9.38 Torque tool producing medium torque for ROV classes 1, 2, 3, and 4. *(Courtesy: Ocean Scan.)*

7. It should be noted that the weight of the torque tool is reduced in the water compared to air; the weight of the torque tool in the water should be within the weight capacity range of the ROV.
8. A torque tool analyzer (Fig. 9.39) is used to show and set the output torque of the torque tool during operation.
9. For linear valves such as slab gate, the stem of the valve does not come out of the ROV bucket; the ROV torque tool is not compatible with rising stems.

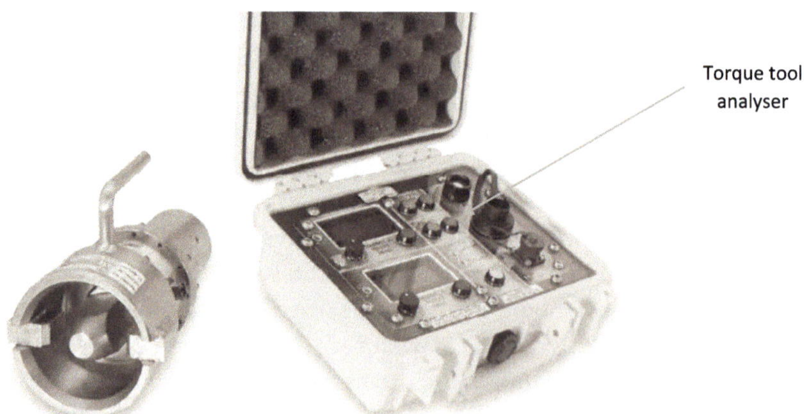

Fig. 9.39 Torque tool analyzer.

10. The speed of the ROV torque tool should be given either by the end user or purchaser in the specifications and could be between 8 and 20 revolutions per minute (rpm) as an example. RPM or rpm is a way of measuring rotation frequency. The number of ROV turns required to operate the valve is normally given by the valve manufacturer along with its tolerances and minimum, average and maximum values. In addition, the valve manufacturer can provide the maximum torque tool speed in rpm that the ROV bucket is capable of withstanding.

Question: An ROV tool has a minimum speed of 8 rpm and a maximum speed of 20 rpm. A 16″ ball valve with a fail-safe close actuator and ROV override requires minimum, average, and maximum ROV turns for opening to closing or vice versa equal to 100, 110, and 120 turns, respectively. How many seconds does it take to open the valve from closed position (minimum and maximum)?

Answer: The maximum speed for the valve operation occurs when the ROV torque tool is at its maximum speed, equal to 20 rpm (20 turns or rotations in 1 min or 60 s) and the number of turns for the ROV torque tool to open the valve from closed position is equal to 100 turn. Thus the 100 turns necessary for opening the valve from closed position take approximately 5 min minimum at a speed of 20 rpm. The minimum speed for the valve operation occurs when the speed of the torque tool is 8 rpm, equal to 8 turns in 1 min or 60 s, and the number of turns of the ROV torque tool to open the valve is 120 turns. Thus the 120 turns needed to open the valve from closed position take $120/8 = 15$ min in this case. Thus the minimum and maximum operation times are 5 and 15 min, respectively.

Question: An ROV tool has a minimum speed of 8 rpm and a maximum speed of 20 rpm. A 10″ ball valve with a fail-safe close actuator and ROV override requires minimum, average, and maximum ROV turns for opening to closing or vice versa equal to 30, 35, and 40 turns, respectively. What is the maximum speed of the ROV operation in 1″ per second?

Answer: The maximum speed of the ROV or valve operation is achieved when the ROV tool has a speed of 20 rpm, and the number of turns for operating the ROV bucket on the valve is 30 turns. It is possible to rotate the valve with the ROV bucket 20 times (turns) in 1 min. Therefore 30 turns of ROV operation require 1.5 min equal to 90 s. The valve is 10″; considering the linear motion of the ROV torque tool and the valve internal, it is possible to achieve a maximum rotation of 1″ per 9 s for this valve.

The following points should be considered while operating the valve with the ROV bucket and torque tool:

- The position of the valve should be checked with the position indicator;
- The interior of the ROV bucket should be checked to ensure that it is free from particles or debris;
- The ROV torque tool should be applied to the ROV bucket;

- Nominal torque value can be applied through the torque tool to the bucket;
- The number of turns can be observed to be according to the values given by the valve supplier;
- The valve position indicator should be observed to change from open to closed or from closed to open position.
- The ROV torque tool should be removed at the end of the operation.

Other important points regarding the operation of valves with a ROV:

- If the actuated valve is opened with a ROV torque tool, it should be closed with the same tool;
- If the valve is opened by the hydraulic fluid supplied to the actuator, it should be closed by the spring. A ROV should not be allowed to close a valve that was opened by the hydraulic force supplied to the actuator.
- If ROV override cannot operate (open) the valve at the provided differential pressure, the differential pressure across the valve should be reduced little by little while continuing to try to operate the valve with the ROV torque tool. In general, operating a valve under lower amounts of differential pressure is easier. If it is still not possible to operate the valve, the best recommendation is to contact the valve supplier.
- It is not recommended to exert torque above the maximum input torque.

9.9 ROV and valve interface

1. All sensitive equipment and components located in the area of ROV operation should be protected against a ROV impact of 1000 kg and 0.5 m/s speed based on the NORSOK standard. These values may be changed for different projects based on the end users' requirements and the type and size of the ROV used for subsea operation. A ROV impact of 3000 kg and 1.5 m/s could be considered as an example. Normally, valves are protected by a steel structure, so this requirement does not place any restriction on valve and actuator manufacturers to modify their design.
2. The ROV system and its surroundings, as well as its components and structures should be designed to minimize the adverse impact of undesirable environmental conditions such as currents. In addition, the ROV interface and its surroundings should have very good visibility.
3. The ROV low torque paddle should be suitable for direct ROV manipulator operation without disconnecting from any other parts of the valve.
4. The low torque paddle for ball valves should have mechanical end stops to keep the valve closure member in fully open and closed positions.

5. Regarding the layout and space requirement for an observation ROV, an operational envelope equal to 1.9 m × 1.5 m × 1.5 m (L × W × H) should be considered as clearance. The given values are general and should be evaluated case by case.
6. Regarding the layout and space requirement for a work ROV, an operational envelope equal to 3.5 m × 2.9 m × 2.9 m (L × W × H) should be considered as clearance. The given values are general and should be evaluated case by case.
7. The distance from the front end of a ROV to the work location should be between 500 and 1500 mm.
8. The surroundings of the structures and components under operation of the ROV should be clear of any object that could disturb the operation of the ROV. In fact, the valve stem and top work should allow for ROV operation and a circle of 300 mm around the valve should be clear from any obstruction.
9. The ROV interface (valve bucket) could be tested or undergo design verification through finite element analysis (FEA) to remain operational under 3000 N m bending and 3000 N m axial loads.
10. There is a safety factor associated with the design of ROVs. As an example, a ROV could have safety factor of 2 which means that it should be designed to withstand at least twice the torque or thrust needed to operate the valve under the most severe operation conditions.
11. The stem of the valve could be damaged due to an excess amount of torque exerted by the ROV torque tool. Some of the valve stem protections against excess torque are summarized in API 17H/ISO 13628-8; these include identifying the maximum torque for safe operation of the valve. The valve supplier provides different torque values such as breakout, running, and damage torque. Normally, damage torque has the highest value compared to breakout and running torque and could be 25% above breakout torque as an example. A ROV could work with hydraulic power, so in this case, the appropriate hydraulic motor and safe hydraulic transmission with proper hydraulic pressure should be used to avoid overtorquing the valve. Using mechanical stops in the valves and actuators can prevent overtorquing, but they are not sufficient. The API standard proposes the proper shape of the end effector, which is defined as the end part of the robot arm or finger that has interface with the torque tool.
12. Temperature class selection for ROVs is similar to that of actuators. Table 9.7 presents the temperature classes provided in API 6A/17D. The minimum temperature class for the ROV could be compatible with the minimum temperature range for one of the classes given in the table (e.g., −18°C correspondents to temperature class S). The maximum temperature class of the ROV should not be less than 65°C. Therefore the maximum temperature associated with temperature class S is not applicable for ROVs. It is possible to select 65°C for the maximum temperature

Table 9.7 Temperature classes according to API 6A/ISO 10423.

Temperature class	Temperature range °C min.	°C max.	°F min.	°F max.
K	−60	82	−75	180
L	−46	82	−50	180
N	−46	60	−50	140
P	−29	82	−20	180
S	−18	60	0	140
T	−18	82	0	180
U	−18	121	0	250
V	2	121	35	250

Note: Minimum temperature is the lowest ambient temperature to which the equipment can be subjected. Maximum temperature is the highest temperature of the fluid that can directly contact the equipment.

of the ROV. However, the minimum and maximum temperature class for a ROV may be given in the project documents such as specifications and data sheets. The given temperature rating in the project documents could be completely different from the standard temperature classes given in the table based on API standards (e.g., −10°C to 70°C).

9.10 ROV and structure interface

The orientation of the subsea structure could affect the ease of ROV operation. There is no certain rule about the direction and orientation of the subsea structure to facilitate ROV operation. Each potential structure orientation should be addressed and evaluated case by case. Other considerations should be addressed during the subsea structure orientation in addition to ROV operation such as visibility, currents, and field layout requirements.

9.11 ROV material selection

The purchaser/end user of the ROV is responsible for selecting the material for the ROV. Frequency of operation in seawater is one of the main parameters of material selection. Therefore considering possible corrosion types due to seawater such as pitting and chloride stress cracking corrosion (CLSCC) are important. In addition to corrosion, the following mechanical and material parameters should be considered for ROVs:

- Mechanical stress and loads that affect the mechanical properties of materials such as yield and tensile strength;

- The fatigue properties and resistance of the material under frequent loads;
- Hardness to prevent possible erosion and wearing;
- Susceptibility to and prevention of marine fouling; this is an important parameter for all subsea components as explained in Section 9.2.

9.12 ROV intervention philosophy and function

The following requirements should be defined and clarified for ROV intervention according to ISO 13628-8/API 17H:

1. Definition of the task: is it really needed or could it be avoided?;
2. Definition of the required type of ROV;
3. Definition of the host facility control system responsible for moving and controlling the ROV;
4. Definition of the required torque tool, ROV interface on the valve or actuator, power supply and control requirements for the ROV;
5. Choice of docking and tool deployment;
6. Relevant documentation for operation or maintenance jobs such as guidelines and procedures;

Note: Different loads should be taken into account in ROV design such as loading coming from:

1. The torque tool during engagement with the ROV bucket.
2. Environmental loads such as waves, seawater pressure, depth, and current.
3. Load of the umbilical connected to the ROV; this load is affected by waves and other actions such as the movement of the umbilical as a type of drag force.
4. Possible clashes of the ROV with subsea structures should be prevented as much as possible.

9.13 Self-assessment questions

1. What is the main reason for using ROVs in underwater subsea valve operation?
   - **A.** This equipment is large and robust.
   - **B.** ROVs can maneuver in deep areas and handle valve operations at depth.
   - **C.** ROVs have the capability to observe, work, and provide information to a surface pilot or operator.
   - **D.** The ROV arm and manipulator have enough flexibility to carry out the underwater job.

   Answer: ROVs are large and expensive equipment but this is not a main reason for using them for subsea operation so option A is not correct. Option B is correct since humans could not have access to valves at such depths without ROV's assistance.

Option C is correct in general but it is not a main reason for using ROVs for subsea valve operation. Again, Option D is not the main reason for using ROVs in subsea. Thus option B is the answer to the question.

2. Which temperature range could be applicable for ROV interface on the valve?

A. −18°C to 82°C.

B. −18°C to 70°C.

C. −10°C to 70°C.

D. All of above.

Answer: Option A is compatible with temperature class T as per API 6A/17D standards and the maximum temperature range is above 65°C so it is correct. Option B has a minimum temperature value compatible with temperature classes of S, T, and U and the maximum given temperature is not compatible with any temperature class but it is higher than 65°C so option B is correct as well. The temperature range given in option C is not compatible with any temperature classes but could be a correct temperature range for the ROV interface (bucket). Therefore option D is correct.

3. Which item is not correct regarding ROV interfaces?

A. According to API 17H/ISO 13628, a ROV class 4 interface requires higher torque for operation than class 3.

B. A ROV class 3 interface is not as common as class 4 for subsea valves.

C. A ROV low torque paddle should be robust enough not to be removed from the top of the valve during ROV operation.

D. ROV operation for cycling the valve is done only through a ROV manipulator and fingers.

Answer: Option A is correct. Since the higher ROV class is used for higher torque requirements, ROV class 4 requires higher torque compared to class 3. Option B is also correct; ROV class 3 is not common for valves in the subsea industry. Option C is correct and it should be borne in mind that a low torque paddle should be connected to the valve during ROV operation. Option D is wrong since the ROV arm and fingers carry a torque tool for valve operation.

4. Which statements are not correct regarding the ROV bucket extension?

A. The extension is limited to the ROV bucket.

B. The height of a subsea manifold is always affected by the height of gate valves, and ball valves always require an extension of the ROV bucket.

C. The position indicator extension should be at the same level as the ROV bucket base in general.

D. The ROV bucket extension can be done through a self-supporting extension.

Answer: Option A is not correct since the extension is not limited to the ROV bucket; the position indicator is also extended. Option B is not correct since gate valve's ROV bucket and position indicator may require extension in some cases. As an example, a large ball valve with a double acting hydraulic actuator could have a higher height compared to a medium sized gate valve with ROV operation in the same pressure class. In this case, if these two valves were inside the same structure or manifold, the gate valve ROV bucket and position indicator should be extended. Option C is correct since the position indicator is extended in order to reach the same level of the ROV bucket base. Option D is also correct since a self-supporting bucket extension can be used for the valves.

5. Which statements about subsea intervention are not correct?

A. ROVs are normally used for installation and replacement of light and small objects;

B. The retrieval actuators of the valves are normally large and should be handled by ART as a type of ROT for intervention.

C. Intervention could cover different tasks such as repair, maintenance, installation, and retrieval.

D. For handling a large subsea component, ROT is used as a backup for ROV and the task is mainly performed by ROV.

Answer: Options A, B, and C are all correct. Option D is wrong since ROT performs the main intervention task for large and heavy items and ROV is used as a backup.

6. Which statements about the materials used for ROVs and ROV buckets are correct?

A. ROV buckets could be made in both soft and metallic materials.

B. Parameters that should be considered for the material selection of ROVs include corrosion, wearing, mechanical strength, and fatigue.

C. Fatigue and load consideration are important for ROV materials since different loads are applied to ROVs such as waves, interaction with the torque tool, etc.

D. All options are correct.

Answer: All options are correct so option D is the correct choice.

7. Which sentence is correct about ROV bucket design?

A. If the maximum output torque of an ROV is 2711 N m, it can be damaged at 2700 N m.

B. The maximum output torque for ROV bucket class 3 is less than that of class 2 and 1.

C. A low torque paddle could be applicable for small valves such as ½″.

D. The access to ½″ ball valves located on the vertical line with the stem on the horizontal plain is from the top.

Answer: Option A is wrong because the damage torque is always higher than the maximum output torque, whereas the damage torque in option A is lower than the output torque. Option B is not correct since the maximum output torque for class 3 is higher than that of class 1 and 2 (refer to API 17H/ISO 13628-8). Option C is correct: ½″ ball valves with ROV operation could be supplied with a low torque paddle. Option

D is not correct because ball valves on the vertical line with the stem on the horizontal plain have a side access for ROV, and no access from the top.

8. Which sentences are correct about operating a valve through the ROV bucket with the torque tool?

A. If the valve is opened by the hydraulic supply to the actuator, it can be closed with a ROV torque tool.

B. A ROV torque tool should be capable of operating a valve even if debris has accumulated in the bucket.

C. The number of ROV rotation turns should be given by the valve supplier in the drawings of the valve and in the IOM manual.

D. The maximum speed of the ROV could be given in the valve specifications, but it can also be advised by the valve supplier.

Answer: Option A is not correct because if the valve is opened by hydraulic supply to the actuator, it should be closed by spring force in the actuator and not by ROV. Option B is wrong since it is not recommended to operate valves with debris inside the bucket. Option C is correct; normally the minimum, average, and maximum number of ROV turns are given in the GA drawings and IOM manual by the valve supplier. Option D is correct as well because the maximum speed of ROV operation is given in the valve specification and could be advised by the valve supplier in the IOM manual.

9. Which sentence is not applicable about position indicators?

A. Position indicators maybe extended along with ROV buckets.

B. Position indicators should be readable from far away.

C. Position indicators require dual sealing for moving parts.

D. Antifouling materials are proposed for position indicators to prevent marine growth and promote readability.

Answer: Option A is correct because the position indicator should be located close to the ROV bucket—more precisely at the same level as the ROV bucket base so that extending the bucket leads to the extension of the position indicator. Option B is not correct since the position of the valve should be readable by the ROV or diver during operation at close range. Option C is correct; dual sealing is required for moving parts. In fact, moving parts create more friction with the sealing so one extra sealing is required as a backup. Option D is correct since antifouling materials prevent marine growth and assist readability of the position indicator by ROV or diver.

10. Which sentence is not correct about ROV key components?

A. Light is required for ROVs to show the way and remove the darkness of the underwater environment.

B. A frame is a structure that protects internal components and provides installation space for ROV components.

C. Pilot control is required for the control and operation of the ROV from the surface.

D. A tether is used to provide electrical or hydraulic power to maneuver the ROV.

Answer: Options A, B, and C are correct. However, option D is not correct since thrusters are the correct component that receive electrical or hydraulic power for maneuvering the ROV.

Further reading

American Petroleum Institute (API) 6A, Specification for Wellhead and Tree Equipment, twenty-first ed., API, Washington, DC, 2018.

American Petroleum Institute (API) 17D, Design and Operation of Subsea Production Systems, Subsea Wellhead and Tree Equipment, second ed., API, Washington, DC, 2011.

American Petroleum Institute (API) RP17H, Remotely Operated Tools and Interface on Subsea Production Systems, third ed., API, Washington, DC, 2019.

International Organization of Standardization (ISO) 10423, Petroleum and Natural Gas Industries—Drilling and Production Equipment—Wellhead and Christmas Tree Equipment, fourth ed., 2009. Geneva, Switzerland.

International Organization of Standardization (ISO) 13628-4, Petroleum and Natural Gas Industries—Design and Operation of Subsea Production Systems—Part 4: Subsea Wellhead and Tree Equipment, second ed., 2010. Geneva, Switzerland.

International Organization of Standardization (ISO) 13628-8, Petroleum and Natural Gas Industries—Design and Operation of Subsea Production Systems—Part 8: Remotely Operated Vehicle (ROV) on Subsea Production Systems, first ed., 2006. Geneva, Switzerland.

Norsok U-007, Subsea Intervention Systems. Rev. 02, 1998. Lysaker, Norway.

Norsok U-102, Remotely Operated Vehicle (ROV) Services. Rev. 01, 2003. Lysaker, Norway.

Rusty, What is An Underwater ROV, [online] Available at: https://bluerobotics.com/learn/what-is-an-rov/. (Accessed 28 August 2020).

Total Marine Technology, ROV and Tooling Handbook, 2017.

CHAPTER TEN

Subsea actuators

Contents

10.1 Introduction

An actuator is a machine or component installed on the top of an industrial valve to automatically or remotely move and control the operation of the valve. The quality of a valve depends on many parameters, such as material selection, mechanical strength, quality control, machining, etc., whereas the performance of a valve is largely dependent on the actuator. Three types of actuators—hydraulic, pneumatic, and electrical—are used

Subsea Valves and Actuators for the Oil and Gas Industry
https://doi.org/10.1016/B978-0-323-90605-0.00007-4

for industrial valves in various sectors of the oil and gas industry, such as offshore (topside) as well as onshore refineries and petrochemical plants. Hydraulic and electrical actuators are applicable for subsea valves, but pneumatic actuators are not, since the typical air pressure in pneumatic actuators ranges only from 5 to 9 bar. This relatively low amount of air pressure is not sufficient to produce enough force for subsea actuators and valve cycling, or a very huge and extremely heavy pneumatic actuator would be required. But it should be noted that the API 6A/ISO 10423 standards, which are used for subsea actuator design, manufacturing, testing, inspection, storage, and transportation, do mention pneumatic supply pressure to subsea actuators even if it is not practical. Subsea valves that operate (open and close) frequently are normally supplied with an actuator. On the other hand, subsea valves that operate infrequently are opened and closed manually by an ROV or diver.

10.2 Subsea hydraulic actuators

10.2.1 Actuator types

Different types of actuators, such as linear, scotch yoke, rack and pinion, and helical spline are explained in this section.

10.2.1.1 Linear actuators

Linear actuators, which are used for subsea gate valves, create motion in a straight line with a linear motion of the stem and closure member (disk or gate). Linear actuators are spring return with a fail-safe closed (FSC) function in most of cases, as illustrated in Fig. 10.1.

A linear actuator could be made with a double acting function, but this type of configuration is not common; most linear actuators have a FSC function with single acting spring return action. Terms such as single and double acting, spring return, fail-safe closed, ROV override bucket, and compensation will be explained later in this chapter.

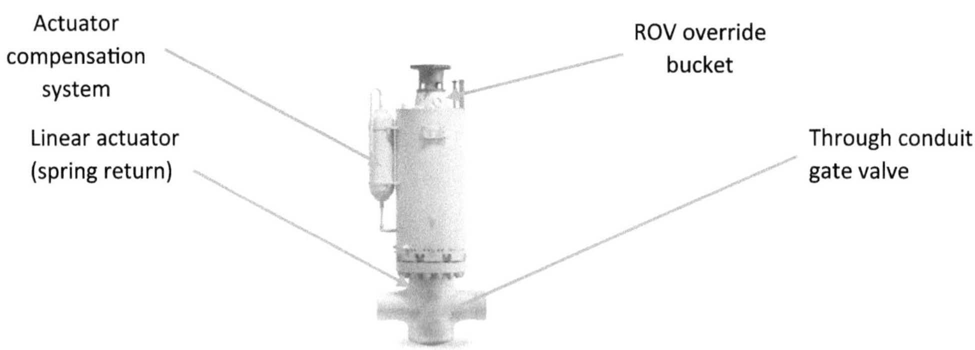

Fig. 10.1 Linear actuator for a gate valve.

Fig. 10.2 illustrates a FSC single acting linear actuator for a through conduit gate (TCG) valve. The hydraulic supply pressure, which is typically 5000 psi, pushes the stem of the valve downward and compresses the spring. The gate valve, which has a reverse port as per the configuration provided and explained in Chapter 4, goes to open position with hydraulic fluid supply. Upon stoppage of the hydraulic fluid or failure of the hydraulic supply, the spring expands, the stem of the valve moves upward, and the valve gets closed. Linear actuators for gate valves can be used for water depths as deep as 2400 m. The actuator illustrated in Fig. 10.2 has a product safety level of 3 (PSL3) as per API 6A and 17D. PSL is defined and explained in detail in Chapter 2. Linear actuators like this one can be made in different materials such as carbon steel, duplex and super duplex, as well as nickel alloys such as Inconel 625. Six hundred cycles based on performance requirement 2 (PR2) are applied on such actuators, and the ROV bucket is made according to the requirements of API 17H/ISO 13628 part 8.

10.2.1.2 Scotch yoke actuators

Scotch yoke actuators (see Fig. 10.3) are common for subsea ball valves with quarter turn or 90-degree motion of the ball for cycling (opening and closing). These actuators can be made in single acting or double acting versions and are suitable for water depths as deep as 3 km. Fig. 10.3 illustrates a scotch yoke actuator on a ball valve. Transformation of the linear shaft or piston movement to rotary movement is achieved through the yoke

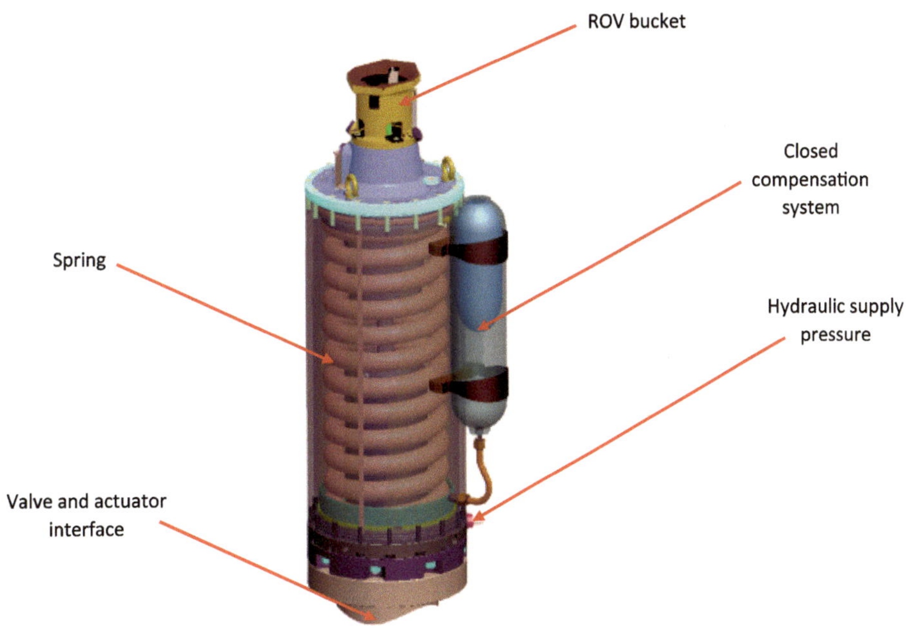

Fig. 10.2 Linear actuator suitable for a gate valve.

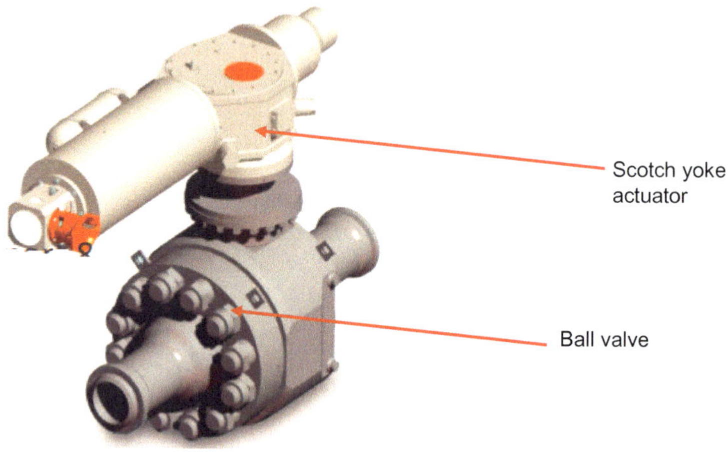

Fig. 10.3 Scotch yoke actuator on a ball valve.

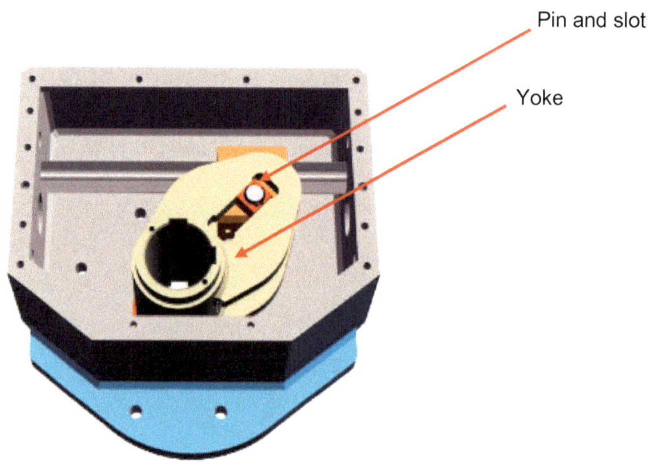

Fig. 10.4 Scotch yoke mechanism. *(Courtesy: Flow Control.)*

mechanism (see Figs. 10.4 and 10.5) inside the actuator, as well as the slot or pin illustrated in Fig. 10.4. Three main advantages of scotch yoke actuators are that they have fewer moving parts, produce high output torque, and have smooth operation.

Scotch yoke operation and function is illustrated in Fig. 10.6. In part 1, the oil moves the piston back and the yoke moves 90 degrees so the ball valve is opened. The spring is consequently compressed due to the oil pressure. In part 2, the oil supply pressure stops, so the spring is expanded and pushes the piston back, causing movement of the yoke 90 degrees to close the valve. Alternatively, the actuator can be opened via an ROV bucket with an ROV override instead of hydraulic pressure. The ROV moves

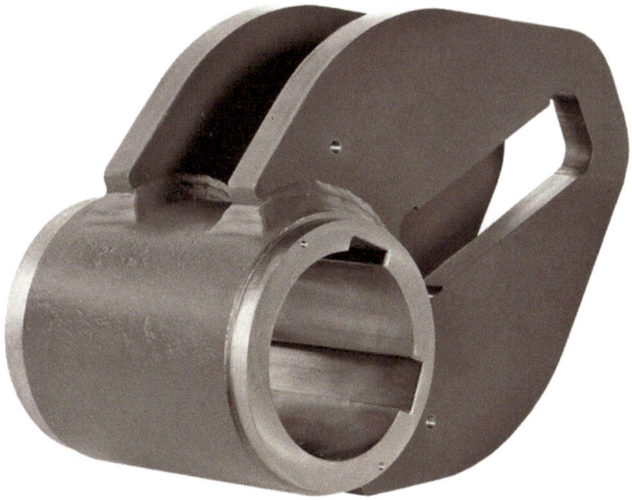

Fig. 10.5 Yoke. *(Courtesy: Pro Control.)*

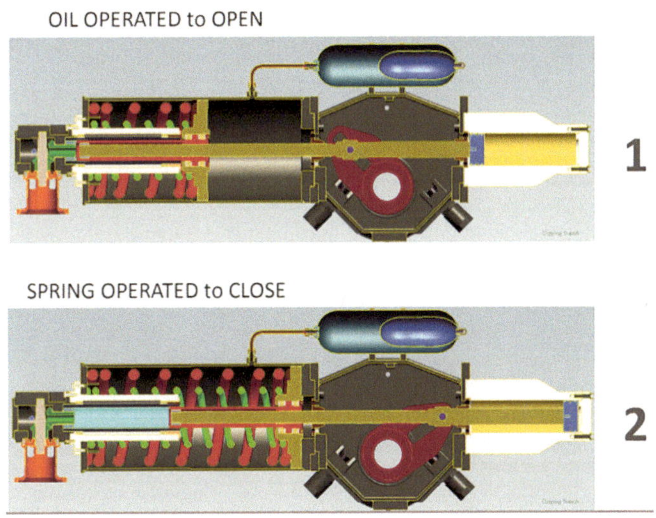

Fig. 10.6 Scotch yoke spring return (single acting)hydraulic actuator operation.

counterclockwise to open the actuator. When the actuator is opened by an ROV, it should be closed by an ROV as well through clockwise movement of the ROV in the bucket.

Normally, the spring of the actuator is not compressed and expanded during ROV override operation. This has two main benefits: first, the spring is exposed to less fatigue since it undergoes no cycle (expansion and compression) during ROV override; second, the actuator can be operated with an ROV override tool even in the case of spring failure. It should be borne in mind that, in general, ROV override operation of a valve requires a low amount of torque depending on the ROV interface class defined in API 17H/ISO 13628-8.

10.2.1.3 Rack and pinion actuators

Rack and pinion actuators (see Fig. 10.7) are common for subsea ball valves with quarter turn or 90-degree motion of the ball for cycling (opening and closing). The name of this actuator refers to its two gears: the rack, which has linear motion, and the pinion, which has rotary motion. The linear motion of the rack is achieved through hydraulic force that is transferred to the pinion, which moves the rotary and cycles the valve 90 degrees to open or closed position. Rack and pinion actuators can be fabricated in both spring return (single acting) and double acting versions. Symmetrical spring return rack and pinion actuators contain two spring housings and two piston and cylinders rather than one, a design that can be selected for its advantages of fast operation and less fatigue. In symmetrical design, the racks are moved in opposite directions to ensure balanced forces on the pinion from both sides. The bottom of the pinion is connected to the valve, so the rack force is transferred to the pinion and consequently to the valve internals. Rack and pinion actuators are a very reliable actuator solution for subsea valves, as they can be relatively more compact than scotch yoke and provide higher torque for valve operation. Double acting rack and pinion actuators with high torque input and more compact design

Fig. 10.7 Rack and pinion spring return (hydraulic) actuator for a 30″ subsea ball valve.

compared to single acting actuators can be used for subsea ball valves in large sizes and high-pressure classes such as 16″, 20″, and 22″. Noticeably, 22″ is not a common size for piping and valves in the oil and gas industry, except for subsea applications. The simple mechanism of the rack and pinion design is illustrated in Fig. 10.8.

10.2.1.4 Helical spline actuators

Helical spline or helical rotary (see Figs. 10.9 and 10.10) is another type of actuator that can be used, especially for ball valves, to save space horizontally, instead of using rack and pinion actuators installed on ball valves with a horizontal bore and vertical stem. In fact,

Fig. 10.8 Rack and pinion.

Fig. 10.9 Subsea helical spline actuator with ROV override. *(Courtesy: ATV.)*

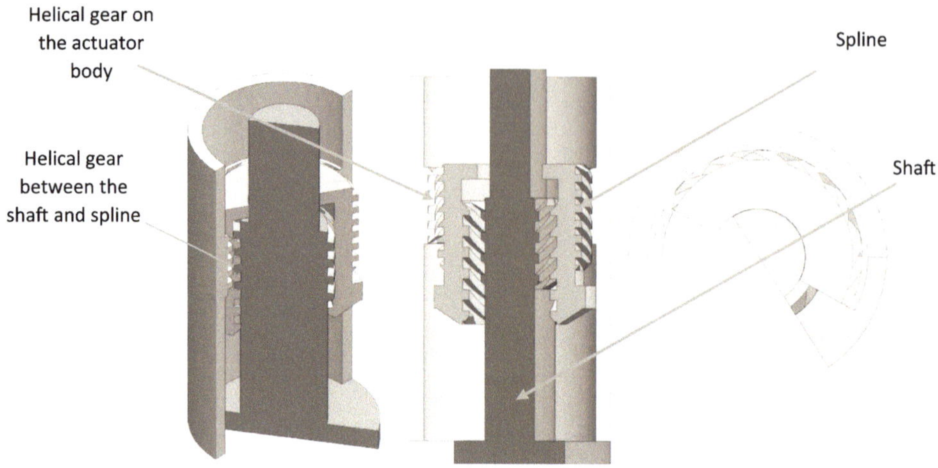

Fig. 10.10 Helical rotary actuator internals.

helical spline actuators have the advantages of compact design as well as weight reduction, but can be more expensive than rack and pinion actuators. In some cases, where there is a space limit sidewise on the manifold where the rack and pinion actuator is located, one solution is to tilt the rack and pinion actuator some degrees, so that the actuator does not stand parallel to the pipe bore. Fig. 10.11 illustrates a rack and pinion actuator tilted 20 degrees and located on the top of a subsea ball valve to save space sidewise, compared to positioning the actuator parallel to the piping and valve bore. The top work of the

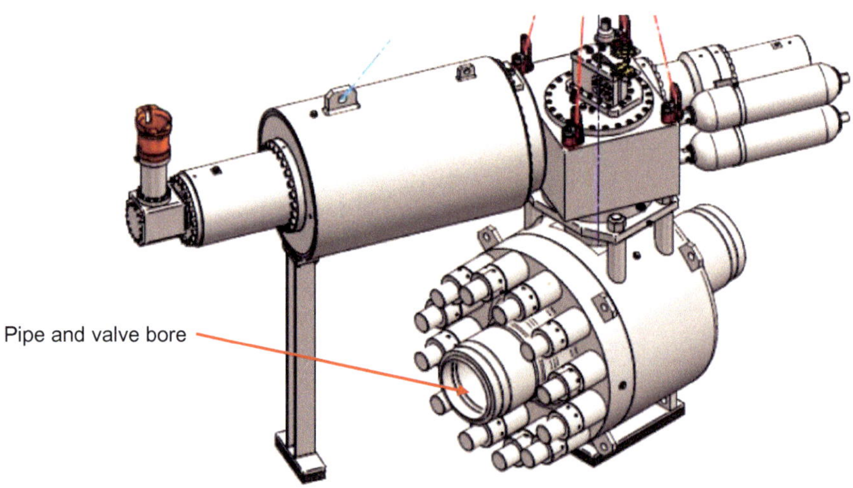

Fig. 10.11 Tilted rack and pinion actuator. *(Courtesy: ATV.)*

valve, such as the stem key grooves and the holes for connecting the valve to actuator, must be adjusted accordingly.

As illustrated in Fig. 10.10, the housing of the actuator has internal gears connected to the external section of the spline. The hydraulic fluid enters from the top and pushes the spline down, while the helical gearing causes the spline and the shaft to rotate at the same time. There are two helical gears: one is between the external part of the spline and the internal section of the actuator body or casing, and the other is between the external section of the shaft and the internal section of the spline. Rotation of the bottom of the shaft will be transferred to the valve.

10.2.1.5 Stepping actuators

Stepping actuators are applicable for choke valves, which are explained in more detail in Chapter 6. A stepping actuator provides linear force to move the choke valve stem up or down. This actuator has many steps in order to operate the valve, and includes one or two hydraulic cylinders and gears.

10.2.2 Pressure-temperature rating

A hydraulic actuator is a cylinder that converts hydraulic fluid power to mechanical work or force. Comparing subsea actuators with those used in other sectors of the oil and gas industry, subsea actuator hydraulic pressure can be much higher compared to topside and onshore actuators. Hydraulic pressure values to the subsea actuators are 207 bar, 345 bar, or 380 bar. A hydraulic pressure of 345 bar equal to 5000 psi can be considered a standard hydraulic pressure for subsea actuators in deep-water applications. Therefore it is more efficient to use hydraulic actuators rather than pneumatic for subsea application. It should be noted that increasing hydraulic pressure in general provides higher force for operation of the actuators, which is a good solution for deeper water actuation applications. The other benefit of increasing the hydraulic oil pressure is to modify the design of the actuator to be more compact by reducing the actuator swept or displacement volume as well as its piston size. Swept volume is defined as the volume of oil that is swept or displaced by the actuator piston during an opening or closing cycle. The other important consideration about actuator pressure value is that 345 bar should be considered a working pressure. The design pressure of an actuator can be calculated at 1.1 times of the operating pressure, which is around 380 bar for a 345 bar working or operating pressure. The actuator should have a temperature rating in addition to a pressure class or rating. Eight temperature ratings are given in API 6A/ISO 10423; Table 10.1 is extracted from these standards.

Actuators have a pressure rating that affects metallic and nonmetallic sealing material selection, as well as an actuator qualification program, which will be explained in more detail later in this chapter. The minimum rating temperature for the actuator should be selected from the minimum temperature values given in Table 10.1, unless another value

Table 10.1 Temperature classes according to API 6A/ISO 10423.

Temperature class	Temperature range °C min.	Temperature range °C max.	Temperature range °F min.	Temperature range °F max.
K	−60	82	−75	180
L	−46	82	−50	180
N	−46	60	−50	140
P	−29	82	−20	180
S	−18	60	0	140
T	−18	82	0	180
U	−18	121	0	250
V	2	121	35	250

Note: Minimum temperature is the lowest ambient temperature to which the equipment can be subjected. Maximum temperature is the highest temperature of the fluid that can directly contact the equipment.

is indicated by the end user or in the project specification. The maximum upper working temperature of the actuator is typically standardized as 65°C and should not be less than 65°C.

10.2.3 ROV override

Actuated valves are normally supplied with ROV override facilities for operating the valves independently of the actuators' source of power. ROV override as a backup for valve actuation is beneficial, especially in the case of actuator failure to function. In the event of ROV override, the valve is operated by a torque tool with rotary motion, which is normally clockwise to close and counterclockwise to open. The design of an ROV override is according to API 17H/ISO 13628-8, "Remotely operated vehicle (ROV) interfaces on subsea production systems." ROV override classes could be 2, 3, or 4, depending on the required torque for operating the actuator and valve. The important point is that if the actuator and the valve are opened with mechanical ROV override, the valve and actuator should be closed with mechanical ROV override and not with actuator spring force. If the valve and actuator are opened with hydraulic supply power, then the valve and actuator should be closed with spring force and not ROV override. Each ROV interface has a specific maximum input torque according to API 17H/ISO 13628-8. As an example, the maximum input torque of an ROV override with an interface of ISO 13628-8 class 4 is 2711 N m. Thus the dedicated torque tool for ROV bucket operation in class 4 should be of the same class. Using a torque tool of a higher class can raise the input torque above the maximum allowed input torque of 2711 N m, which can damage the actuator as well as the valve stem and closure member. Actuators are normally supplied with an ROV override bucket to be able to operate the valves independently of the actuators. Fig. 10.2 illustrates an ROV bucket or receptacle on the top of a linear actuator. If the hydraulic supply to the FSC linear actuator stops, the spring closes the valve and actuator. In this case, the ROV can rotate counterclockwise and open the valve through the ROV bucket, instead of using hydraulic supply to open the valve and actuator. The valve can be closed by rotating the ROV bucket clockwise. Different parameters are essential in ROV design and selection, such as interface and class, minimum and maximum speed in Rpm (revolutions per minute), minimum and maximum ROV input torque, ROV orientation (rotary or linear), and ROV load resistance according to the API 6DSS/ISO 14723 standards for subsea pipeline valves. ROV override on the actuator maybe extended to reach the top of the structure roof. In fact, it is practically impossible for an ROV to be moved inside the manifold for the operation of valves and actuators. Thus the ROV override should be extended to the structure roof to enable ROV operation. The ROV override extension is normally installed through a self-supporting piece of pipe (spool), as illustrated in Fig. 10.12. A nonsupporting extension spool for raising both the ROV bucket and position indicator is shown in Fig. 10.13.

The position indicator for the actuator and valve, which will be explained later in this chapter, should be extended to be close to the ROV bucket and readable by the ROV. Fig. 10.14 illustrates subsea valves with ROV buckets in orange color located at the structure roof (top) level.

The extension of the ROV bucket and position indicator can be supplied with a compensation system to prevent failure due to seawater head. Relief valves can be installed on the extensions to adjust the compensation pressure. Fig. 10.15 illustrates a schematic for extending the position indicator to the manifold roof through a self-supporting extension

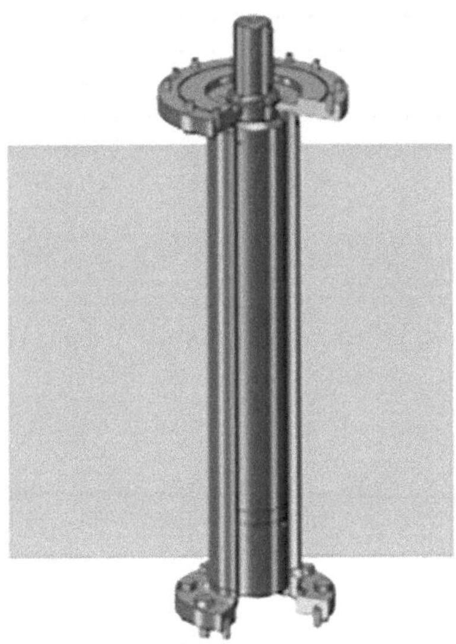

Fig. 10.12 Subsea self-supporting ROV bucket extension. *(Courtesy: PetrolValve.)*

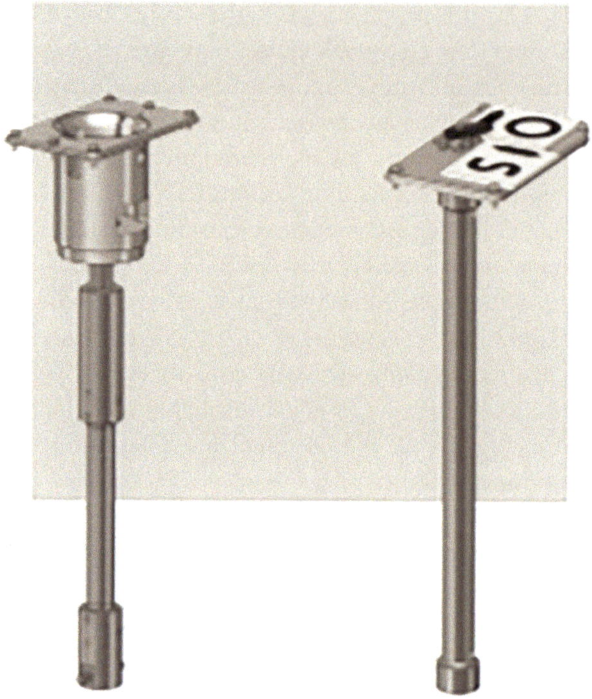

Fig. 10.13 Nonsupporting ROV bucket and position indicator extension. *(Courtesy: PetrolValve.)*

Fig. 10.14 Subsea valves in a structure.

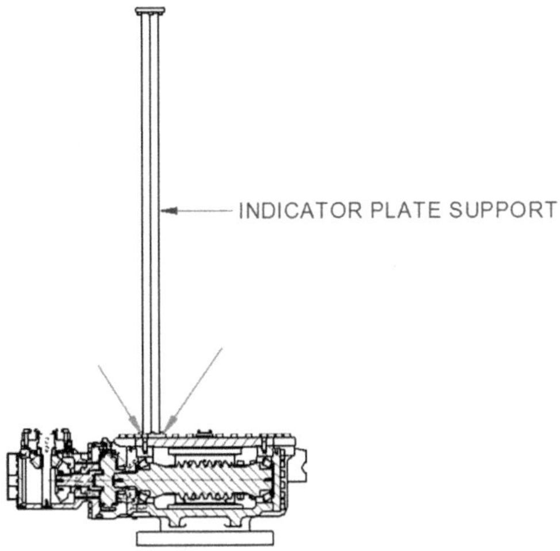

Fig. 10.15 Position indicator extension on a rack and pinion actuator.

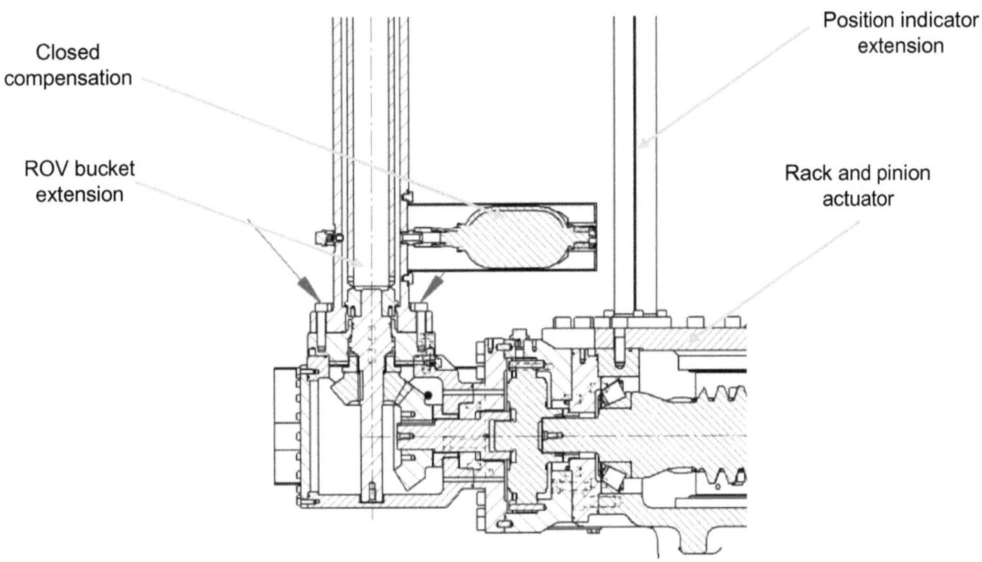

Fig. 10.16 ROV bucket extension with a closed compensation system on a rack and pinion actuator.

installed directly on a rack and pinion actuator. Fig. 10.16 illustrates the installation of a ROV bucket extension that is pressure compensated with a closed compensation system. O-ring sealing, such as Buna N or hydrogenated Buna N, can be used to seal between the actuator and stem extension.

10.2.4 Single vs double acting actuators

Both spring return (single acting) and double acting actuators are used for subsea valves. Spring return actuators could have either a fail-safe closed (FSC) or fail-safe open (FSO) mode of operation. The fail mode of a spring return actuator is always achieved by spring force (see Fig. 10.17). This means that spring force always closes a FSC actuator and opens a FSO actuator. The opposite action is done by the hydraulic fluid, which always opens FSC actuators and closes FSO actuators. Double acting actuators (see Fig. 10.18) have a fail-as-is (FAI) failure mode, meaning that they stand still upon stoppage of the hydraulic fluid supply. Both the opening and closing of double acting actuators are done by hydraulic oil pressure. Double acting hydraulic actuators are generally more compact than spring return (single acting) actuators and produce higher force or torque for the valve and actuator operation. Generally, double acting actuators are more suitable for large, high-pressure class valves (e.g., 22″/517 bar) compared to single acting, since they are more compact and produce higher force.

10.2.5 Sealing

All of the sealings inside the actuators around the shaft and position indicator should be dual to provide higher safety barriers against leakage. In addition, the actuator should be protected against any possible leakage from the valve stem. Failure of a valve's stem sealing could lead to release of the valve's fluid service (e.g., hydrocarbon) to the actuator internals. Therefore a pressure relief valve (as an example) could be installed on the valve

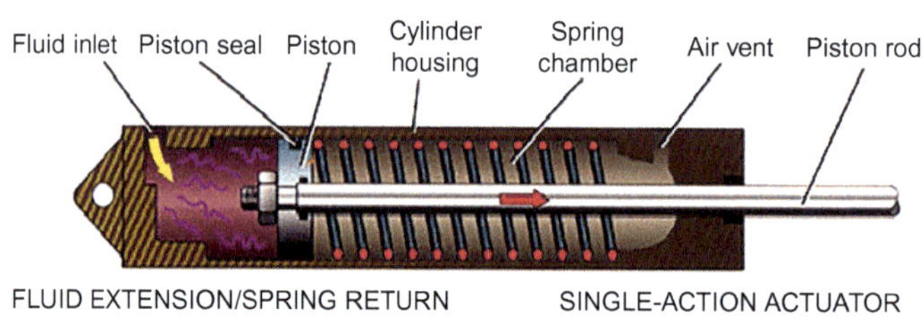

Fig. 10.17 Spring return hydraulic actuator. *(Courtesy: Springer.)*

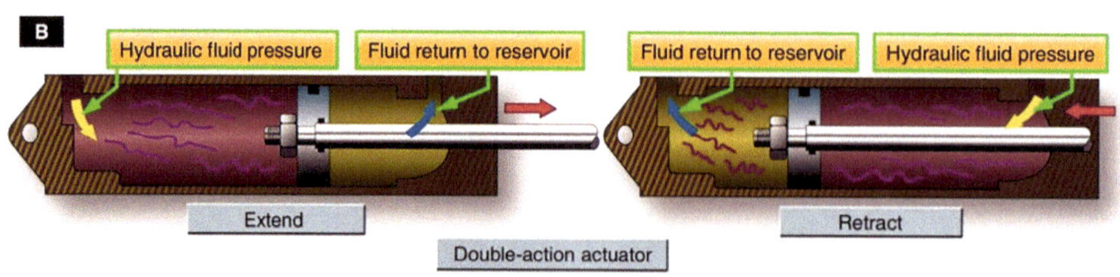

Fig. 10.18 Double acting actuator. *(Courtesy: Springer.)*

body or bonnet to release the pressure due to leakage from the valve sealing to outside of the valve. A pressure relief valve on the valve body creates an extra leakage point from the seawater to the valve internals, so it should be provided with adequate sealing to prevent external seawater leakage to the valve's internals. The actuator sealing material, given by the valve manufacturer and/or valve supplier, should be compatible with the hydraulic fluid. The sealing should be capable of providing adequate sealing to withstand the supply pressure of the hydraulic fluid at the rated temperature.

10.2.6 All-electrical actuators

The other important topic in subsea actuation is the shift from hydraulic to electrical actuators, known as a state-of-art modification in the subsea sector of the oil and gas industry. All-electrical subsea technology is not something new and has been experienced by many end users. This change of actuator operation to electrical power brings simplification and cost reduction, and is environmentally friendly. It is well known that spillage of hydraulic oil is harmful to the environment, threatening marine life and damaging the ecosystem. Moreover, removing the hydraulic oil from the actuators prevents the exposure of personnel and operators who are involved in hydraulic actuator manufacturing, testing, inspection, transportation, etc. Thus using electrical actuators is beneficial from a Health, Safety, and Environment (HSE) point of view. There are other advantages associated with using electrical actuators, such as improving actuator response time, reliability and efficiency, and the functionality of the subsea system. Electrical power provides smooth operation with accurate positioning, less possibility of actuator failure, and fewer maintenance requirements.

A hydraulic system is comprised of a hydraulic power unit (HPU), an umbilical, and a subsea control module (SCM). Subsea umbilicals are cables and hoses that are used to supply electrical and hydraulic power to subsea equipment such as valves and actuators. All subsea electrical actuator strategies lead to the elimination of the HPU, SCM, and the hydraulic oil-transmitting umbilical, which is a good cost-saving strategy. The space saving benefit due to the usage of an all-electrical subsea system is not limited to subsea; using an all-electrical system reduces the amount of required space topside due to the elimination of the topside HPU. In fact, shifting to an electrical subsea system reduces both operation and capital cost (OPEX and CAPEX). Some case studies have proven that using an all-electrical subsea system can save costs in the range of 10%–30%. Fig. 10.19 illustrates a subsea electrical actuator.

10.2.7 SIL/SAR requirements

One of the main questions that arises due to the shift from hydraulic to electrical actuator usage is whether there is any need to consider Safety Integrity Level (SIL) for electrical actuators. It is important to know the accurate meaning of SIL before answering this

Fig. 10.19 Subsea electrical actuator. *(Courtesy: Baker Hughes.)*

question. SIL is part of international standards such as International Electro technical Commission IEC 61508 and IEC 61510 that provides suppliers and end users with a common framework to design products and systems for safety-related applications. SIL provides a scientific and numeric approach to specifying and designing safety systems, enabling the risk of failure to be quantified. Four SILs are defined in IEC 61508. SIL 1 represents the lowest level of safety and SIL 4 represents the highest safety level. The probability of failure on demand (PFD) and the probability of dangerous failure per hour (PFH) on safety functions for each SIL are presented in Table 10.2.

Table 10.2 Failure probabilities based on SIL grade according to IEC.

SIL	Average PFD (*low demand mode of operation*)	Average PFH (*high demand or continuous mode of operation*)
1	$\geq 10^{-2}$ to $<10^{-1}$	$\geq 10^{-6}$ to $<10^{-5}$
2	$\geq 10^{-3}$ to $<10^{-2}$	$\geq 10^{-7}$ to $<10^{-6}$
3	$\geq 10^{-4}$ to $<10^{-3}$	$\geq 10^{-8}$ to $<10^{-7}$
4	$\geq 10^{-5}$ to $<10^{-4}$	$\geq 10^{-9}$ to $<10^{-8}$

Normally, subsea valves on manifolds and Christmas trees do not have a safety function such as emergency shutdown (ESD) or high integrity pressure protection (HIPPS). SIL and a safety analysis report (SAR) are applicable for safety critical valves with ESD or HIPPS functions. Changing the actuators from hydraulic to electrical does not change the SIL requirement. Thus SIL is not applicable for manifolds and Christmas tree valves, regardless of whether they use a hydraulic or electrical actuator.

Fig. 10.20 illustrates a 7 1/16″ slab gate valve in a pressure class of 517 bar equal to 7500 psi with a spring return linear actuator used for subsea manifolds in normally open position that gets closed through actuator spring force to isolate the manifold branch from the connected wellhead. The manifold should be isolated from the wellhead in the event of wellhead operational problems or during maintenance.

10.2.8 Hydraulic control systems

Clean hydraulic fluid supply is provided for subsea actuators installed on the seabed in subsea production systems. The hydraulic fluid is transported from the topside facilities through an umbilical to the subsea distribution unit (SDU) for actuator and valve operation. As discussed earlier in this chapter, the hydraulic fluid should have sufficient pressure to be able to fully open and close the valves and actuators. In early developments of subsea fields, each subsea valve and actuator was connected directly to the topside host facilities through a unique umbilical. This type of hydraulic supply system is called a "direct hydraulic control system" and is shown in Fig. 10.21. The main component of the hydraulic supply unit is the HPU. The HPU contains a motor, a reservoir or storage tank, a hydraulic pump, and an accumulator. A reservoir is a storage tank used to store the hydraulic fluid. The pump provides sufficient pressure to move the hydraulic fluid, and the accumulator compensates for the pressure damping of the pumps. In fact, accumulators are the components that store the hydraulic energy or power, so accumulators

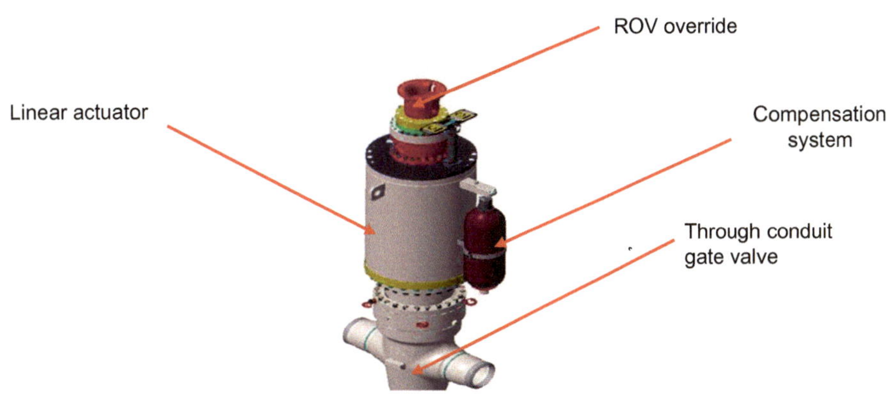

Fig. 10.20 A 7 1/16″ slab gate valve with a linear actuator on a subsea manifold.

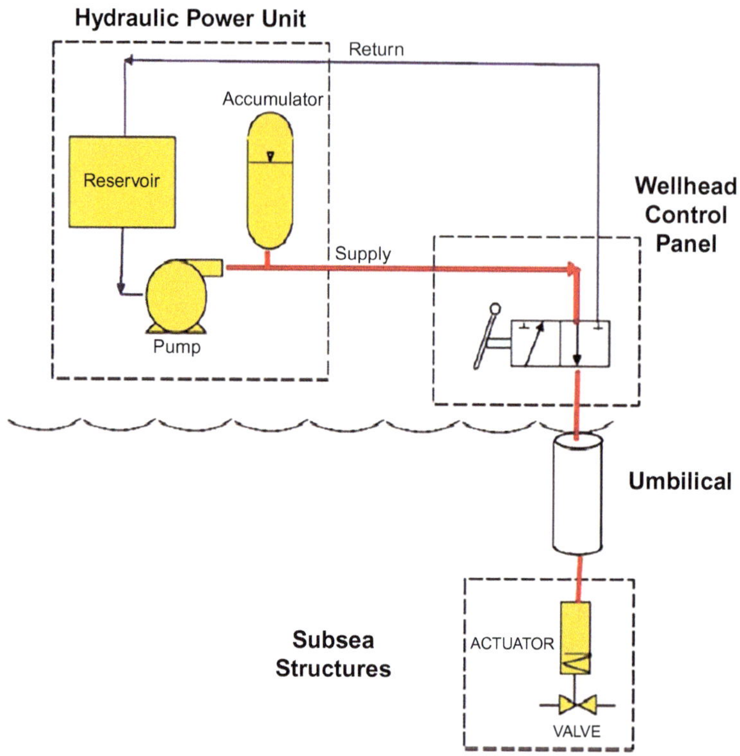

Fig. 10.21 Direct hydraulic control system. *(Courtesy: Elsevier.)*

can save energy for the pumps and leads to considerably less powerful pumping. The main advantage of a direct hydraulic control system is its reliability. The reliability of the system is due to the location of the HPU system topside, so it is easier to have access to the HPU for different purposes such as maintenance. In addition, this system is relatively economical, since there is no SDU or SCM. However, the actuator response time could be very slow, since there is a long distance from the HPU to the subsea actuator. The other disadvantage of the system is its large number of umbilicals. Therefore a more complex distribution system was invented by adding an SCM to the hydraulic supply system as illustrated in Fig. 10.22.

The development of a piloted hydraulic control system led to a reduction in the number of umbilicals and in actuator response time.

10.2.8.1 Open- vs closed-loop hydraulic systems

The other main decision in the development of subsea fields is the choice between an open- or a closed-loop hydraulic distribution system. Hydraulic oil is transmitted from the topside facilities or floating production storage and offloading (FPSO) unit to the

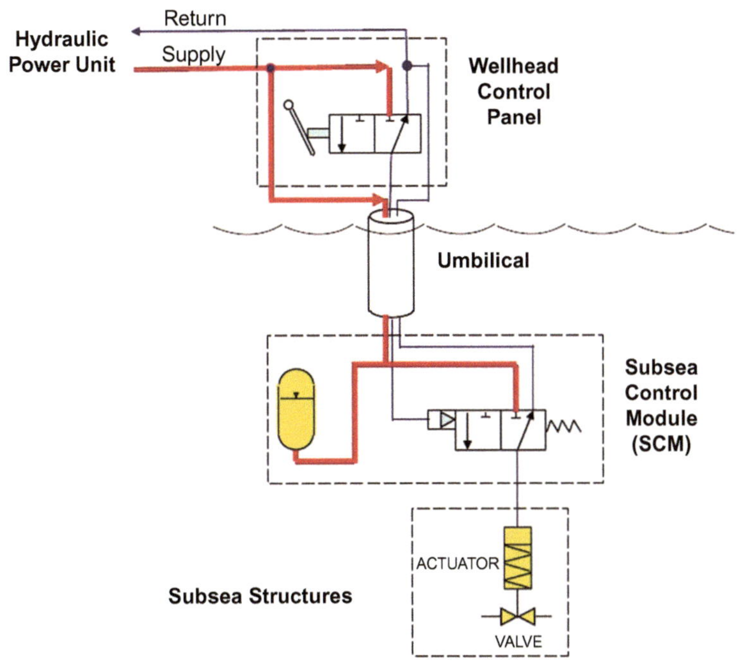

Fig. 10.22 Piloted hydraulic control system. *(Courtesy: Elsevier.)*

SCMs in both open- and closed-loop hydraulic distribution systems. The main feature of an open-loop system is that the hydraulic fluid is discharged to the sea after usage in the subsea actuator. Therefore this system is not environmentally friendly, although it is cheaper than a closed-loop hydraulic system. In a closed-loop system, the hydraulic fluid is returned to the topside facilities through an umbilical after being used in the hydraulic actuator, rather than being dumped into the sea. This makes a closed-loop system more HSE friendly, but more expensive due to the larger number of umbilicals required. The details of both closed and open-loop systems are illustrated in Fig. 10.23.

There is a subsea electronic module (SEM) in subsea control module (SCM), refer to Fig. 10.23. SEM contains the electrical supply and communications channels between the subsea and topside facilities. A direct control valve (DCV) is used to direct and change the route of the hydraulic supply to the actuated valves. In fact, an SCM contains different software and valves for routing and directing the hydraulic fluid to the different types of actuators. The actuators in Fig. 10.23 are spring return hydraulic actuators that open when the hydraulic oil supply enters the piston and cylinder side of the actuator through the green line. When hydraulic supply to the actuator stops, spring force closes the valve and actuator. The hydraulic fluid moves out of the actuator through the same green line that directed the hydraulic fluid to the actuator. The outlet or returned hydraulic fluid enters the DCV and is directed to the sea in an open-loop system. In a closed-loop

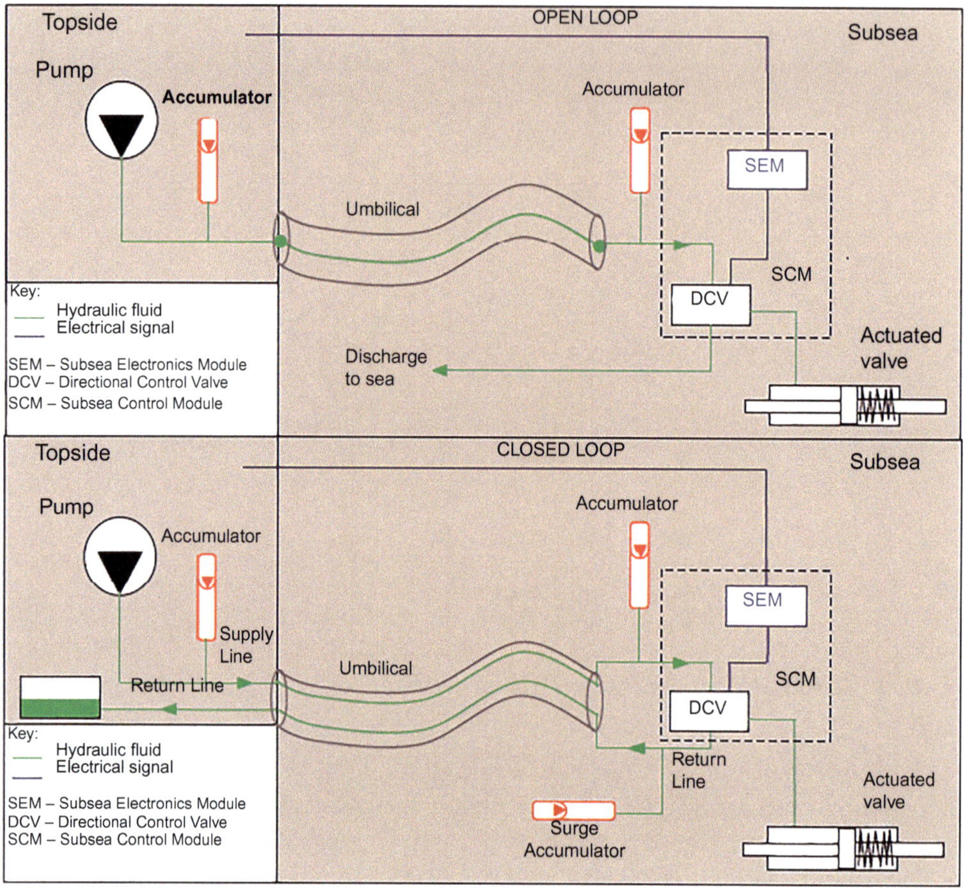

Fig. 10.23 Comparison between open- and closed-loop hydraulic systems. *(Courtesy: URS Cooperation Limited.)*

system, the discharged hydraulic fluid is directed to the DCV and then to the return umbilical line instead of being discharged to the sea. There is only one hydraulic supply line for open-loop systems, whereas closed-loop systems have two: a hydraulic supply line and a hydraulic return line. In short, in an open-loop system, discharge to the sea occurs when the spring within the actuated valve returns the valve to a closed position, forcing the hydraulic fluid to be discharged into the marine environment. There is a delay in closing the actuators and the connected valves in a closed-loop system, since the spring force in the actuator must overcome the hydraulic friction in the return hydraulic line. Although using a closed-loop hydraulic system necessitates spring return actuator design change and increases both response time and the volume of the umbilical, it is more environmentally friendly.

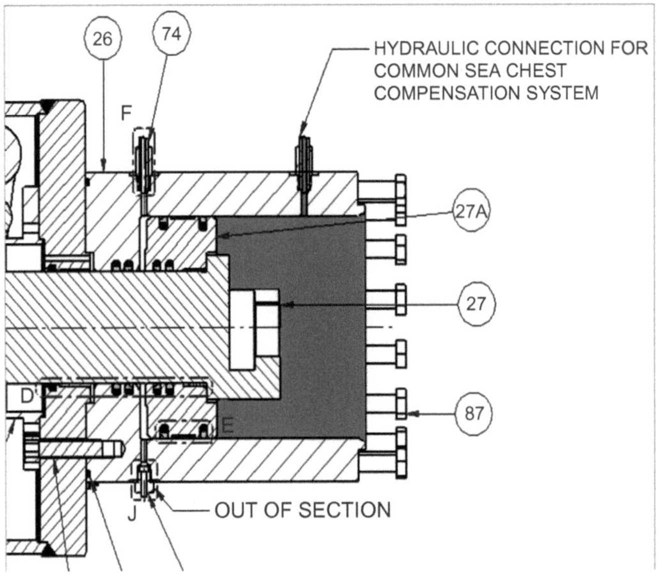

Fig. 10.24 Bottom part of a linear hydraulic actuator connected to a closed-loop system.

A closed-loop system and its connection to an actuator is not always like the schematic shown in Fig. 10.24, which illustrates the bottom part of a linear actuator connected to a closed-loop system installed on a slab gate valve in a manifold. Part #74 shows the hydraulic supply port, which is plugged. The other plugged connection under part #74 will be connected to the common return line tubing that transports the hydraulic fluid to the topside facilities as part of a closed-loop system. The important point is that this hydraulic fluid affects the opening and closing of the actuator and valve. Let's suppose that the hydraulic supply is coming from port #74 and puts the actuator shaft (part #27) down. The hydraulic fluid that accumulates at the bottom of the actuator prevents the shaft from moving downward and opening the valve. It should be noted that the gate valve connected to this actuator has a reverse port that gets opened when the stem and gate move downward. Therefore an actuator with higher opening torque compared to a standard actuator should be designed in some cases to neutralize the effect of the hydraulic pressure at the bottom of an actuator connected to a common return line. On the other hand, the hydraulic pressure at the bottom of the actuator helps to close the valve and actuator when the stem moves upward.

Fig. 10.25 illustrates a spring return FSC rack and pinion actuator connected to a common return line or closed-loop system through a "recirculation hydraulic connection." The hydraulic fluid enters the port labeled "main hydraulic connection" in Fig. 10.25 and moves the piston and rack from left to right to open the actuator

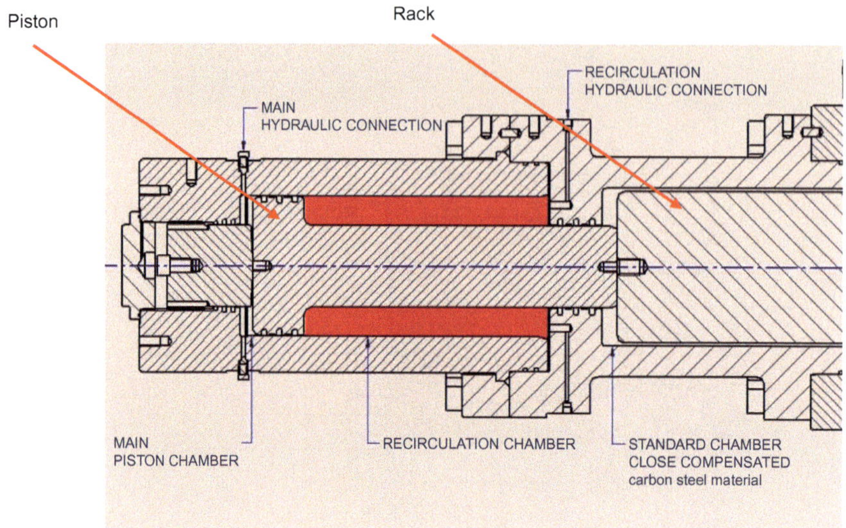

Fig. 10.25 Rack and pinion hydraulic actuator connected to a closed-loop system.

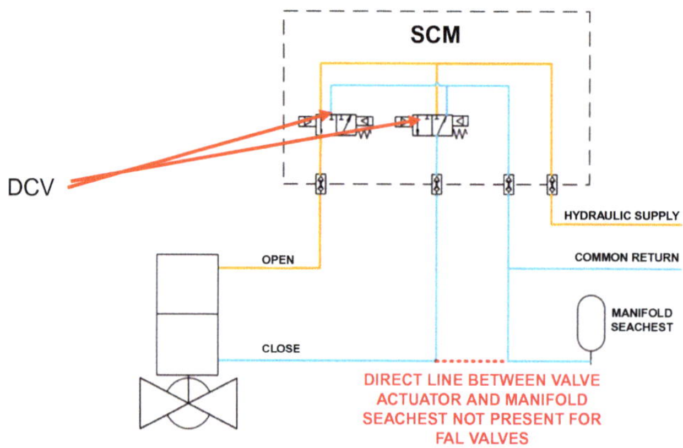

Fig. 10.26 Double acting hydraulic actuator connected to a common return line in a closed-loop system.

and valve. Two areas around the shaft, highlighted in red in the figure, are connected to the closed-loop system. The hydraulic fluid in a recirculation connection acts against opening the valve and actuator and assists the closing action.

Fig. 10.26 illustrates a double acting actuated ball valve connected to a closed-loop system. There are two DCVs in the SCM in the figure. DCVs are important components in hydraulic systems that direct the fluid flow in different paths. Each hydraulic cylinder is connected to one line. The bottom part of the actuator in the schematic is connected to a blue line and the upper part is connected to an orange line. Each of the lines connected to the actuator is used as both supply and return hydraulic lines. Once the hydraulic pressure is supplied to the actuator through the orange line, it opens the valve and actuator and the hydraulic fluid moves out of the bottom part of the actuator through the blue line to the common return line. Then the DCV in the middle of the figure can change in position and allow the hydraulic supply to enter the blue line connected to the bottom part of the actuator. Then the actuator gets closed and the hydraulic fluid on the top of the actuator moves back from the orange color line into the common return line. The key point is that in a double acting actuator connected to a closed-loop system according to the schematic given in Fig. 10.26, there is no need to change the configuration of the actuator, unlike the two previous cases of the FSC rack and pinion and linear actuators. However, it should be noted that connecting a common return line to a double acting actuator affects the effective produced torque/force of the actuator as well as its sizing.

10.2.9 Compensation systems

Compensation systems, including both open and closed types are explained in Chapter 11.

10.2.10 Design and qualification

Subsea actuators are designed, manufactured, and tested according to the requirements of API 6A/ISO 10423 and API 17D/ISO 13826-4. API 6A and ISO 10423 are international standards that provide requirements and recommendations for the design, testing, inspection and quality control, marking, storage, and shipment of wellhead and Christmas tree equipment. API 17D/ISO 13826-4, like API 6A/ISO 10423, are international standards for subsea wellhead and Christmas tree equipment. In addition to these standards, specific project and client requirements are applicable for actuators. According to the earlier mentioned standards, subsea actuators and valves should undergo a series of qualifications for design validation. These qualifications include temperature-pressure cycles, a hyperbaric test, and an endurance test. These tests and qualification programs are normally followed by disassembly, visual inspection, and nondestructive tests. The number of cycles could be 200 times for each type of qualification test, equal to 600 cycles as an example. API 6A annex F and 17D define two types of cycles called "performance requirement levels" (PR1 and 2) for valve and actuator qualification. PR1 requires only three pressure-temperature cycles for valve qualification, which is not common. PR2 requires 200 pressure-temperature cycles, which is more applicable in the industry for both valves and actuators. In addition, subsea actuators should be designed for a specific operational life, which could be 20, 25, or 30 years as examples, and for the water depth specified by the end user. The design or operational life, as well as the valve and actuator installation water depth, is normally provided in the data sheets by the end user.

10.2.11 Materials

Materials used for subsea actuators include carbon steel or low-temperature carbon steel, low alloy steel with high mechanical strength for some of the internals and some seawater corrosion-resistant alloys such as 25Cr super duplex and nickel aluminum bronze (NAB) materials. As an example, for a spring return rack and pinion actuator, the internal gears, rack, pinion, and spring could be made in a high-strength, low alloy steel (e.g., 39NiCrMo3 according to the EN 10083-3 standard). This low alloy steel contains between 0.35% and 0.43% carbon, between 0.5% and 0.8% manganese (Mn), maximum phosphor of 0.025%, maximum sulfur of 0.035%, between 0.6% and 1% chromium, between 0.15% and 0.25% molybdenum, as well as between 0.7% and 1% nickel with a quench-temper heat treatment. This material has almost the same chemical composition and mechanical strength as AISI 9840. Quenching and tempering are two processes that are done to strengthen materials like steel. The quench-temper process involves heating the material and then cooling it in cold water. In fact, quenching means cooling the material to achieve a very hard structure. Tempering is done after quenching in order to achieve higher toughness and ductility. A low alloy steel could have very high mechanical strength as strong as 160 ksi yield and 180 ksi tensile. Alternatively, AISI 4340 is popular for actuator components such as the internal gear, rack and pinion; it is known as a high mechanical strength material containing low alloy steel with a yield strength of approximately 103 ksi and a tensile strength of 161 ksi. This material, like AISI 9840, contains between 0.36% and 0.43% carbon and a small percentage of other alloys such as nickel, molybdenum, manganese, and chromium. The American Society for Testing and Materials (ASTM) A29 standard specification for general requirements for steel bars, carbon and alloy steels, hot-wrought should be followed for AISI 4340 materials.

The override shaft that transfers the ROV load to the actuator gear, as well as the rack and pinion, should be in a strong material such as 25Cr super duplex ASTM A182 F55. The cylinder and piston of the actuator could be in contact with seawater, so a pitting-resistant material such as 25Cr super duplex ASTM A182 F55 could be selected for the cylinder and NAB in grade ASTM B124 C63000 for the piston. These parts can be called "subsea trim"; trim refers to valve internals that are in contact with the process fluid. These are also called "wetted parts." Wetted parts or trim in actuators are the actuator internals that are in contact with the hydraulic fluid. These parts could be in contact with seawater during wet storage when the actuator is in the sea before installation and operation. But it should be borne in mind that it is recommended to store actuators at ambient temperature, typically from 5°C to 40°C, in a dry, nonhumid environment. The actuator hydraulic connections, such as hydraulic supply and return, are normally supplied with two plugged connections. C63000 NAB contains approximately 82% copper and has a relatively high mechanical strength. Forged materials are the preferred choices for subsea actuators, since forging produces a better quality than casting with less possibility of defect, and provides a stronger structure.

The pressure-containing parts of actuators, such as the cylinder, piston, and housing should have full material traceability, normally according to EN 10204 type 3.1. When purchasing steel materials, they come with mill test certificates (MTC) containing data related to quality of the product, such as chemical composition, mechanical strength, etc. EN 10204 defines the different types and levels of these certificates. The certification levels for steel products are organized into four main categories in the oil and gas industry based on the EN 10204 standard:

1. EN 10204 2.1: This is the lowest level of certificate, in which the manufacturer of the steel products just gives a declaration of compliance with the purchase order without any test report. This level does not provide a good assurance for the quality of the product.
2. EN 10204 2.2: This category indicates that the manufacturer of the steel provides both a declaration of conformity and a test report. However, since the test report is prepared and validated by the producer of the steel, it is still not very reliable.
3. EN 10204 3.1: This level means that the manufacturer of the steel products provides a declaration of conformity and verifies that a specific inspection has been done on the products by a mill (producer) inspector who is independent from the manufacturing department.
4. EN 10204 3.2: This is the highest and the most reliable inspection level; it includes the EN 10204 3.1 requirements and, in addition, a test report that is prepared and validated by the mill's authorized inspector and validated by either the purchaser inspector or a regulatory organization.

10.3 Actuators on valves

Actuators are installed on valves after complete valve assembly, including all parts of the valves such as the body, bonnet, stem, closure member, sealing, and gaskets. It is recommended to start pressure testing the valves after actuator assembly. However, in practice, actuated valves are sometimes tested bare stem without actuators due to the late arrival of the actuators from the actuator supplier when valve and actuator are supplied by two different manufacturers. Dual pins between the valve's top work and the bottom of the actuator ensure correct alignment and assembly of the actuator on top of the valve. Fig. 10.27 illustrates an actuator with four tie-rods that are used to correctly align the actuator on top of the valve.

10.4 Actuator tubing interface

As mentioned before, the actuator suppliers should provide hydraulic actuators with plugging for the hydraulic connections. One type of these fittings could be

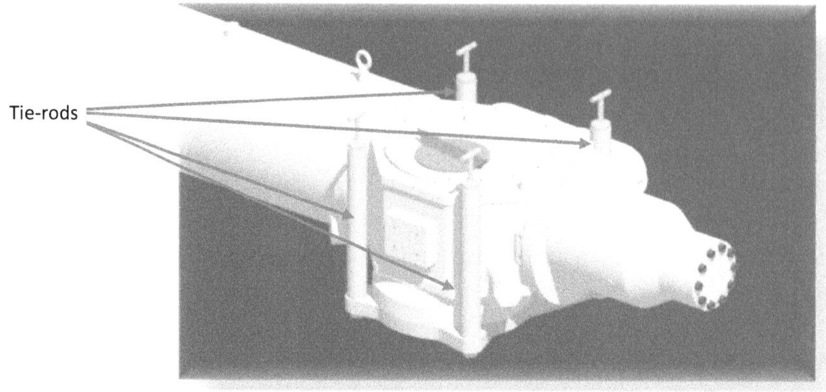

Fig. 10.27 Tie-rods for correct alignment and installation of the actuator on the valve.

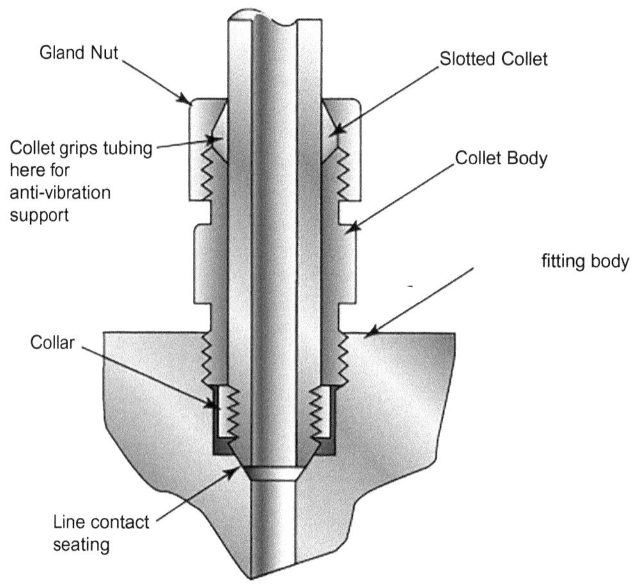

Fig. 10.28 Autoclave assembly for the actuator's hydraulic connection.

autoclaved plugs, also called "cone and threaded assembly," illustrated in Fig. 10.28. The main components in this type of connection are tubing, a gland, and a collar, illustrated in Fig. 10.29.

This connection is attached on one side to the actuator hydraulic connection and welded to the tubing from the other side. The connection could be as small as 3/8″ in size. It should be noted that the hydraulic connections to the actuator are mainly tubes

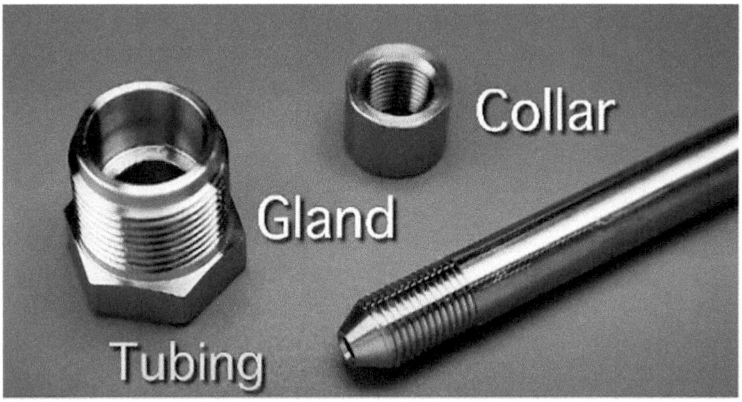

Fig. 10.29 Main components of the autoclave assembly.

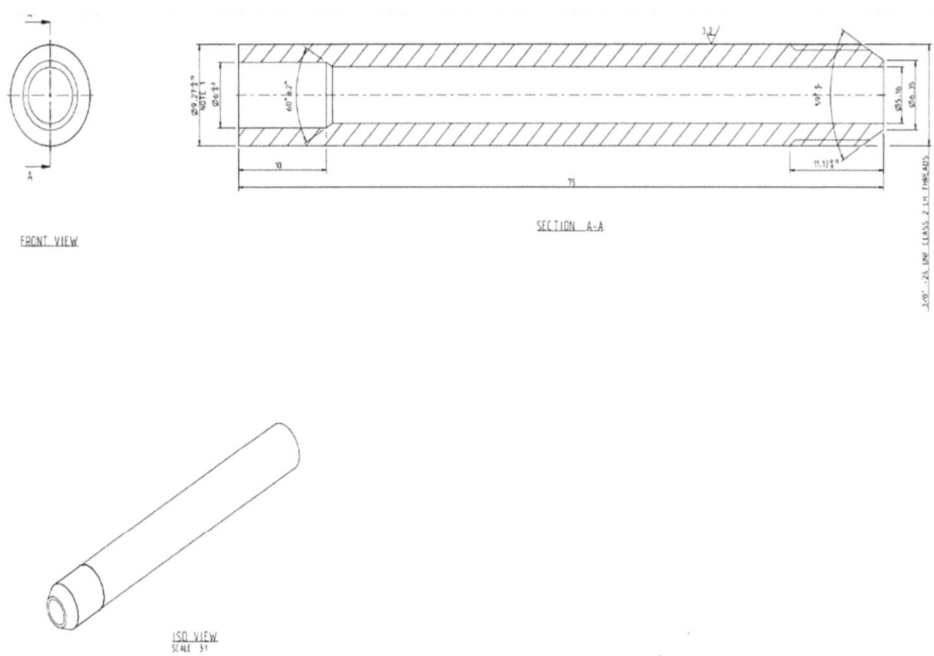

Fig. 10.30 Cone and threaded nipple for blinding the supply hydraulic actuator connection.

that are welded together—except for the tube connection to the body actuator, which could be a cone and threaded connection. Fig. 10.30 illustrates a cone and threaded nipple component used for blinding the hydraulic actuator connections.

In general, threaded tubing connections are not allowed for subsea use. The tube has male threads at the top that engage into the female threads of the collar. One or two threads on the top of the tube can remain free without any engagement to the collar threads. Lubricating the threads during the assembly of components is essential to prevent galling and friction between the metal contacts. The gland or collet provides an anti-vibration feature for the tube assembly, which enters the actuator body (shell) through its male threads. In general, there is no need to design and apply an antivibration gland for tubing in the subsea sector, since the tubes are butt welded together which can provide robust joint. An antivibration gland or feature is applicable for threaded connections that are not welded. It is important to note that fillet weld and socket weld are not acceptable for subsea tubing connections. Loctite, as a sealant, and threaded locking can be used for assembling the actuator fitting to the body of the actuator. Teflon or PTFE tape is another material that could be used for inserting the actuator tube connection to the body of the actuator. However, some valve and actuator specifications may not allow the usage of PTFE. The other important point is that the hydraulic connection plugs should be designed based on 1.5 times of the hydraulic working pressure.

10.5 Lifting the actuator

Both valves and actuators should have a suitable lifting device, either a pad eye or eyebolt, in a suitable size and of dimensions to support a sufficient safe working load (SWL). SWL is defined as the maximum safe load that the lifting equipment can lift, move, suspend, and lower down to the ground without any problem of breaking. API 17D/ISO13628-4 has specified a safety factor of 2.5 times the SWL for an individual pad eye or permeant lifting points that should be designed according to API 17D/ISO 13628-4 Annex K. All of the required dimensions for pad eyes as a type of permeant lifting lug are given in the above-mentioned standard and annex. Lifting lugs could be eye bolts, also called lifting eyes (Figs. 10.31 and 10.32), that are screwed into the valve or actuator body, or they could be pad eyes or another type of welded lifting component welded to the actuator and valve body/bonnet or integral liftings. Fig. 10.33 illustrates a lifting lug welded to the body of a valve. All of the welds for lifting lugs such as pad eyes are considered critical, which means that they require 100% both volumetric

Fig. 10.31 Screwed-in lifting lug.

Fig. 10.32 Lifting eyes installed on a valve.

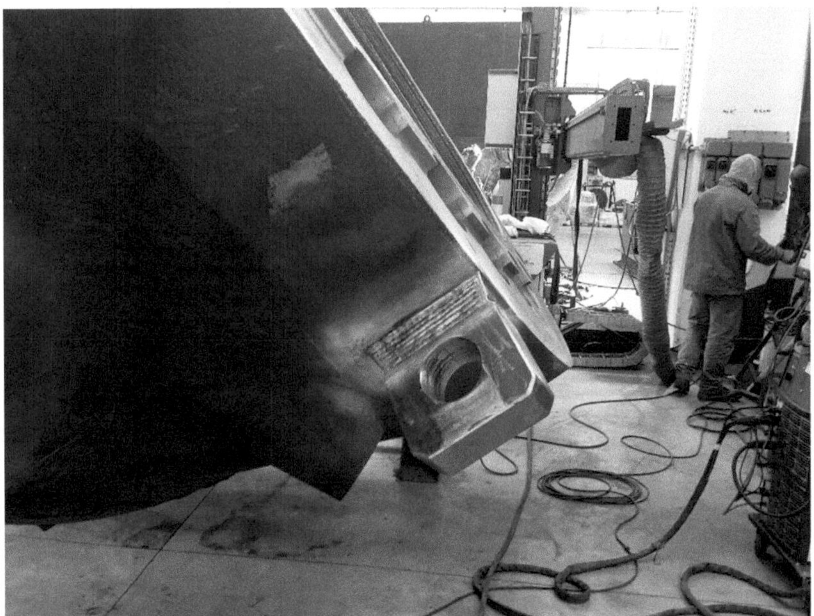

Fig. 10.33 Lifting lug welded to the body of a large-bore ball valve.

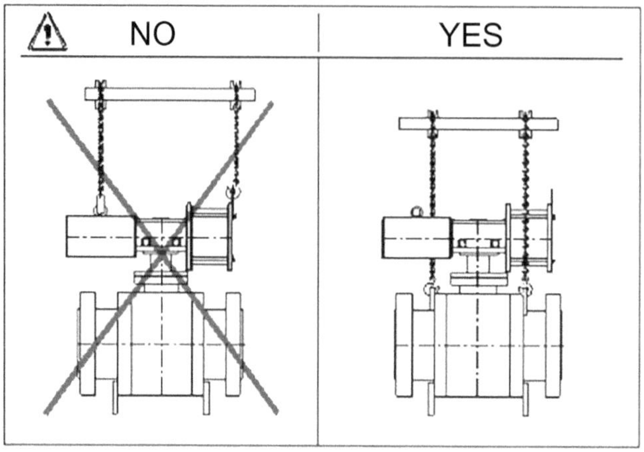

Fig. 10.34 Valve and actuator assembly lifting. *(Cameron.)*

nondestructive test examination such as a radiography test or ultrasonic test, as well as a surface examination such as a liquid penetration or magnetic particle test.

Actuators normally have lifting lugs for lifting and transportation purposes. However, lifting the assembly of the valve and actuator with the actuator lifting lugs is not allowed (see Fig. 10.34 left side). The picture on the right illustrates the correct lifting method, namely lifting the valve and actuator assembly with the valve lifting lugs. Therefore the lifting lugs or points provided on the valves should be designed and considered for the whole assembly of the actuated valves.

If the actuator is light and/or the connection points between the valve and actuator (actuator and valve integrity) are strong enough, it is fine sometimes to lift the valve and actuator assembly with the actuator lifting lugs, as illustrated in Fig. 10.35.

A procedure for lifting the actuator as a single unit, as well as lifting the actuator and valve assembly, including sketches and calculations (if applicable) should be provided by the valve and actuator manufacturer. This procedure could be included in the actuator's Installation, Operation, and Maintenance (IOM) manual, or could be a part of the handling, transportation, and storage procedure. As an example, Fig. 10.36 illustrates a sketch for lifting a double acting actuator from the valve lifting lug with slings or a chain provided by the actuator supplier.

The preferred choice for actuated valve transportation is to ship both the valve and the actuator as a single unit from the valve supplier factory to the fabrication yard, as separation of the actuator from the valve could jeopardize the testing of the valve and actuator as a single unit. However, the actuator can be removed from the top of the ball after the factory acceptance test (FAT) and shipped separately in the case of huge and heavy actuated valves that might encounter some road limitations during transportation. If the actuator is required to be shipped separately, this should be clearly stated in the actuator's IOM manual.

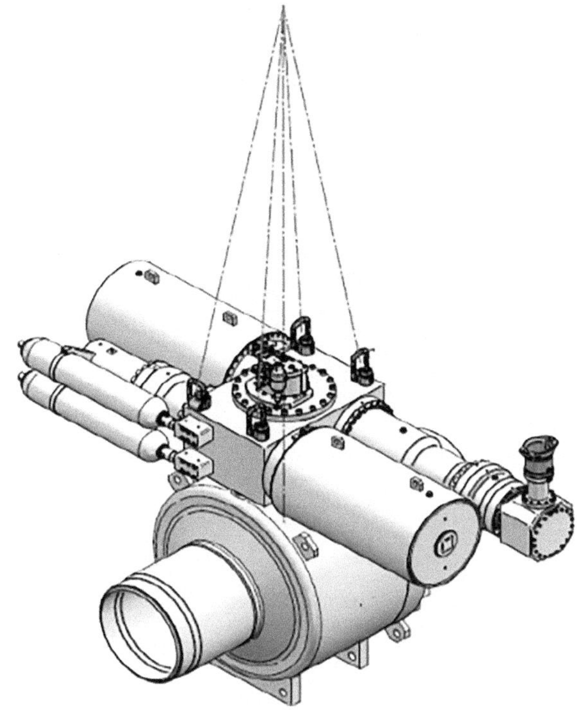

Fig. 10.35 A large ball valve and symmetrical actuator assembly being lifted from the actuator. *(Advanced Technology Valve.)*

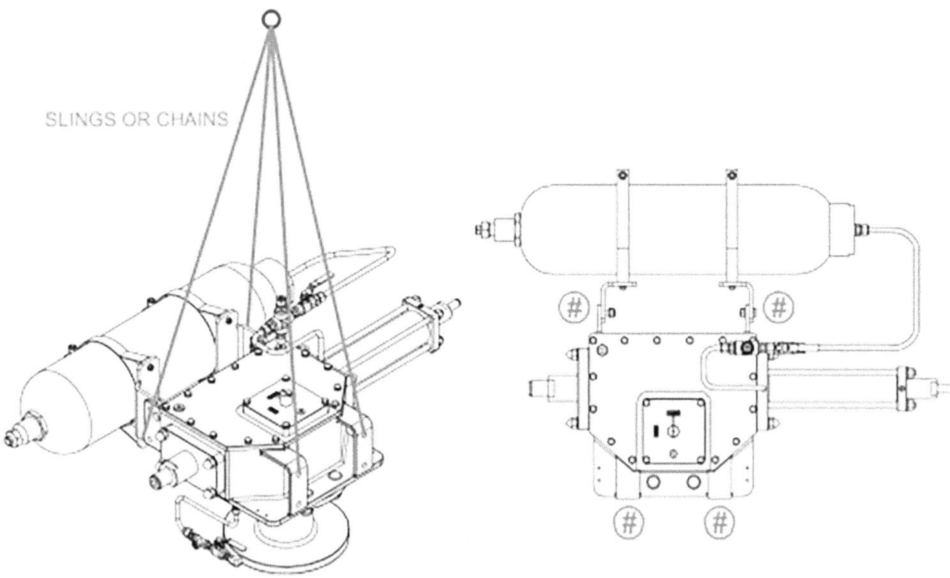

Fig. 10.36 Double acting actuator lifting sketch. *(Cameron.)*

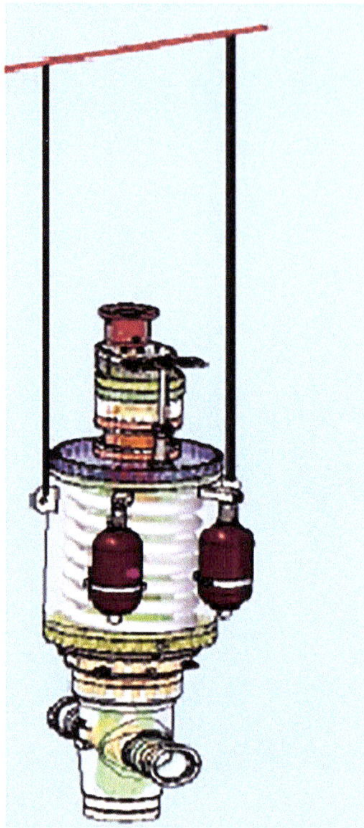

Fig. 10.37 Lifting an actuated slab gate valve from the linear actuator. *(Courtesy: ATV.)*

Question: A 7 1/16″ actuated slab gate valve in a pressure class of 517 bar has four lifting eye bolts. Its linear actuator has two permeant lifting points. The valve weighs 1400 kg, the actuator weighs 2500 kg, and the assembly is thus 3900 kg. The IOM manual shows that the whole assembly of the valve and actuator can be lifted from the actuator (refer to Fig. 10.37). What working load limit could be considered for the valve and actuator lifting lugs in order to satisfy API 17D/ISO 13628-4?

Answer: The valve has temporary lifting lugs, so they have no assigned working load limit (WLL). Normally all four valve lifting lugs are used to lift the valve, so the practical WLL for each valve lifting lug could be equal to the whole weight of valve divided by four, i.e., a 350 kg WLL for each valve lifting eye bolt. However, the valve supplier may consider a higher safety factor for valve lifting lugs and designate a 1400 kg WLL for each valve lifting eyebolt. The actuator has two permanent lifting lugs, and each of these should be able to be used to safely lift the entire 3900 kg valve and actuator assembly. In fact, the SWL of each permanent actuator lifting lug is 3900 kg. Having a safety factor of 2.5 according to API 17D/ISO 13628-4 gives a WLL of 2.5 multiplied by 3900 and equal to 9750 kg for each individual actuator lifting point.

The valve in Fig. 10.37 is a 7 1/16″ slab gate valve in pressure class 7500 equal to 517 bar. The valve has four lifting lugs and its linear actuator has two lifting lugs.

10.6 Valve and actuator installation

The complete assembly of the valve and actuator should be capable of sealing and operating in the direction of installation. In addition, strong and sufficient integrity should be achieved for the assembly of the valve and actuator at the specific direction of installation. There are three typical installation directions for valve and actuator assemblies: horizontal bore with vertical stem, vertical bore with horizontal stem, and horizontal bore with horizontal stem. The orientation of the bore and stem of the valve should be clearly specified in the valve data sheets, as it could affect the connection design and the strength of the connection between the valve and actuator as well as the sealing capability of the valve. As an example, most large bore valves above 2″ in the manifold have a horizontal bore and vertical stem. Fig. 10.38 illustrates three 10″ ball valves on the manifold header, all with a horizontal bore and vertical stem. These ball valves are not actuated; they are operated manually via ROV. All four slab gate valves on the branches are 7 1/16″ with hydraulic actuators with a horizontal bore and vertical stem.

Fig. 10.39 illustrates one of the 7 1/16″ slab gate valves in the previous picture with a horizontal bore and vertical stem located on a production manifold during the fabrication activities in a yard. Half inch ball valves with ROV operation and a low torque paddle installed at the top of the valve are located on the side of manifolds and could

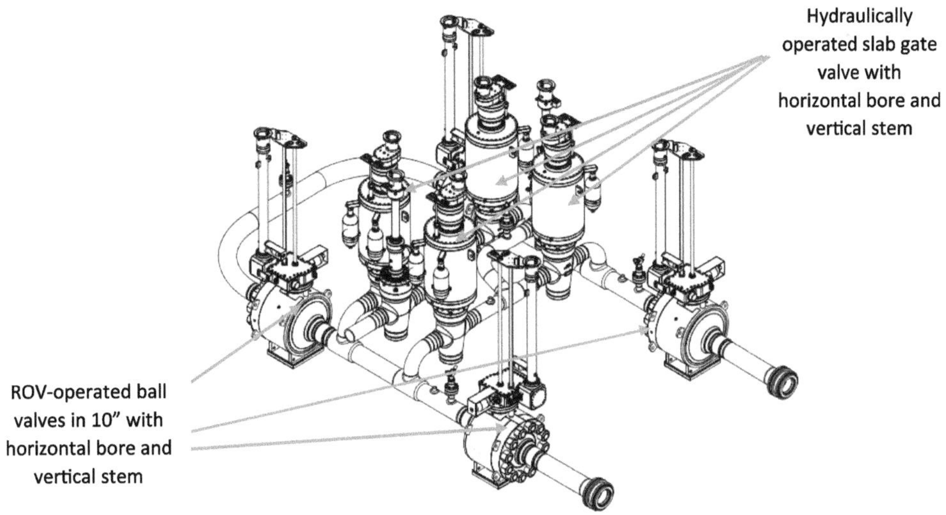

Fig. 10.38 Large bore slab and ball valves with horizontal bore and vertical stem on subsea manifolds with horizontal bore and vertical stem.

Fig. 10.39 A 7 1/16″ subsea slab gate valve with horizontal bore and vertical stem for a subsea manifold. *(Courtesy: Baker Hughes.)*

have a vertical bore with horizontal stem, as illustrated in Fig. 10.40. These ½″ ball valves are used for hydraulic fluid supply to the actuators for automatic operation of the valves. Fig. 10.41 illustrates a side view from outside of the manifold where all of these nonactuated ½″ ball valves with low torque paddles are located. These ½″ low torque paddle ball valves can be installed in a third position, horizontal bore with horizontal stem, depending on the piping arrangement and routing. The preferred solution is to standardize the direction of installation for similar valves of the same size, pressure class, and installation depth with the same actuator types.

10.7 Effects of actuation on valve design

Compared to manually operated valves, higher load (force) or torque (force × distance) is applied from the actuator to the valve drive train including the valve stem. The drive train is defined as all the parts of the valve drive between the operator and closure member, including the closure member but excluding the valve operator, according to the API 6DSS/ISO 14723 standards. The drive train for a ball valve includes the stem, stem key, ball and stem joint, as illustrated in Fig. 10.42. The stem key is located at the top of the stem, which is not included in the figure.

It is essential to make sure that the actuator output torque or thrust value on the maximum hydraulic supply pressure or spring action or maximum electricity voltage (applicable for electrically operated actuators) does not exceed the maximum allowable stem

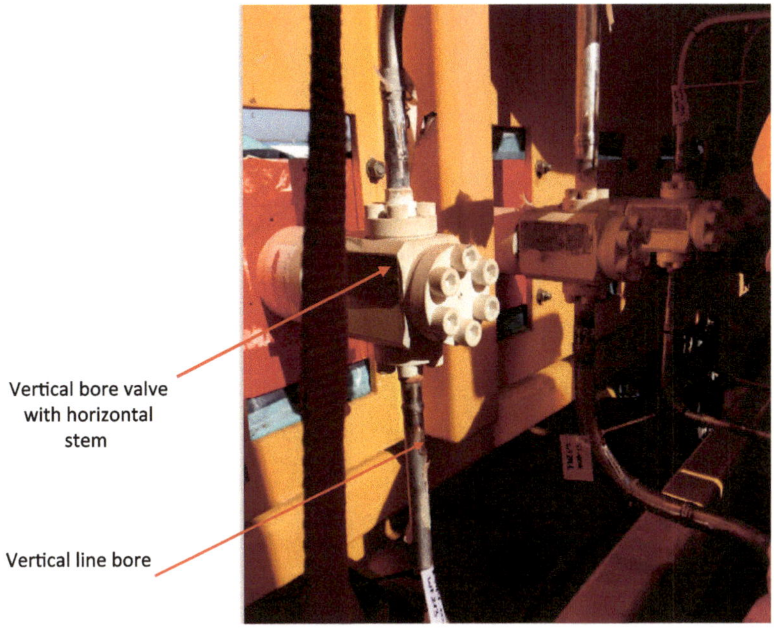

Fig. 10.40 A ½″ ball valve for hydraulic supply to the actuators on a subsea manifold with vertical bore and horizontal stem.

Fig. 10.41 Side view of a subsea manifold with low torque paddle ball valves with ROV side operation.

Fig. 10.42 Drive train of a ball valve excluding the stem key.

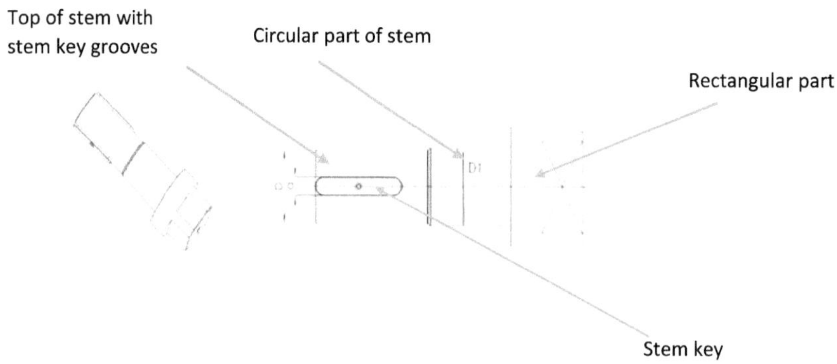

Fig. 10.43 Ball valve stem. *(Courtesy: Springer.)*

torque (MAST) during the whole travel distance of the actuator. If the actuator output exceeds the valve's MAST, the stem material could be upgraded to a higher mechanical strength material, or the diameter of the stem could be increased to improve the MAST. MAST can be calculated according to ASME boiler and pressure vessel code section VIII Div. 02. All four sections of the stem, including the top part of the stem with two groves, the circular part of the stem, the rectangular section at the bottom, as well as the stem keys are illustrated in Fig. 10.43. Additionally, it is essential to design the valve stem to withstand the actuator force/torque in case the valve obturator or closure member gets stuck. Normally, five types of actuator torques should be considered for MAST evaluation, depending on the position of the closure member or the beginning/ending stroke of the actuator spring or hydraulic oil; these are defined as follows:

1. *Break to open (BTO):* This torque is measured when the valve is closed and the ball opens against just one seat under pressure. This torque, also called breakaway torque, is the largest torque in this case.
2. *Running torque (RT):* The torque of the valve when the ball opens at approximately 35–45 degrees.
3. *End to open (ETO):* The torque of the valve when the ball opens at the 80-degree position very close to fully open position of the valve.
4. *Break to close (BTC):* When the valve is in fully open position, the torque required to break the open position of the valve to close the valve.
5. *End to close (ETC):* The torque required to fully close the valve when the valve is about to close.

These torques are all illustrated in Fig. 10.44 according to the position of the closure member (ball). It should be borne in mind that the actuator should supply enough force for complete cycling (opening and closing) of the valve under full differential design pressure. Generally, more force is required to open the valve against full differential pressure. As an example, when the valve has a design pressure of 7500 psi equal to 517 bar and it is closed, the actuator BTO and ETO torque should be considered when the valve is pressurized by 517 bar from one seat with zero pressure in the cavity for a standard valve design (not double block and bleed DBB). Any internal or external fluid pressure that

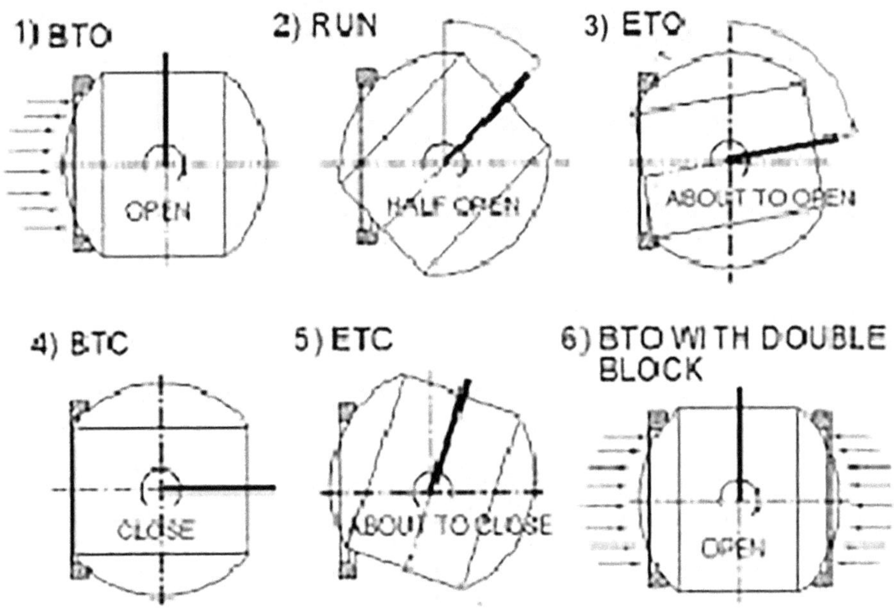

Fig. 10.44 Different torque description according to the position of the ball.

Fig. 10.45 Movement of the ball inside a ball valve.

could facilitate cycling the valve should not be taken into account in actuator design or in calculating the produced torque. As an example, the valve internal fluid helps to close a reverse port gate valve and reduces the BTC and ETC forces. However, this force reduction should not be taken into account in the actuator's design, selection, and sizing.

Fig. 10.45 illustrates the movement of the ball during a ball valve function test. The purpose of this test is to make sure that the ball valve operates smoothly without stick-stuck, and that the ball can close the valve completely and align to the pipe bore. The valve is reduced bore since it has bore reduction compared to the connected flange. The question is, does the ball position show which type of torque is needed for valve operation? The ball position could be either during BTO or ETC torque. If the direction of ball movement is from upward to downward position, then it is related to ETC torque. Conversely, if the valve is moved from downward to upward, its movement is connected to BTO torque.

10.8 Actuator sizing

Actuator sizing is an important and challenging task. Selection of the wrong size actuator could damage the valve, or the actuator might not be able to operate the valve. Actuator sizing is mainly affected by four main parameters, which are summarized as follows:

1. Valve torque/force: The force or torque for operating a valve is given by the valve manufacturer. Five or six force/torque values are associated with each type of valve depending on the position of the closure member, as discussed in the previous section. The actuator should produce higher torque than the required torque for opening and closing the valve. Normally, safety factor values such as 1.5 or 2 are considered for actuator sizing and selection. A safety factor of 2 means that the actuator should provide at least twice the required torque necessary for operating the valve from any position of the closure member. As an example, if a valve's required opening torque is 200 N m, the actuator should provide at least 400 N m for valve operation. The safety factor could depend on the type of valve or the function of the valve. As an example, some end users or EPC contractor companies may require a safety factor of a minimum of 1.5 for gate valves and 2 for ball valves. In some cases, valves with high safety functions, such as subsea safety isolation valves (SSIV) or HIPPS valves, should have a minimum safety factor of 2. Normally, the most severe operating condition, such as full pressure on one side of the valve with zero pressure in the cavity for opening the valve, and full pressure across the valve with zero pressure in the cavity for closing the valve should be considered when determining actuator sizing.
2. Actuator type: The type of the actuator affects its size. As an example, hydraulic actuators are more compact compared to pneumatic actuators with the same output torque. Another example is to use a helical spline actuator instead of a rack and pinion actuator, which is more compact.
3. Failure mode: FSC and FSO actuators with one side spring and one side hydraulic piston and cylinder are larger than FAI double acting actuators with two side hydraulic piston and cylinder. Additionally, double acting actuators provide higher torque compared to single acting actuators.
4. Minimum and maximum actuator pressure: The other important point is that the actuators should be able to open and close the valves in all operating and with minimum hydraulic pressure. As an example, the minimum and maximum hydraulic supply pressure to a hydraulic actuator is 290 bar and 345 bar, respectively. The valve is double acting with FAI function. There are three important considerations, as follows:
   - The minimum hydraulic supply pressure equal to 290 bar should be able to fully stroke the actuator for opening and closing.
   - The maximum hydraulic supply pressure equal to 345 should not generate any torque exceeding MAST, which can damage the stem and/or drive train.
   - Normally, valves and actuators are designed according to a maximum hydraulic supply pressure of 345 bar so the hydraulic supply pressure to the actuator during the operation should not exceed 345 bar.

Safety factor is also applicable for ROV override. As an example, ROV override should be designed based on 2.0 times the torque or force required for operation of the valve. Again, the most severe operation condition should be considered during ROV override sizing.

It should be noted that connecting the actuator to common return line affects its sizing and safety factor. As an example, the assumption is that a FAI actuator is connected to an open-loop system with a safety factor of 2. If the hydraulic transportation system is changed from open loop to closed loop as illustrated in Fig. 10.23, the hydraulic pressure in the common return line resists against closing the actuator so a safety factor of less than 2 (e.g., 1.7) could be achieved instead. In this case, a bigger hydraulic actuator is required to maintain safety factor of 2 in connection with a closed-loop system.

10.9 End stops

The other important design consideration is to have end stops on both ends of the actuator (generally applicable for ball valves) to provide complete alignment of the ball hole with the internal diameter of the connected pipe to provide full bore access. Rotating the end stops clockwise or counterclockwise adjusts the closure member of the valve for full closing and opening positions. Fig. 10.46 illustrates spring return scotch yoke actuators with end stops at one end. Let's suppose that the ball in a ball valve rotates just 88 degrees instead of 90 degrees. The end stop can be used to rotate the ball two more degrees and adjust the ball. However, the ball in the ball valve could have a small amount, such as 0.25 or 0.5 degrees of freedom from the fully open or closed position according to the specifications. The end stop could be achieved through the valve instead of the actuator.

10.10 Position indicators

In general, there are two types of position indicators: local and remote.

10.10.1 Local position indicators

It is possible to recognize the position of a valve and actuator (closed/open) by means of a position indicator installed on the top of the actuator beside the ROV override to be visible to the ROV during operation. Position indicators should have an Aquasign trademark; Aquasign uses an antifouling silicon marking system that prevents marine growth and provides high visibility for position indicators and other subsea components during their design life. Additionally, the Aquasign should be ultraviolet (UV) stable and resistant to prevent it from being damaged by sunlight. Fig. 10.47 illustrates a local position indicator labeled with "O" (Open) and "S" (Shut/Closed). When the valve is open, the indicator points to the letter "O" and when the valve is closed, the indicator shows the letter "S." Position indicators are applicable for ball, gate, and choke valves and are not used for axial or nozzle check valves. The requirement in the API 6DSS standard is to have the

Fig. 10.46 Spring return actuators.

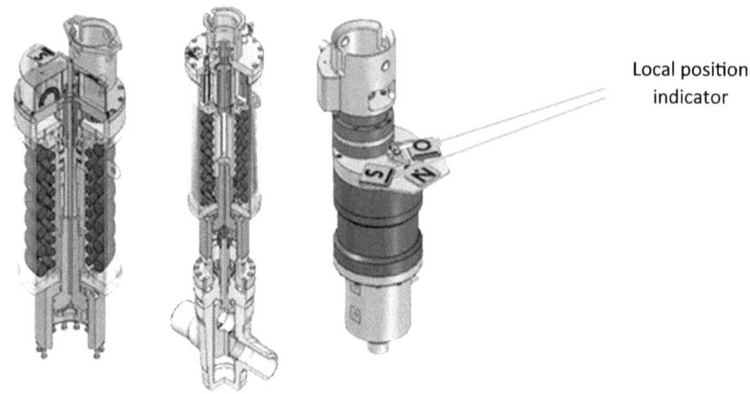

Fig. 10.47 Actuator with position indicator. *(Perar.)*

Fig. 10.48 Ball valve body and stem key groove.

position indicator parallel to the pipe when the valve is open and perpendicular to the pipe when the valve is closed for quarter turn valves such as ball valves.

It should be possible to recognize the position of valves without actuators as well as position indicator on the top. For quarter-turn valves such as ball valves, there is a key or slot on the top of the valve stem that shows the direction of the hole inside the ball or bore of the valve. Fig. 10.48 illustrates the body of a ball valve with grooves on the stem that facilitates placing the stem key bars in the grooves. When the stem groove stands perpendicular to the valve bore, it indicates that the hole inside of the ball is perpendicular to the fluid passage so the valve is in closed position. The valve standards do not specify any way to recognize the position of TCG valves when there is no position indicator and the actuators are removed from the valve. Gate valves with a rising stem and hydraulic actuator have a reverse port that is open when the stem is down and closed when the stem is up. ROV-operated gate valves are supplied with normal ports, so the valve is closed when the stem is down and open when the stem in up.

10.10.2 Remote position indicators

Remote, also called smart position indicators, provide accurate and reliable data about the open, closed, and intermediate positions of valves and actuators. Proximity switches are used to detect the movement and position of the valve closure member via electrical signals and without any direct contact. The switches can act as position transmitters that give feedback from 4 to 20 mA to a remote location such as a control room. Other communication protocols could be used. Remote position transmitters are more common for ESD and HIPPS valves, as well as those on the wellhead such as choke valves. Remote position indicators are applicable for all of the types of actuators mentioned in this chapter.

10.11 Fixed or retrievable actuators

Actuators can be categorized as fixed or retrievable. Fixed actuators remain on the valves during their whole design life; retrievable actuators can be removed from one valve and transferred to another for operation, or transported to the surface for maintenance. A retrievable actuator can be separated from a valve during normal operation without any need to stop the production line. Pressure caps (Fig. 10.49) are installed on the valves after the removal of actuators to prevent possible valve stem leakage to the environment.

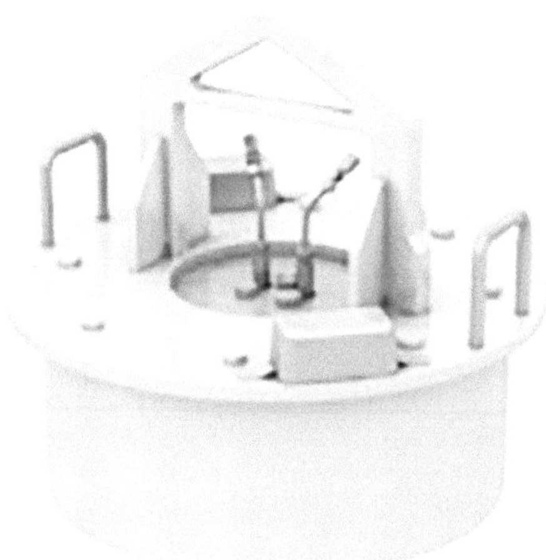

Fig. 10.49 Pressure cap.

10.12 Conclusion

In general, the following points should be considered for the design of subsea actuators:

- Subsea actuators should be fully designed and compensated for the design water depth;
- They should be designed in accordance with the design of the plant without the need for any maintenance in the case of fixed actuators;
- They should have a compact design, especially those intended for use in subsea manifolds and other structures;
- Their design, qualification, and FAT should be compatible with relevant API and ISO standards;
- They should meet special requirements for fast operation in the case of ESD and HIPPS applications;
- They may be either hydraulic or electric based on end user or EPC contractor requirements;
- The actuator type and safety function (mode) should be as per the end user or EPC contractor requirements;
- They should have the possibility of being assessed for SIL as well as the requirements of IEC 61508 and 61511 in the case of ESD or HIPPS application;
- They should have a clear and visible local or remote position indicator;
- They should be capable of being operated by ROV override bucket or by a diver;
- They should have the possibility of being retrievable in the case of end user requirements, and include a pressure cap installation on the bare stem valve after actuator removal.

Further reading

American Petroleum Institute (API) 6A, Specification for Wellhead and Tree Equipment, 21st ed., API, Washington, DC, 2018.

American Petroleum Institute (API) 17D, Design and Operation of Subsea Production Systems, Subsea Wellhead and Tree Equipment, second ed., API, Washington, DC, 2011.

International Organization of Standardization (ISO) 10423, Petroleum and Natural Gas Industries—Drilling and Production Equipment—Wellhead and Christmas Tree Equipment, fourth ed., 2009 Geneva.

International Organization of Standardization (ISO) 13628-4, Petroleum and Natural Gas Industries—Design and Operation of Subsea Production Systems—Part 4: Subsea Wellhead and Tree Equipment, second ed., 2010 Geneva.

K. Sotoodeh, Safety integrity level in valves. J. Fail. Anal. Prev. 19 (2019) 832–837 Springer. https://doi.org/10.1007/s11668-019-00666-2.

K. Sotoodeh, The importance of maximum allowable stem torque in valves. SN Appl. Sci. 1 (2019) 433, Springer Switzerland. https://doi.org/10.1007/s42452-019-0445-0.

K. Sotoodeh, Actuator sizing and selection. SN Appl. Sci. 1 (2019) 1207 Springer Switzerland. https://doi.org/10.1007/s42452-019-1248-z.

CHAPTER ELEVEN

Compensation systems

Contents

11.1 Introduction

Subsea actuators are installed in harsh, corrosive seawater environments at different water depths, at times as deep as three kilometers in some cases. Thus a high level of quality and integrity is expected from subsea actuators to ensure long life and reliable performance in harsh working conditions. Therefore actuators should be selected in materials suitable for seawater applications, and subsea actuators should be compensated for the selected depth. The seawater column exerts external pressure on both subsea valves and actuators. As a rule of thumb, each 10-m water depth applies 1 bar of pressure to subsea components. Therefore an actuator installed at a water depth of 3000 m receives approximately 300 bar of seawater external pressure. One solution to compensate for this pressure is to increase the thickness of the actuator, which makes it much heavier. Alternatively, it is common to connect a capsule filled with pressurized hydraulic oil almost equal to the seawater external pressure head to the actuator spring housing: this is called a compensator or compensation system. Using a compensation system is not only limited to the actuators' spring housing but also compensation is used on valve gear boxes and close to ROV buckets (see Fig. 11.1) to protect these components from collapse due to the external pressure.

11.2 Types of compensation systems

There are two main types of compensation systems: closed compensation and open compensation.

Subsea Valves and Actuators for the Oil and Gas Industry
https://doi.org/10.1016/B978-0-323-90605-0.00008-6

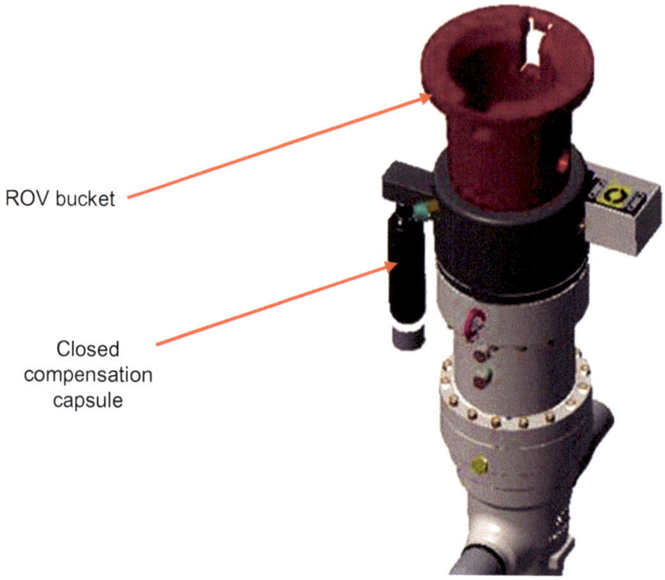

Fig. 11.1 Closed compensation capsule on an ROV-operated gate valve close to the ROV bucket.

11.2.1 Closed compensation systems

Fig. 11.2 illustrates a closed compensation system that contains hydraulic oil compensation fluid in a rubber diaphragm or bladder located at the top of the compensator. The bladder's function is to prevent seawater from entering the actuator. It is common to make the bladder or diaphragm from nitrile rubber, which is known as Nitrile rubber (NBR) or Buna N. NBR is classified as an elastomer polymer sealing that returns to its original shape after being stretched to a high extent. Seawater enters the compensation system from the bottom as the main pressure source. The compensation hydraulic fluid, which is normally different from the actuator hydraulic fluid, then moves to the actuator spring housing.

Fig. 11.3 provides an overview of a capsule-shaped closed compensation system installed on a spring return actuator. The closed compensation system is filled with compensation hydraulic fluid, illustrated in orange color, which is connected and distributed in the spring chamber of the actuator. The compensation oil is normally different from the actuator supply fluid, which is shown in green color in the figure. The schema on the right side of Fig. 11.3 illustrates the actuator in open position and shows the seawater at the bottom of the compensation system in blue color at the lowest level; the schema on the left illustrates the actuator in closed position and shows the seawater in blue color

Fig. 11.2 Closed compensation system.

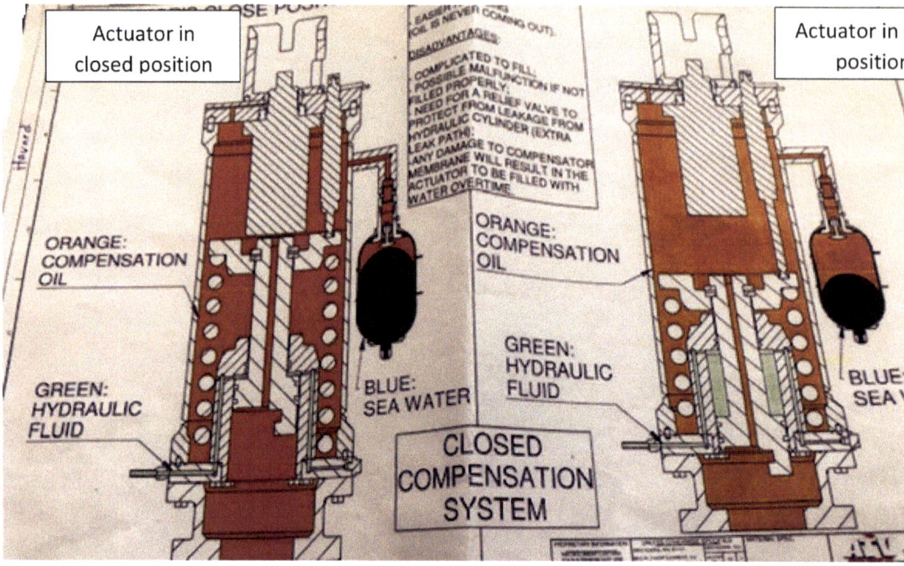

Fig. 11.3 Closed compensation system installed on a spring return actuator.

in the compensation system at the highest level. The main advantage of a closed compensation system is that it is environmentally friendly, as the oil is not discharged into the seawater. The HSE-friendly nature of this system could be the main reason to choose a closed compensation over an open one. However, there are some disadvantages associated with this system, such as the difficulty of changing the compensation oil. There is no easy access to the capsule internals for this purpose. A closed compensation system requires two relief valves, one inlet and one outlet, in order to balance the internal pressure with that of the seawater. If the pressure inside the compensation system increases and exceeds the seawater pressure by more than 5 bar, as an example, the pressure relief should release the excess pressure from the compensation system to the sea. On the other hand, if the pressure at the seawater depth exceeds the compensation system internal pressure, the extra pressure in the seawater should be released to the compensation system through another pressure relief valve to balance the two pressure values. A pressure relief valve is a type of safety valve used in the oil and gas industry to release extra liquid pressure to the environment. Making two holes in the compensation cavity to insert the pressure relief valves adds additional leakage paths to the compensation system, reducing its reliability. The other disadvantage of this system is related to the potential failure of the diaphragm or rubber or membrane, which leads to mixing of the compensation oil and seawater. This allows seawater to enter the actuator and cause corrosion. Closed compensation systems are normally made of carbon steel.

11.2.2 Open compensation systems

An open compensation system (see Fig. 11.4) is like a storage tank connected to the actuator casing and the seawater directly enters the oil tank. The compensation oil in an open compensation system should have a higher density than the seawater and not be mixable with it. These two requirements entail a separation between the oil and the seawater and prevent the seawater from entering the actuator.

Fig. 11.5 illustrates a spring return actuator in closed position compensated with an open tank located at the top left and filled in with seawater (highlighted in blue). The main advantages of open compensated systems are summarized as follows:

- There is no need for a pressure relief valve;
- It is easy to change the compensation oil.

Open compensation systems do have a major disadvantage, however: Tilting the actuator leads to the release of compensation fluid into the sea, so this system is less environmentally friendly compared to closed compensation systems.

Open compensation systems should be made of a material compatible with seawater, such as 25Cr super duplex or any material with a pitting resistance equal number (PREN) higher than 40, to avoid pitting corrosion. PREN is a measurement method used to evaluate the resistance of stainless steel against localized corrosion such as pitting, which

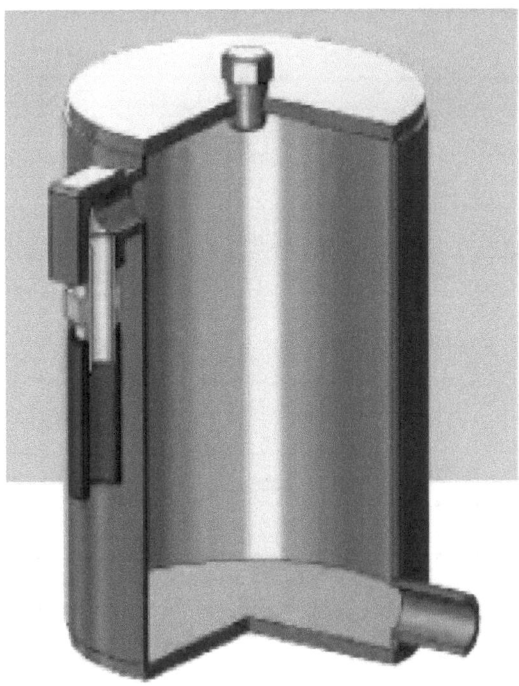

Fig. 11.4 Open compensation system.

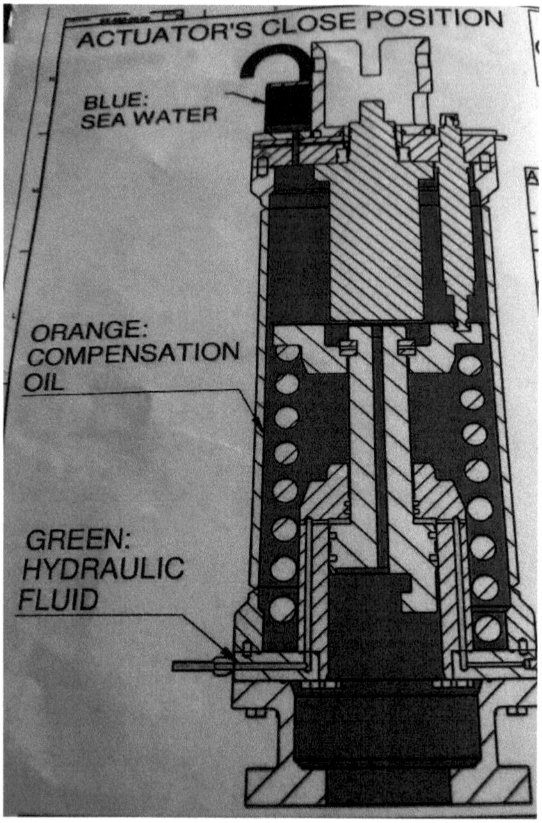

Fig. 11.5 Open compensation system on a spring return actuator.

depends on different chemical composition content such as chromium (Cr), molybdenum (Mo), and nitrogen (N) and can be calculated through the equation as follows:

$$\text{PREN} = \%\text{Cr} + 3.3 \times \%\text{Mo} + 16 \times \%\text{N}$$

$$25\ \text{Cr super duplex PREN} = 25 + 3.3 \times 3.5 + 16 \times 0.22 = 25 + 11.55 + 3.52 = 40.07$$

11.3 Compensation system benefits

The main benefits of compensation systems are summarized as follows:

- Provide isolation for the valve gear box and actuator from ingress of seawater inside the gear box, hydraulic cylinder (in the case of a double acting actuator), or spring chamber (in the case of a single acting (spring return) actuator). It should be noted that the ingress of seawater into an actuator's spring chamber and gear box causes crevice and pitting corrosion and consequently leads to valve operator failure.

– Provide protection against changes in pressure inside the actuator spring chamber or gear box caused by different factors such as solar radiation, subsea deployment, or operation of the actuator or gear box. The essential fact is that extremely high and low pressure inside the spring chamber could prevent the actuator from fully moving the valve. High pressure inside the spring chamber could result in less spring torque for closing the valve and low pressure in the spring chamber could cause ingress of seawater into the spring chamber, resulting in corrosion and malfunction of the spring. Therefore because the actuator and valve could fail due to pressure changes, the compensation provides a critical benefit to the system that goes beyond corrosion protection.

Figs. 11.6 and 11.7 illustrate a worm screw gear box with both closed and open compensation systems. This type of gear box can be used for ball valves. Figs. 11.8 and 11.9

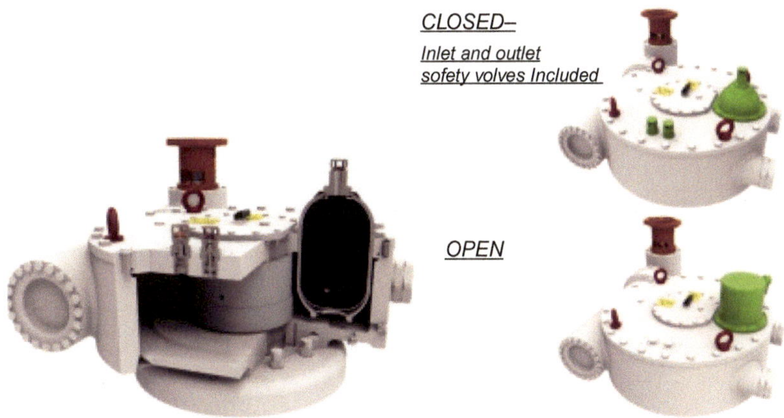

Fig. 11.6 Worm screw gear box with closed and open compensation system. *(Courtesy: Galperti Engineering.)*

Fig. 11.7 Worm screw gear box with closed and open compensation system. *(Courtesy: Galperti Engineering.)*

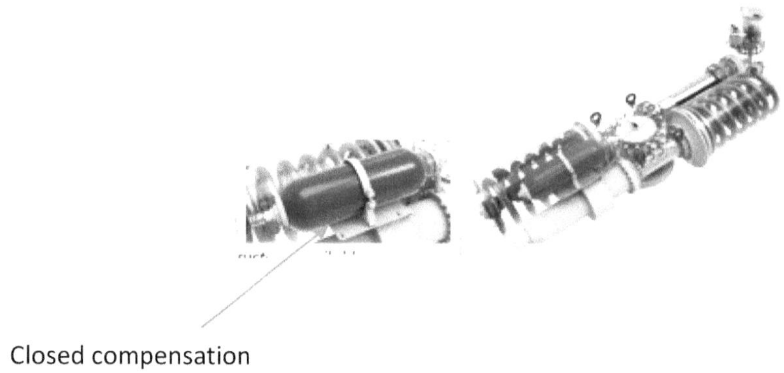

Fig. 11.8 Symmetrical spring return hydraulic actuator with a closed compensation system. *(Courtesy: Galperti Engineering.)*

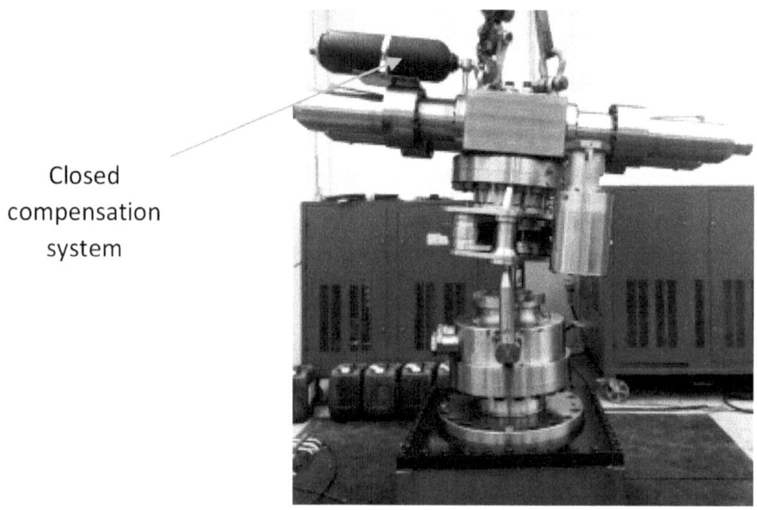

Fig. 11.9 Symmetrical spring return hydraulic actuator with a closed compensation system. *(Courtesy: Galperti Engineering.)*

illustrate a closed compensation capsule on the spring chamber of a spring return rack and pinion symmetrical hydraulic actuator. A worm gear (or worm) as illustrated in Fig. 11.10 is a specific gear composition in which a screw (worm) meshes with a gear/wheel similar to a spur gear which can be used for ball valves.

Fig. 11.10 Worm screw gear.

11.4 Compensation system requirements

The following requirements should be considered in compensation system design and operation:

- There should be a proper sealing between the hydraulic supply piston and cylinder and the compensated spring chamber. In fact, the hydraulic supply fluid should not enter the spring chamber and the compensation fluid in the spring chamber should be completely isolated from the piston and cylinder area of the actuator. Seals on the piston rod of the actuator play an important role in the isolation of these two areas.
- Filling the compensation system is very essential. The valve supplier should advise for the safe and optimal approach to filling the compensation system in the installation, operation, and maintenance (IOM) manual for the actuators and gear boxes.
- At certain depths, valves and actuators may need an extension on the ROV bucket to reach to a certain height to provide access to the remote-operated vehicle (ROV) so that they can be operated from the top of the manifold roof. In some cases, a pressure compensation system should be placed on the valve extension. Typically, the compensation system is filled with hydraulic fluid at the fabrication yard.
- The type of compensation (open or closed) should be stated in the valve data sheet. Closed compensation systems should be prioritized due to their more health-, safety-, and environmentally-(HSE) friendly nature compared to open compensation systems.
- Both of the pressure relief valves used on closed compensation systems should be tested to make sure they are capable of relieving the extra pressure at the desired, set pressures. The valve supplier is responsible for proving a procedure for testing the pressure relief valves. The relief valves should have serial numbers for full traceability.

- Qualification testing in a hyperbaric chamber should be performed while the compensation cavity or system is filled with the compensation fluid specified in the valve data sheet.
- The procedure given as follows can be used for the qualification of compensation systems. This test can be called a speed test and is applicable for both open and closed compensation systems. The valve and actuator should be operated at the speed given in the valve data sheet. The operation of the valve and actuator at normal speed should not cause the operation of any relief valves on the compensation system to release extra pressure in the compensation to the sea or vice versa. For an open compensation system, no fluid should escape from the compensation tank.
- A material data sheet that includes all the physical, chemical, and other relevant characteristics of the hydraulic fluid to be used for the compensation system should be given to the purchaser by the valve and actuator supplier.

CHAPTER TWELVE

Material selection and corrosion

Contents

12.1 Introduction

The selection of materials used in the oil and gas industry, especially in the offshore sector, is a challenging and complex task. In fact, material selection is an aspect of engineering design that should be approached with extra care in order to prevent the failure of components such as valves due to corrosion, loads, or any other causes. The role of the material department in selecting appropriate materials goes beyond the study and evaluation of corrosion; many other parameters, such as weight, strength, and compatibility with process conditions, should be considered. Chart 12.1 shows most of the effective parameters of the material selection process for subsea valves. Considering all of these parameters adds a high level of complexity to the material selection process.

When it comes to the process of selecting materials to prevent corrosion or any other cause of failure, many factors should be taken into account; these are listed as follows:

- Type of fluid service, including its chemical composition, such as carbon dioxide (CO_2) and hydrogen sulfite (H_2S) which cause sweet and sour corrosion, respectively. These corrosion mechanisms are explained later in this chapter.
- Service fluid pressure and temperature; generally higher pressure increases the effect of $CO_2$and (H_2S) corrosion. Some types of corrosion, like carbon dioxide corrosion and pitting, can be more severe at higher temperatures.
- Galvanic corrosion, which can occur in an electrolyte environment when two metals with different electrical potentials have contact.
- Crevice corrosion, which can happen in the grooves of carbon steel and low alloy steels where fluid may get trapped under the sealing element.

Subsea Valves and Actuators for the Oil and Gas Industry
https://doi.org/10.1016/B978-0-323-90605-0.00004-9

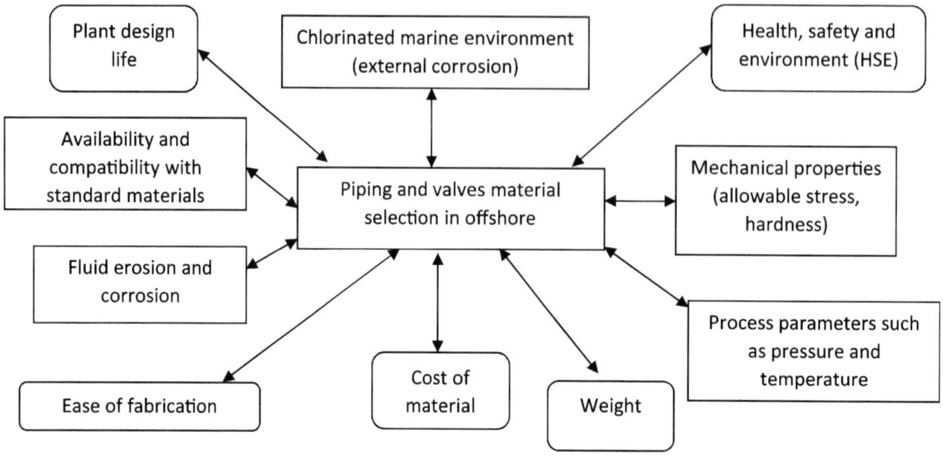

Chart 12.1 Effective parameters on subsea valves material selection.

- Design life, since some types of corrosion, such as carbon dioxide corrosion, are time dependent and can increase in rate over time.
- Compatibility of the material not only with the service fluid, but with all the types of fluid passing through the valve, such as injection fluid.
- Compatibility of the nonmetallic sealing materials with the pressure, temperature, and fluid type.
- External protection of materials to prevent external corrosion by adding coating and cathodic protection.
- Strength of material is important, especially for the pressure-containing parts of valves such as the valve stem. The pressure-containing parts are those whose failure to function leads to the leakage of valve fluid to the environment.
- Availability of material is a very important parameter in the material selection process. Many young engineers may select an ideal material for a specific part of the valve that is not common in the industry. As an example, hard and high mechanical strength nickel alloys such as Inconel 718 and 725 are not common for the body and bonnet of the valves, which are pressure-containing parts.
- Cost is an important consideration for material selection. As an example, carbon steel with full Inconel 625 overlay for the body and bonnet of the valve in sizes such as 8″ and above could be more cost effective than selecting a super duplex body valve. However, it should be noted that material failure during operation imposes high operation cost (OPEX) on the project. Thus, in some cases, it is better to select more expensive materials in order to reduce OPEX risk. It can be concluded that the cost of the material depends on both the cost of expenditure (CAPEX) and OPEX.

Valve material failure is a very undesirable event that can have negative consequences such as loss of asset; loss and stoppage of production; plant shutdown; loss of human life (personnel); environmental pollution; other health, safety, and environmental (HSE) problems, etc. One of the main challenges involved in selecting valve materials for the subsea environment is that this environment is very corrosive and harsh. Seawater contains chloride, which is highly corrosive and can cause different types of corrosion such as pitting and chloride stress cracking corrosion (CLSCC), which will be explained in more detail later in this chapter. Valves used in production normally deal with untreated oil and gas, which contain considerable amounts of undesirable corrosive compounds such as carbon dioxide and hydrogen sulfide. In addition, a large amount of sand is typically produced with the production fluid, which can cause erosion in the valves. It should be noted that eroded metal is more susceptible to corrosion.

12.2 Corrosion definition and types

Corrosion has several definitions. One of the definitions is the destruction or deterioration of material through interaction with the environment and/or fluid. This definition is not limited to metals and can be applicable for other kinds of materials such as plastics and rubbers, etc. Corrosion in a metal can also be defined as losing electrons.

Corrosion is a very costly phenomenon that causes approximately 100 billion dollars per year in damage in the United States alone. Corrosion prevention is normally achieved by material and corrosion engineers through different strategies, such as proper material selection, using corrosion inhibitors, cathodic protection, selecting the most suitable coating, etc.

Corrosion is divided into two types: internal and external. Internal corrosion is caused by the fluid inside the valve, and external corrosion is caused by the seawater located outside the valve boundary.

Fig. 12.1 illustrates a schematic of corrosion occurring in a metal in an anodic area where iron in the form of Fe^{2+} and electrons $2e^-$ are released.

$$Fe \rightarrow Fe^{2+} + 2e^-$$

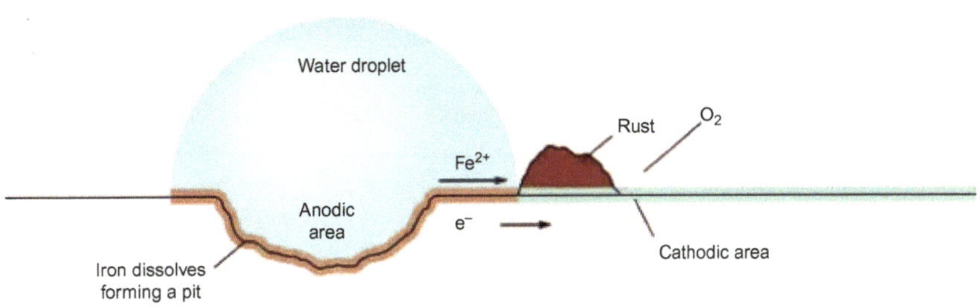

Fig. 12.1 Corrosion mechanism.

Some corrosion types, like the one illustrated in the figure, require water to occur. In this case, Fe^{2+} is reacting with oxygen and water to make iron oxide in the form of rust.

$$Fe^{2+} + O_2 + H_2O \rightarrow Fe_2O_3 + H_2$$

12.2.1 Internal corrosion types

12.2.1.1 Erosion corrosion

Erosion corrosion is the degradation of a material surface due to a combination of erosion and corrosion. Different factors cause erosion, such as particles, high fluid velocity, turbulent flow, sudden pressure drop of the liquid, abrasion, etc. Normally, carbon and low alloy steels are highly sensitive to erosion corrosion in cases where high fluid velocity can accelerate the combination of erosion and corrosion. Generally, high fluid velocity in corrosion-resistant alloys (CRAs) does not produce or increase corrosion degradation.

Cavitation is a type of erosion corrosion in which the liquid flow undergoes a significant pressure drop that causes bubbles to be produced out of the liquid. The bubbles can burst forcefully and put high loads on materials, which in turn create pits and irregularities in the material. Fig. 12.2 illustrates cavitation in a globe valve plug, representing considerable mechanical damage. If the high velocity of fluid is combined with corrosion, this type of erosion corrosion can be called impingement corrosion. Adding erosion allowance to steels is one way to mitigate the effect of erosion. DNV-RP-0501 is a standard developed by Det Norske Veritas for managing sand production and erosion. The standard provides a guideline for the evaluation of erosion and wear in piping systems, including industrial valves, and a procedure for calculating erosion rate and allowance. Cavitation is also briefly addressed in this document.

Fig. 12.2 Cavitation in the plug of a globe valve.

12.2.1.2 Fatigue corrosion

Subsea valves are subject to frequent cycles under a combination of loads during the qualification program, which causes fatigue to different valve components. The damaged surfaces, or any parts of the material that are damaged due to frequent applied mechanical loads or fluid pressure, are susceptible parts that can be corroded easily. While the initiation of fatigue in the metal can cause corrosion, the opposite action is also possible, meaning that corroded sections can be damaged by fatigue stress. Fatigue failure can be divided into three phases: initiation, propagation, and final rupture. Finite element analysis (FEA) can be performed to validate the design of subsea valves against fatigue or cycle loads.

12.2.1.3 Fretting corrosion

Corrosion caused by friction/metals sliding over each other is called fretting corrosion. Some approaches to preventing this type of corrosion involve using harder materials, using lubrication or applying hardfacing materials such as stellite or tungsten carbide to the contacting metal surfaces. As an example, it is common in the valve industry to apply tungsten carbide hardfacing to the closure member and contact seats of gate and ball valves to prevent galling or fretting corrosion. Fig. 12.3 illustrates the ball of a ball valve in 13 chromium, 4 nickel martensitic stainless steel hardfaced with tungsten carbide.

Fig. 12.3 Ball of a ball valve hardfaced with tungsten carbide.

12.2.1.4 Biological or microbial corrosion

Various biological agents or microorganisms exist in the reservoir and in the oil and gas production environment; these can cause corrosion in different ways, such as by forming corrosive compounds like hydrogen sulfide, or making the environment more corrosive. Several types of bacteria can produce toxic and corrosive hydrogen sulfide in the reservoir to change the hydrocarbon characteristic from sweet to sour crude, which contains a considerable amount of hydrogen sulfide. Materials that are at risk of sour corrosion or hydrogen sulfide should be manufactured according to the requirements of ISO 15156. The important and interesting fact is that materials for nonsour oil and gas fields

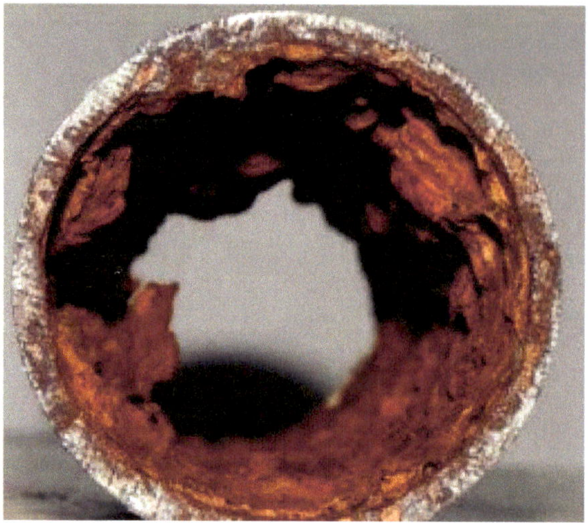

Fig. 12.4 Microbial corrosion on a piece of pipe.

could be designed based on the sour service requirements provided in the above-mentioned standard, due to the possibility of the crude oil naturally changing from sweet to sour over time due to bacterial activity. Fig. 12.4 illustrates microbial corrosion on a piece of pipe:

12.2.1.5 Carbon dioxide (sweet) corrosion

"Acid gases" refer to two of the undesirable by-products of oil and gas, namely CO_2 and H_2S. This name is chosen for them simply because of their acidic chemical properties in the presence of water. CO_2 corrosion of carbon and alloy steels is called "sweet corrosion," which is defined as the deterioration of metal components resulting from contact with gas or solutions including both CO_2 and water, since CO_2 is noncorrosive in the absence of water. In the presence of water, carbon dioxide can form carbonic acid, which reduces the pH of the fluid and causes corrosion. This type of corrosion is known as a uniform corrosion, which causes metal loss. Thus the combat strategy to address this type of corrosion is to add some millimeters of additional thickness to carbon steel and low alloy steels (e.g., 3 mm corrosion allowance). Corrosion-resistant alloys (CRAs) such as stainless steels and nickel alloys are immune to carbon dioxide corrosion. Fig. 12.5 illustrates carbon dioxide corrosion as a type of metal loss on susceptible material. Different models have been developed to predict the carbon dioxide corrosion rate based on process parameters such as pressure, temperature, amount of carbon dioxide and hydrogen sulfide in the fluid, pH value, fluid velocity, etc. The NORSOK standard developed for Norwegian petroleum industry has standardized the corrosion allowance on carbon and low alloy steel to 3 mm. Adding more corrosion allowance to the

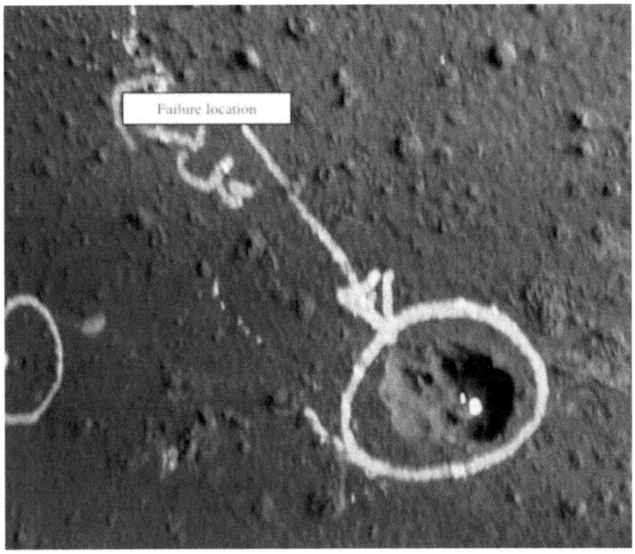

Fig. 12.5 Carbon dioxide corrosion.

thickness of piping and valves could make the component heavier. In addition to adding corrosion allowance, there are other strategies to prevent this type of corrosion, such as injecting corrosion inhibitor into the piping system, including the valves, using CRAs, etc.

12.2.1.6 Hydrogen sulfide (sour) corrosion

According to some estimates, sour corrosion is responsible for 40% of all corrosion in the oil and gas industry. Hydrogen sulfide (H_2s) is the main compound that leads to sour corrosion in its different forms, which are sulfide stress cracking (SSC), stress cracking corrosion (SCC), hydrogen-induced cracking (HIC), stress-oriented hydrogen-induced cracking (SOHIC), and stepwise cracking corrosion (SWC). These will be discussed in detail as follows. All of these kinds of damage require a wet or aqueous environment to occur. Sour corrosion is one of the most common and dangerous types of corrosion in the oil and gas industry due to the toxic nature of H_2S. H_2S is a chemical compound with the smell of a rotten egg, and even a small concentration of this substance, such as 100 parts per million (ppm), can kill a human. The corrosion mechanism of hydrogen sulfide can be very aggressive; material exposed to H_2s can crack, and the crack can lead to leakage of the internal fluid in just hours. Therefore adding corrosion allowance is not applicable for hydrogen sulfide corrosion prevention. Hydrogen sulfide can produce hydrogen atoms and sulfuric acid in reaction with water. ISO15156, previously merged with the National Association of Corrosion Engineers (NACE) standard MR0175, provides a guideline for metallic material selection in sour services in the oil and gas industry. Despite its name, NACE is an international organization. The above-mentioned standards mention some important parameters for metallic materials in sour service, such as chemical composition control, hardness control and reduction within the given limits in the standard, applying proper heat treatment and methods of fabrication, etc.

Sulfide stress cracking corrosion (SSC) is the cracking of metallic material under the combined effect of hydrogen sulfide and tensile stress. The important point is that compressive load cannot cause this type of corrosion. This type of corrosion is not classified as a general or uniform corrosion type; it can happen locally on special areas such as weld joints and heat-affected zones located around the welding area. For this reason, welded joints in sour services require postweld heat treatment (PWHT) to release the residual stress in the material. It should be noted that the type of stress could be either residual or applied. Hydrogen sources can diffuse inside the material and initiate cracks in localized areas. Different approaches such as PWHT and controlling the hardness and stress level of the material are important measures to reduce the risk of SSC. Fig. 12.6 illustrates SSC in a piece of piping.

Stress cracking corrosion (SCC): Information about SCC is provided in Section 12.2.1.9.

Hydrogen-induced cracking (HIC): HIC is another mechanism of sour service corrosion and is a major risk in carbon steel and low alloy steels. Notably, HIC is not a risk on exotic, corrosion-resistant alloys such as stainless steels and nickel alloys. The other important characteristic of HISC is that it does not require any stress for cracking. However, a number of hydrogen-induced cracks stacked together due to applied or residual stress can lead to another sour corrosion mechanism known as stress-oriented hydrogen-induced cracking (SOHIC).

Fig. 12.6 SSC on a piece of piping.

Stepwise cracking (SWC): This type of corrosion is a general term that refers to the blistering and cracking of metal surfaces without any applied stress to the material.

12.2.1.7 Crevice corrosion

Crevice corrosion is a localized attack that occurs inside or close to a crevice or gap that is in contact with a sealing or other metal. This type of corrosion can occur in sealing areas where a gasket or soft sealing is placed, and in the seat pockets of industrial valves. Crevice corrosion can have the same mechanism as pitting corrosion and can lead to pitting corrosion if left untreated. Carbon steel and low alloy steels are more prone to this type of corrosion, but it can also occur in stainless steel materials. Factors that can affect the severity of crevice corrosion include crevice type, such as metal-to-metal or metal-to-soft-material contact, the geometry of the crevice, such as its size, depth, and roughness as well as environmental factors such as pH, temperature, and oxygen level.

12.2.1.8 Pitting corrosion

Pitting corrosion is a type of localized corrosion in which cavities or pits form on a metallic surface. This type of corrosion is very difficult to predict, unlike uniform corrosion types. Seawater and chloride can cause pitting corrosion in susceptible materials, meaning that two factors—environment and metallurgical or material factors—contribute for HISC to happen. Fig. 12.7 illustrates pitting corrosion on steel. In general, chloride can damage the passive or protective layer of the material. Different steels without a high pitting resistance equivalent number (PREN), such as martensitic stainless steel and Inconel X750, which are very common valve spring materials, are not recommended for subsea valves. PREN is a number that represents the material's level of resistance against pitting; it is dependent on certain chemical compositions in the material such as chromium, nitrogen, and molybdenum. Eq. (12.1) is used to calculate a material's PREN.

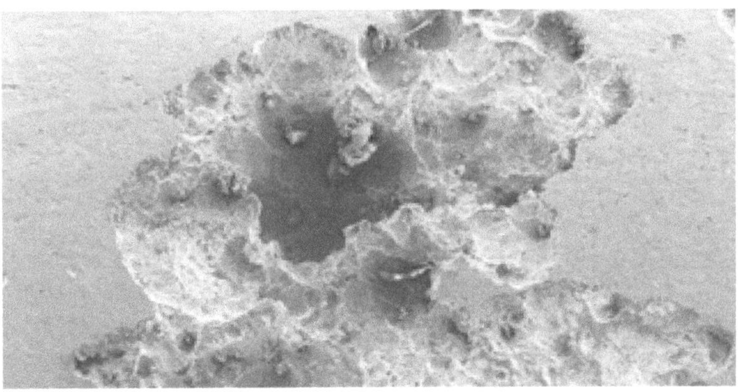

Fig. 12.7 Pitting corrosion in steel.

PREN calculation

$$\text{PREN} = \text{Chromium} + 3.3\,\text{Molybdenum} + 16\,\text{Nitrogen} \tag{12.1}$$

Different approaches to preventing pitting corrosion are proper selection of materials with a high PREN (above 40), controlling pH, chloride content, and temperature, using cathodic protection, coating the material in seawater, and applying a pitting corrosion test according to ASTM G48. This standard provides test methods to assess both the pitting and crevice corrosion resistance of materials in chloride solution.

12.2.1.9 Stress cracking corrosion

The mechanism of SCC is same as that of SSC; SCC can occur due to the combination of stress and a corrosive compound. As an example, chloride content in the fluid service can cause stress cracking corrosion, which can be exacerbated due to the impact of hydrogen sulfide. Fig. 12.8 illustrates SCC in a carbon steel pipeline.

Thus stress cracking corrosion is defined as the creation and extension of cracks inside a material due to such factors as a corrosive environment and stress. The type of stress could be either applied or residual. Applied stressors are those generated by mechanical loads. The applied stress must be tensile (tension) and not compression for corrosion to occur. Residual stress could be due to welding and metallurgical changes in areas next to the welding, i.e. the heat affected zone (HAZ). Residual stresses due to welding, which

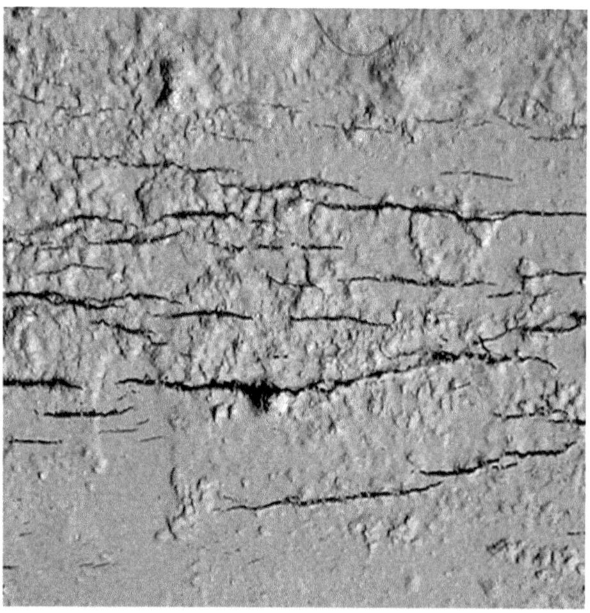

Fig. 12.8 Stress cracking corrosion of a carbon steel pipeline in sour service.

do exist in the HAZ, can be released through PWHT. Different media can cause SCC, such as chloride, amine, hydrogen sulfide, etc. SSC caused by chloride is called chloride stress cracking corrosion (CLSCC), which can be created after the development of pitting corrosion. CLSCC is very common in austenitic stainless steel bolting. Chloride cracking typically occurs in marine areas such as offshore. Fig. 12.9 illustrates the failure of a bolt due to CLSCC. Higher temperatures increase the possibility and occurrence of this type of corrosion.

12.2.1.10 Intergranular stress cracking corrosion

Intergranular stress cracking corrosion is a type of corrosion that occurs at grain boundaries inside the material structure, specifically in austenitic stainless materials. The migration of carbide and formation of chromium carbide in the grain boundaries lead to the depletion of chromium and the corrosion of the vulnerable austenitic stainless steel. Depletion of chromium typically occurs at high temperatures during welding, so intergranular corrosion is also known as "weld decay." Using austenitic stainless steels with grade "L" indicating low carbon as well as using austenitic stainless steels in grades 321 and 347 can prevent this type of corrosion. Fig. 12.10 illustrates ISCC corrosion in an austenitic stainless material.

12.2.1.11 Corrosion under insulation

Insulation is sometimes applied on piping and valves for different purposes, such as maintaining a warm or cold temperature inside the pipe or valve without any effect from environmental temperature. The insulation could sit directly on the external part of the piping and valve or have some gap. If moisture becomes trapped between the pipe/valve external part and the insulation, this type of corrosion can occur.

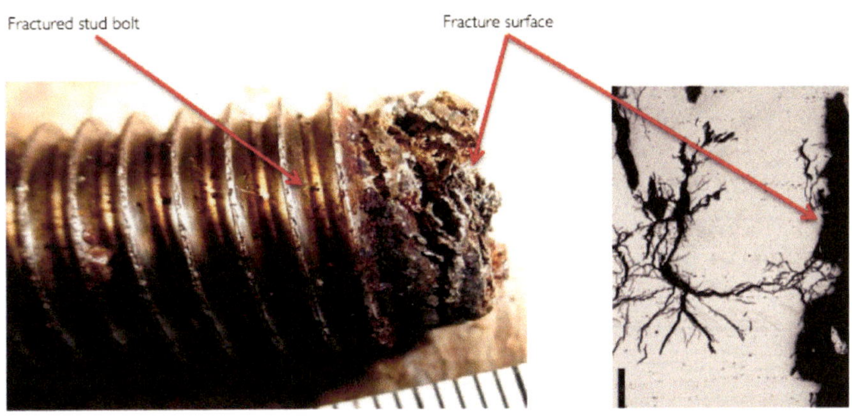

Fig. 12.9 Chloride stress cracking of a bolt in an offshore environment. *(Courtesy: Valve World/ Springer.)*

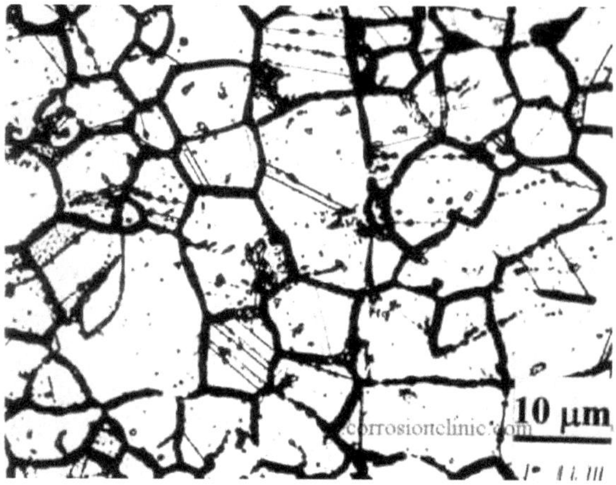

Fig. 12.10 Intergranular stress cracking corrosion.

12.2.2 External corrosion types

12.2.2.1 Hydrogen-induced stress cracking corrosion

The most common type of external corrosion in the subsea environment is called hydrogen-induced stress cracking corrosion (HISC). Historically, some duplex and super duplex subsea components such as piping and valves have failed due to this type of corrosion. HISC corrosion has impacted subsea facilities in the oil and gas industry for the past 15 years. In fact, both 22Cr duplex and 25Cr super duplex, which are known as ferritic-austenitic stainless steels, have been listed as materials susceptible to HISC. Although some studies, especially those by Det Norske Veritas have been done on HISC, this phenomenon is not completely clear. Det Norske Veritas (DNV) has developed a guideline for the design of duplex components based on the DNV-RP-F112 standard to avoid HISC. The primary causes of this type of corrosion are identified in the DNV standard. The main reason for HISC failure has been attributed to a combination of stress coming from loads (e.g., tension in the pipeline) and ingress of hydrogen formed at the steel surface due to cathodic protection, as well as the use of the previously mentioned susceptible materials. Fig. 12.11 illustrates the contributions of these three factors for HISC occurrence.

Cathodic protection, along with coating, is an external seawater corrosion mitigation measure for piping, valves, and structures mainly in the subsea sector of the oil and gas industry. Corrosion, in simple language, means losing electrons from a metal. Three main parameters are required for corrosion to occur: one is the presence of two dissimilar metals, an electrolyte such as water, and a path between the two metals. Cathodic protection provides electrons from an active metal that can be corroded easily, called an anode, to the protected component, called a cathode, to prevent seawater corrosion.

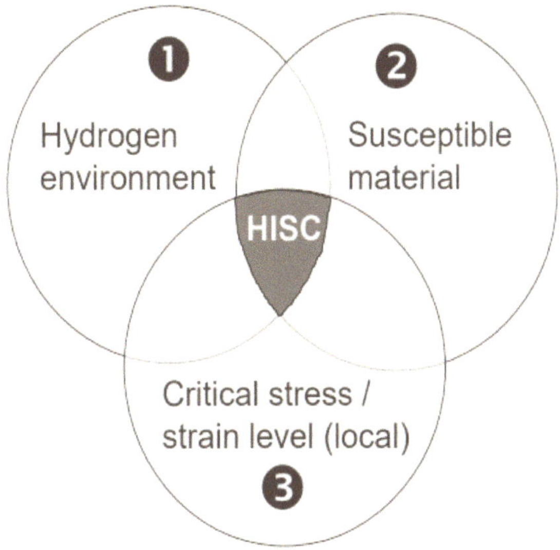

Fig. 12.11 HISC occurrence and contributing factors. *(Courtesy: DNV.)*

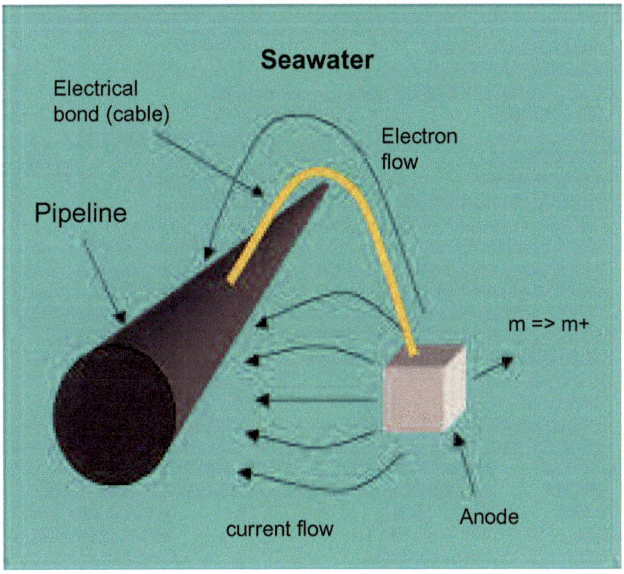

Fig. 12.12 Cathodic protection of a subsea pipeline. *(Courtesy: Elsevier and Springer.)*

The anode is a material that can easily lose an electron, such as aluminum, magnesium, or zinc, for the protection of the cathode. Fig. 12.12 illustrates a sacrificial anode connected to a piece of pipe for the purpose of cathodic protection. The basic concept of cathodic protection is to keep the electrical potential of the metal below its corrosion threshold so that it will be incapable of corroding.

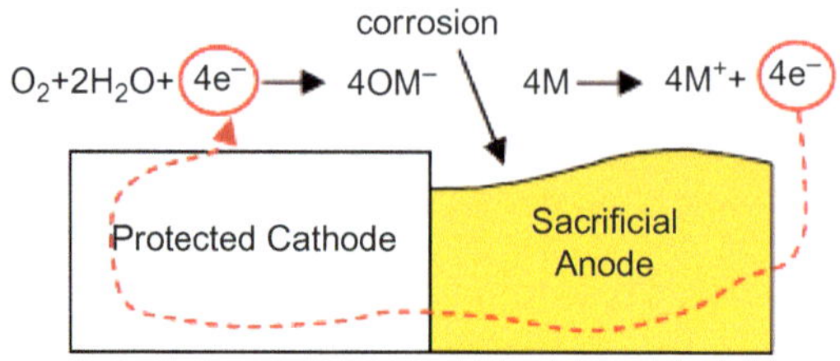

Fig. 12.13 Chemical reaction between anode and cathode.

The reaction in anodes is given in Eq. (12.2):

$$Fe \rightarrow Fe^{2+} + 2e^{-} \tag{12.2}$$

The extra electrons released from anodes will be transferred to cathodes (see Fig. 12.13) and will react with oxygen and water according to Eq. (12.3).

$$O_2 + 2H_2O + 4e^{-} \rightarrow 4\,OH^{-} \tag{12.3}$$

Fig. 12.13 illustrates the chemical reaction between the anode and cathode.

As illustrated in Fig. 12.13, the pH of seawater increases due to the formation of OH^-, and the reaction of OH^- is beneficial; it makes a protective layer of $CaCo_3$ in reaction with the calcium and magnesium of the seawater. This deposit or protective layer protects the cathode and increases its life. But the important point is related to a situation in which the electrical potential of the anode is too negative. In this case, extra electrons are produced from the anode and are transmitted toward the cathode. The extra electrons can cause a second cathodic reaction, leading to the possibility of HISC by creating hydrogen, as per Eq. (12.4).

$$H_2O + e^{-} \rightarrow H + OH^{-} \tag{12.4}$$

Fig. 12.14 illustrates HISC in a welding joint between a duplex pipe and a low alloy hub buttered with Inconel 725. Buttering is the addition of material, by welding, on one or both faces of a joint, prior to the preparation of the joint for final welding, for the purpose of providing a suitable transition weld deposit for the subsequent completion of the joint. Low alloy steel is buttered with Inconel 725, and a PWHT is applied on buttered low alloy steel with Inconel 725 before welding it to a duplex pipe. It is important to exclude duplex pipe from PWHT since the heat produced during PWHT can make sigma brittle phase in duplex, which causes duplex failure. Fig. 12.15 illustrates buttering between carbon steel and duplex stainless steel piping to facilitate welding.

The exact HISC corrosion mechanism is not fully understood in detail. Duplex and super duplex materials have two main types of crystalline structure: austenitic and ferrite. Ferritic structure is more prone to cracking than austenitic structure. Hydrogen absorbed inside material vulnerable to HISC can embrittle ferritic phase and cause cracking,

Fig. 12.14 HISC failure on Inconel 725 buttering used to connect a pipe and hub. *(Courtesy: General Electric.)*

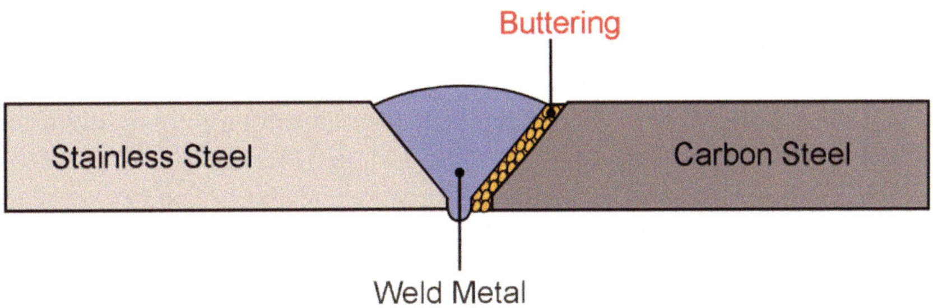

Fig. 12.15 Buttering between duplex stainless steel and carbon steel pipe.

whereas austenitic phase is resistant and acts as a crack barrier. It should be noted that the model proposed by the DNV standard to mitigate HISC considers material quality a factor, inasmuch as fine austenitic spacing has higher HISC resistance compared to materials with coarse spacing. However, materials with excessive austenitic spacing and grain flow perpendicular to the main loading direction cannot be evaluated for HISC by the methodology given in the DNV standard, since the given model and load limits for these materials are nonconservative. Failure of materials with a large grain size and adverse orientation has been widely experienced in the industry. Therefore DNV-RP-F112 suggests reviewing the manufacturing routes and processes for materials that failed due to HISC in order to achieve an acceptable grain size and orientation. Forged components are widely used for valves in the subsea sector of the oil and gas industry; they should be close to their final shape, and the direction of grains should follow the forging direction to reduce the risk of HISC.

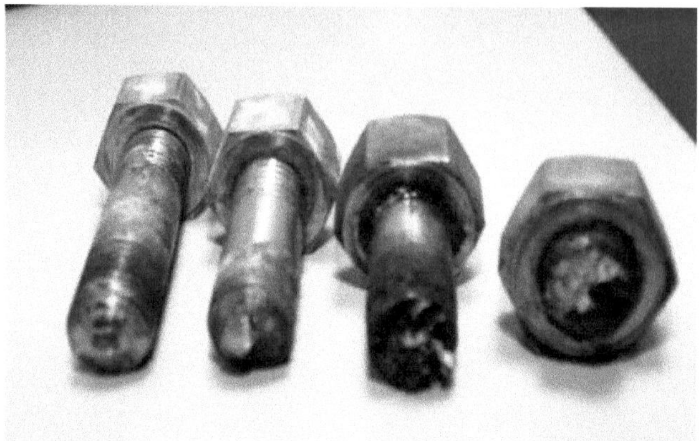

Fig. 12.16 HISC failure of low alloy steel bolts.

The other important point related to HISC is that this type of corrosion is not limited to duplex and super duplex: According to the API 20E standard, hard and high mechanical strength nickel alloys, such as Inconel 718, X750 and 725, as well as low alloy steel bolting are also subject to failure due to HISC, as illustrated in Fig. 12.16.

The DNV-RP-F112 standard provides a stress and strain model, or criteria for HISC evaluation, for the purpose of HISC failure prevention for duplex and stainless steel subsea pipeline. The model is applicable for subsea valves. HISC evaluation according to DNV includes a formula as well as a finite element load analysis. The first thing to check in stress calculations is to make sure that both membrane and bending-plus-membrane stresses are within the limitation given in Eqs. (12.5), (12.6):

$$\delta_m < \alpha_m \cdot \gamma_{HISC} \cdot \text{SMYS} \tag{12.5}$$

$$\delta_{m+b} < \alpha_{m+b} \cdot \gamma_{HISC} \cdot \text{SMYS} \tag{12.6}$$

Both limitations given before should be met. Coefficients of α_m and δ_{m+b} are allowable stress factors given in Table 12.1, according to Table D1 of the DNV standard.

γ_{HISC} is the material quality factor; its relevant values are given in Table 12.2 according to Table D2 of the DNV standard.

Table 12.1 Allowable SMYS factors for duplex and super duplex stainless steels.

Allowable SMYS factors for duplex and super duplex stainless steels	
Membrane α_m	0.80
Membrane plus bending stress α_{m+b}	0.80[a]

[a]The membrane plus bending stress could be 1.00 or 0.90, alternatively, depending on the location selected for HISC analysis.

Table 12.2 HISC material quality factors.

HISC material quality factors	
Fine austenitic spacing	100%
Coarse austenitic spacing	85%

Table 12.3 Strain limit for HISC analysis.

	Strain limit	
Location in the component	**Within L_{res} from weld**	**Outside L_{res} from weld**
Outside 5% wall thickness from any surface	Min. [0.3%; 0.5%$-\varepsilon_{res}$]	0.3%
Within 5% wall thickness from surface	1.00%$-\varepsilon_{res}$	1.00%
	0.6%$-\varepsilon_{res}$	0.6%

The second criterion (strain limit) based on FEA analysis for HISC is given in Table 12.3 with reference to Table E1 of DNV RP-F-112.

L_{res} is defined as the distance from the weld in which residual strain and stress should be taken into account. Residual stress in the material is induced by the manufacturing process or by welding, which is totally different from the type of stress placed on the materials from applied loads. Residual strain is a type of strain associated with residual stress. In general, strain is defined as the measurement of the material in response to applied force. Strain (ε) is defined as per Eq. (12.7).

$$\varepsilon = \frac{\Delta L}{L} = \frac{-\text{Change in length}}{\text{Original length}} \tag{12.7}$$

12.2.3 Types of internal and external corrosion

Some types of corrosion can be categorized as both internal and external. These types of corrosion can occur due to either internal fluid in the piping and valves or external corrosion from fluid in the environment.

12.2.3.1 Galvanic corrosion

Galvanic corrosion can be classified as an external type of corrosion; it occurs when two dissimilar metals with different electrical potential are in contact in an electrolyte. An electrolyte is a substance that produces electrically conducted solution such as water. The electrical potential difference between two dissimilar metals leads to electron migration from the less Noble material to the more Noble one. The greater the electrical potential difference between the two metals, the higher the corrosion rate for the less Noble material acting as an anode. The electrical potential of different materials is given in a table called "galvanic series," shown in Fig. 12.17. According to the galvanic series table, materials such as magnesium, zinc, and aluminum have the most negative electrical

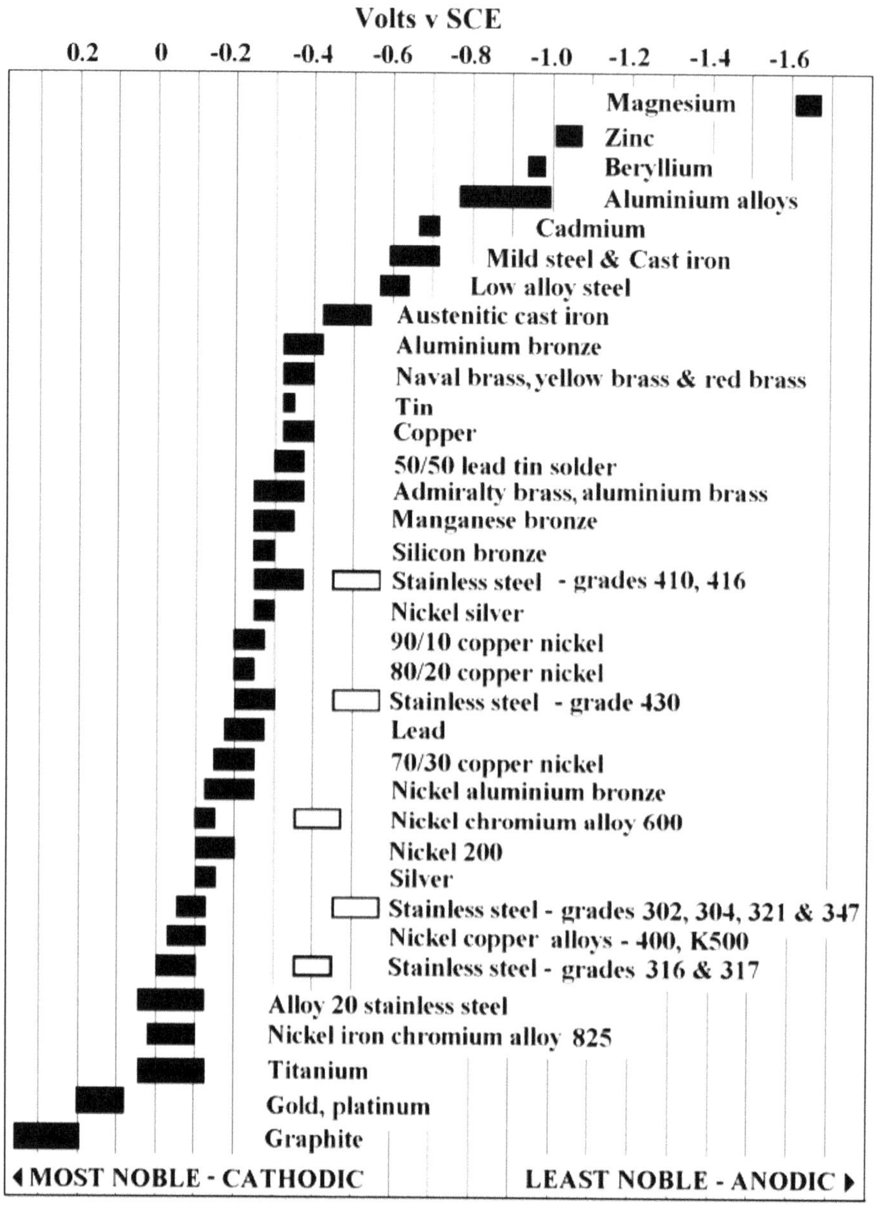

Fig. 12.17 Galvanic series of materials.

potential, meaning that these materials can lose electrons easily and be corroded. Thus these are the most suitable materials to be chosen for anodes. At the other end of the spectrum, graphite, gold, platinum, and titanium are the most Noble materials with the most positive electrical potential, so they have a cathodic character. Graphite, which can be used for stem sealing and packing for valves in seawater service, can cause galvanic

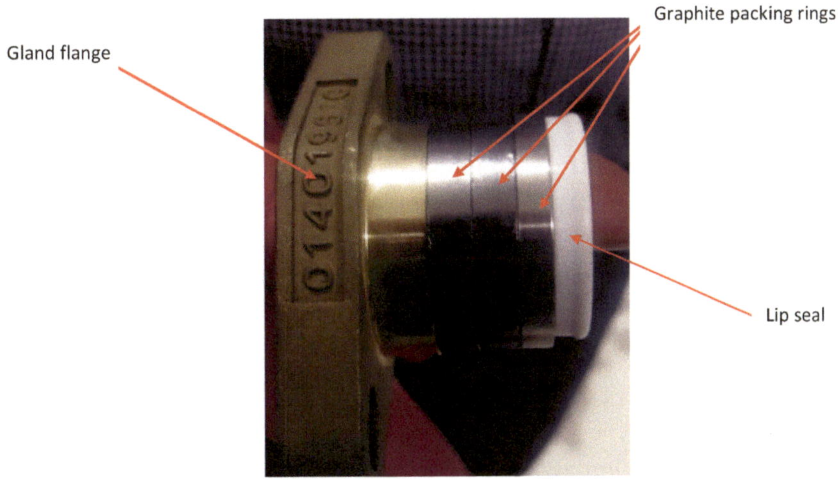

Fig. 12.18 Isolation of graphite packing rings from seawater service to prevent galvanic corrosion.

corrosion of valve stems in materials such as 22Cr or 25Cr super duplex. Thus graphite packing could be isolated by means of other sealings to prevent it, along with the connected stem area, from coming into contact with the seawater to prevent galvanic corrosion. Fig. 12.18 illustrates the use of a lip seal to isolate three graphite packing rings and a stem in duplex from contact with the seawater, thus preventing galvanic corrosion of the stem.

Fig. 12.19 illustrates galvanic corrosion between the stem of a valve in 22Cr duplex material and a connected key in AISI 4140. As a result of galvanic corrosion, the less Noble low alloy AISI 4140 is corroded in the offshore environment. "Offshore environment" refers to a humid, chloride-containing, corrosive atmosphere.

Fig. 12.19 Galvanic corrosion in the stem key of a valve in AISI 4140 in an offshore environment.

12.2.4 Review of materials for subsea valves

12.2.4.1 Carbon steel

Carbon steel, a carbon and iron alloy, is the cheapest choice of metal for piping and valves. This material is classified as a noncorrosion-resistant alloy that can be corroded by corrosive compounds such as carbon dioxide, hydrogen sulfite, etc. Other elements, such as silicon, copper, and manganese, can be added to carbon steel. Carbon steel can be divided into different categories, such as low carbon with a carbon content less than 0.25%, medium carbon with a carbon content between 0.25% and 0.5%, and high carbon with a carbon content between 0.5% and 1.25%. It should be noted that the classification of carbon steel based on carbon content can differ from what is mentioned here. If the carbon content of an iron-carbon alloy is more than 2%, the material is called cast iron. Cast iron was a popular choice for piping swage and water pipe before the advent of plastic piping. Carbon steel can be corroded in the form of metal loss due to carbon dioxide CO_2. Therefore 3 mm corrosion allowance as an average can be added to carbon steel pipe and valve bodies according to Norsok M-001, "material selection" and Norsok L-001, "piping and valves." In addition, hydrogen-induced cracking (HIC) is a big risk for carbon steel; hydrogen atoms can accumulate in the void spaces in the microstructure of carbon steel; after collision they get bigger and cause cracking or other types of failures in the metal (see Fig. 12.20). One approach to mitigating HIC corrosion is to used killed carbon steel. Killing carbon steel process involves deoxidating the steel with aluminum or silicon, which makes the steel homogenous and free of porosity.

The other important parameter for carbon steel is carbon equivalent (CE), which affects the weldability of carbon steel. Increasing the percentage of some of the alloys, such as carbon and manganese, in the carbon steel improves the mechanical strength of the carbon steel, but also increases its hardness and brittleness. Some piping and valve specifications put limitations, such as 0.39 or 0.40 or 0.42, on CE to prevent cracking the

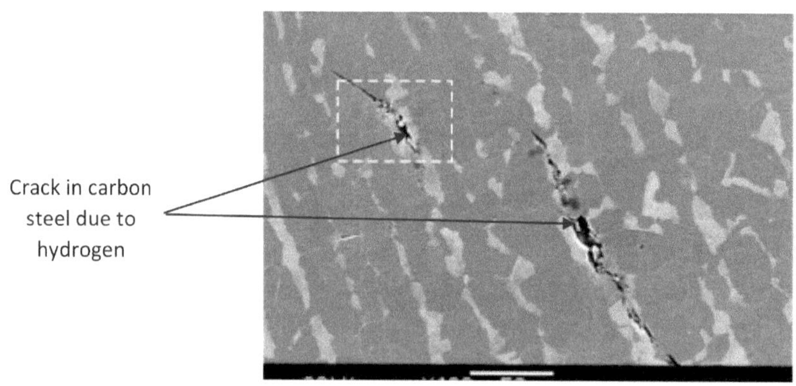

Fig. 12.20 Hydrogen-induced cracking corrosion in carbon steel.

carbon steel due to welding or sour corrosion such as HIC. CE can be calculated through Eq. (12.8):

Carbon equivalent calculation

$$CE = C + \frac{Mn}{6} + \frac{Cr + Mo + V}{5} + (Ni + Cu)/15 \quad (12.8)$$

where:

C: carbon; Mn: magnesium; Cr: chromium; Mo: molybdenum; V: vanadium; Ni: nickel; Cu: copper.

Carbon steel, according to ASME B31.3, "process piping code," can be used for design temperatures between −29°C and approximately 400°C. Alternatively, low-temperature carbon steel (LTCS) can be used if the design temperature of the piping and valve will be between −46°C and −29°C. LTCS is probably more suitable for lower temperature than carbon steel due to the addition of some alloys such as nickel and magnesium. A Charpy impact test is applied on LTCS at −46°C to evaluate the resistance of the materials against breakage. A V-notch is created on a sample of the material (see Fig. 12.21) and the required energy to break the material through the V-notch is measured. The mechanism of low temperature failure in the material is cracking, which is evaluated by the impact test. Adding 3 mm of corrosion allowance is also applicable for piping and valve bodies in LTCS material.

The common, low-temperature carbon steel for the body and bonnet of subsea valves is American Society for Testing and Materials (ASTM) A350 LF2. ASTM is an international standard. One of the common heat treatments for this material is quenching and tempering, the process of heating and cooling a material to achieve certain mechanical and physical properties. Quenching means rapid cooling of a metal to adjust to the mechanical properties of a solid state. Quenching can be done in water as the cooling medium and results in a very hard structure. Tempering, which is done after quenching, is a heat treatment for improving the toughness of a material. Tempering is normally done after a hardening heat treatment to reduce the brittleness of the material. Carbon steel can be corroded easily in seawater if unprotected; however, coating and cathodic protection prevent the corrosion of carbon steel material in deep-water applications. Carbon steel is

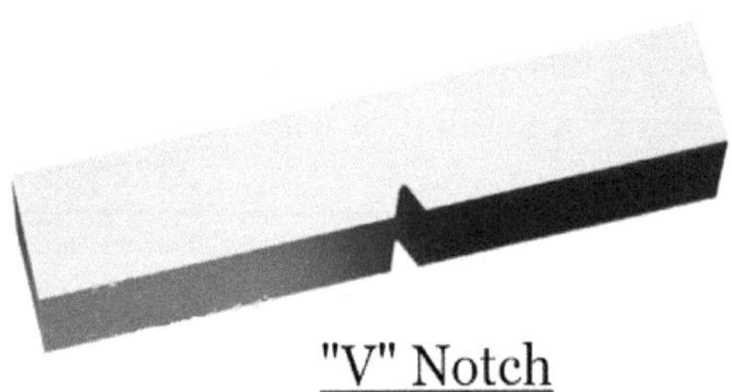

Fig. 12.21 V Notch on metal for the Charpy impact test.

subject to erosion and corrosion internally when used for oil and gas services, so a full Inconel 625 overlay should be applied internally on the carbon steel. An Inconel 625 overlay on carbon steel prevents the risk of both CO_2 and H_2S corrosion, especially HIC on the carbon steel. The thickness of the Inconel 625 should be at least 3 mm, according to API 6A/API 17D standards, and applied in one or two layers depending on the welding technology. Gas Tungsten Arc Welding (GTAW), also known as tungsten inert gas (TIG), is an arc welding process that uses a nonconsumable tungsten electrode to weld. Applying an Inconel 625 overlay through TIG or GTAW requires two layers of welding. An alternative process, cold metal transfer (CMT) welding, may require one layer of Inconel 625 weld overlay on the carbon steel. Carbon steel material can be used in noncorrosive fluid such as MEG (mono-ethylene-glycol). MEG is injected into gas to dehydrate the gas, removing water droplets and preventing hydrate formation. Valves in MEG service could have body and bonnet material in LTCS (ASTM A350 LF2) or low alloy steel since MEG is not corrosive. However, the seat pocket, sealing areas, and any possible grooves should be overlaid with Inconel 625 to prevent crevice corrosion. In fact, fluid can be trapped in the grooves under the sealing and cause corrosion, so a partial Inconel 625 overlay on groove areas on carbon steel or low alloy steel body and bonnet of the valves is recommended. The qualification of the Inconel 625 weld overlay shall be done according to ASME BPVC, Section IX or ISO 15614-7 according to API 6A/API 17D standards. The other important parameter in applying a weld overlay of Inconel 625 on carbon and low alloy steels is the amount of iron diffusion in Inconel 625, which can be limited to 5%. It should be noted that material selection normally begins with the piping; in some cases, the body and bonnet of the valve can have the same material as the piping, but this rule is not always correct. For example, carbon steel valves with a full Inconel 625 overlay may be welded to duplex or super duplex piping through a pup piece of the same material as the pipe. Welding a carbon steel valve with super duplex piping is challenging, especially when the thickness of the pipe is more than 20 mm, because carbon steel welding in thickness values above 20 mm and in sour services requires PWHT to release the residual stress from the material after welding according to ASME B31.3, "process piping code requirement." However, super duplex is sensitive to PWHT due to the formation of sigma phase, which can cause brittleness and cracking. Therefore the solution is to apply an Inconel 625 buttering on the carbon steel pipe and then apply PWHT. Afterward, the carbon steel buttered with Inconel 625 can be connected to super duplex with Inconel 625 filler (see Fig. 12.22). ERNiCrMo-3 is the weld wire for Inconel 625. In this technique of welding, super duplex is not subject to PWHT, so there is no risk of sigma phase formation. Fig. 12.23 shows a LTCS body valve in ASTM A 350 LF2 with a full Inconel 625 overlay being welded to 25Cr super duplex piping. A pup piece in 25Cr super duplex is being welded to a LTCS body valve from one side and to a 25Cr super duplex pipe from the

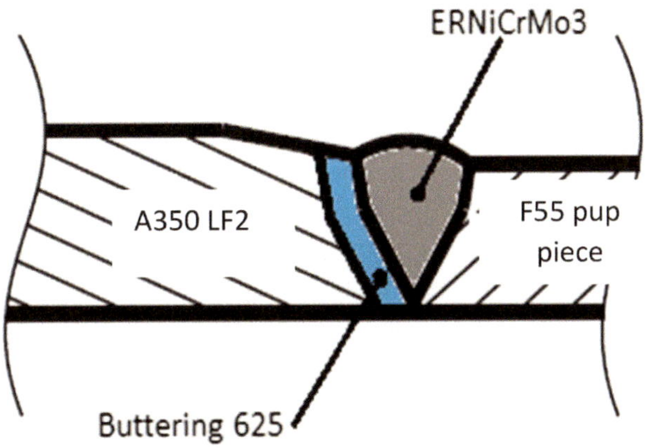

Fig. 12.22 Buttering of a valve body in LTCS welded to a super duplex pup piece with an Inconel 625 filler. *(Courtesy: Valve World/Springer.)*

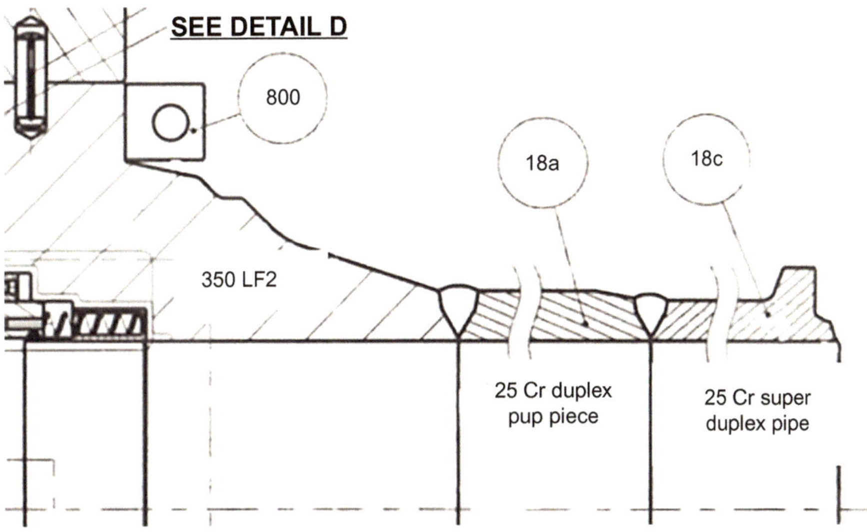

Fig. 12.23 Welding a LTCS carbon steel to 25Cr super duplex piping.

other side. The pup piece is normally in forged material; in this case it is ASTM A182 F55. Refer to the illustration as follows—the challenging weld is between the body of the valve in carbon steel and the pup piece in super duplex. The super duplex pup piece is welded to the piping in the fabrication yard without any challenge since the two materials are identical.

12.2.4.2 Low alloy steel

ASTM A694 F60, F65, and F70, including alloys such as nickel, molybdenum, and magnesium, are widely used for the bodies and bonnets of valves as an alternative to carbon steel. This grade of low alloy steel has higher strength compared to carbon steel. The letter "F" in the ASTM grade designation indicates that the material is forged, and the number after the letter "F" indicates the yield strength of the material. As an example, F60 has a yield strength of 60,000 psi. This material, when used for the body and bonnet of a valve, can be either partially or fully overlaid with Inconel 625, just like carbon steel, since low alloy steel is not classified as a CRA. The chemical composition of the low alloy steel should be controlled to make sure that its carbon equivalent is within the accepted range given before for carbon steel to improve weldability and prevent cracking due to sour corrosion. Fig. 12.24 illustrates a 7 1/16″ through conduit slab gate valve in pressure class 10,000 psi to be used in a subsea manifold with a body in low alloy steel grade F60.

Fig. 12.24 7 1/16″ class 10,000 psi slab gate valve in F60 low alloy steel body material.

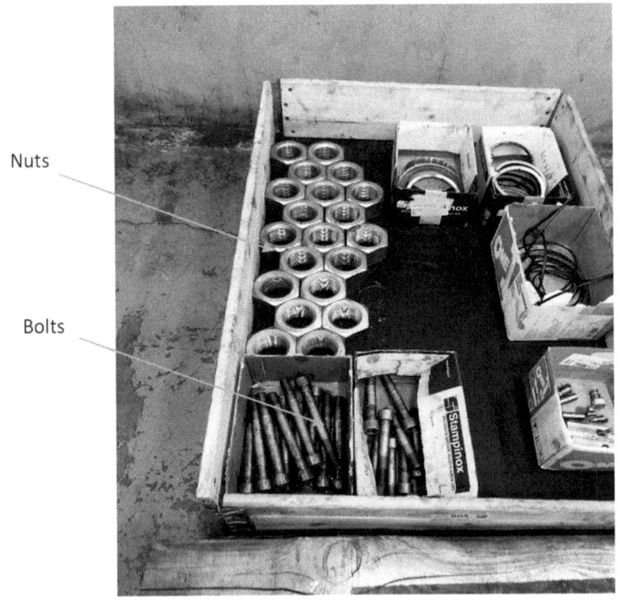

Fig. 12.25 Low alloy steel bolting with zinc-nickel coating for a subsea ball valve.

Low alloy steel bolts, based on ASTM A193 and A194, are very common for subsea valves as per API 20 E standard or a specification applicable for carbon steel and low alloy steel bolting. These noncorrosion-resistant alloy materials are provided with coating and cathodic protection. Different types of coating, such as zinc-nickel plate, PTFE, or zinc flake coating, which is a mixture of aluminum and zinc, are applicable in the subsea sector. Because subsea valve bolting is provided with cathodic protection, the use of corrosion-resistant alloys for bolting, such as super duplex and nickel alloys covered by API 20F, is not common in the subsea sector. The hardness level of the bolting material should be controlled based on NACE MR 0175 if the bolts will be used for sour service fluid. Super duplex is a common CRA bolting material that can be used to mitigate the risk of corrosion under insulation (CUI). Fig. 12.25 illustrates low alloy steel bolting, including both bolts and nuts with zinc-nickel coating for a subsea ball valve.

12.2.4.3 Stainless steels

Stainless steels are a large group of materials; they can be defined as corrosion-resistant alloys that contain iron as the main steel element and more than 10.5% or 11% chromium. The main passive corrosion protective layer in stainless steel is called chromium oxide Cr_2O_3, which is stable with a minimum 10.5% chromium content. Therefore the amount of chromium in stainless steel affects its corrosion resistance, high temperature resistance, pitting corrosion resistance, etc. In addition to chromium, nickel has a high influence on the properties of stainless steels. Nickel is the main component that improves the stability of austenite phase in stainless steel. Molybdenum is another element in stainless steel that improves its resistance to the pitting and crevice corrosion that can happen in seawater service. Nitrogen, like molybdenum, improves the resistance of stainless steel to pitting and crevice corrosion. In general, stainless steels can be split into four categories: ferritic, martensitic, austenitic, and duplex, which are explained in the next sections. Ferritic stainless steel types are not explained since they are not common in the oil and gas industry.

Martensitic stainless steels

Martensitic stainless steel grades that are used for industrial valves in the oil and gas industry are normally 13Cr, 13Cr-4Ni, or 17-4 PH. These materials have very high mechanical strength due to their chemical properties as well as the heat treatments that are used for valve stems. Valve stems are categorized as a pressure-containing part, so they should be made from high mechanical strength materials especially in the case of actuation. Fig. 12.26 illustrates a large ball valve stem in 13Cr-4Ni martensitic stainless steel for use in the offshore sector of the oil and gas industry on a topside platform. In this case, not only the stem of the valve but also the ball material is in martensitic stainless steel, as the force from the stem is transmitted to the ball.

Generally, quench and temper heat treatments are used for martensitic stainless steel, which gives them high mechanical strength. The main weak points of these materials are their susceptibility to stress cracking corrosion and hydrogen attack corrosion types such as HISC and hydrogen embrittlement. Thus martensitic stainless steels, including the grades mentioned before, are not used and not recommended for subsea valves. 13Cr contains 13% chromium and 0.15% carbon; it possesses very good ductility, good general corrosion resistance, and high mechanical strength. 13Cr-4Ni contains 13% chromium with an addition of 4% nickel, which improves the ductility, weldability, impact, and

Fig. 12.26 Ball and stem of a large ball valve in the offshore industry (nonsubsea sector) in 13Cr-4Ni. *(Courtesy: Springer.)*

low-temperature resistance of the material. 17-4 PH has 17% chromium, 4% nickel, and 4% copper which gives it a very hard characteristic achieved via participation hardening or age hardening heat treatment.

Austenitic stainless steels

Austenitic stainless steels, like martensitic stainless steels, are not used for subsea valves. Austenitic stainless steels are known as 300 series, which contains mostly 18% chromium and 8% nickel. Other chemicals, such as carbon, molybdenum, sulfur, phosphor, etc., exist in austenitic stainless steels. Austenitic or austenite phase is a solid solution of carbon and iron. Austenitic stainless steels do not have high mechanical strength, unlike duplex and martensitic stainless steels. Different grades of this material such as 304, 316, 321, and 347 are more common. Grade 316 contains approximately 2% molybdenum, which improves the resistance of this grade against pitting and stress cracking corrosion. That is the reason why the Norsok standard only allows using stainless steel grade 316 in the offshore environment. However, stainless steel grade 316 is not resistant to seawater corrosion and is not used subsea for the main components of valves. In general, one of the main weak points of austenitic stainless steels is its low resistance against chloride. In fact, austenitic stainless steel grades 300 in general are at high risk of pitting and chloride stress cracking corrosion (CLSCC) in the offshore environment. It should be noted that the NORSOK standard allows the use of uncoated austenitic stainless steel grade 316 for operating temperatures less than 60°C. Austenitic stainless steels may be coated for use in offshore environment in some parts of the world, especially in higher temperature ranges, to prevent corrosion. However, even the coating of austenitic stainless steels may not be sufficient to prevent corrosion in the offshore environment. Figs. 12.27 and 12.28 illustrate pitting corrosion of austenitic stainless steel tubes in the offshore environment.

Fig. 12.27 Pitting corrosion of a 316 tube in the offshore environment.

Fig. 12.28 Pitting corrosion of a 316 tube in the offshore environment.

Grades 321 and 347 are stabilized by certain chemicals, so they are more resistant against intergranular stress cracking corrosion. Other important grades of austenitic stainless steel are those with a letter "L" at the end of their number, such as 304L and 316L. "L" means low carbon, which improves some of the properties of stainless steel, such as increasing weldability and reducing the chance of weld decay or intergranular stress cracking corrosion, and increasing the resistance of the material in high-temperature applications. One of the most important properties of austenitic stainless steels is their resistance in temperature applications as low as −196°C. Thus austenitic stainless steel materials are common for extremely low-temperature applications, also called cryogenic applications. In addition, austenitic stainless steels are used onshore, in refineries and petrochemical plants, instead of carbon steel where carbon steel is at a high risk of carbon dioxide corrosion. Fig. 12.29 illustrates a cryogenic valve made in austenitic stainless steel after the cryogenic test. One of the main cryogenic applications is in liquefied natural gas (LNG) plants where heavier gases are liquefied and separated from the oil.

Duplex stainless steels

Duplex and super duplex stainless steels are the most common materials for subsea valves. This type of stainless steel is called duplex because it has two kinds of microstructure: ferritic and austenitic. Duplex has 22% chromium and super duplex has 25% chromium, which is higher than the chromium content in austenitic stainless steels. Higher chromium improves the corrosion resistance of duplex stainless steels compared to austenitic grades. However, common grade 300 austenitic stainless steels in general have higher nickel content, which improves the toughness of these grades. Better toughness improves the resistance of austenitic stainless steels in low-temperature applications. One of the main limitations of duplex stainless steels is low-temperature resistance; they should be used at or above a minimum temperature of −46°C. In addition, duplex stainless steel can be more susceptible to cracks in hydrogen sulfide-containing environments. But

Fig. 12.29 Austenitic stainless steel body valve for cryogenic application after the cryogenic test.

duplex and super duplex are more resistant against pitting and chloride stress cracking corrosion compared to austenitic stainless steels. In addition, the mechanical properties of duplex and super duplex are higher than austenitic stainless steels. As an example, yield and tensile strength of a 316 pipe in UNS S31600 could be about 30,000 and 75,000 psi, respectively. The yield and tensile strength of 22% Cr (chromium) duplex stainless steel UNS 31803 are 65,000 and 90,000 psi, respectively. Super duplex UNS 32750 can have a mechanical strength of approximately 90,000 psi yield and 118,000 psi tensile. Increasing the level of chromium in super duplex compared to duplex improves mechanical strength, pitting, and crevice corrosion resistance and sour corrosion resistance. As outlined in ISO 15156 and NACE MR 0103, super duplex has a higher corrosion resistance than 22Cr duplex in an environment containing H_2S. However, it should be noted that duplex is sensitive to high temperatures, especially welding, due to sigma brittle phase made of chromium and molybdenum; increasing the chromium content increases the risk of sigma phase formation. Thus super duplex is more susceptible to sigma phase formation at high temperatures and during welding in comparison to duplex stainless steel. Fig. 12.30 illustrates the formation of sigma phase in duplex stainless steel after 9 min of exposure to a temperature of 850°C. Normally, pitting and crevice corrosion tests according to ASTM G48 are applicable for duplex and super duplex to be used in seawater or for seawater fluid service. Fig. 12.31 illustrates subsea gate valve internals, including the stem, disks, and seats, in F55 grade, which is the forging of super duplex. Figs. 12.32 and 12.33 illustrate the seats and disk of a gate valve in super duplex material, respectively.

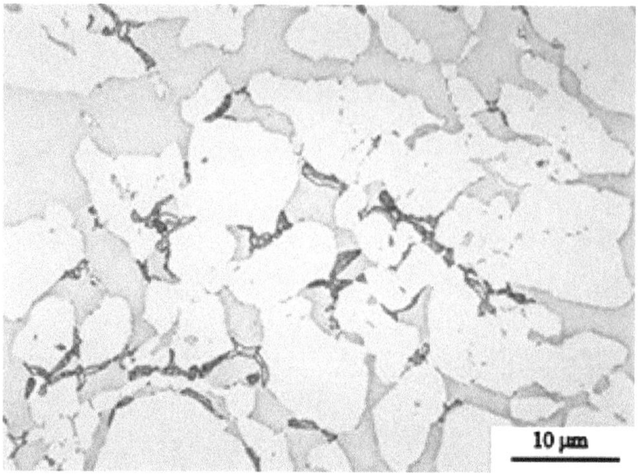

Fig. 12.30 Sigma phase formation in duplex stainless steel after 9 min exposure to 850°C. *(Courtesy: Sandvik.)*

Fig. 12.31 Super duplex stem (F55) for a subsea gate valve.

12.2.4.4 Nickel alloys

Inconel 625 or alloy 625 has approximately 65% nickel, 22% chromium, 9% molybdenum, and 4% niobium. Inconel 625 is a nickel-based super alloy with high strength properties and resistance to corrosion, including H_2S and CO_2, and elevated temperatures. This material can be used for the body, bonnet, seats, stem, and metallic sealing

Fig. 12.32 Super duplex seats F55 of a subsea gate valve.

Fig. 12.33 Super duplex disk (F55) for a subsea gate valve.

Fig. 12.34 Inconel 625 ring type joint gasket.

of subsea valves. Alloy 625 can be used as an internal overlay for valve bodies, bonnets, and internals in carbon steel or low alloy steel to prevent different types of corrosion, such as carbon dioxide, hydrogen sulfide, and pitting. Fig. 12.34 illustrates a ring type joint gasket in Inconel 625 used for sealing between the body and bonnet of a through conduit gate valve.

Inconel 725 is another nickel alloy containing nickel, chromium, and molybdenum alloys. It is a more expensive nickel alloy than Inconel 625 with the same corrosion-resistant properties. Inconel 725 can demonstrate higher mechanical strength compared to Inconel 625, so it is a good choice for valve stems. Age hardening is a very common heat treatment for Inconel 725 that provides very high mechanical strength. As discussed before, however, Inconel 725 is vulnerable to HISC corrosion.

Inconel 718 is another type of nickel alloy with nickel, chromium, and molybdenum that can provide almost the same strength as Inconel 725. Inconel 718 does not have as high a corrosion resistance as Inconel 725, and is not suitable for seawater service. The heat treatment for Inconel 718, like Inconel 725, is age hardening, and this material can be used as an option for valve stems. If the mechanical strength of 25Cr super duplex for the stem is not high enough, the stem material can be upgraded to an alloy with higher mechanical strength, such as Inconel 718 or 725.

Inconel X750 is a popular material for valve springs due to its high fatigue properties and resistance against frequent loads. However, this material does not have a sufficient pitting resistance equivalent number so it is not recommended for subsea valve applications. An alternative spring material in seawater is Elgiloy UNS R30003. The heat treatment for Inconel X750 is precipitation hardening, which adds some elements like aluminum and titanium.

12.2.4.5 Hardfacing alloys

Hardfacing is a method for improving the wear resistance of a metal by depositing hardfacing alloys such as Stellite, Colmonoy, Ultimet, Chromium Carbide, etc. In fact, hardfacing is a metal working process in which a harder material is applied on a base metal through different processes such as weld overlay, spray, or laser to mitigate the risk of erosion, galling, and cavitation. A variety of tests should be performed on hardfacing alloys to ensure the quality of the hardfacing overlay, such as a thickness test, an adhesion test, a corrosion test, a nondestructive test (NDT), a chemical composition test, etc. The more common types of hardfacing materials in the subsea industry are tungsten carbide and Ultimet. Stellite is a very common hardfacing material in the oil and gas industry. These materials are explained in detail as follows.

Tungsten carbide, with a chemical formula of WC, contains tungsten and carbon atoms. Tungsten carbide is overlaid on the ball and seat surfaces of metal seat ball valves, as mentioned before. Some end users may opt not to apply a sealant injection fitting on metal seat ball valves, since the seats are strengthened and hardfaced with tungsten carbide. Sealant injection fittings are used to inject sealant on the ball valve seats as a means of emergency maintenance. However, it should be noted that sealant injections are used for the industrial valves in all sectors of oil and gas industry except for subsea. Subsea valves should be maintenance free so they do not have sealant injection fitting. Fig. 12.35 illustrates the ball of a ball valve after tungsten carbide coating and lapping. High Velocity Oxygen Fuel (HVOF) thermal spray coating is the common method of applying the tungsten carbide (see Fig. 12.36). This process pushes a tremendous amount of kinetic

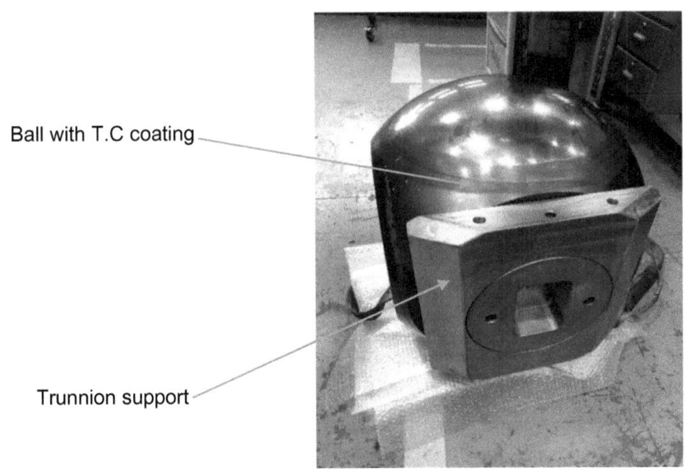

Fig. 12.35 Ball of a subsea ball valve in super duplex with tungsten carbide hardfacing.

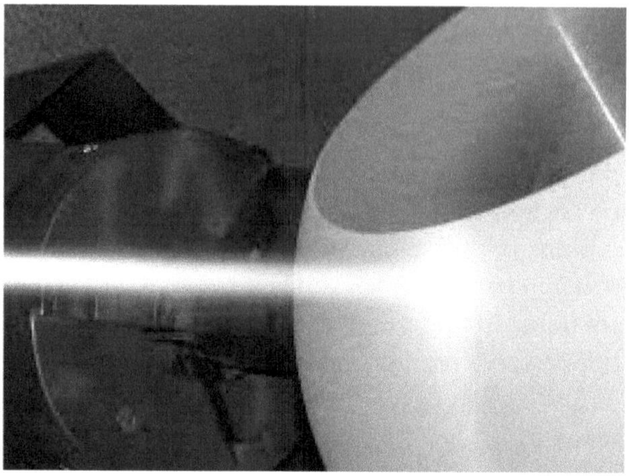

Fig. 12.36 HVOF tungsten carbide coating being applied to a valve.

energy into the powder particles. When these highly energetic particles strike the substrate, their velocity drops to zero and they absorb their kinetic energy, which causes them to bond instantly to the workpiece. The thickness of tungsten carbide for subsea valves could be between 0.2 and 0.4 mm, based on project specifications.

In some cases, the subsea valve internal materials should be selected by considering the wet storage in which the valve internals are in contact with seawater. There are some debates about using tungsten carbide in seawater. Another hardfacing material that can be used in seawater as an alternative to tungsten carbide is Ultimet or Tribaloy.

Ultimet (UNS R31233) is a cobalt alloy, like stellite 6 and tungsten carbide, that provides very good wearing resistance. This material contains other compounds like chromium, nickel, molybdenum, iron, and tungsten. Based on industry experience, overlaying this material is not as easy for valve manufacturers as applying stellite 6. Sufficient adhesion of Ultimet with the core material can be a challenge in some cases, and can necessitate removing the weld overlay after some years of operation. This type of overlay is very common for 25Cr duplex valves in seawater service in the Norwegian offshore industry (nonsubsea sector). The next paragraph provides some detailed information about Tribaloy hardfacing, which, like Ultimet, is suitable for seawater service.

Tribaloy alloys with either a nickel or cobalt base have been developed for applications of extreme wear, high temperature, and corrosive media. As mentioned before, this material is a good alternative for stellite in seawater service. Tribaloy 800 is a cobalt-chromium-molybdenum alloy with high erosion, corrosion, and galling resistance.

Tribaloy 800 has a minimum of 50% cobalt, 17.5% Cr, and 28.5% Mo. The high amount of molybdenum in Tribaloy 800 increases its resistance to pitting corrosion.

Stellite 6 UNS R30006 is the most common hardfacing alloy used in the valve industry, displaying outstanding hardness and toughness. This material is a cobalt alloy that contains other compounds, such as chromium, carbon, and tungsten, with high resistance to wear, galling, and corrosion due to the hard carbide phase in the CoCr alloy matrix. Stellite 21 is another grade of stellite that is not as popular as stellite 6 in the oil and gas industry. Applying stellite on 22Cr duplex and 25Cr super duplex is possible, but it is challenging due to the formation of sigma phase in these two types of duplex materials during welding. Sigma phase is a hard and brittle crystal structure that is formed at high temperatures and can lead to cracking and failure. Alloy elements such as chromium and molybdenum in the duplex promote the formation of sigma phase at the high temperature of welding and the slow cooling down, which increases the hardness and reduces the toughness of the material.

12.2.4.6 Cobalt alloys

Elgiloy is a cobalt alloy that is proposed for subsea valve springs due to its good resistance to seawater service. More precisely, Elgiloy is a cobalt, chromium, and nickel alloy containing approximately 40% cobalt, 20% chromium, 15% nickel, 7% molybdenum, and the rest iron and manganese. This material is used for springs because of its resistance to corrosion and fatigue, and its high strength.

12.3 Conclusion and recommendations

Material selection for subsea valves should address the different valve components such as the body, bonnet, closure member, seats, stem, pup piece, spring, etc., and consider the different parameters reviewed in this chapter. Table 12.4 summarizes and proposes different materials for subsea valve components, and lists their advantages and disadvantages.

Table 12.4 Summary of material selection for different valve components.

Valve component	Proposed materials	Pros and cons
Body/bonnet	Carbon/alloy steel	Economical, but not corrosion resistant Inconel 625 can be overlaid partially or fully
	Super duplex	Good resistance to pitting corrosion but limited resistance to H_2S
	Inconel 625	Expensive but highly corrosion resistant

Continued

Table 12.4 Summary of material selection for different valve components—cont'd

Valve component	Proposed materials	Pros and cons
Stem	Super duplex	Good resistance to pitting corrosion but limited resistance to H_2S; limited strength can be valid in case of actuation
	Inconel 625	Good resistance to pitting, resistant to H_2S corrosion; limited strength can be valid in case of actuation
	Inconel 718	High mechanical strength but poor pitting corrosion resistance
	Inconel 725	Expensive choice but high mechanical and corrosion resistance
Trim (seats and closure member)	Super duplex	Good resistance to pitting corrosion but limited resistance to H_2S
	Inconel 625	Good resistance to pitting and H_2S corrosion
Bolting	Low alloy and carbon steel/zinc nickel electroplated	Economic, but not suitable under insulation
	Super duplex	More expensive for bolting but suitable under insulation
Pup piece	Carbon steel	No comment
	Super duplex	No comment
	Inconel 625	No comment
Spring	Inconel X750	High mechanical strength and fatigue resistance but poor pitting and corrosion resistance
	Elgiloy	High mechanical strength and fatigue resistance and suitable for seawater service
	Inconel 625	Suitable for seawater service but low mechanical strength and fatigue resistance
Hardfacing materials	Tungsten carbide	No comment
	Stellite	No comment

Further reading

ASB Industries, Innovative Surfacing Technologies, [Online]. Available from: https://www.asbindustries.com/tungsten-carbide-coatings, 2018. Accessed 12 September 2020.

H.E. Amaya, A.D. Onesubsea, Hydrogen embrittlement stress cracking (HISC) in the oil and gas industry, in: Conference Paper. Materials Science Technology, A Cameron & Schelumberger Company, Columbus, OH, 2015.

American Society for Testing and Materials (ASTM) G48. 2003. Standard Test Methods for Pitting and Crevice Corrosion Resistance of Stainless Steels and Related Alloys by Use of Ferritic Chloride Solution. USA.

American Petroleum Institute (API) 6A, Specification for Wellhead and Tree Equipment, twenty-first ed., API, Washington, DC, 2018.

American Petroleum Institute (API) 17D, Design and Operation of Subsea Production Systems, Subsea Wellhead and Tree Equipment, second ed., API, Washington, DC, 2011.

American Society of Mechanical Engineers (ASME), Process Piping, (2012). New York, NY.
American Petroleum Institute (API) 20E, Alloy and Carbon Steel Bolting for Use in the Petroleum and Natural Gas Industries, first ed., API, Washington, DC, 2012.
American Petroleum Institute (API) 20F, Corrosion Resistant Bolting for Use in the Petroleum and Natural Gas Industries, first ed., API, Washington, DC, 2015.
A. Bahrami, P. Woollin, Hydrogen induced stress cracking of duplex stainless-steel subsea components, in: Conference paper, Offshore Mechanics and Arctic Engineering (OMAE) Conference, Shanghai, China, TWI Ltd, 2010.
Y. Bai, Q. Bai, Subsea Engineering Handbook, first ed., Elsevier, Atlanta, GA, 2012.
Det Norske Veritas (DNV) RP-F-112, Design of Duplex Stainless Steel Subsea Equipment Exposed to Cathodic Protection, Det Norske Veritas, Hovik, Norway, 2008.
Det Norske Veritas (DNV) RP-O-501, Erosive Wear in Piping Systems, Det Norske Veritas, Hovik, Norway, 2007.
D.W. Green, R.H. Perry, Perry's Chemical Engineers' Handbook, ninth ed., McGraw-Hill, 2018.
International Organization of Standardization (ISO) 10423, Petroleum and Natural Gas Industries—Drilling and Production Equipment—Wellhead and Christmas Tree Equipment, fourth ed., (2009)Geneva, Switzerland.
International Organization of Standardization (ISO) 13628-4, Petroleum and Natural Gas Industries—Design and Operation of Subsea Production Systems—Part 4: Subsea Wellhead and Tree Equipment, second ed., (2010)Geneva, Switzerland.
International Organization of Standardization (ISO) 15156, Petroleum and Natural Gas Industries—Material for Use in H_2S Containing Environments in Oil and Gas Production, third ed., (2015). Geneva, Switzerland.
K.L. Jepsen, P. Grumsen, Failed austenitic SS bolts in valves, Valve World Mag. (2014). [online] Available at: https://www.valveworld.net/pdf/failed_austenitic_bolts_in_valves.pdf. Accessed 12 September 2020.
B. Nesbitt, Handbook of Valves and Actuators, first ed., Elsevier, New York, 2007.
NORSOK M-001, Material Selection, fifth ed., (2014). Lysaker, Norway.
NORSOK L-001, Piping and Valves, (2017). 4th revision, Oslo, Norway.
M. Solnordal, S. Wasberg, G. Heiberg, E. Odd, Hydrogen Induced Stress Cracking (HISC) and DNV-RP-F112, Det Norske Veritas, Hovik, Norway, 2009.
K. Sotoodeh, Analysis and improvement of material selection for process piping system in offshore industry. Am. J. Mech. Eng. 6 (1) (2018) 17–26, https://doi.org/10.12691/ajme-6-1-3.
K. Sotoodeh, Welding a carbon steel body to 22Cr duplex pipe: a case study, Valve World Mag. 24 (5) (2019) 38–41.
K. Sotoodeh, Application of hardfacing alloys for valves, Valve World Mag. 24 (8) (2019) 63–64. September.
K. Sotoodeh, Requirement and calculation of corrosion allowance for piping and valves in oil and gas industry. J. Bio Tribo Corros. 6 (2020) 21, https://doi.org/10.1007/s40735-019-0319-4. Springer.
K. Sotoodeh, Material selection for subsea valves, Stainless Steel World Mag. 32 (2020) 47–50.
K. Sotoodeh, Application of austenitic stainless-steel butterfly valves in NGL recovery and cryogenic applications, Stainless Steel World Americas Mag. 3 (2020) 14–15.
K. Sotoodeh, Optimized material selection for subsea valves to prevent failure and improve reliability. J. Life Cycle Reliab. Saf. Eng. (2020). https://doi.org/10.1007/s41872-020-00152-x. Springer.
WebCorr, Different Types of Corrosion, Crevice Corrosion, [online] Available at: https://www.corrosionclinic.com/types_of_corrosion/crevice_corrosion.htm#:~:text=Crevice%20Corrosion%20refers%20to%20the,metal%20and%20non%2Dmetallic%20material, 2020. Accessed 11 September 2020.
Workshop agreement NORSOK M-WA-01, HISC Guideline for Duplex Stainless Steel. Revision 01, NORSOK, Lysaker, Norway, 2005.

CHAPTER THIRTEEN

Valve and actuator sealing

Contents

13.1 Introduction to subsea valve reliability

Subsea valves face the unique challenge of being designed for zero maintenance and leakage for a relatively long period of time, such as 20—and in some cases even 50—years of design life. The main reason to achieve zero leakage and maintenance for subsea valves is that many new subsea projects are located at water depths between 1 and 3 km, far from shore where maintenance is not possible. In fact, the challenge of conducting subsea projects at deeper depths is coupled with other difficulties, such as increased distance from shore and harsh environments like that of the icy arctic. It should be added that subsea valves are operated in very sensitive marine environments where the spillage of oil and gas has extremely negative consequences for the environment, its biological resources, and human and animal life. As an example, one of the world's largest oil spillage disasters occurred on April 20, 2010, in the Gulf of Mexico on a platform operated by British Petroleum (BP). This spill was caused by various factors, including the failure of a subsea safety valve located on the sea floor at the wellhead; this valve, also called a blowout preventer, was designed to prevent blowouts by shutting down the wellbore. The disaster killed approximately 82,000 birds of 102 species, approximately 6165 sea turtles, and up to 25,900 marine mammals such as whales and

Subsea Valves and Actuators for the Oil and Gas Industry
https://doi.org/10.1016/B978-0-323-90605-0.00010-4

(A) (B)

Fig. 13.1 (A) Explosion of the BP platform in the Gulf of Mexico. (B) Death of a whale due to the BP platform explosion. *((A) Courtesy: Niva. (B) Courtesy: Environmental Pollution Center.)*

dolphins. Fig. 13.1A and B illustrate the catastrophic explosion of the BP-operated platform and the death of a whale due to this disaster.

The other adverse effects of oil spillage are increasing operational costs due to the loss of assets and decreased production. Fig. 13.2 summarizes the challenges confronting subsea valves addressed in this chapter.

The existing standards for subsea valves are reviewed and discussed in Chapter 2. Valves can emit hydrocarbons into the atmosphere due to different factors, such as improper fit, wear and tear, and corrosion. Fugitive emission (FE) is defined as unintentional leakage from sealed surfaces such as packing and gaskets resulting from a faulty or corroded

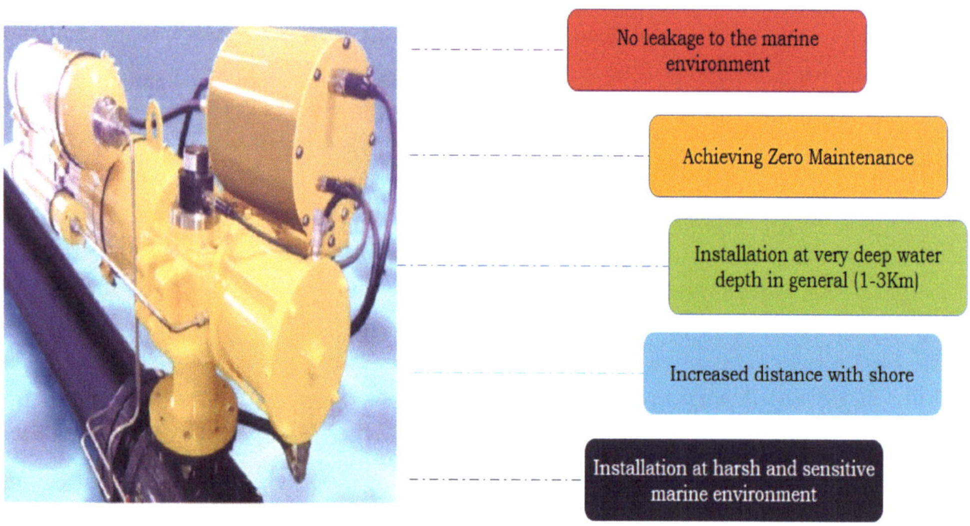

Fig. 13.2 Subsea valve challenges during installation and operation. *(Courtesy: Springer.)*

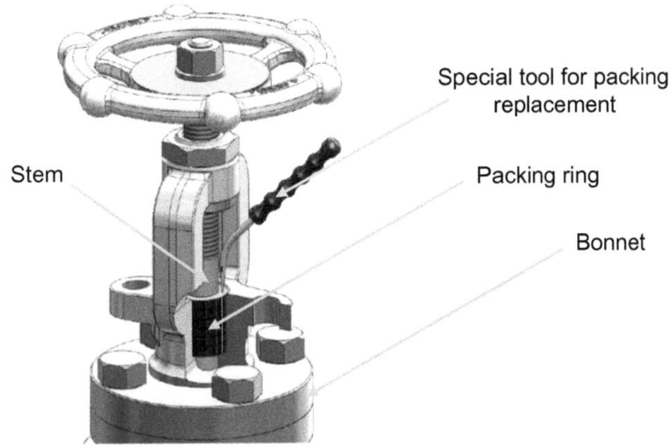

Fig. 13.3 Packing replacement on a wedge gate valve.

connection. Packing in the valve industry refers to ring-shaped components made of resilient material such as graphite or Teflon (PTFE) installed between the valve stem and the bonnet or stuffing box to prevent leakage from the valve stem to the environment. Fig. 13.3 shows the replacement of packing rings from a wedge gate valve.

Many standards and specifications have been established to maintain a very low level of fugitive emissions from valves; these include:

- ISO 15848: Industrial valve fugitive emissions measurement test and qualification procedure;
- Shell MESC SPE 77/312: Industrial valve fugitive emission (FE) measurement, classification system, qualification procedures and valve prototype and production tests;
- API 622: Type testing of process valve packing for fugitive emissions;
- API 624: Type testing of process valve leak performance for fugitive emissions and seat leakage applicable for rising stem valves such as wedge gate valves with graphite packing;
- API 641: Type testing of quarter-turn valves for fugitive emissions applicable for ball valves.

The main technical issue addressed in this chapter is that none of the FE standards are used for *subsea* valves. And, looking at the subsea valve standards mentioned before, there is no single sentence about FE tests and standards. The main technical question is, then, how can subsea valves achieve zero leakage over a long period of design life and in a harsh environment? The aim of this section is to provide a detailed discussion of the parameters needed in order to improve the safety and reliability of safety valves in regard to the zero-leakage requirement. In general, three unique factors—design, manufacturing, and testing—contribute to ensuring the zero leakage/zero maintenance expectation. Sealing plays a very important role in all three stages of valve design, manufacturing, and testing in order to achieve the goal of zero emissions to the environment. The reliability of a valve's sealing is dependent on tests and finite element analysis (FEA). The reliability of sealings is evaluated and estimated during both qualification tests and factory acceptance tests (FAT), which are explained in Chapter 14. In general, testing every single stem sealing is recommended during the FAT. Hyperbaric testing can validate the performance of external sealing, which is designed to prevent the leakage of seawater into the valve internals, especially the stem areas. FEA can evaluate sealing performance and assess the possibility of failure under different load conditions, as well as fluid pressure, temperature, and pressure values. Sealing requirements and functionality are more complicated in high-pressure and high-temperature applications. Fig. 13.4 illustrates a stress analysis for a high-pressure, high-temperature stem seal in an extremely high-pressure condition of 20,000 psi and 177°C. In addition, the sealing supplier should apply relevant tests to the sealing based on the governing project standards.

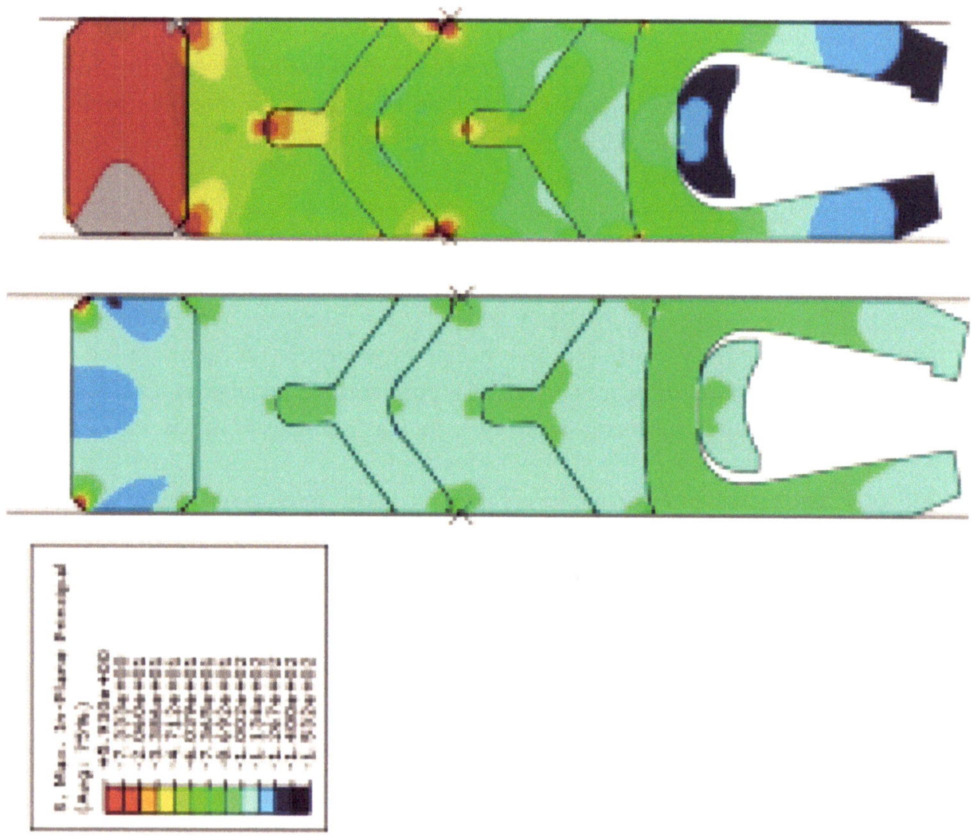

Fig. 13.4 High-pressure/high-temperature stress analysis for a stem seat at 20,000 psi and 177°C. *(Courtesy: Valve World.)*

13.2 Introduction to valve sealing

The subsea oil and gas application represents one of the most challenging environments due to extreme pressure, relatively high temperatures in some cases, and harsh erosive and corrosive fluid, which demands excellent sealing selection and extensive testing and qualification. Valves sealings are an essential component for safe valve operation and are critical in terms of achieving the leakage-free expectation. The sealing element around the valve stem could consist of a combination of different seals, such as O-rings, lip seals, chevron V-packs (see Fig. 13.5), and bellows (see Fig. 13.6). Stem sealing prevents leakage from the stem area during valve operation. The other unique challenge of stem sealing for subsea valves is that it should prevent the leakage of seawater to the valve internals, including the stem seal area. The sealing materials' capability is not limited to providing sealing; seals should also resist wearing, friction, and contact damage caused by moving parts. Moreover, sealing components in valves are not limited to stem

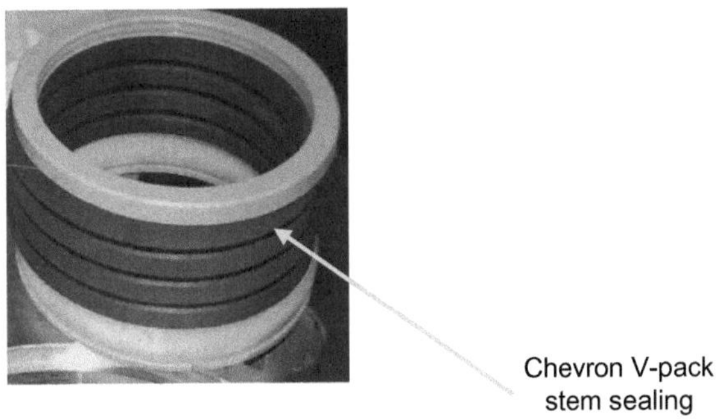

Fig. 13.5 Chevron V-pack stem sealing. *(Courtesy: Galperti Engineering.)*

Fig. 13.6 Stem sealing bellows. *(Courtesy: Galperti Engineering.)*

Fig. 13.7 Sealing of a seat and body, as well as a seat (retainer) and ball in a ball valve.

sealing. Other metallic and soft materials in the form of gaskets, O-rings, or thermoplastic materials can be used to provide sealing between different components of valves, such as the seat and body, seat (retainer) and ball, etc., as illustrated in Fig. 13.7. Different parameters affect valve sealing selection, such as sealing purpose or application, pressure, temperature, etc. As an example, elastomer sealing is mainly used for stem sealing, but can also be used between two valve components such as the body and bottom cover in rare cases. Thermoplastic sealings are normally used for sealing between the seat and closure member in soft seat valves. Sealings can be categorized as dynamic or static; static sealing is used between two nonmovable components, while dynamic sealing is used between at least one movable component. As an example, stem sealings located between the body or bonnet of a valve and its stem are considered dynamic, since the stem of the valve is a moving part. The stem of a valve could rotate 90 degrees in a ball valve, or move linearly up and down in a gate valve; a gate valve stem with linear movement produces relatively more friction and wearing on stem sealings.

13.3 Elastomeric polymer sealing

13.3.1 Introduction

NORSOK M-710 addresses the qualification of nonmetallic sealing materials and manufacturers, which will be explained in more detail later in this chapter. According to NORSOK M-710, a polymer is "a high molecular weight organic compound, natural or synthetic, whose structure can be represented by a repeated small unit. Polymer includes both thermoplastic and thermoset materials, of which elastomers are subcategory." A polymer, as illustrated in Fig. 13.8, is a chemical compound with molecules bonded together in long, repeating chains. The term polymer is widely used in the plastic

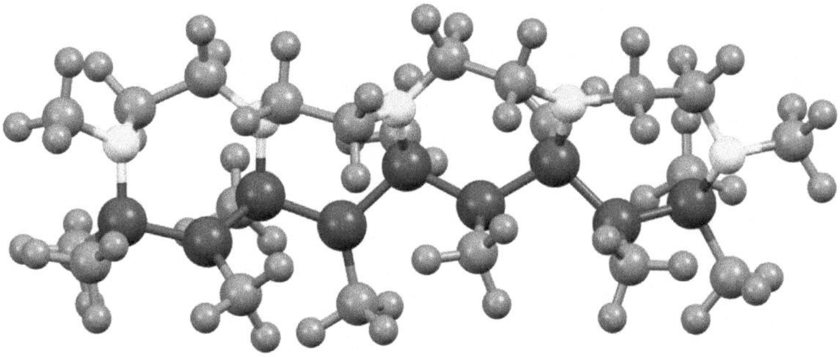

Fig. 13.8 Polymer chain.

industry. One of the main characteristics of a polymer is elasticity. Thus, as it was mentioned before, elastomers are a type of polymers. Rubber and natural rubber are considered as a type of elastomers.

The history of using rubber materials can be traced back many years; rubber has been found at archeological sites of ancient civilizations such as the Aztecs and Mayans. The usage of rubber in European industry dates back to the 18th century. In the 20th century, rubber was used for industrial valves, including those in the oil and gas industry. Elastomers, like plastics, are made of polymers. As an example, C_5H_8 is called isoprene (Fig. 13.9) and natural rubber is built up of regular chains of isoprene, as illustrated in Fig. 13.10.

13.3.2 Properties

Elastomers, also colloquially called rubbers, are elastic substances that can be made naturally, like natural rubber, or produced synthetically, like neoprene. Elastomers are so named because they can change their shapes under applied force and regain their original

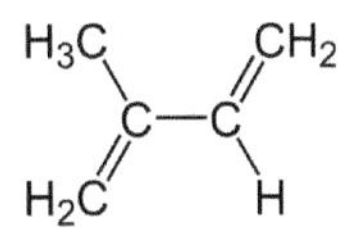

Fig. 13.9 Isoprene.

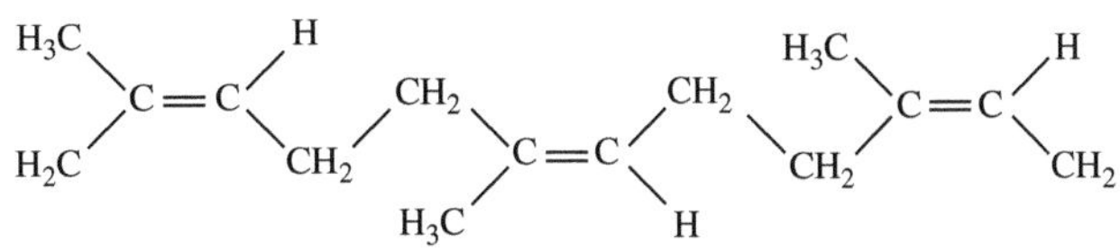

Fig. 13.10 Natural rubber made of chains of isoprene.

shape after removal of the force. The main characteristics of elastomers are high flexibility, high elongation, resilience, and damping. Damping is a property of elastomers that causes them to convert mechanical energy to heat during deflection. Elongation is defined as the percentage of increase in shape or length of a material as a result of applying force. Elongation can be calculated by dividing the length increase by the original length in percentage. Elongation is dependent on different factors, such as hardness and tensile strength; higher elongation means more flexibility against cracking. An elastomer can often stretch 700%, meaning six times its original shape; if it stretches to twice its original shape, then the elongation is 100%. Generally, higher elongation means higher ductility. Other important properties of rubber are listed as follows:

- *Elasticity:* Elasticity is one of the major properties of elastomers, especially O-rings; it means that they can be stretched up to 10 times their original length and still return to their original shape.
- *Contraction by heat:* Unlike most materials, which expand when heated, elastomers do the opposite and shrink when heated. This phenomenon is caused by the behavior of the material's molecules and their reaction toward each other.
- *Specific gravity:* The specific gravity (SG) of a liquid or solid is defined as the ratio of its density to the density of water. The specific gravity of rubber is normally between 0.9 and 0.91 g/m^3. Some sources assign its SG as 0.96 g/m^3. Fig. 13.11 illustrates a scientific specific gravity measurement tool or tester for rubber materials.

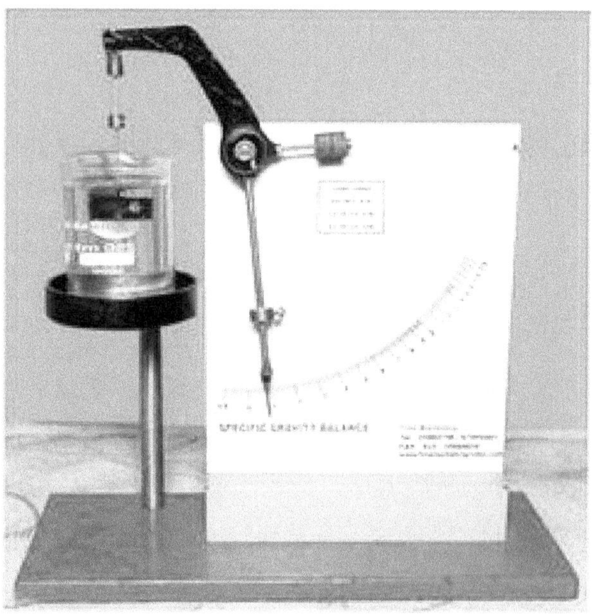

Fig. 13.11 Specific gravity measurement of rubber materials.

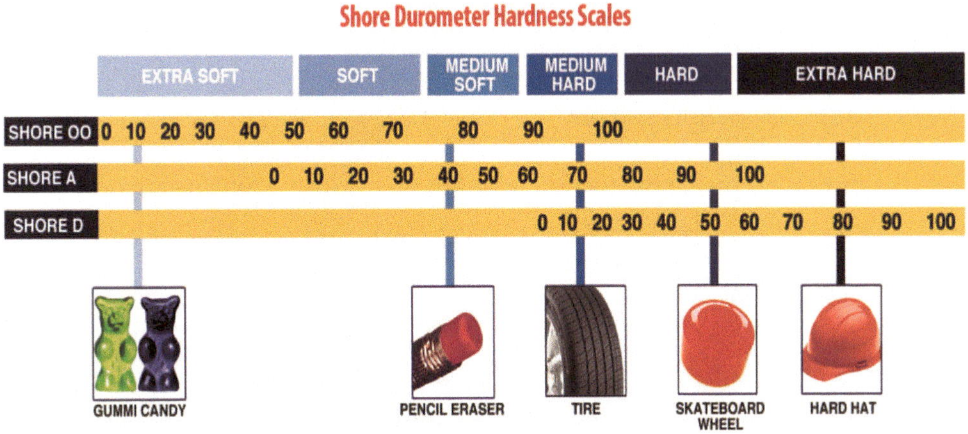

Fig. 13.12 Shore hardness values for different materials.

- *Hardness:* The term hardness is the measurement of a material's resistance to a deformation force applied for a specific duration and its resistance against plastic deformation. Plastic deformation refers to a kind of deformation that changes the shape of the material completely without any chance for the material to return to its original shape. Normally, the hardness of rubber materials is expressed as a value of shore hardness, which is measured by a shore instrument available in different types such as A and D (see Fig. 13.12). The shore hardness of Viton could be almost 70 based on shore A. The structure of elastomers provides them with different hardness values. Shore hardness A, as can be seen in the figure, is more common for soft and medium-soft elastomers, while shore hardness D is more common for medium-hard, hard and extra-hard sealing.
- *Abrasion:* Abrasion refers to the wear resistance of a material. Generally, hydrogenated nitrile rubber provides very good wear resistance. Other types of rubber, such as NBR, FKM, and EPDM, have good wear resistance. Abrasion resistance is connected to hardness such that having higher abrasion resistance leads to higher hardness.
- *Tensile strength:* Tensile strength or ultimate tensile strength refers to the point at which the applied stress or force ruptures the material during the test at a specific rate of elongation, which is expressed as the force per unit area. The tensile strength of a rubber material could be around 10–25 MPa; during tensile test the material is pulled until the point of failure. Tensile strength is the key factor for buyers and makers of elastomers, as it focuses on the point of failure due to stretching.
- *Gas permeability:* Elastomers allow gas to permeate their structures. To reduce permeability and minimize the entrance of gas into elastomers, one or more of following points should be considered: using a larger O-ring section, applying more compression, optimizing the surface finish, and selecting high-density elastomers. The essential

Fig. 13.13 AED HNBR.

consideration for rubber materials made in the shape of an O-ring is to have antiexplosive decompression (AED) in the application of high-pressure gas services. Rapid reduction of the gas service pressure leads to ingress of gas into the elastomer seal, which expands it and finally causes sealing failure. Therefore it is recommended by NORSOK M-710 to apply a rapid gas decompression (RGD) test for O-rings. Fig. 13.13 illustrates an O-ring made from hydrogenated Buna N (HNBR) with an AED characteristic.

- *Operating Temperature:* There is an operating temperature range for each type of sealing, which is typically defined in the project documents and specifications. The operating temperature of a sealing depends on certain factors, such as the duration of cold or hot conditions, the application of the sealing as a static or dynamic seal, etc. Normally, the operating temperature (not the design temperature) is used for soft materials temperature limit including elastomers. One of the weak points of elastomers is that they become harder and less resilient in low-temperature applications. Some elastomer grades perform better at low temperatures, but they are generally soft and, at high temperatures, softness can reduce the strength of an elastomer.
- *Aging:* A heat aging test is performed to record the changes in the different properties of elastomers, such as hardness, elongation, and tensile strength due to heating for a specific time.
- *Color:* Rubber materials normally contain some black carbon fibers, so most are black in color. The addition of carbon fiber increases the strength and heat resistance of the rubber.

- *Compression set:* This is defined as the limit at which the elastomer fails to return to its original shape after releasing the compression load. Rubber materials can lose elasticity under frequent compression forces.
- *Corrosion resistance:* Some chemicals and compounds such as MEG, methanol, carbon dioxide, and hydrogen sulfide can affect the performance of elastomers. If some of these products enter the polymer structure, the mechanical strength of the elastomer can deteriorate. This can lead to tearing and cracking of the elastomer. Different standards address sealing in sour services, as follows:
- NORSOK M-710: This standard covers the requirements and procedures for the qualification of nonmetallic polymer materials for use in sour service.
- ISO 23936: This standard covers nonmetallic materials in contact with media related to oil and gas production, and explains the requirements and procedures for assessing the sealing capability of elastomeric materials to withstand static exposure to a vapor phase sour gas environment (gas hydrocarbon with hydrogen sulfide at elevated temperatures).
- API 6A: This standard covers wellheads and Christmas tree equipment, including valves. It describes the testing of elastomers in sour service in section F.1.13.5.2 for material classes DD-EE-FF-HH. Different test requirements, such as temperature, pressure, exposure period, and test fluids are provided in this section of the standard.
- NACE TM 0187: The name of this standard is "Evaluation of elastomeric materials in sour gas environments." It establishes a method to measure the capability of elastomeric materials to withstand static exposure to a vapor phase sour gas environment.

  Sour service refers to a fluid media with a considerable amount of hydrogen sulfide. The failure of elastomers due to hydrogen sulfide is not limited to cracking; other failures could occur due to increasing hardness and loss of elasticity. Some of the failures are visible and some are not. The severity of elastomer failure due to hydrogen sulfide depends on other parameters, such as time, temperature, hydrogen sulfide concentration, and partial pressure. The partial pressure of hydrogen sulfide depends on the operating or working pressure of the system. Fig. 13.14 illustrates a hydrogen sulfide crack in a Viton FKM O-ring. The failure of O-rings in sour service is not limited to Viton FKM. Other elastomeric O-rings, such as NBR, HNBR, and EPDM, are at risk of sour corrosion and should be tested. Different types of elastomeric O-rings are explained in the next section.

13.3.3 Types of elastomers

Different types of elastomers are used in the valve industry. These are listed as follows:

1. **Buna N/Nitrile rubber/NBR:** This type of synthetic rubber is very common for valve sealing in all sectors of the oil and gas industry. It can be used in an operating temperature range between −20°C and 90°C. Some specifications may put more

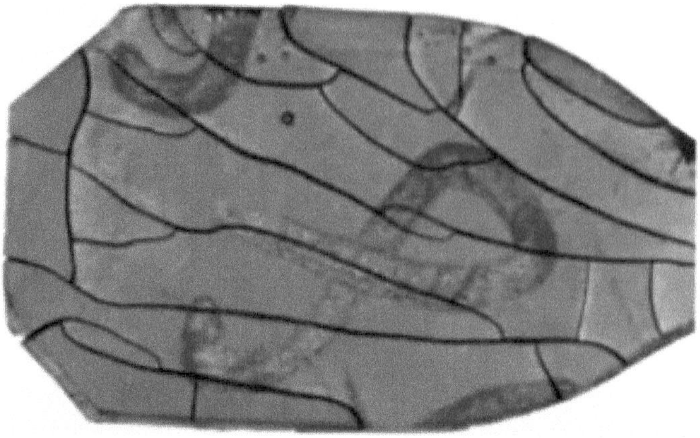

Fig. 13.14 Sour corrosion crack in Viton FKM.

limitations on the operating temperature of this material, such as 0–80°C. The duration of heat and high temperature exposure affects the maximum operating temperature of Buna N in a way that exposure time of this rubber to 90°C or even 100°C for a short time may not damage this material. The application type, i.e., dynamic or static, can also affect the temperature range. Normally, dynamic seals that provide sealing for moving parts are more exposed to wearing, and are thus less able to bear extreme temperatures compared to the conditions in which they are used as static sealing. Buna N has very good resistance to chemicals and hydrocarbon fluids such as oil and hydraulic oil, as well as very good abrasion resistance. However, the resistance of Buna N to ozone, weather, and UV rays is poor. Fig. 13.15 illustrates a nitrile

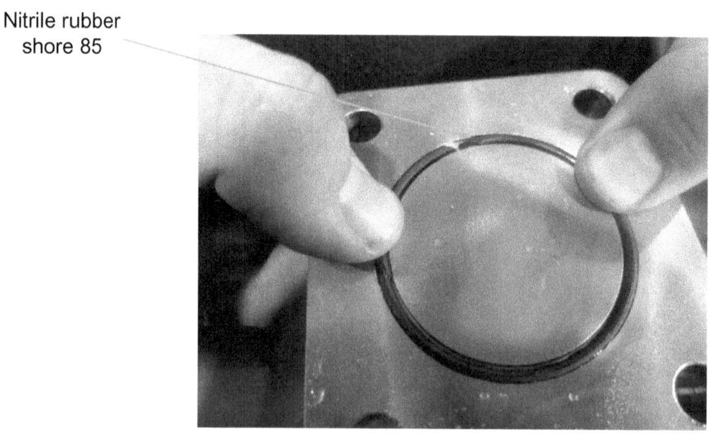

Fig. 13.15 NBR (Nitrile Rubber) used for sealing the bottom cover to the body of a butterfly valve.

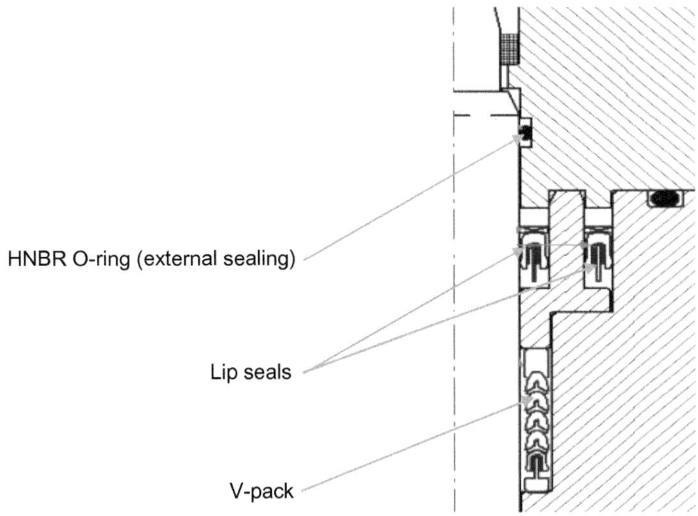

Fig. 13.16 Stem sealing for a slab gate valve.

rubber seal with shore hardness equal to 85 that is used to seal the body and bottom cover of a butterfly valve in the topside sector of the oil and gas industry.

2. **HNBR**, also called hydrogenated Buna N, has higher temperature resistance and ozone resistance compared to Buna N. This type of synthetic rubber is more common for subsea valves in such applications as actuator sealing and valve primary external sealing due to its high resistance to chemical and hydrocarbon fluid and relatively low cost. Hydrogenated Buna N or Nitrile can be used in a temperature range between −40°C and 150°C. Fig. 13.16 illustrates the stem sealing area of a slab gate valve. Two different types of sealing are used for the stem sealing—one internal and the other external. Internal sealings prevent leakage of the fluid from inside the piping and valves to the external environment (seawater). External sealing (HNBR in this case) prevents leakage of seawater to the valve internals. The primary internal sealing is a V-pack and the secondary stem sealing consists of two lip seals. V-packs and lip seals will be explained later in this chapter.

   Fig. 13.17 illustrates the bottom part of a linear hydraulic actuator used for a subsea slab gate valve. There are four sealings between the shaft and the cylinder, all made of HNBR.

3. **EPDM:** This is a synthetic rubber made by the copolymerization of ethylene and propylene. The full name of this material is Ethylene propylene diene monomer. EPDM can be used in a variety of temperatures, ranging from −50°C to 150°C, and has excellent resistance to weather, ozone, and heat. However, it has poor resistance to oil and gas and petroleum products. Thus this type of rubber is not popular for valves in the oil and gas industry.

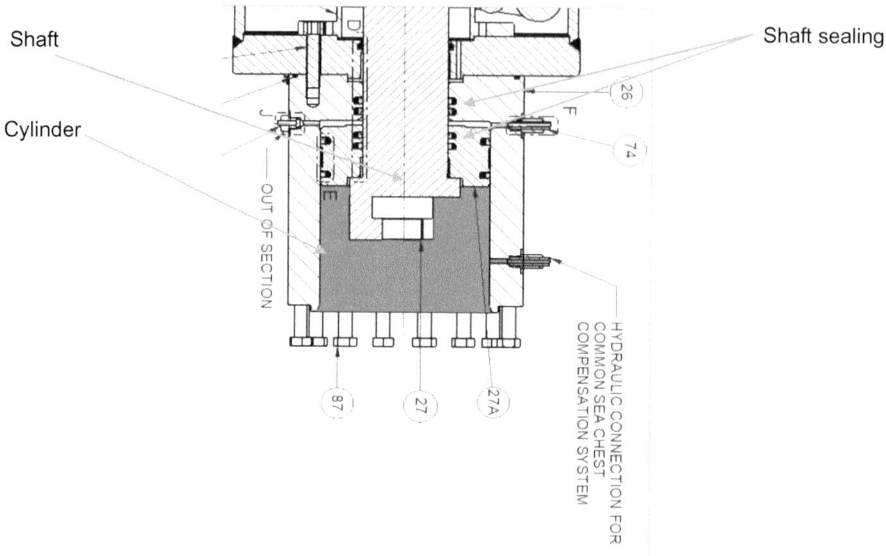

Fig. 13.17 Bottom part of a linear hydraulic actuator with highlighted shaft and shaft sealing.

4. **Neoprene:** This type of rubber is not common for the valves and actuators in the subsea sector of the oil and gas industry. It is a synthetic rubber produced by the polymerization of chloroprene. Neoprene has very good abrasion, water, ozone, UV ray, and heat resistance. However, it has poor resistance to acids and chloride.
5. **Silicon:** This type of rubber is excellent in high-temperature conditions and can be used in a wide range of temperatures from −60°C to 220°C. It provides very good UV, ozone, and weather resistance. It has poor resistance to oil, so it is not common in the oil and gas industry, especially the subsea sector.
6. **Viton:** Viton is the trademark name of fluoroelastomers, also known as FKM. Viton is very common for the sealing of industrial valves (mainly stem sealing) in the oil and gas industry especially subsea, and can be used in temperature ranges of −40°C to 180°C. Viton provides outstanding resistance to weather, UV, and ozone. In general, FKM does not provide as high a corrosion resistance as HNBR to chemicals and hydrocarbon. But Viton provides very good pressure, temperature, and corrosion resistance to hydrocarbon fluid and its cost is relatively low.
7. **Glyd Ring:** Glyd ring (see Fig. 13.18) is the trademark name of a double acting O-ring with an energized piston seal used for dynamic applications such as stem sealing. This sealing is used in some slab gate valves as the primary barrier against seawater ingress to the valve stem areas. This sealing has many advantages, such as low friction, minimum breakout force, and high wear resistance. Glyd rings can also be used for actuator cylinder sealing.

Fig. 13.18 Glyd ring.

13.4 Thermoplastic sealing

Thermoplastic materials are a type of plastic polymer material that becomes moldable at certain high temperatures and then becomes solid after cooling. Compared to elastomers, thermoplastics have a greatly reduced elasticity capability and can easily become permanently deformed. Thermoplastics generally have high chemical and corrosion resistance. Most types of thermoplastics are relatively heavy. Different types of thermoplastics are mentioned as follows:

1. **Teflon (PTFE):** PTFE is the shortened name of the chemical polytetrafluoroethylene, and Teflon is the trade name of PTFE. This material is very flexible and has a nonstick property. Teflon was invented in 1938 and the Teflon trademark was registered in 1945. Teflon was used as a coating for frying pans in the 1960s. PTFE is used as a soft seat material in ball valves in low-pressure classes, normally class 300 up to 50 bar. It should be noted that PTFE can be reinforced with graphite or glass fiber, which leads to higher pressure and temperature resistance. ISO 17292, which is a standard for ball valves in petroleum, petrochemical, and other aligned industries, provides a table for the maximum pressure resistance of PTFE and RPTFE (reinforced PTFE) in floating ball valves (see Table 13.1). Floating ball valves are ball valves without any support plate or other means of support under the ball (closure member) of the valve. Although the table only addresses floating ball valves, it can be used for trunnion ball valves as well. Trunnion ball valves are normally larger ball valves with a support under the ball that facilitates opening and closing the valve with optimum force or torque. Higher temperature decreases the maximum operating pressure at which PTFE and RPTFE seats can be used for ball valves. In general, RPTFE seats have better pressure-temperature resistance compared to virgin PTFE. It should be noted that Table 13.1 is just an example and should not be used as a

Table 13.1 Maximum pressure values (bar) based on temperature for PTFE and RPTFE seated ball valves.

Temperature °C	PTFE seats				Reinforced PTFE seats			
	Floating ball			Trunnion	Floating ball			Trunnion
	DN ≤ 50	50 < DN ≤ 100	DN > 100	DN > 50	DN ≤ 50	50 < DN ≤ 100	DN > 100	DN > 50
	NPS ≤ 2	2 < NPS ≤ 4	NPS > 4	NPS > 2	NPS ≤ 2	2 < NPS ≤ 4	NPS > 4	NPS > 2
–29 to 38	69,0	51,0	19,7	51,0	75,9	51,0	19,7	51,0
50	63,6	47,1	18,2	47,1	70,4	47,8	18,4	47,8
75	53,3	39,2	15,2	39,2	59,9	40,4	15,6	40,4
100	43,0	31,3	12,1	31,3	49,4	33,1	12,8	33,1
125	32,7	23,3	9,1	23,3	38,9	25,8	10,0	25,8
150	22,4	15,4	6,1	15,4	28,3	18,4	7,2	18,4
175	12,1	7,5	3,0	7,5	17,8	11,1	4,4	11,1
200	–	–	–	–	7,3	3,7	1,6	3,7
205	–	–	–	–	5,2	2,3	1,0	2,3

(Courtesy: ISO 17292.)

guideline for projects. Normally, the pressure and temperature range of soft materials should be based on the project specifications or the governing standard. In general, PTFE and RPTFE can be used from −80°C to 200°C. In conclusion, since PTFE and RPTFE are not suitable for high-pressure classes, such as 517 bar and 690 bar, which are the case in subsea, this material is not applicable for subsea valve seats. Instead, PTFE is normally used as an internal lining for bearings. Bearings are installed around the stem to prevent lateral movement of the stem. Normally, bearings are made of a metallic material such as Inconel 625 with an internal lining of PTFE to prevent friction and galling between the stem and bearings. If for some reason PTFE lining is not used for the bearings, tungsten carbide hardfacing should be applied on the stem contact area with the bearing to prevent galling and friction between the stem and bearings. Fig. 13.19 illustrates PTFE seats for ball valves in the petroleum industry.

2. **PEEK:** Polyether ether ketone or PEEK is an organic, thermoplastic material normally used as a soft seat or a soft seal between the backseat of the valve and the valve body. PEEK is compatible with chemicals and process fluid and is suitable for high-pressure classes in API 6A/17D. PEEK can be used in a wide range of operating temperatures from −80°C to 200°C. Fig. 13.20 illustrates the seat of a slab gate valve with a PEEK insert used as a seal between the seat and body of a valve.
3. **Devlon:** Devlon is a thermoplastic material developed by James Walker with some additives that improve resistance against both temperature and pressure compared to

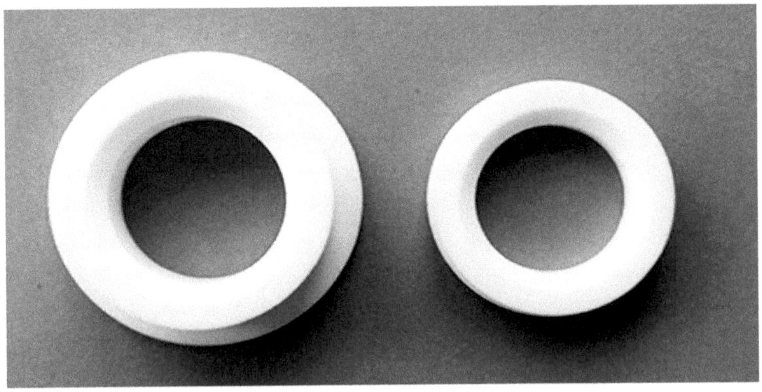

Fig. 13.19 PTFE seats for ball valves.

Fig. 13.20 Seat of a slab gate valve with a PEEK seat insert.

PTFE and nylon. Although PEEK can provide better pressure resistance than Devlon, Devlon is a more cost-effective choice. In addition, Devlon provides less friction compared to PEEK. Devlon can be used for valve seats in a wide temperature range from −50°C to 176°C. The other advantage of Devlon is its low moisture absorption, which improves the seal's dimensional stability. The other important parameter in seat sealing is the ability to provide sealing in low-pressure fluid conditions. Devlon is ideal for this purpose. It should be noted that different grades of Devlon are available on the market, such as T-100, S-Grade, V-API, Devlube, and A153. Fig. 13.21 illustrates Devlon seats for a ball valve. Devlon may be used for ROV buckets for manual subsea valves.

4. **Nylon:** This thermoplastic was discovered in 1935 by DuPont, an American company. Nylon is used in different products such as textiles, car tires, and mechanical

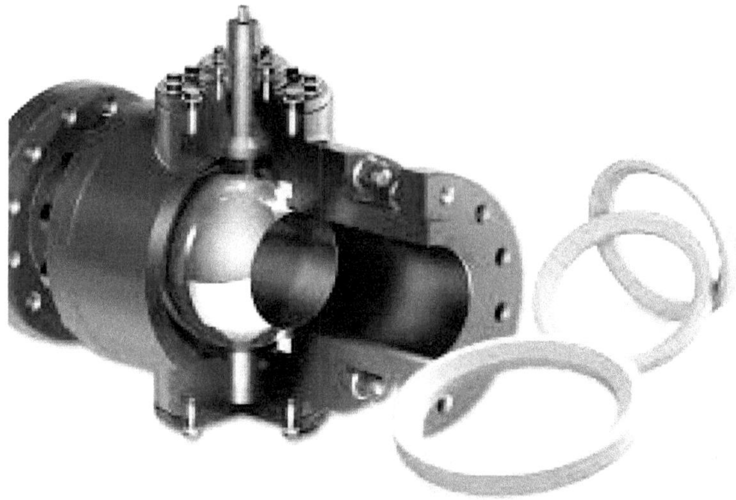

Fig. 13.21 Devlon seats for a ball valve. *(Courtesy: James Walker.)*

equipment. Nylon is strong and has relatively good temperature resistance and chemical compatibility. Nylon valve seats can be used from −20°C to 100°C. This material is a very good choice due to its relatively low cost, but it is not compatible with some chemicals such as MEG and Methanol. MEG and methanol are injected into subsea valves used on production lines; their seat material (in the case of soft seats) should be compatible with the injection fluid. Thus nylon is not generally recommended. It should be noted that soft seat ball valves are not typically used in subsea.

5. **Turcite:** This thermoplastic has excellent operating temperature resistance; it can be used for operating temperatures from −200°C to 250°C. One of the applications of this material for subsea valves is for sealing the stem as a wear ring installed below other types of stem sealing, such as V-pack and lip seal.
6. **Vespel:** This is a very robust stem sealing that, like Turcite, can be used in extreme temperature ranges and can withstand high friction and heavy loads.
7. **Kelf:** This elastomer has very good low-temperature resistance and can be used for temperatures as low as −150°C. Kelf is also called PCTFE (PolyChloroFluoroEthylene) thermoplastic, and can tolerate low-temperature extremes with very good corrosion resistance against chemicals and very strong mechanical proprieties.

13.5 Thermoplastic spring-energized sealing

Thermoplastic spring-energized seals which are lip seals and spring seals are explained in this section:

1. *Lip seal:* Lip seals are made of PTFE, which is energized by a metallic material such as Elgiloy (see Figs. 13.22 and 13.23). More precisely, Elgiloy (Co, Cr, Ni) is a cobalt alloy containing 40% cobalt, 20% chromium, 15% nickel, and 7% molybdenum. A lip

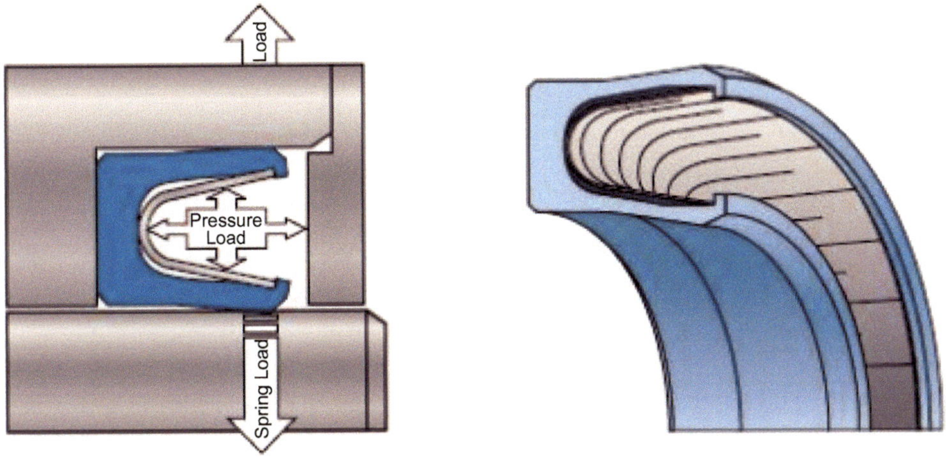

Fig. 13.22 Effect of a spring on lip seal sealing performance.

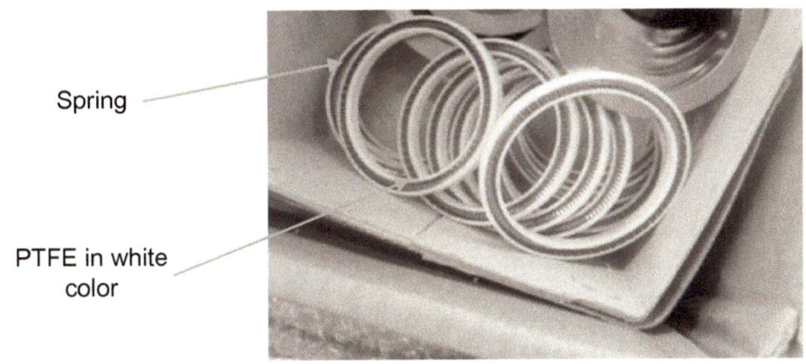

Fig. 13.23 Lip seal made of PTFE and an energized spring in Elgiloy.

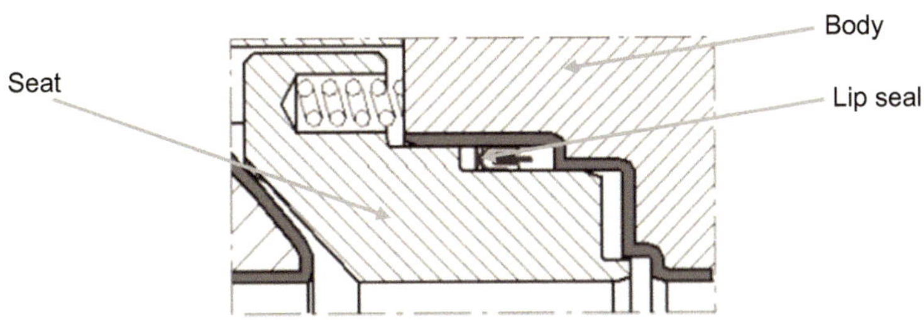

Fig. 13.24 One lip seal between the seat and body of a valve (SPE seat). *(Courtesy: Galperti Engineering.)*

seal can be used for sealing between the seat and body of a valve (see Figs. 13.24 and 13.25) or for stem sealing (see Fig. 13.16). The important point is that the lip seal used between the seat and body in Fig. 13.25 prevents leakage from the flow on the left side of the valve to the seat and body cavity of the valve. However, if pressure accumulates in the valve cavity, it can be released to the left side piping. This type of

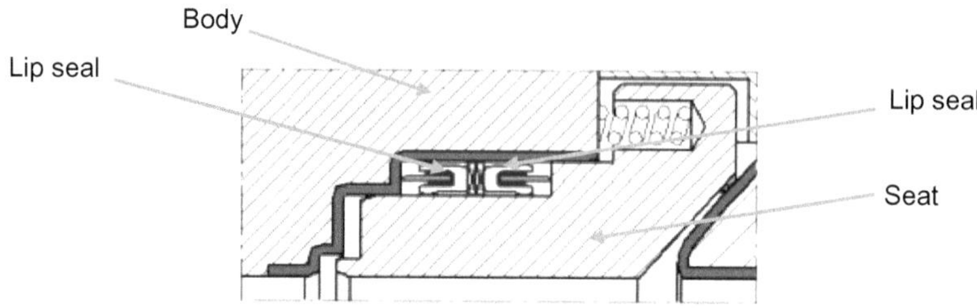

Fig. 13.25 Two lip seals between the seat and body of a valve (DPE seat). *(Courtesy: Galperti Engineering.)*

sealing for the seat of a valve is considered single piston effect (SPE). Alternatively, two lip seals could be placed in the area between the valve body and the seat to prevent any leakage from both sides—from the seat area and cavity to the line on the left side and as from the line on the left side to the body cavity. The type of seat design illustrated in Fig. 13.25 is known as double piston effect (DPE).

NORSOK M-710 accepts metallic springs for lip seals in Inconel 625, alloy C276, Elgiloy (UNS R30003), and super alloy 35N (UNS R30035). Inconel and Hastelloy are nickel alloys that are explained in more detail in Chapter 12. The unified numbering system (UNS) is a numbering system that is widely accepted in North America. Each UNS number represents a unique material with a specific chemical composition. Super alloy 35N is a cobalt alloy that has undergone an age-hardening heat treatment and possesses super high mechanical strength, toughness, ductility, and outstanding corrosion resistance, especially in seawater. This material could consist of approximately 30% cobalt, 35% nickel, 20% chromium, and 10% molybdenum. The assistance of a spring helps to provide a load for sealing the lip seal, especially when media pressure is too low to fully actuate the lips of the sealing (see Fig. 13.22).

Lip seals have two main advantages: they can be used in a wide range of temperatures and pressures, and can be used for high pressure gas. In fact, lip seals have an antiexplosive decompression (AED) property. The important point is that a lip seal can provide sealing in only one direction, so it is not a bidirectional sealing like an O-ring. The fluid can be sealed with lip seals from the direction where the metallic springs are located or the "lip side." The temperature range of the lip seal depends on the temperature range of both its metallic and soft materials. A lip seal with a PTFE and Elgiloy spring could be used for a temperature range of approximately −196°C to 250°C, so it can be also used for cryogenic or very cold temperatures.

2. *Spring seals:* Spring seals are energized with a U-cup shape (see Fig. 13.26) and are not common in the valve industry. Spring seals have advantages over other types of sealing, such as better resistance against higher temperatures and chemicals, and less friction.

Fig. 13.26 Spring seals.

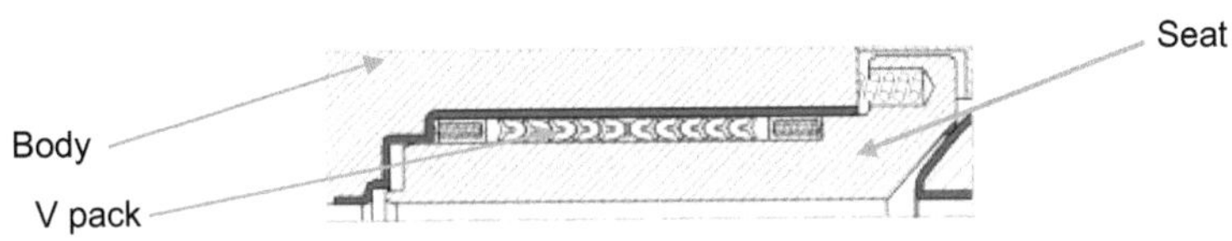

Fig. 13.27 Single V-pack between a valve seat and body (SPE design). *(Courtesy: Galperti Engineering.)*

13.6 Thermoplastic combined sealing

V-pack/Vee pack: This type of sealing is a very high technology, advanced sealing that can be used for stem sealing as well as sealing between the valve closure member and body (see Fig. 13.27), like lip seals. The design illustrated in Fig. 13.27 is a single piston effect design. It is possible to have two V-packs for a double piston effect seat design, as illustrated in Fig. 13.28.

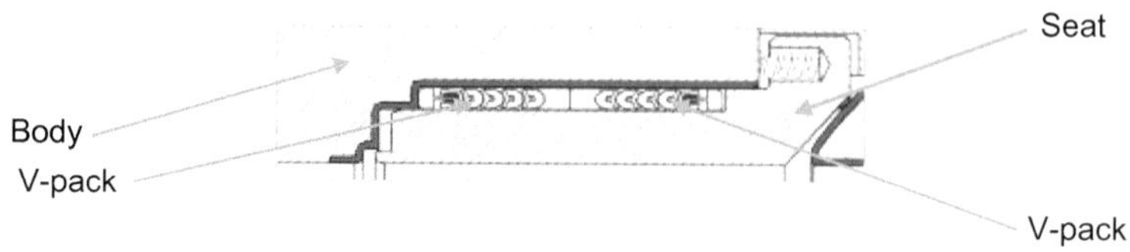

Fig. 13.28 Two V-packs between a valve seat and body (DPE design). *(Courtesy: Galperti Engineering.)*

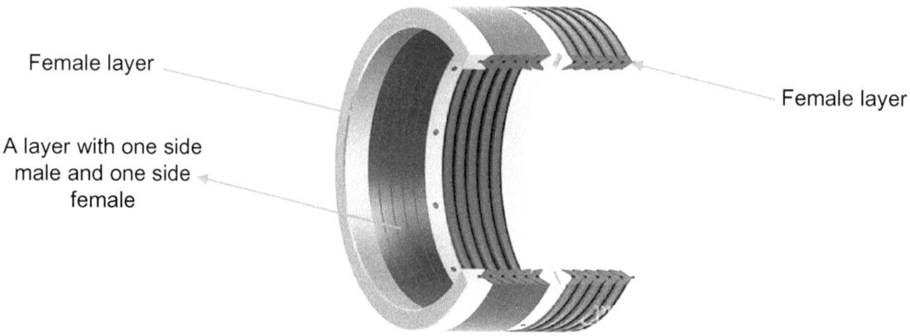

Fig. 13.29 V-pack with male and female layers made of PTFEE and PEEK.

The vee pack illustrated in Fig. 13.29, also called Chevron V-pack sealing, includes male and female adaptors or layers, which could be a group of three or five layers or even more. This type of sealing can withstand very high pressure and can be used for heavy-duty applications like the valve stem sealing or body and closure member sealing of subsea valves. V-packs are also suitable sealing for rods and pistons. Fig. 13.29 shows a V-pack made of PEEK and PTFE thermoplastic seals. V-packs can be made from layers of elastomers such as Viton FKM, Neoprene, EPDM, etc. The bottom and top layers have one flat side surface on both ends, rather than the V shape, and other V-shaped layers are located in between. The combinations of construction materials define the resistance of the V-pack against pressure, temperature, and fluid.

13.7 Graphite sealing

Graphite sealing or packing is widely used for valves in the oil and gas industry due to its high temperature and pressure resistance. Graphite packing can be used for extreme temperatures, which could be as low as −250°C and as high as 500°C. Graphite can be used in two forms: dye-formed and braided. Graphite is fire resistant and is a very Noble material with very good corrosion resistance against oil, gas, and chemicals. Graphite is very common for the compression packing of valves to be used in the onshore and offshore sectors of the oil and gas industry—excluding subsea valves. Compression packing refers to the soft materials placed in an area around the valve stem, known as the "stuffing box," and compressed by a valve part called the "gland or gland follower" to create stem

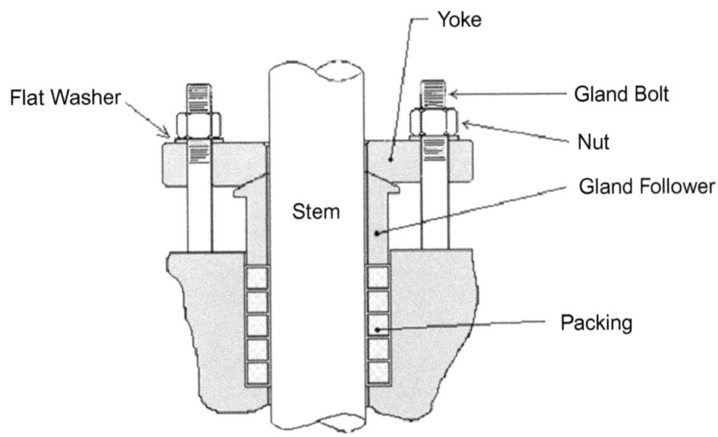

Fig. 13.30 Graphite packing.

sealing, as illustrated in Fig. 13.30. Graphite packing is not common for valves in the subsea sector of the oil and gas industry, although it is common for other valves used topside, and in refineries and petrochemical plants. The high purity of graphite (more than 98% or 99% pure) is a key consideration for graphite packing design and selection. Graphite may be required to have less than 50 ppm chloride and less than 700 ppm sulfur. Probably the reason for putting limitation on the chloride content is that it can form acid and lead to corrosion of the packing. Sulfide should be limited since it can oxidize and make SO_2, which disturbs the tightness of the packing and causes corrosion.

13.8 Metallic sealing

Metallic gaskets are common for sealing the body and bonnet or body pieces of subsea valves. These gaskets are normally ring type joint (RTJ) and BX types, as per the API 6A/API 17D standard, and provide bidirectional sealing to prevent leakage from the valve internals to the sea, as well as leakage from the sea to the valve internals. Fig. 13.31 illustrates different kinds of sealing on a ball valve, including a metallic BX gasket between body pieces. Inconel 625 is a very common material for BX gaskets, but other materials such as Hastelloy C-276, duplex, and other types of stainless steel are available on the market.

The other type of metallic sealings could be bellows sealing, which is the primary internal sealing around the valve stem as illustrated in Fig. 13.6. A completely different approach to packing is a device called a bellows seal—an accordion-like metal tube fastened to the valve stem and to the bonnet, forming a leak-proof seal with negligible friction. The metallic gasket is made of Inconel 625 in many cases. Inconel 625 provides higher corrosion and fatigue resistance compared to stainless steel bellows. Hastelloy

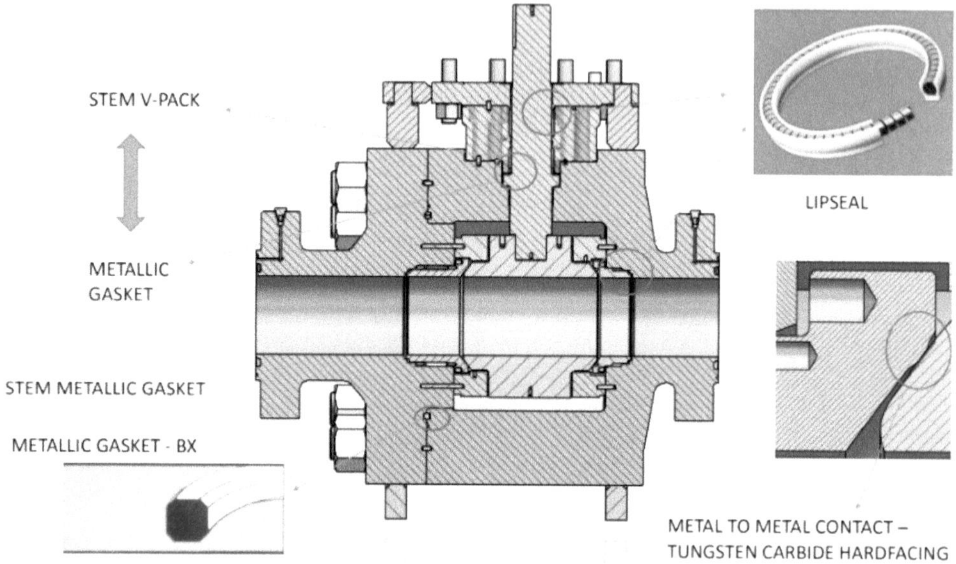

Fig. 13.31 Ball valve sealing including body pieces and metallic gaskets.

C-276 could be a better choice than Inconel 625, due to its higher fatigue and corrosion resistance, especially in seawater service. Another way to increase the fatigue resistance of bellows is to use two of them instead of one. The mechanical strength and fatigue resistance of bellows are important, because they are the primary stem sealing that could be exposed to high fluid pressure inside the valves. In general, different parameters can lead to bellows fatigue failure, such as the material of construction, pressure and temperature, fabrication method, etc.

Figs. 13.32 and 13.33 illustrate a combination of a metallic BX gasket with an O-ring in elastomeric material such as HNBR or Viton FKM for sealing between the body and bonnet of a slab gate valve. The Inconel 625 weld overlay on the sealing areas, indicated in red (dark gray in print versions) in the figure, is mainly to prevent crevice corrosion which can occur when fluid is trapped under the sealings. The O-ring provides sealing

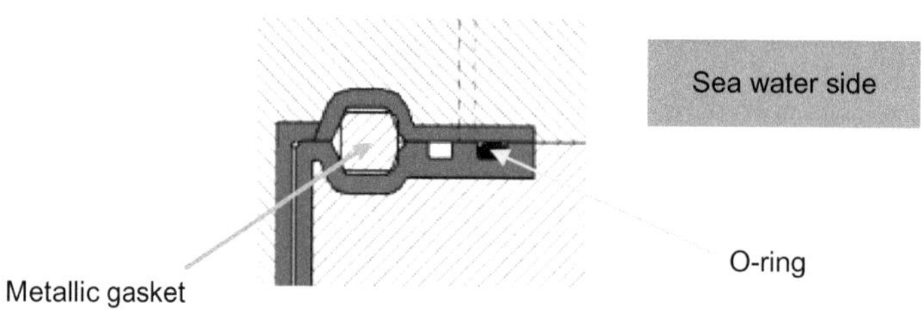

Fig. 13.32 Combination of a metallic gasket and O-ring for sealing the body and bonnet of a gate valve. *(Courtesy: Galperti Engineering.)*

Fig. 13.33 Combination of a metallic gasket and O-ring for sealing the body and bonnet of a gate valve. *(Courtesy: Galperti Engineering.)*

against the external fluid, which is seawater. The metallic ring is the primary sealing against valve internal pressure but it provides bidirectional sealing so, in case of the failure of the O-ring, the metallic gasket acts as a second barrier against external pressure.

Fig. 13.34 illustrates a complete stem sealing, as well as sealing for the body pieces of a subsea ball valve. Item #1 is the metallic bellows stem sealing, which is used for the large bore ball valve installed in the deep-water area. Bellows are typically omitted for small ball valves in sizes such as ½″ and ¾″ and for ball valves used in shallow water depths. Bellows can be in Inconel 625 material. Items #2 and #3 are metallic gaskets plus O-ring located between the valve body pieces, as in the arrangement explained before regarding sealing

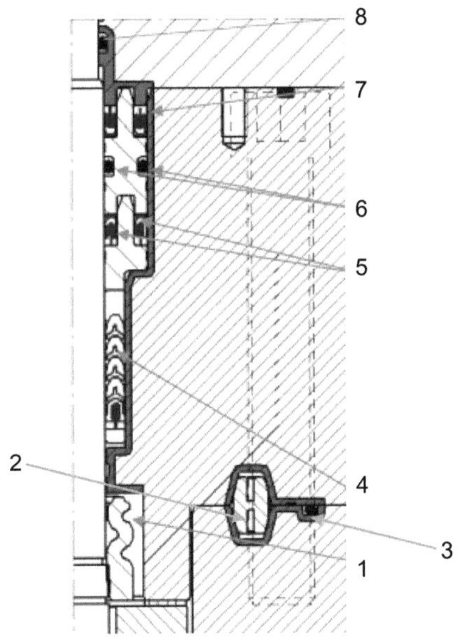

Fig. 13.34 A ball valve stem sealing arrangement. *(Courtesy: Galperti Engineering.)*

Table 13.2 Sealing summarizing for ball, gate, and check valves.

	Ball valve	Through conduit gate valve	Check valve
Body and bonnet/ Body pieces	Primary: one metallic bidirectional gasket Secondary: one soft O-ring such as HNBR	Primary: one metallic bidirectional gasket Secondary: one soft O-ring such as HNBR	Not applicable
Seat and closure	Mainly metallic but soft thermoplastic is also possible	Mainly metallic	Metallic
Seat and body	Nonmetallic materials such as lip seal/vee pack	Nonmetallic materials such as a lip seal/vee pack	Not applicable
Stem	Large size valves like 8″ and above: Primary internal: metal such as bellows; secondary internal: V-pack made of soft rings, third internal: lip seal made of metallic spring and thermoplastic material External primary and secondary sealing: O-ring and lip seal Small size valves like less than 2″: Primary internal: V-pack made of soft rings Secondary internal: lip seal made of metallic spring and thermoplastic material External primary and secondary sealing: O-ring or lip seal	Primary internal: V-pack Secondary internal: lip seal Note: The backseat of the valve is also used as the secondary stem sealing mechanism. External sealing: lip seal and O-ring	Not applicable

between the body and bonnet of a gate valve. Item #4 is a vee-pack, which acts as the secondary internal sealing; in case of failure and leakage from the bellows, the vee-pack acts as the second barrier against the valve internal fluid pressure and flow. Item #5 is a lip seal that acts as the tertiary fluid barrier. Item #7 is another lip seal that acts as the primary barrier against the external seawater fluid. Item #6 is an O-ring, which could be in Viton FKM or HNBR; it acts as the secondary barrier against external seawater pressure so, in case of leakage or failure of the lip seal (Item #7), the O-ring prevents ingress of seawater to the stem area. Item #8 is an O-ring, which can be HNBR or Viton FKM as examples, used to prevent leakage from the valve stem area to the valve operator.

Table 13.2 provides a standard or summary model of the different types of sealing used for subsea valves covered by the API 6DSS standard, "Pipeline subsea valves."

In general, the following soft sealings are common for subsea valves and actuators: HNBR or Viton O-ring, lip seal, V-pack, Glyd ring, PEEK.

13.9 NORSOK M-710 review of soft materials

NORSOK M-710 covers the requirement procedures for the qualification of non-metallic or soft sealing and the manufacturers of soft materials. Different standards, such as the American Society for Tests and Materials (ASTM), British Standard (BS), International Organization for Standardization (ISO), and Deutsches Institut für Normung (DIN) as per Table 13.3, extracted from NORSOK M-710, second edition in 2001, are listed for polymeric sealing testing, including both thermoplastic and elastomer materials.

The selection of soft sealing materials is made through an evaluation of the soft material properties and the functionality of the soft material for a specific requirement. The following parameters are of important for soft materials and should be taken into account during the soft sealing selection process:

- Compatibility with fluid and environment: it is the responsibility of the purchaser to define the fluid condition and type in the relevant engineering documents; additionally, the resistance of the sealing to aggressive fluid such as hydrogen sulfide is important;
- Adequate physical and mechanical properties, such as hardness, tensile strength, elongation at break, modulus of elasticity, compression set, tear resistance, etc.;
- Absorption;
- Thermal resistance and flexibility at low temperatures;
- Aging resistance;
- Explosive decompression resistance;
- Extrusion resistance: Sealings are placed in metal grooves; extrusion occurs when the pressure in the system applies force to the sealing and pushes it off the grooves. Proper groove dimensions and seal selection are important to prevent this type of failure;
- Creep resistance: Creep resistance is important for materials that should withstand loads in high-temperature conditions. Creep failure can cause deformation due to the application of stress at high temperatures;
- Field-proven solutions;
- Resistance to thermal cycling and dynamic movement;
- Suitability of long-term sealing without any replacement: Unlike valves in other sectors of the oil and gas industry, no sealant injection is designed on the seat and stem sealing of subsea valves to repair the damaged soft material as an emergency action before the next maintenance. Fig. 13.35 illustrates sealant injection ports on the seats of a ball valve used in the nonsubsea sector;
- Low friction, especially in the case of stem sealing. High friction of the sealing with the stem increases the force or torque required for the operation of the valve. Using

Table 13.3 List of applicable testing standards for thermoplastics and elastomers.

Standards		Thermoplastics	Elastometers
ASTM D 395	Standard Test Method for Rubber Property—Compression Set (method B).		X
ASTM D 638	Test Method for Tensile Properties of Plastics.	X	
ASMT D 695	Test Method for Compressive Properties of Rigid Plastics.	X	
ASTM D 746	Test Method for Brittleness Temperature of Plastics and Elastomers by Impact.	X	X
ASTM D 790	Test method for Flexural Properties of Unreinforced and Reinforced Plastics and Electrical Insulating Materials.	X	
ASTM D 792	Test Methods for Specific Gravity and Density of Plastics by Displacement.	X	X
ASTM D 1414	Methods of Testing Rubber O-rings.		X
ASTM D 1415	Standard Test Method for Rubber property—International hardness (IRHD).		X
ASTM D 1525	Test Method for Vicat Softening Point for Plastics.	X	
ASTM D 2240	Test Method for Rubber Property—Durometer Hardness (Shore A/D).	X	X
ASTM D 2990	Test methods for Tensile. Compressive and Flexural Creep and Creep Rupture Test of Plastics.	X	
BS 1806	Standard Inch Sizes of O-rings.		X
BS 6442	Specification for limits of surface imperfections on elastomeric toroidal sealing rings (O-rings).		X
DIN 53453	Testing of Plastics. Impact Flexural Test.	X	
ISO 868	Determination of indentation hardness by means of a Durometer (Shore A/D hardness).	X	X
ISO 1432	Rubber vulcanized—low temperature stiffening (Gehman test).		X
ISO 1817	Vulcanized rubbers—Resistance to liquids—methods of tests.		X
ISO D34	Tear resistance, method A.		X
ISO R 812	Method of test for temperature limit of brittleness for vulcanized rubbers.		X
Description of Arrhenius method: ASTM D 3032; Method of testing hook-up wire insulation.		X	X

(Courtesy: NORSOK M-710, 2nd revision, 2001.)

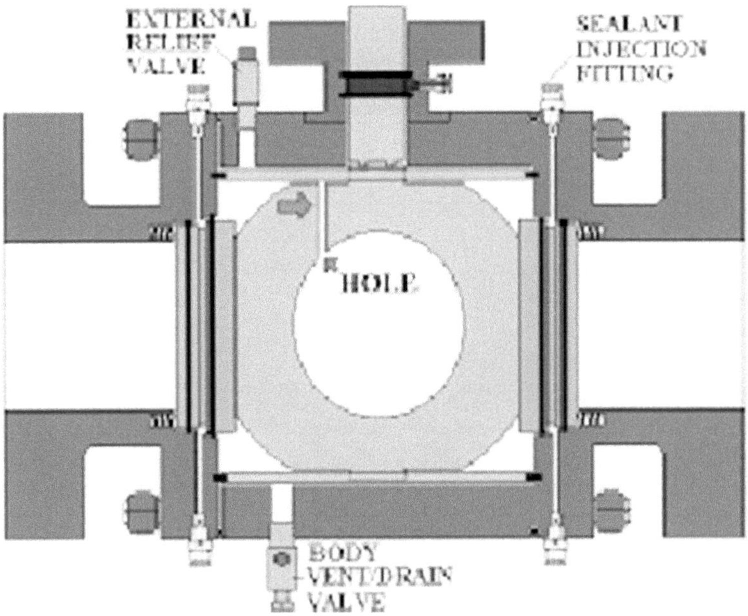

Fig. 13.35 Sealant injection ports on the seats of a ball valve.

multiple, heavy sealings can reduce the risk of leakage and emission; however, it has the drawback of necessitating high force/torque to operate the valve.

In general, soft materials, such as thermoplastics and elastomers, provide class A, zero leakage performance, but their performance is limited by certain parameters that affect the valve sealing selection, such as media type, pressure, temperature, AED and RGD requirements, cycling, roughness, etc., which are summarized in Fig. 13.36. More challenging operation conditions lead to the selection of more robust and complicated sealings for industrial valves. Generally, lip seals require less roughness for sealing compared to O-rings, and subsea valve sealing should be resistant to cycling and frequent operation of the valves.

The required tests and documentations for thermoplastic and elastomer materials based on the specified tests in NORSOK M-710 are given in Tables 13.4 and 13.5, respectively.

The proposed tests for thermoplastic sealing according to Table 13.4 provided in NORSOK M-710 are explained in more detail as follows:

1. *Softening temperature measurement:* This test is done based on ASTM D1525 to measure the temperature at which a flat needle, called a Vicat needle, penetrates the specimen of thermoplastic to the depth of 1 mm under a specific load of either 10 N or 15 N, for test type A or B, respectively. This temperature is considered the elevated softening temperature. The test specimen is lowered into oil for the test and the oil

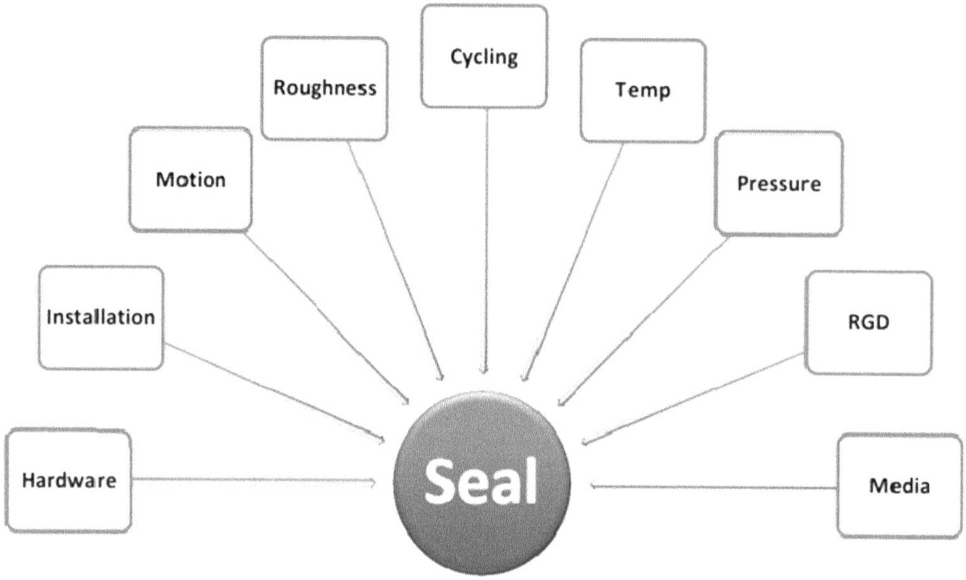

Fig. 13.36 Seal selection operational parameters. *(Courtesy: Valve World.)*

temperature is elevated to between 50°C and 120°C. Fig. 13.37 illustrates a softening temperature measurement test.

2. *Specific gravity test:* This test is performed according to the requirements of ASTM D792. Relative density, or specific gravity, is the ratio of the density of a selected material to the density of water at an ambient temperature of 23°C. Density is defined

Table 13.4 Required tests and documentations for thermoplastic material properties.

Properties	Documentation	QC test
• Softening point (ASTM D 1525).	D	
• Specific gravity (ASTM D 792).	D	B
• Hardness (ISO 868/ASTM D 2240. Shore D).	D	B
• Tensile properties and elongation (ASTM D 638).	D	B
• Compression properties (ASTM D 695).	D	
• Impact strength (DIN 53453).	D	
• Measurement of glass transition temperature by DSC.	D	
• Resistance to creep under permanent tensile and compressive loads (ASTM D 2990).	D	
• Aging characteristics (Annex C).	D	

D: Properties to be documented for each supplier and for each type of material. Nominal values with tolerances shall be given.
B: Properties to be documented on a batch-wise basis, minimum of three samples per test, per batch; the acceptance criteria shall be established prior to the test and based on the qualification test results.

Table 13.5 Required tests and documentations for elastomer material properties.

Properties	Documentation	QC tests
• Specific gravity (ASTM D 792)	D	B
• Hardness (Shore A/IRHD) (ISO 868, ASTM D 2240/ASTM 1415).	D	B
• Tensile and elongation properties (ASTM D 1414).	D	B
• Compression set (ASTM D 395/ASTM D1414) at 100°C for 7 days in Nitrogen atmosphere.	D	
• Low temperature characteristics by DSC or DTMA.	D	
• Tear resistance (ISO D 34).	D	
• Gehman plot (ISO 1432) +20°C to −20°C.	D	
• Aging/RGD characteristics (Annex A/B).	D	

D: Properties to be documented for each supplier and for each type of material. Nominal values with tolerances shall be given.
B: Properties to be documented on a batch-wise basis, minimum of three samples per test, per batch; the acceptance criteria shall be established prior to the test and based on the qualification test results.

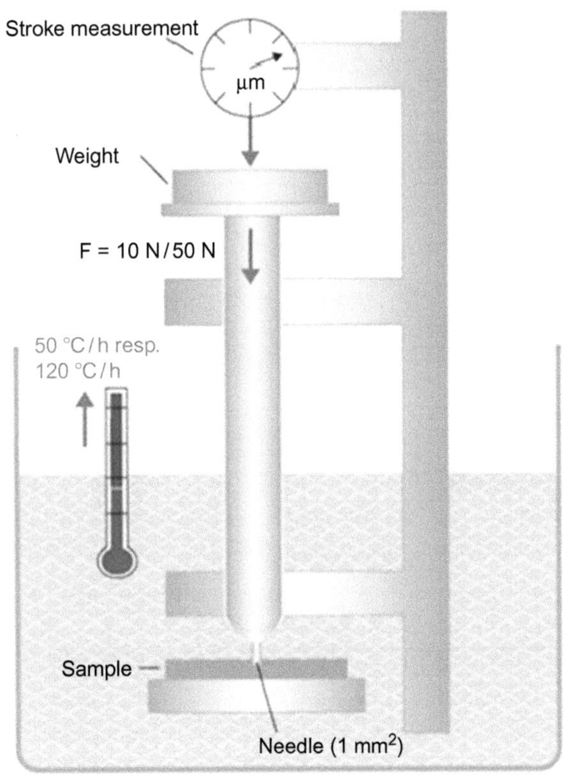

Fig. 13.37 Softening temperature measurement test.

Fig. 13.38 Specific gravity test measurement.

as the ratio of the mass to unit volume of the material. Two types of specific gravity tests (A and B) are performed according to the relevant ASTM standard. Method A, in which the specimen is weighed in air and then weighed while submerged in water, is more common (see Fig. 13.38).

3. *Hardness test:* A hardness test is conducted according to ASTM D2240/ISO 868 shore D. A durometer is typically used to measure the hardness of rubber and other soft materials. The penetration of a rubber material under a specific hardness indenter is performed during the hardness test. The specimen is placed on a flat surface and the indenter is pressed into the specimen (see Fig. 13.39).
4. *Tensile properties and elongation:* These two properties of rubber materials are measured according to ASTM D 638. The test involves applying a tensile force to a sample of rubber material and measuring different mechanical properties, such as tensile strength break, which is defined as the amount of force applied to a plastic before it breaks. The other property measured during this test is tensile modulus, defined as the degree to

Fig. 13.39 Hardness testing of elastomers.

which a material can be deformed before it breaks. Elongation and Poisson's ratio are two other key values. Poisson's ratio is the relationship between how far a material is stretched and how thin it gets during the stretching process.

5. *Compression properties:* Compression properties describe the behavior of a material when it receives a compressive load. Compressive modulus is another property of rubber material measured by this test, according to the ASTM D695 standard. Compressive strength is defined as the capacity of a material to withstand loads that are intended to compress the material and reduce its size. Compression properties are exactly the opposite of elongation values.
6. *Impact strength:* DIN 53453 provides methods for performing the Charpy impact test. This test is used to measure the amount of energy required to break a material. The Charpy impact test, also called the V-notch test, is illustrated in Fig. 13.40.
7. *Measurement of glass transition temperature by differential scanning calorimetry (DSC):* Glass transition is one of the most important properties of materials such as polymers; it refers to the temperature at which the properties of the material change from hard

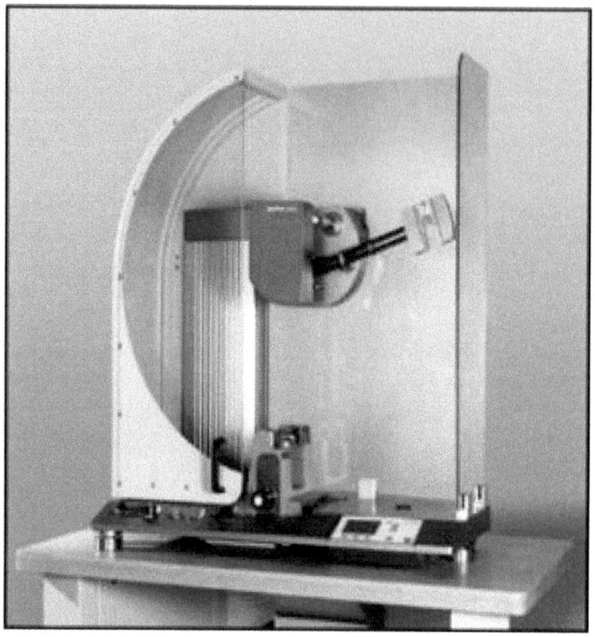

Fig. 13.40 Charpy impact test.

to brittle to soft. The glass transition temperature depends on the stiffness of the polymeric material, as illustrated in Fig. 13.41. The glass temperature is typically lower than the melting temperature.

8. *Resistance to creep under permeant tensile and compressive loads:* This test is performed according to ASTM D 2990 to measure the creep characteristic of the rubber material under loads and high-temperature conditions. Creep, in material science, is defined as

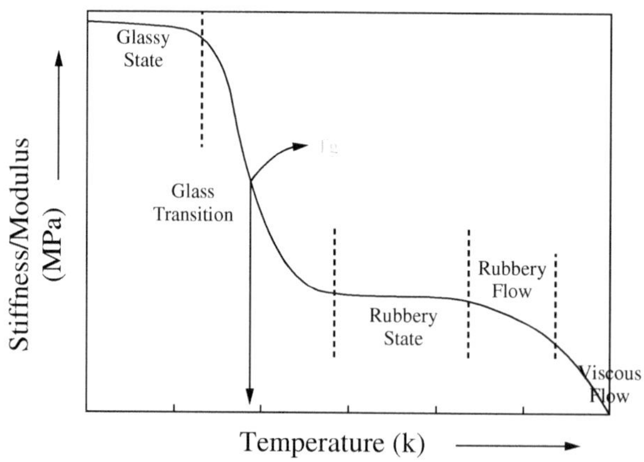

Fig. 13.41 Glass transition temperature.

the tendency of a material to move slowly and deform permanently, typically due to being placed under constant mechanical stresses at high temperatures.

9. *Aging characteristic:* NORSOK M-710 defines a procedure for predicting the progressive degradation of thermoplastic and elastomer sealing exposed to high pressure and temperature over a long period of time.

The proposed tests for elastomeric sealing, according to Table 13.5 provided in NORSOK M-710, are explained in more detail as follows. It should be noted that the specific gravity test, hardness test, and tensile and elongation properties test, which are explained for elastomeric sealing, are also applicable for elastomers:

1. *Compression set:* Compression set is defined as the permeant deformation of a material after the removal of the applied force. A compression set test is performed according to ASTM D1414 at 100°C for 7 days in a nitrogen atmosphere. ASTM D1414 is used to measure some of the mechanical properties of rubber O-rings, such as tensile strength, density, and compression set.
2. *Low temperature characteristic by DSC:* DSC is the most common thermal analysis technique. Endothermic and exothermic flow refers to the heat flow input to the sample and the heat flow output from the sample. DSC measures the temperature transition between the input flow to the sample and the output flow from the sample. The heat input produced or absorbed by the elastomeric material can be measured by DSC. Low-temperature DSC in conjunction with electrical resistance is implemented to assess the physical and chemical behavior and transition of an elastomer during freezing and warming. An elastomer sample 10 mg in mass on average is placed in an aluminum pan with a lid for sealing purposes. Nitrogen purging is used to purge the sample chamber constantly. During cooling and warming, the absorbed or produced heat is measured from the sample and the different properties of the elastomer, such as melting point and crystallization, are measured. A low-temperature test for elastomers is important since these materials can be harder and less resilient.
3. *Tear resistance*: This test is implemented according to ISO 34. Tear strength or tear resistance is defined as the maximum force required to tear a test specimen in two different directions of stress and perpendicular to the stress direction. Tear strength represents the level of a rubber material's resistance to cracking. As illustrated in Fig. 13.42, the rubber sample is placed in a special machine that pulls it apart until crack and rupture take place. The maximum force required to initiate or expand a tear is recorded as force per thickness.
4. *Gehman plot:* This test is performed to determine the relative stiffness of elastomers according to ISO 1432 requirements. ASTM D1053 is another standard that is used to measure a material's relative stiffness. In fact, rubber materials lose their resiliency at low temperatures and can lose their main characteristic behaviors when frozen. Thus a Gehman plot test is performed to measure the material's torsional stiffness at low temperature.

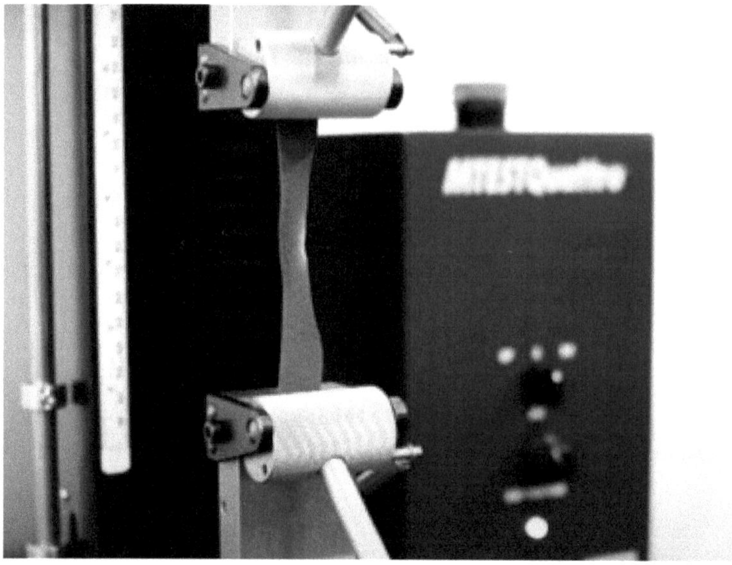

Fig. 13.42 Tear test.

13.10 Self-assessment questions

1. Which sentence is not correct about subsea valve sealing?
   - **A.** Spiral wound gaskets are not used for subsea valves.
   - **B.** A V-pack can be used for subsea valve stem sealing and is comprised of thermoplastic materials.
   - **C.** Bellows are not recommended for linear gate valves due to their high friction with the valve stem.
   - **D.** A lip seal is only applicable for subsea valve stem sealing.

Answer: Spiral wound gaskets are made of metallic rings and graphite fillers, according to ASME B16.20 (see Fig. 13.43). This type of gasket is suitable for low-pressure classes such as class 150 equal to a nominal pressure of 20 bar and class 300 equal to a nominal pressure of 50 bar. Subsea valves based on API 6A /17D have much higher pressure classes, so spiral wound gaskets are not applicable. Therefore option A is correct. Option B is also correct, since V-packs are used for the stem sealings of subsea valves, which could contain male and female layers of PEEK and PTFE as elastomers. It is essential to remember that a V-pack can be used for sealing the seat and body of gate and ball valves. Therefore the usage of V-packs is not limited to stem sealing in subsea valves. Option C is correct, since a bellows seal can create high friction with the stem, especially in a gate valve in which the valve stem moves in a linear line and creates a high amount of friction with a metallic sealing such as a bellows. Option D is wrong, since lip seal usage is not limited to stem

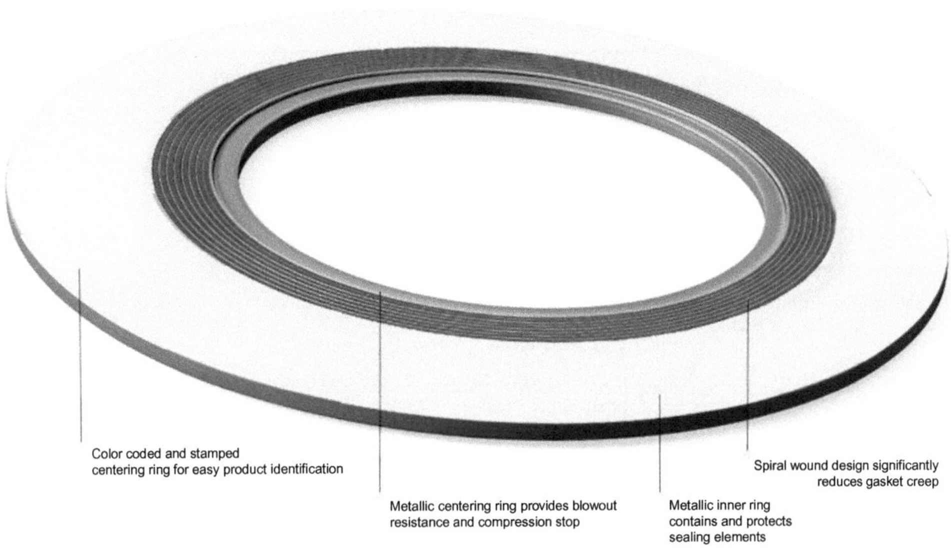

Fig. 13.43 Spiral wound gasket.

sealing, and a lip seal, like a V-pack, can be used for sealing between the seat and body of ball and gate valves.

2. Which sentence is not correct about elastomer sealing?

  A. Adding hydrogen to NBR and forming HNBR does not have any advantage.

  B. EPDM is not resistant to hydrocarbons, so this type of elastomer is not common in valves used for subsea purposes.

  C. Viton FKM is a more common sealing in the valve industry than neoprene and silicon.

  D. HNBR is a common sealing material for both valves and actuators.

Answer: Option A is wrong because adding hydrogen to NBR, or more accurately the hydrogeneration of NBR, improves its temperature and ozone resistance. Option B is correct, since EPDM is not compatible with oil and gas and is not common for subsea valves. Viton FKM is common in the valve industry and is more popular compared to neoprene and silicon. D is also correct, because HNBR is common for the sealing of both valves and actuators.

3. Which sentence is correct about the NORSOK M-710 standard?

  A. This standard is used only for the qualification of soft sealing materials. Manufacturers of soft materials are evaluated through a different standard.

  B. Lip seals as well as metallic gaskets are covered by this standard.

  C. Different sets of tests for both elastomers and thermoplastic materials are defined in this standard.

  D. NORSOK M-710 is only applicable to projects in Norway.

Answer: Option A is not correct, since NORSOK M-710 is used for the qualification of both soft sealing materials and manufacturers. Option B is wrong because metallic gaskets are not covered by the NORSOK M-710 standard. Option C is correct; different tests are defined for both elastomers and thermoplastics in NORSOK M-710. NORSOK M-710 is used by end users and companies outside Norway so it is not limited to Norway. Thus Option D is wrong as well.

4. Which parameters are important for the selection of soft materials?
   - **A.** Appearance and color, low friction, strength and heat resistance.
   - **B.** Compatibility with fluid, resistance to galvanic corrosion in the case of graphite selection, life expectancy, and field-proven efficiency.
   - **C.** Direction of sealing, abrasion resistance, and corrosion resistance, especially in the case of hydrogen sulfide.
   - **D.** Strength, heat, and pressure resistance.

Answer: Option A is not completely correct, because the appearance and color of the sealing is not important for the selection of soft materials. Option B is not completely correct, since galvanic corrosion does not affect or interfere with the selection of graphite material selection, due to graphite's high Nobility. Option C is wrong since the direction of sealing does not affect the sealing selection. As an example, a lip seal provides unidirectional sealing, but could be preferred over an O-ring due to its higher pressure–temperature resistance, RGD property, and high corrosion resistance. Option D is correct, since the strength and heat and pressure resistance of soft materials are key factors during the selection process.

5. Which sentence is not correct about elastomeric seals?
   - **A.** Temperature, pressure, and fluid type are essential factors in selecting elastomeric seals.
   - **B.** HNBR and Viton FKM are the most popular elastomeric seals for valves in the oil and gas industry.
   - **C.** Elastomer seals have less resilience at low temperatures.
   - **D.** Graphite which is an elastomeric seal should be pure and some limitations of chloride and sulfide content.

Answer: All of the sentences are correct statements. But graphite is not an elastomeric seal, so option D is not correct.

6. Which sentence is not correct about lip seals and V-packs?
   - **A.** Lip seals' PTFE part is energized through spring force.
   - **B.** The spring material in a lip seal could be Elgiloy, as a popular choice for subsea applications.
   - **C.** A V-pack cannot be made more than six rings.
   - **D.** A V-pack can be used as the primary internal stem seal in gate valves.

Answer: Option A is correct, since a lip seal contains an elastomeric part (normally PTFE) and a spring. The spring energizes the PTFE or elastomeric part for sealing. Option B is correct; Elgiloy, which is a cobalt alloy, is a very common material for valve springs and lip seal springs. Option C is wrong, since there is no limitation on the number of V-packs layers. The number of layers in a V-pack could exceed six rings. Option D is correct, since a V-pack is normally selected as the primary internal pressure steam sealing for through conduit gate valves.

Further reading

American Petroleum Institute (API) 6A, Specification for Wellhead and Tree Equipment, 21st ed., API, Washington, DC, 2018.

American Petroleum Institute (API) 6DSS, Specification for Subsea Pipeline Valves, third ed., API, Washington, DC, 2017.

S. Burftscher, A. Derfmann, K. Bergmeister, Mechanical aspects of high damping rubber, in: University of Applied Sciences, Institute of Structural Engineering. 2nd International PhD Symposium in Civil Engineering. Budapest, 1998.

L. Cannistraci, In the ocean's depths, valves face unique challenges, Valve Mag. (2010). [online] Available at: https://www.valvemagazine.com/magazine/sections/features/4419-in-the-oceans-depths-valves-face-unique-challenges.html?showall=1&start=0. (Accessed 24 July 2020).

Craftech Industries Inc., Top 5 Elastomers for Gasket and Seal Applications, 2020, [online] Available at: https://www.craftechind.com/top-5-elastomers-gasket-seal-applications/. (Accessed 3 September 2020).

Explore the world of piping, Introduction to Bellow Sealed Valves, 2020, [online] Available at: http://www.wermac.org/valves/valves_bellows_sealed.html. (Accessed 9 September 2020).

International Organization of Standardization (ISO) 12792, Metal Ball Valves for Petroleum Petrochemical and Aligned Industry, 2004.

International Organization of Standardization (ISO) 23936, Petroleum, Petrochemical and Natural Gas Industries—Non-Metallic Materials in Contact With Media Related to Oil and Gas Production. Part 2: Elastomers, first ed., 2011.

J. Walker, Devlon V-API Thermoplastic Valve Seats, 2020, [online] Available at: https://www.jameswalker.biz/nb/pdf_docs/184-devlon-v-api-thermoplastic-valve-seats. (Accessed 4 September 2020).

A. Kruijer, Exchanging elastomer knowledge. Shell company, Valve World Mag. (n.d.) [online] Available at: https://www.valve-world.net/pdf/vw10ce_emissions_shell.pdf.

Lyphilization Technology Inc., Low Temperature Thermal Analysis, 2020, [online] Available at : https://www.lyotechnology.com/thermal-analysis.cfm. (Accessed 9 September 2020).

J. Mullins, The Eight Failures That Caused the Gulf Oil Spill, 2010, [online] Available at : https://www.newscientist.com/article/dn19425-the-eight-failures-that-caused-the-gulf-oil-spill/. (Accessed 24 July 2020).

NORSOK, Qualification of Non-metallic Sealing Materials and Manufacturers. NORSOK M-710, second ed., 2001. Oslo.

National Association of Corrosion Engineers (NACE) (TM0187), Evaluating Elastomeric Materials in Sour Gas Environments, 2011.

J. Praveen, M. Pathan, K. Ansari, Hyperbaric pressure testing of a subsea valve to validate deep water condition, Int. J. Mech. Prod. Eng. Res. Dev. 8 (2) (2018) 1011–1022.

Precision Polymer Engineering, Elastomer for Sour Gas, 2020, [online] Available at:https://www.prepol.com/industries/sour-gas-h2s-resistant-elastomers. (Accessed 9 September 2020).

Reading plastic machining & Fabrication, PTFE vs Teflon: How Are These Plastics Related? 2020, [online] Available at:https://readingplastic.com/ptfe-vs-teflon/#:~:text=Essentially%2C%20the%20only%20difference%20lies,look%20no%20further%20than%20PTFE. (Accessed 4 September 2020).

K. Sotoodeh, A review on subsea process and valve technology, J. Mar. Syst. Ocean Technol. 14 (2019) 210–219 Springer.

Superior Sealing Technology, Elastomer Properties, [online] Available at:https://superiorltd.com/material-science/elastomer-properties/. (Accessed 4 September 2020).

CHAPTER FOURTEEN

Testing

Contents

14.1 Introduction

Today, subsea valves and actuators are used in very deep-water depths—as deep as 3 km for example—and they face the challenge of receiving zero maintenance for a long time, typically between 20 and 30 years. For this reason, different aspects of valve design, such as sealing, machining tolerance, material selection, and corrosion resistance may be more complicated for subsea valves compared to valves used in other sectors of the oil and gas industry. Indeed, more stringent tests in the form of Factory Acceptance Tests (FATs) and qualification tests are conducted on subsea valves to ensure their reliability and achieve the zero-maintenance expectation. FATs include different series of tests, mainly pressure tests, that should be conducted on each valve and actuator after final assembly and prior to shipment. In general, qualification tests are applied to newly designed valves and actuators and include three main test categories as a minimum: hyperbaric testing, endurance testing, and pressure and temperature cycling. The main aim of this chapter is to explain and review the most important FATs and qualification tests based on industrial standards such as API 6DSS/ISO 14723, "Specification for subsea pipeline valves"; API 17D/ISO 13628-4, "Subsea wellhead and Christmas trees equipment"; and API 6A/ISO 10423, "Specification for wellhead and Christmas trees equipment" as well as

Subsea Valves and Actuators for the Oil and Gas Industry
https://doi.org/10.1016/B978-0-323-90605-0.00015-3

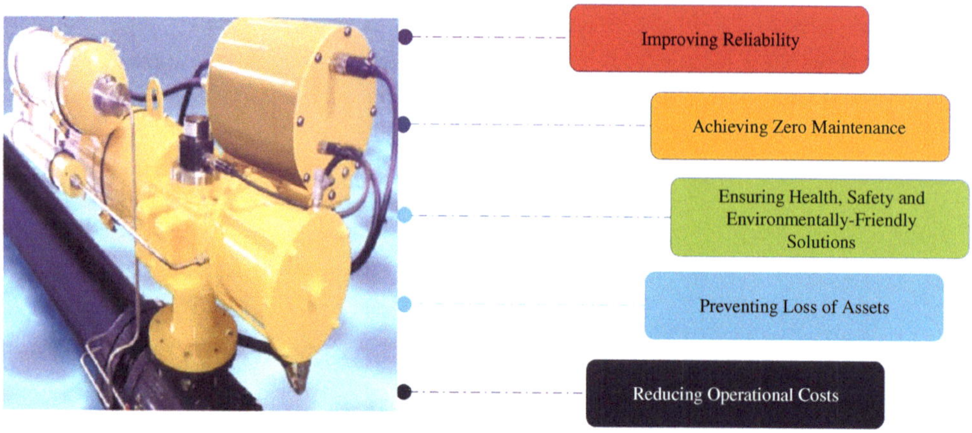

Fig. 14.1 Benefits of subsea valve and actuator testing.

industrial practices and experience. Fig. 14.1 illustrates the general benefits that can be achieved through accurate testing of subsea valves and actuators.

Subsea actuators could work with hydraulic or electrical power; this chapter mainly addresses the testing of hydraulic actuators. The main reason is that hydraulic actuators are still more common than electrical actuators in author of the book experience. The main qualification tests mentioned earlier are followed by disassembly and visual inspection of the valves, as well as a nondestructive test in many cases make sure that there is no sign of wear on the valves or actuators.

14.2 Factory acceptance test

14.2.1 Safety concerns

Performing FATs involves risk and hazards due to dealing with high-pressure fluid that could cause mechanical failure of the valves and actuators. Therefore strict safety consideration and implementation based on company risk procedures and regulations are essential. Extreme caution should be applied during the testing of both valves and actuators to prevent injury. Wearing safety glasses, a safety hat, and safety shoes in some areas of the factory are essential practices. It should be noted that the high-pressure gas test requires an especially high level of safety; the valve should be tested in a pit at a sufficient distance from personnel, and the pit should be equipped with gas detector alarms.

14.2.2 Test preparation

Valves are normally pressurized during the pressure test with water, air, and gas. The gas could be nitrogen or nitrogen helium. Testing is normally done at room temperature, which is between 2°C and 38°C, with some limitation on the chloride content of the water such as a maximum of 20 or 30 parts per million (ppm). Limitation of chloride is more important for testing valves in carbon steel or austenitic stainless steels that are more sensitive to corrosion due to chloride. However, limiting the chloride content of the test fluid in many specifications is generally applicable to all applicable valve materials. Adding some percentage of corrosion inhibitor (e.g., from 1% to 3%) to the test fluid is recommended to prevent corrosion. Fig. 14.2 illustrates a storage tank used for supplying the test medium that contains 97% water and 3% corrosion inhibitor.

It is important to dry the valves after the test and remove all water as a part of the preservation requirements. Preservation is defined as series of actions that are taken to protect the valve against harmful environmental factors such as humidity, wind, snow, low temperature, etc. Protection of the valve through preservation is not limited to after the test; rather, preservation can be applied at different stages of the valve cycle, such as storage, lifting, handling, and installation. Preservation is the responsibility of the valve and actuator manufacturer/supplier based on the project specifications provided by the end user and/or purchaser company. Fig. 14.3 illustrates some rust inside a valve that was not drained properly and dried after the FAT and pressure test. Thus the test water remained in the valve for a long time and caused corrosion.

All test equipment, especially pressure gauges, should be calibrated and certified before the test is conducted. The calibration of the test equipment should be done within 3 months before the test implementation. Pressure gauges are used to measure the pressure during the test; the test pressure applied during the test should be between 20% and

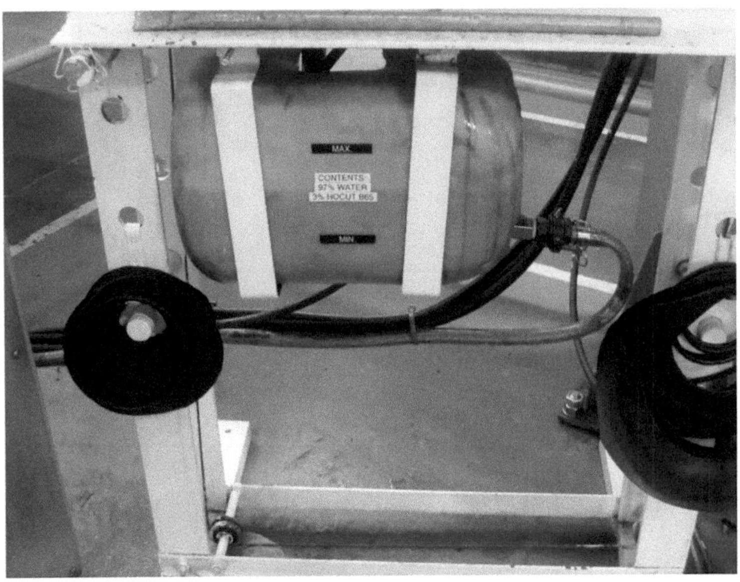

Fig. 14.2 Test pressure fluid storage tank containing 97% water and 3% corrosion inhibitor.

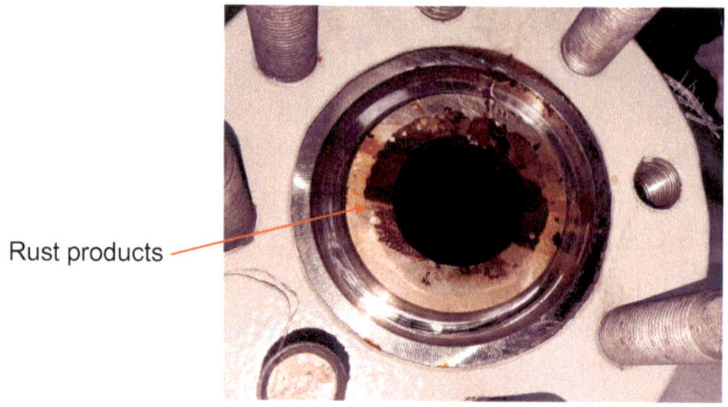

Fig. 14.3 Rust in the valve due to test water that remained in the valve after the pressure test.

80% of the full scale of the pressure gauge. Fig. 14.4 illustrates a pressure gauge with a dedicated serial number used for the valve pressure test. The label shows that the pressure gauge was calibrated on the 17th of October 2013, so the test should not take place after the 17th of January 2014. After calibration, a unique calibration or certification number is given and labeled on the pressure gauge. All of the test results should be documented and

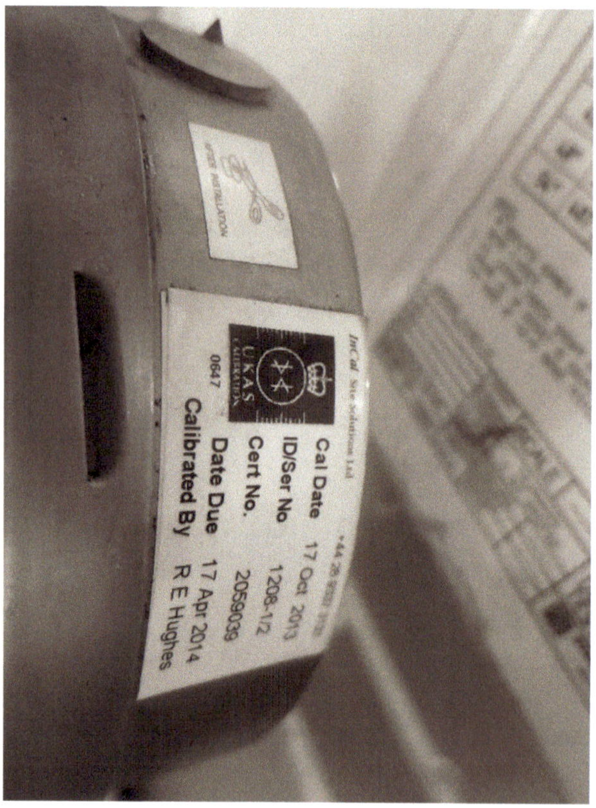

Fig. 14.4 Calibrated pressure gauge.

available for a third-party inspector or end user representative upon their request. Digital test equipment can provide charts and test records, unlike analogue/more traditional test equipment. The other important thing is that the valve is normally tested without painting or coating. The presence of coating on the valve body or shell during the body or shell test can prevent detection of possible leakage from defective areas of the valve. The valve should be clean and free from grease or any contamination during the tests. The applied test pressure should not be less than the nominal test pressure. But it is possible to increase the test pressure to 5% above the specified or nominal test pressure. It is also important to wait until the pressure is stabilized during the test. The pressure test is considered stabilized when the change rate of the pressure is not more than 3% of the test pressure per hour or 34.5 bar per hour, whichever is less.

14.2.3 Valve tests

14.2.3.1 Valve pressure tests

Valve pressure tests are normally performed on both the seats and body of the valves. The tests are typically high-pressure hydrotests (1.5 times and 1.1 times of design pressure on the body and seats, respectively), low-pressure air/gas seat tests (5.5–6.9 bar) as per API 6DSS, or 300 psi equal to 20 bar as per API 6A/17D and high-pressure gas tests (1.1 times of design pressure or simply design pressure on both body and seats). The seat of the valves could be tested with a pressure of 1.2 or 1.25 times of the design pressure as an alternative to 1.1 times of design pressure due to subsea pipeline testing requirements as per the Det Norske Veritas (DNV) requirement. Body or shell test acceptance is proof of the structural integrity and robustness of the valve body pieces or shell, as the main pressure-containing part, against the internal fluid pressure. Valve stem sealings may be tested during the body pressure test. The valve is fully filled from pressurized media, which could be water or gas, and the external area of the valve is monitored for any possible leakage during the shell test. Normally no leakage is allowed from the body or shell of the valve during the body or shell test. Leakage from the body of the valve has very negative consequences, such as seawater pollution, so it is not tolerated. Ball valves are body pressure tested in partially or half open position to allow pressure fluid to circulate in all areas of the valve. The other reason is that the seats on ball valves would be damaged at 1.5 times the design pressure at both fully open and fully closed conditions. Fig. 14.5 illustrates a 14″ reduced bore top entry ball valve with a 10″ bore in a pressure class of 2500 equal to 428 bar. Fig. 14.6 illustrates a crack that occurred during a shell pressure test inside the bolt hole in the body of a valve. Although the crack was identified during the test, which prevents severe failure of the valve during operation in the field, the detection of this problem during the test has a negative impact on the project schedule and delivery because the valve is useless and a new one will need to be manufactured from scratch. The duration of the body test is very long; as per API 6DSS it is at least 4 h. Typically, API 6A can be used for the test duration of both seats and body depending on the

Fig. 14.5 14″ × 10″ Class 2500 top entry ball valve.

production specification level (PSL) as per Table 14.1. PSL3G is more common PSL for subsea valves. The PSL depends on the valve pressure and material class and defines the level of design, quality, and test of the subsea products. Gas test requirements are applicable for PSL3G and PSL4. More information about PSL is given in Chapter 2. A seat test, also called a leakage test, is performed to evaluate the sealing capability of the valve between the valve closure member and the seats. A gas test could be applied on both the body and seats of the valves and is implemented after the hydrostatic test. In fact, any possible leakage from the valve seats that could not be detected during the hydrostatic test can be detected during the gas test. As stated in the table, the shell and seat tests are performed in more than one stage for subsea valves and actuators, unlike industrial valves in other sectors of the oil and gas industry, which are subject to each type of test in one stage. Performing double or triple tests on the hydrostatic shell or body, and seat tests for subsea valves and actuators improves the safety and reliability of the products. ISO 10423 provides the gas test durations for the body of the valves in a single stage and not less than 15 min. The gas seat test should be done in two stages: the first stage

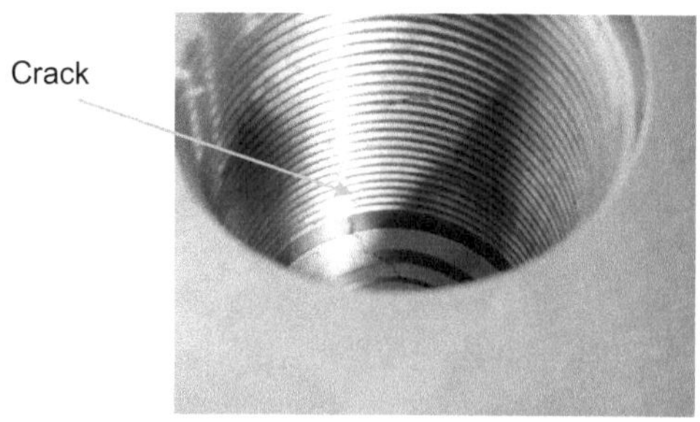

Fig. 14.6 Crack detected in the bolt hole of a valve body during the shell test.

Table 14.1 Hydrostatic shell and seat test duration as per API 6A.

		PSL1 (min)	PSL2 (min)	PSL3 (min)	PSL3G (min)	PSL4 (min)
Shell test	Primary test duration	3	3	3	3	3
	Secondary test duration	3	3	15	15	15
Seat test	Primary test duration	3	3	3	3	3
	Secondary test duration	3	3	15	3	15
	Tertiary test duration	NA	3	15	3	15

at design pressure or 1.1 times the design pressure (high-pressure gas test) and the second stage with 20 bar pressure (low-pressure gas test); the proposal in this chapter is to have each stage last not less than 15 min. If the seat test requires different steps such as three stages, the valve should be cycled to fully open and closed positions after the completion of each seat test and before performing the next stage seat test.

The accepted leakage rate from the seats could be rate A (zero leakage), B, C, or D as per ISO 5208. In some cases, the allowable leakage rate from a through conduit gate (TCG) valve could be less than that of a ball valve since the sealing between the seats and disk of a gate valve is achieved through the contact of two flat surfaces. However, sealing between the ball and seat of a ball valve is achieved through contact of two spherical areas. Valves can be tested on a test bench, but in some cases, this can apply excessive force on the valve, which could jeopardize the valve sealing; in such cases it is better to test the valve on the ground or in a pit. Valves may also be tested off the test bench when they are very large. When testing valves on the ground or in a pit, they are blinded with two caps or a blind hub at both ends of the valves (see Fig. 14.7). API 6DSS mentions

Fig. 14.7 Testing a very large valve in a pit.

another reason for testing the valves off of the test bench. When a pressure test is applied on the test bench, it does not allow full pressure force to be transmitted to the valve, so at least one valve per type, size, and pressure class should be tested with the pressure caps or hubs to ensure that full axial test pressure is transmitted to the valve.

The amount of leakage from the valve could be counted with a bubble counter (in the case of gas) or a drop counter (in the case of liquid). Fig. 14.8 illustrates a bubble counter. The bubbles are released to the water inside the counter which makes the bubbles countable. The bubble counter could be connected to one side of the valve seat to measure the bubbles leaking from the seats.

There are three differences between API 6DSS and the API 6A and 17D standards regarding the high- and low-pressure gas tests:

1. API 6DSS specifies at least 1.1 times of the rating pressure for high-pressure gas tests on both body and seats. However, both API 6A and 17D specify working pressure as the high-pressure gas test for both body and seats. Working pressure could be considered equal to operating pressure, which is normally less than design pressure. The proposal of the author is to apply the high gas test pressure as per API 6DSS, which is equal to 759 bar for a valve in 10,000 psi equal to 690 bar. The reason to apply higher pressure gas test equal to 1.1 times of the rating pressure is for improving the safety and reliability of the valves.
2. API 6DSS recommends a low-pressure gas test with a pressure value between 5.5 and 6.9 bar, whereas the nominal pressure for low-pressure gas tests recommended in API 6A and 17D is equal to 20 bar. It should be noted that a low-pressure gas test is only applicable for valve seats. In practice, it has been observed that low-pressure gas tests

Fig. 14.8 Liquid in the bubble counter.

with both pressure values in two stages have been applied to valves. However, the proposal in this chapter is to choose between one of the low-pressure gas tests due to the fact that this test is expensive, relatively time consuming, and involves a high number of Health, safety and environment (HSE) precautions.

3. API 6DSS allows a high-pressure gas test to be done out of a water bath. However, API 6A and 17D emphasize submerging valves in a water bath during gas tests in order to detect possible leakage in the form of bubbles in the water bath, probably to ensure higher safety. The proposal in this chapter is to avoid submerging the valve in water for the gas test, as long as enough safety precautions can be applied, in order to reduce the time required for testing.

14.2.3.2 Valve function tests

Function testing of the valves could be done during or after the pressure tests. The valve and operator (actuator or gear box) shall be assembled during this test. The functionality of the valves and actuators, such as their speed of cycling, smooth cycling, required torque for opening and closing, as well as the complete alignment and correct placing of the closure member are measured and observed during this test. The test could be done with water as a fluid medium and/or without any fluid media. It can be repeated three or four times through 3–4 opening and closing cycles, and can be done after the hydrostatic test pressure implementation on the seats. The first important observation is that the operator should operate the valve smoothly without any chattering or stick-slip. This test can show that the actuators have sufficient force to fully open the valves through hydraulic fluid supply to the actuators. The measured torque produced by the operator should not exceed the maximum torque provided in the valve data sheets. Also, other torque values during valve travel, such as break to open, end to open, running, break to close, and end to close, could be measured and compared with the values provided by the valve and actuator supplier. The other important task during the test is to monitor and record the hydraulic supply pressure to the actuator and the different torque measurements. The most challenging torque scenario for opening the valve is to open it against full differential pressure (maximum valve rating) and the most challenging scenario for closing the valve is to close it when the valve is being operated at the maximum valve pressure rating with zero pressure in the valve cavity.

As an example, the break to open torque for a 7 1/16″ slab gate valve in pressure class 7500 psi equal to 517 bar is given as 1758 Newton meters (N m) and the supply hydraulic pressure during the valve opening is measured at 330 bar during the test. The function test should be done by pressurizing one side of the valve with 517 bar when the valve is closed and measuring BTO and hydraulic pressure during valve opening. The acceptance criteria are to measure BTO at less than 1758 N m. and the capacity to open the valve with a hydraulic pressure of 330 bar which is less than the nominal hydraulic pressure.

Based on industrial experience, the present author offers the following recommendations for improving the execution of the function tests:

1. Two valve cycles are currently required to measure all the valves and actuators operating torques. Six tests including 12 cycles could be done for the subsea valves function test; three tests and 6 cycles with water, and three tests and 6 cycles without any fluid. Repeating the test improves the accuracy of the test data and improves the reliability of the valves for functioning.
2. Any measured value of torque with a variation of more than 5% from the value provided in the valve data sheet should be rejected and the test should be repeated.
3. The valve closing and opening time is mentioned in the valve data sheet and should be compared with the values obtained through the test. Normally, the opening and closing times for manifold valves are not very critical. However, the closing time for emergency shutdown and HIPPS valves is very critical. Thus failure to function and achieve the required closing time means that the valve should be rejected.
4. DBB valves should be capable of provide sealing while they are pressurized from both ends from the connected piping to the valves. For DBB valves that provide sealing from both sides, the function tests should be done while pressurizing the valve from both sides. Chattering of the valve during the test should lead to rejection.
5. The alignment of the closure member with the valve bore to keep the valve in fully open position (e.g., the alignment of the ball with the bore for a ball valve during opening; see Fig. 14.9) and during the complete closing of the valve should be checked during the function tests.
6. Subsea valves are typically equipped with ROV interfaces so that they can be operated independently from the actuator in the case of actuator malfunction or unavailability of hydraulic fluid power supply to the actuator. The same tests for measuring the torque value of the valve could be repeated with the ROV torque tool instead of the actuator.

14.2.3.3 Body cavity test

This test is typically applicable to ball valves and through conduit expanding gate valves. This test is not performed on and is not applicable to axial check valves. The valve cavity is defined as the area between the valve closure member and body in which fluid can be

Fig. 14.9 Ball valve ball alignment to the valve bore during the function test.

trapped and pressurized and thus damage the valve body and internals if it is not released. Therefore the existence of a body cavity for ball valves could be counted as a disadvantage or weak point. The cavity of a ball valve (see Fig. 14.10) has communication with the valve internals in partially open positions; however, it is disconnected from the valve bore when the valve is opened or closed. To avoid damage to the valve due to overpressurized fluid in the cavity, the seat of the valve should be able to release the cavity pressure at a maximum of 33% of the maximum rated pressure according to the API 6DSS requirement. It should be noted that the seats of the valves could be designed as self-relieving, also called single piston effect (SPE), or double piston effect (DPE) types. The self-relieving feature of SPE valves' seats allows the cavity pressure to be released to the valve internal bore, unlike DPE seats. The cavity test is applied when the valve is in closed position. The cavity is used as the test port and upstream and downstream valve bores are used for pressure monitoring. A small amount of pressure is applied to the cavity, such as 1 or 2 bar, and the pressure is increased gradually until the self-reliving seats release the pressure to the line. At that time, the pressure gauges, located upstream or downstream of the valve, show the pressure increase on the upstream or downstream bore of the valve. The releasing pressure value from the cavity to the valve internals is measured and should be a maximum of 33% times the design pressure. However, some valve specifications may ask for lower cavity release pressure such as 10%. However, based on present author industrial experience, the allowable maximum pressure inside the valve cavity is 1.33% of the design pressure or maximum working pressure in some cases.

The spring torque behind the seat has an impact on the release of the pressure from the cavity at a specific pressure. If the valve contains two double piston effect seats, called double isolation and bleed type 1 (DIB1), then a cavity relief test is not applicable. The cavity should be pressurized from the test fluid for DIB1 valves when the valve is in closed position and both the upstream and downstream side of the cavity should be monitored for possible leakage. The amount of leakage from the seats should be compared with the allowable leakage based on the ISO 5208 standard. DIB2 valves with

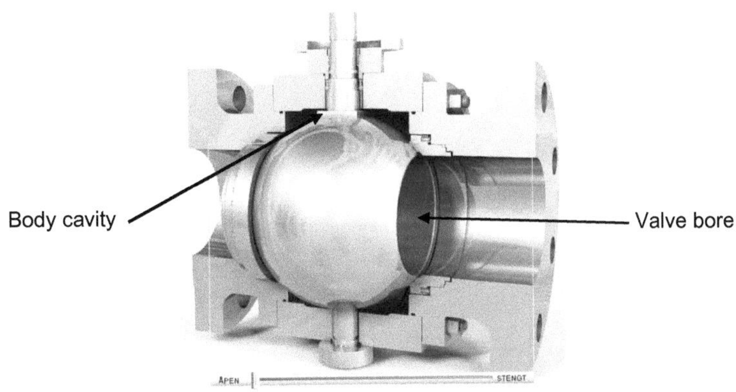

Fig. 14.10 Ball valve in open position highlighting the body cavity.

one seat self-relieving (SR) and one seat double piston effect undergo a different test procedure. The cavity test for DIB2 valves can be done in two stages. The first is to test the self-relieving seat, as in the normal procedure explained before. The second is to pressurize the self-relieving seat side from the valve inside, pressurize the cavity at the same time, and monitor the leakage from the double piston effect seat. The amount of leakage from a double piston effect seat should be same as the allowable leakage rate from the seat (e.g., Leak rate B as per ISO 5208).

14.2.3.4 Backseat test

The backseat test is applicable for TCG valves and not for axial check or ball valves. In fact, other types of subsea valves except for gate valves do not have a backseat feature. Fig. 14.11 illustrates a subsea through conduit slab gate valve with a reverse port gate (hole in the gate) in an open position. According to subsea valve standards, the backseat area can be counted as a secondary stem seal that provides stem sealing when the wider stem area sits on the valve internal bonnet below the stem sealing areas. The backseat test could be performed with design pressure or 1.1 times of the design pressure depending on the design standard. Changing the valve position to closed leads to moving the stem upward, so the stem will be back-seated—meaning that the wider or larger stem area connected to the valve closure member sits on the backseat area and isolates the stem sealing area. Pressure can be applied from each port on both ends of the valve and any possible leakage from the backseat can be monitored from the port located above the backseat. This port

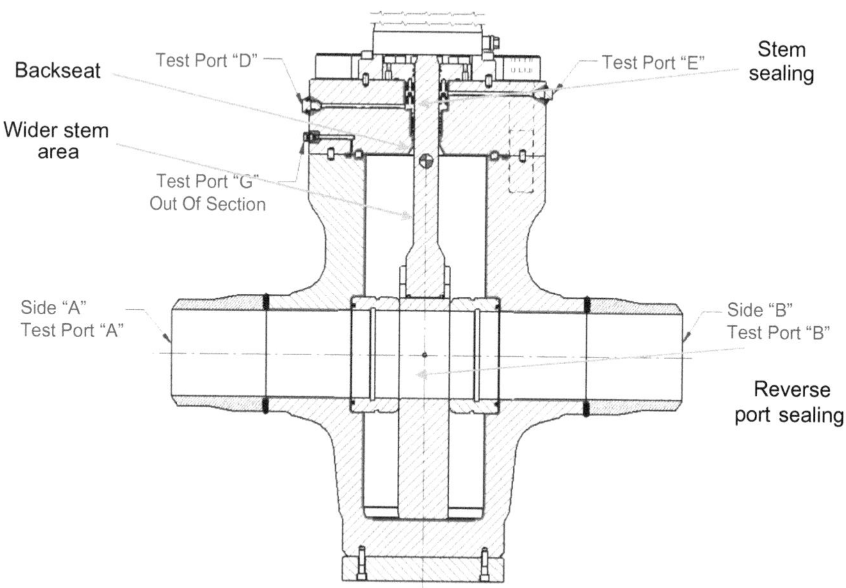

Fig. 14.11 Through conduit slab gate valve.

is not shown in Fig. 14.11. The recommendation of the author based on industrial experience is to apply the backseat test without any stem sealing to more easily monitor any possible leakage from the backseat. No visual external leakage is allowed in the backseat test. The proposed test duration is 15 min according to the API 6A standard. The test is done with water as a test fluid, and should be repeated with nitrogen or nitrogen helium for valves in PSL3G or PSL4.

14.2.3.5 Stem seal tests

This type of test is not reviewed in detail by any of the applicable international standards for subsea valves. The details of the stem seals illustrated in Fig. 14.11 are explained as follows:

1. A V-pack, located at the bottom of the valve stem, acts as the primary internal fluid barrier. This seal, which could be made of soft thermoplastic material layers such as Teflon and PEEK (polyetheretherketone), should prevent the leakage of the internal fluid that is passed from the backseat area outward. Detailed information about types of valve seals, such as elastomeric, V-pack, etc., is available in Chapter 13.
2. A lip seal made of a metallic spring and a soft thermoplastic Teflon seal is located above the V-pack; it acts as a secondary stem sealing against internal fluid. This type of seal prevents leakage from the stem sealing in case the V-pack fails to provide acceptable sealing.
3. An O-ring is located on the top of the stem to prevent ingress of seawater inside the valve, including the stem sealing area. This kind of sealing is called the external primary stem sealing barrier.
4. A lip seal is located under the O-ring and above the other lip seal (secondary internal stem seal barrier). This lip seal can be called a secondary external stem seal barrier; it should prevent ingress of seawater to the valve internals in case of failure of the O-ring to provide sealing.

Port D is connected to the area between the V-pack and lip seal, which are the primary and secondary internal pressure barrier stem seals. Both ends of the valves could be pressurized through Port A and B, and the valve can be kept in open position as illustrated in the Figure so it is not in backseat position. Therefore the pressurized fluid will be in contact with and pressurize the V-pack. Any possible leakage from the V-pack can be monitored and eventually measured through Port D. Testing the V-pack as a barrier against internal pressure inside the valve could be done as a part of the shell or body pressure test. The duration and numbers of the test are the same as the body pressure test and should follow the requirements of Table 14.1. Port E is located between two lip seals: one is the secondary barrier against internal pressure and the other is the secondary barrier against external pressure. Port D can be pressurized by the test fluid, which could be water, and any possible leakage from the bottom lip seal can be monitored and finally measured through port E. One more test port may be required between the O-ring and secondary external stem seal barrier lip seal to be able to test the secondary external seal (the lip seal) against external pressure leakage in case of failure of the O-ring. If that extra test port is provided, it can be pressurized and any leakage from the external seawater secondary barrier, which is a lip seal, can be monitored through port E. The sealing capability of the O-ring can be evaluated by means of a hyperbaric test, which is explained in detail later in this chapter. Any seawater leakage from the O-ring can be monitored through the test port located between the O-ring and the lip seal (secondary external pressure barrier). The important point is that the stem sealing test ports are used just for the purpose of testing the stem sealing and should be plugged and seal welded after the test; these ports are not used during the operation of the subsea valves.

14.2.3.6 Body/bonnet or body pieces sealing tests

A body and bonnet sealing test is applicable to gate and top entry ball valves; body pieces sealing tests are applicable to side entry ball valves. These tests are not covered in detail by the international standards applicable to subsea valves. Fig. 14.11 illustrates a gate valve with two sealings between the body and the bonnet. The details of these two seals are illustrated in Fig. 14.12. The internal one is a metal seal in Inconel 625 material that provides bidirectional sealing. The main purpose of this metal seal is to prevent the leakage of internal fluid to the seawater. The second function of this metallic seal or gasket is to provide sealing against seawater ingress and leakage to the valve internals in case of failure of the other sealing. The other sealing is an elastomeric O-ring designed and selected to provide a barrier against seawater leakage to the valve internals. Any leakage from the elastomeric sealing can be monitored through port G illustrated in Fig. 14.11. Port G is also capable of showing the valve internal pressure leakage from the metallic gasket. Port G is connected to the area between the two body and bonnet seals, which are an O-ring and metallic gasket, typically BX type. The key point is that testing implementation on the O-ring external sealing

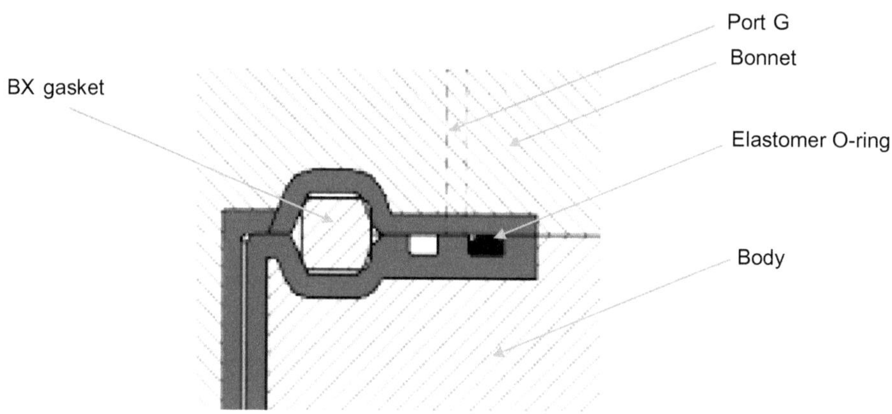

Fig. 14.12 Through conduit gate valve body/bonnet seals.

should probably be done as a part of the hyperbaric test and not the FATs. The reason is that the simulation of seawater head or pressure is done during the hyperbaric test in a hyperbaric chamber, while the FATs are performed mainly out of the water. The exception could be the high-pressure gas tests on the body and seats of subsea valves, which could be performed in the water for safety reasons. However, the water during these two pressure tests is pressurized and does not simulate seawater head pressure at depth.

14.2.3.7 Postpressure valve tests

Some tests performed on subsea valves are not categorized into pressure or functional tests. Three of these tests, namely the drift test, electrical continuity test, and bolt tension or torque verification, are summarized and explained in this section.

Drift test

International standards such as API 6A, API 17D and equivalent International Organization of Standardization (ISO) standards propose drift tests for full bore valves. Bore refers to the internal diameter of the valve; as explained in Chapter 2, a full-bore valve could have a smaller internal diameter than that of the connected piping, also called the "meeting piping." A drift test is performed to make sure that the valve has the defined bore as per the valve data sheet. The drift test should be applied after assembling, testing, and operating the valve during the testing (in functional tests) as per API 6A requirements. That is the reason why this test is categorized as a postpressure test type. This test is required for all product safety levels (PSLs) defined in API 6A. PSL, as explained in Chapter 2, represents the product technical quality requirement. During the drift test, a mandrel with three disks located on both ends and in the middle of the mandrel should be able to pass through the valve without any restriction. API 6A provides the required dimensions of the mandrel, such as the length and diameter; these are provided in Table 14.2, extracted from API 6A.

Fig. 14.13 illustrates a drift test for a top entry ball valve located on a gas export pipeline with an internal diameter equal to 280 mm. One of the disks of the mandrel is shown in the Figure, while the other disk inside the valve is also visible. The material of the disks is in Teflon, which is a nonmetallic or soft material. The reason for using Teflon disks is to prevent cross-contamination between the disk and the internal part of the valve. Cross-contamination occurs due to contact of two dissimilar metals such as carbon steel and stainless steel. The contamination of stainless steel by carbon steel could prevent the formation of chromium oxide in stainless steel, which in turn leads to corrosion and failure of the stainless steel material. It should be noted that the corrosion-resistant property and capability of stainless steel is largely dependent on the formation of chromium oxide.

The drift test is not a time-consuming or complicated test. However, this test is proposed to be limited to piggable valves based on author experiences in the industry. Piggable valves are those that are exposed to running a "pig" or piping injected gadget;

Table 14.2 Dimensions of the drift test tool (mandrel) according to API 6A.

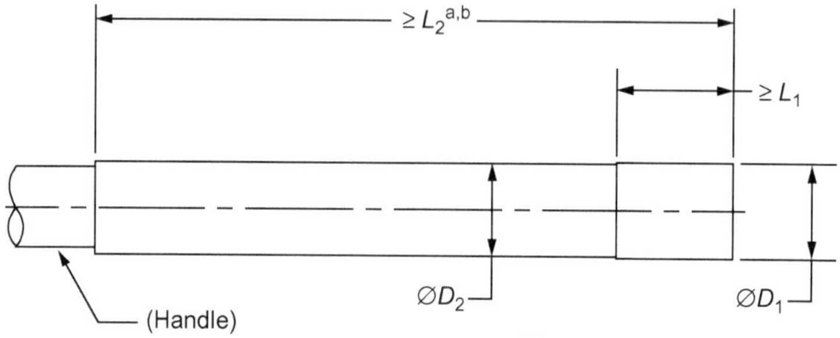

Dimensions in millimeters (inches)[c]

Nominal bore size	L_1 minimum		D_1 $^{+0.69}_{-0}$ mm $\left(^{+0.027}_{-0}\text{ in.}\right)$		D_2 $^{+0.69}_{-0}$ mm $\left(^{+0.027}_{-0}\text{ in.}\right)$	
in.	mm	(in.)	mm	(in.)	mm	(in.)
$1^{13}/_{16}$	76	(3.00)	45.21	(1.78)	38.6	(1.52)
$2^{1}/_{16}$	76	(3.00)	51.56	(2.03)	48.3	(1.90)
$2^{9}/_{16}$	76	(3.00)	64.26	(2.53)	59.7	(2.35)
$3^{1}/_{16}$	78	(3.06)	76.96	(3.03)	73.2	(2.88)
$3^{1}/_{8}$	79	(3.12)	78.49	(3.09)	73.2	(2.88)
$4^{1}/_{16}$	103	(4.06)	102.36	(4.03)	97.3	(3.83)
$4^{1}/_{8}$	105	(4.12)	104.8	(4.09)	100.8	(3.97)
$4^{1}/_{4}$	108	(4.25)	108.0	(4.22)	101.6	(4.00)
$5^{1}/_{8}$	130	(5.12)	129.29	(5.09)	126.2	(4.97)
6	152	(6.00)	151.64	(5.97)	148.8	(5.85)
$6^{1}/_{8}$	156	(6.12)	154.69	(6.09)	151.6	(5.97)
$6^{3}/_{8}$	162	(6.38)	161.04	(6.34)	158.0	(6.22)
$6^{5}/_{8}$	168	(6.62)	167.39	(6.59)	164.3	(6.47)
$7^{1}/_{16}$	179	(7.06)	178.56	(7.03)	175.5	(6.91)
$7^{1}/_{8}$	181	(7.12)	180.85	(7.09)	177.0	(6.97)
9	228	(9.00)	227.84	(8.97)	224.8	(8.85)
11	279	(11.00)	278.64	(10.97)	275.6	(10.85)

[a]For valves, L_2 = valve face-to-face length.
[b]For trees, L_2 = 1065 mm (42 in.).
[c]For USV and BSDV valves with nonstandard bores, see section 11.4.2 in API 6A standard.

they should have a controlled and specific bore equal to that of the connected piping with some tolerances, such as 3%, in order to facilitate the passage of the pig through the valves. It is also proposed to apply this test from both directions to improve the accuracy and reliability of the test implementation. A drift test is not required for check valves, as most types of check valves are not piggable.

Table 14.2 provides the required mandrel dimensions for API 6A standard valves only. Mandrel dimensions for nonstandard bore valves should be according to the requirements given as follows.

Fig. 14.13 Drift test on a top entry ball valve. *(Courtesy: Elsevier and Springer.)*

Where B is the nominal bore size,

F is the maximum face-to-face length of the valve as per API 6A, and

L_1, L_2, D_1, and D_2 are drift dimensions presented in Table 14.2; the following shall apply:

$L_1 \geq B$

$L_2 \geq F$

$D_1 = B - 0.76$ mm ($= B - 0.03$ in.)

$D_2 = D_1 - 3.30$ mm ($= D_1 - 0.13$ in.)

+tolerance for D_1 and $D_2 = +0.69$ mm ($= +0.027$ in.)

−tolerance for D_1 and $D_2 = -0$.

Electrical continuity test

External corrosion of subsea valves and piping due to seawater is discussed in Chapter 12. All subsea valves, piping, structures, and other components are at risk of external corrosion due to contact with seawater. Two main external corrosion prevention measures are coating and cathodic protection. Corrosion of metals in simple language means losing an electron. Metals with less Nobility such as aluminum, zinc, and magnesium have a tendency to act as an anode and lose electrons more quickly and easily to protect the more Noble materials in a cathode. Thus corrosion can be prevented if an electron is transmitted and given to the metallic material. Fig. 14.14 illustrates the protection of a subsea valve from seawater corrosion through cathodic protection in which a less Noble metal as an anode is sacrificed in favor of the valve and transmits electrons to the valve through a cable. All external parts of the valve, such as the body, bonnet, and bolting, are protected from seawater corrosion via cathodic protection as illustrated in the figure. The electrical

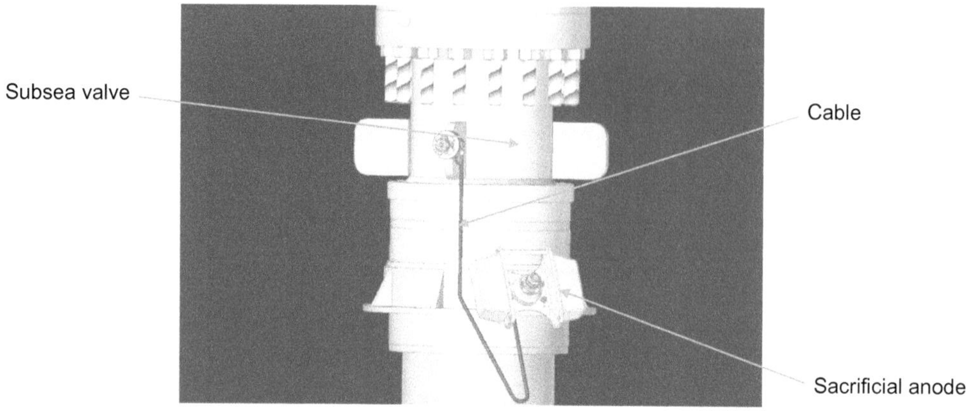

Fig. 14.14 Subsea valve with cathodic protection. *(Courtesy: Imenco AS and Springer.)*

continuity of the body and bonnet, body and bolting, and bonnet and bolting should be tested during the FAT to make sure that these components and the contacts do not provide electrical resistance more than 0.1 Ω while the supply electricity is equal to 12 V. Electrical continuity tests with different criteria are also applied to some valves in the nonsubsea sector of the oil and gas industry to make sure that the valves cannot keep any static electricity inside. An electrical continuity test is very important for valves in the nonsubsea sector to prevent fire. Valves that can keep static electricity inside can catch fire easily due to a combination of electricity, hydrocarbon fluid, and heat.

Bolt tension/torque verification

This type of test is not specified or required according to subsea valve standards. However, it can be performed as per project specifications or end user requirements. Pressure-containing bolts that connect the body and bonnet of gate and top entry ball valves, and the body pieces bolts of side entry ball valves are critical since they join together pressure-containing components. Failure of these critical bolts leads to leakage from the valve to the sea. All the bolts require a certain amount of tension or torque to provide acceptable sealing. This amount of torque or tension is applied by the personnel who assemble the valves during assembly. However, the valves will undergo different pressure tests that may affect the tightness of the valves' bolts. Thus the recommendation is to test the torque or tension of the bolts with a torque measurement tool to make sure that the valve bolting (bolt and nut) provides sufficient tightness. The important point is that this test can be performed on stud bolts and not pressure cap bolts. In addition, some of the valves' bolting components could be hidden after assembly, so they should be excluded from this test. Fig. 14.15 illustrates a ball valve highlighting the body, body closure, and bolting.

Fig. 14.16 illustrates cap screws on a subsea TCG valve; TCG valves are not subject to torque or tension verification tests.

Fig. 14.15 Ball valve highlighting body, body closure, and bolting.

Fig. 14.16 Cap screw bolting for a subsea gate valve.

14.2.4 Actuator tests

14.2.4.1 Actuator pressure tests

An actuator is defined as a mechanical component installed on the top of an industrial valve to control the motion (opening and closing) of the valves. The performance of the valve is very much dependent on the actuator. Actuators can be categorized in general as single acting or double acting. More information about valve actuation is available in Chapter 10. The tests implemented on actuators could be divided into two categories: those applied only on the actuators and those applied to the actuators when they are installed on the valves. In general, pressure tests for the actuators are implemented on the actuator when the actuator is off the valve. This section focuses on tests for hydraulic actuators. As a part of the pressure test, each hydraulic actuator piston and cylinder should be tested separately for validation of structural integrity. The duration and number of pressure tests performed on the actuator depends on the PSL and can be extracted from Table 14.1. PSL 3G is very common for subsea valves and actuators; a PSL 3G actuator should be shell pressure tested twice and leak tested three times. The actuator should be kept in fail-safe closed position. The test pressure, which is 1.5 times of the design pressure, should be applied to the cylinder inlet port. It should be noted that API 17D proposes a test pressure equal to 1.5 times the rated working pressure (RWP), which is less than the proposed value of the pressure test in this text. As an example, if the operating pressure of the hydraulic supply to the actuator is 5000 psi equal to 345 bar, the design pressure would be 1.1 times the operating pressure (345 bar) which is equal to 380 bar. The applied test pressure to the actuator for a shell test is 570 bar. Testing the actuator with the higher pressure proposed before can improve the reliability and safety of actuators. After applying the test pressure, the operator should wait for pressure stabilization. No visible leakage is allowed from the actuator cylinder during the test. In addition, the reduction in hydraulic pressure is a maximum of 20 bar for the actuator shell test. The more restricted requirement for shell test hydraulic pressure reduction could be at maximum 3% of the applied test pressure per hour. Therefore if the pressure test is 570 bar, and the test duration is 15 min, the maximum allowable hydraulic pressure reduction during the shell test of the actuator is equal to $0.03 \times 570 = 17.1$ bar in 1 h. However, since the test duration is 15 min, the maximum allowable pressure reduction during a quarter hour is equal to 17.1 divided by four, which equals 4.275 bar. Therefore if the hydraulic pressure drops below 565.725 bar, this pressure reduction is counted as the leakage rate and the test should be repeated according to the duration given in Table 14.1. The duration of the first and second actuator shell tests is 15 and 3 min, respectively. Fig. 14.17 illustrates and compares single acting (spring return) and double acting hydraulic actuator structures. Both the hydraulic ports and cylinder should be tested for double acting hydraulics actuators.

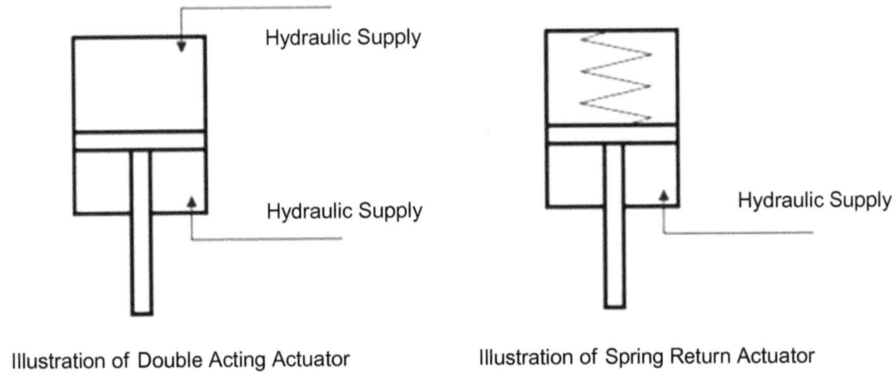

Fig. 14.17 Single acting and double acting hydraulic actuators.

14.2.4.2 Actuator seal test

The other pressure test on the actuators is a seal test, which is done after the shell test with a hydraulic pressure equal to 0.2 times of the hydraulic working pressure or RWP. No leakage is allowed from the piston and cylinder of actuators during the seal test and the allowable hydraulic pressure reduction during this test according to API 17D is 2 bar. The more restricted requirement for allowable pressure drop during the test is at maximum 3% of the test pressure during the test period. Therefore if the hydraulic supply pressure to the actuator were 345 bar, the test pressure would be 0.2 times 345 bar, equal to 69 bar. If the test duration were 3 min, then the maximum allowable pressure drop would be $69 \times 0.03 \times 3/60$ equal to 0.1 bar. The duration and number of tests should be according to the values given in Table 14.1 for the seat of the valves. The test can be applied three times and the duration of the test at each stage should not be less than 3 min. No leakage from the actuator sealings such as the piston sealing or any other internal actuator sealing into the spring chamber is allowed. The seals around the actuator shaft could be considered the most important seals. These seals are considered dynamic since the seal has been placed between at least one moving item, which is the shaft. There are other static seals in the actuator cylinder. Fig. 14.18 illustrates the hydraulic supply port to a single acting hydraulic actuator which should not pass through the dynamic seals around the shaft. Leakage of the hydraulic from the shaft sealing could disturb functioning of the actuator. As an example, actuator may not be able to produce enough force or torque to open the valve completely. Dynamic seals are dual; one works as a backup for the other in case of seal failure. The spring chamber is located at the upper part of the actuator, which is not shown in the figure.

14.2.4.3 Spring chamber hydrostatic test

Double acting hydraulic actuators have two pistons and cylinders, whereas single acting (spring return) actuators have one cylinder and one spring chamber (see the illustration in Fig. 14.17). The spring chamber of a hydraulic actuator is typically pressure compensated

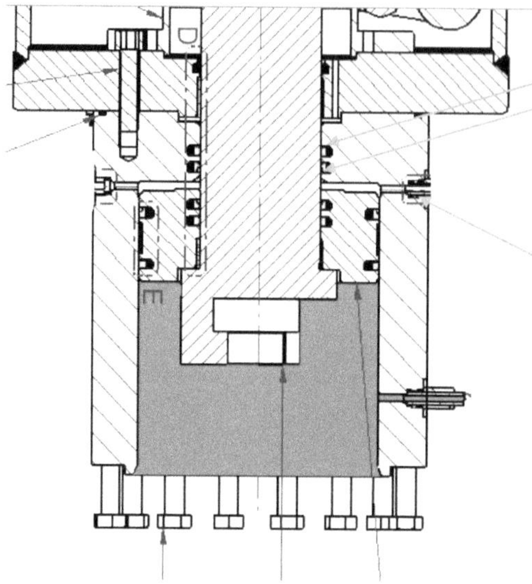

Fig. 14.18 Single acting hydraulic actuator highlighting the dynamic seals around the stem. Note: The *blue arrows* (dark gray in print versions) should be ignored in this figure. *(Courtesy: Springer.)*

with a compensation system and hydraulic fluid, as shown in Fig. 14.19. The compensation fluid in the compensation capsule enters the spring chamber while the valve and actuator are in operation. The compensation capsule is closed type and connected to the actuator casing or shell to provide pressure balance between the casing of the actuator and the seawater head applies as an external pressure on the subsea components including

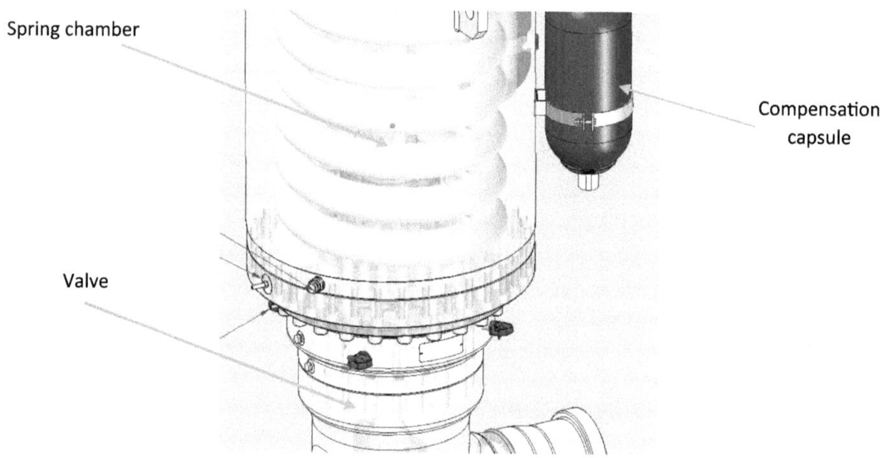

Fig. 14.19 Pressure compensated spring chamber of a single acting hydraulic actuator. *(Courtesy: Springer.)*

the valves and actuators. If no compensation is applied to the actuator installed subsea, the actuator would be imploded due to external seawater head. The pressure of the compensation fluid inside the capsule is dependent on the water depth where the valve and actuator are installed and operated. Let's suppose that the actuator is installed at the depth of 2500 m so the seawater head pressure is roughly equal to 250 bar. For the hydrostatic pressure test of the spring chamber, the compensation cavity should be empty from the compensation fluid, which is typically hydraulic oil. The test fluid should be water including corrosion inhibitor and the test pressure should be 1.5 times of the compensating working pressure which is 250 bar equal to 375 bar in this example. The acceptance criterion is that no leakage is allowed from the spring chamber or the compensation system. The maximum pressure test reduction during the test is 3% of the test pressure during the test period. If the PSL is 3G, then the test can be done in two stages, with a duration of 15 and 3 min, respectively. API 17D does not allow performing this test, like other pressure tests, in less than 3 min.

14.2.4.4 Hydraulic oil cleanliness test

This test is not a pressure test, but is categorized in the pressure test section since it is the main requirement of the hydraulic test fluid. The hydraulic fluid used for testing the actuators should be clean with an acceptable level of particles as per the SAE 4059 or NAS 1638 standard. In addition, all of the hydraulic connections to the hydraulic actuators, such as fittings and tubes, should be clean and the hydraulic actuators should be assembled in a clean room. All components used for assembly, like couplings, pipe work, etc., should be flushed with clean hydraulic fluid before assembly. The flushing should be done with turbulent hydraulic fluid with a Reynolds number higher than 4000 to ensure the cleanliness of the flushed components. Any particles in the actuator system and hydraulic fluid can cause malfunction of the actuators during testing and operation. The class of hydraulic fluid to be used is normally specified by the valve purchaser and should be applied and implemented by the actuator manufacturer. Table 14.3 presents cleanliness levels from 00 to 12 as per NAS 1638; this indicates the number of particles in 10 mL based on different particle sizes. Table 14.4 also presents the cleanliness level from 000 to 12 as per SAE 4059. MTD stands for medium test dust and ACFTD stands for air cleaner fine test dust. These two acronyms are mentioned in Table 14.4.

Fig. 14.20 illustrates actuator flushing after the test or assembly with clean hydraulic fluid according to NAS 1638 level 6. The hydraulic fluid used for flushing should be exactly the same as the type of hydraulic fluid used in the field for the operation of the hydraulic actuators. The cleanliness of the hydraulic fluid to be used for the pressure test of hydraulic actuators should be verified before the test to comply with the required cleanliness level. The cleanliness report of the hydraulic actuator should be attached to the FAT report. Typically, a filtration system is used to achieve the required cleanliness of the hydraulic fluid supply to the actuator.

Table 14.3 Particles limit per 100 mL according to different levels of NAS.

Size range		5–15 μm	15–25 μm	25–50 μm	50–100 μm	>100 μm
NAS classes (based on maximum contamination limits, particles per 100 mL)	00	125	22	4	1	0
	0	250	44	8	2	0
	1	500	89	16	3	1
	2	1000	178	32	6	1
	3	2000	356	63	11	2
	4	4000	712	126	22	4
	5	8000	1425	253	45	8
	6	16,000	2850	506	90	16
	7	32,000	5700	1012	180	32
	8	64,000	11,400	2025	360	64
	9	128,000	22,800	4050	720	128
	10	256,000	45,600	8100	1440	256
	11	512,000	91,000	16,200	2880	512
	12	1,024,000	182,400	32,400	5760	1024

14.2.4.5 Actuator characterization test

This test is performed when the actuator is off the valve and on the test bench, same as the pressure test. The aim of this test is to find a relationship between the hydraulic pressure value and the required operating torque. Therefore the output torque can be measured from the actuator when the hydraulic supply pressure is 100% equal to 345 bar, 75% equal

Table 14.4 Particle limit per mL according to different levels of SAE 4059.

	Maximum contamination limits (particles per mL)					
MTD ACFTD Size code	>4μm(c) >2 μm A	>6 μm(c) >5 μm B	>14 μm(c) >15 μm C	>21 μm(c) >25 μm D	>38 μm(c) >50 μm E	>70 μm(c) >100 μm F
000	195	76	14	3	1	0
00	390	152	27	5	1	0
0	780	304	54	10	2	0
1	1560	609	109	20	4	1
2	3120	1220	217	39	7	1
3	6250	2430	432	76	13	2
4	12,500	4860	864	152	26	4
5	25,000	9730	1730	306	53	8
6	50,000	19,500	3460	612	106	18
7	100,000	38,900	6920	1220	212	32
8	200,000	77,900	13,900	2450	424	64
9	400,000	156,000	27,700	4900	848	128
10	800,000	311,000	55,400	9800	1700	256
11	160,000	623,000	111,000	19,600	3390	512
12	320,000	1,250,000	222,000	39,200	6780	1024

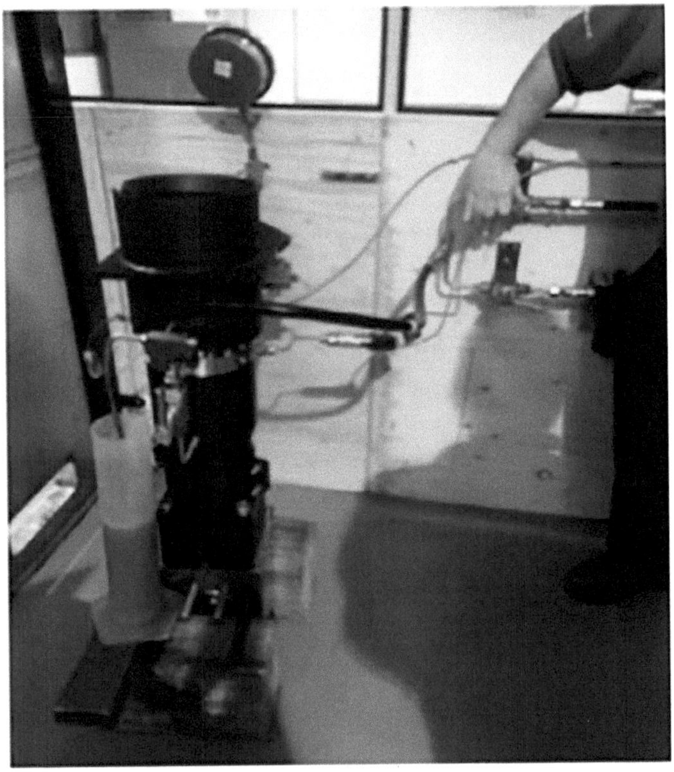

Fig. 14.20 Actuator flushing after the test. *(Courtesy: Pacson.)*

to 259 bar, 50% equal to 173 bar, and 25% equal to 86 bar. The actuator should be in closed position and the above-mentioned pressure values should be applied to the actuator for opening. The opening torque values based on each supply pressure are measured. The test should be repeated after releasing the hydraulic pressure from the piston and cylinder and measuring the spring torque. It should be noted that the spring torque should always be constant.

14.2.4.6 Actuator full opening and closing test

In this test, the maximum supply hydraulic pressure equal to 345 bar is applied and the actuator is checked/tested to open the valve in fully open position. The more critical scenario for keeping the actuator in fully open position is when the hydraulic pressure is dropped down to the minimum hydraulic supply pressure. The minimum hydraulic supply pressure is typically provided to the actuator supplier by the purchasing company. As an example, the minimum hydraulic supply pressure could be 295 bar. This pressure is very important since it should be capable of keeping the valve in fully open position. The other type of test that is recommended is to apply 345 bar hydraulic pressure, then reduce the pressure gradually until the actuator is closed completely. The hydraulic pressure that leads to fully closing the actuator and the valve should be recorded. This pressure should not be higher than 295 bar. The reason is that if 300 bar is registered as the highest pressure that keeps the actuator in closed position, then for sure the minimum supply pressure, which is 295 bar, cannot keep the actuator open. This means that during operation when the pressure of the hydraulic supply fluctuates to 295 bar, the actuator gets closed which is not conductive to operation.

14.2.5 Manual override gear test

14.2.5.1 Gear seawater sealing test

Tests of manual gears are not covered by the relevant API and ISO standards for subsea valves and actuators. The proposed test on the gear box in this section is to make sure that there is no communication between the internal part and external seawater due to any hole or leakage point on the gear box casing. Air with low pressure such as 0.5 or 1 bar should be injected in to the gear box and no leakage should be observed from outside. Soap containing water can be placed on the external surface of the gear box casing for better observation and detection of any possible leakage. When there is no leakage from inside the gear box to the outside, we can then make sure that there should not be any leakage from the outside to the inside of the gear box. It should be noted that the gear box is normally supplied with a compensation system, which should be empty during the gear box test.

14.2.5.2 Gear functional test

This test is a part of the valve function test explained in Section 14.2.3.2.

14.2.6 Relief valve test

A relief valve may be installed on the valve bonnet or upper part of the valve body to release possible leakage from the valve toward the actuator to the sea and prevent fluid pressure accumulation in the actuator. Another application of the pressure relief valve could be on the cavity of valves in liquid service with double isolation and bleed type 1 (DIB1) with two double piston effect seats. Furthermore, the compensation system connected to the spring chamber of a spring return actuator or gear box could have two relief valves to control the pressure inside the compensation system. If the pressure inside the compensation cavity exceeds the seawater head pressure by some amount, such as 3 bar, for example, the relief valve can release the overpressure in the compensation system to the sea. If the seawater head pressure exceeds the compensation system internal pressure by more than 3–5 bar, as examples, the relief valve should release some external pressure from the sea to the compensation system. All pressure relief valves should be tested to make sure that they are capable of relieving pressure at 1 bar above the set pressure values. A detailed explanation of relief valves is provided in Chapter 7.

14.2.7 Factory acceptance test conclusions

Subsea valves and actuators should be in service and operation for a long time—25 or even 30 years—without any maintenance. Therefore very restricted tests in comparison to the tests applied to industrial valves and actuators in the nonsubsea sector of the oil and gas industry should be applied to validate the product's safety and reliability during its life cycle. The testing of subsea valves is not limited to FAT and qualification tests, which should be done to validate newly designed valves, and are explained in the next section. Some of the FAT tests are based on international standards and some are proposed in this book based on industrial experiences. Table 14.5 summarizes all of the proposed FATs and indicates whether their requirements are coming from an international standard or an industrial practice.

All of the new tests proposed in this book to improve the quality and reliability of valves are summarized in Fig. 14.21.

Table 14.5 List of the reviewed tests based on standards and industrial experience.

Test name	Section (Application)	Standard recommendation	Industrial practice
Valve hydrostatic body test	14.2.3.1 Valve	✓	
Valve hydrostatic seat test	14.2.3.1 Valve	✓	
Valve body gas test	14.2.3.1 Valve	✓	
Valve seat gas test	14.2.3.1 Valve	✓	
Valve function test	14.2.3.2 Valve and gear		✓
Cavity test	14.2.3.3 Valve	✓	
Backseat test	14.2.3.4 Valve	✓	
Stem sealing tests (All)	14.2.3.5 Valve		✓
Body/bonnet and body pieces sealing tests	14.2.3.6 Valve		✓
Drift test	14.2.3.7.1 Valve	✓	
Electrical continuity test	14.2.3.7.2 Valve	✓	
Bolt tension/torque verification	14.2.3.7.3 Valve		✓
Actuator shell test	14.2.4.1.1 Actuator	✓	
Actuator seal test	14.2.4.1.2 Actuator	✓	
Spring chamber hydrostatic test	14.2.4.1.3 Actuator	✓	
Hydraulic oil cleanliness test	14.2.4.1.4 Actuator	✓	
Actuator characterization test	14.2.4.1.5 Actuator		✓
Actuator full opening and closing test	14.2.4.1.6 Actuator		✓
Gear seawater sealing and function test	2.5 Gear box		✓
Relief valve test	2.6 Compensation system and valve		✓

Fig. 14.21 Proposed tests based on industrial practices.

14.3 Qualification tests

Components that are tested during qualification testing along with valves and actuators include the compensation system, ROV bucket, and position indicator. Four main qualification tests—pressure-temperature cycling, endurance life cycle test, hyperbaric test, and disassembly and inspection—are explained in this section.

14.3.1 Pressure-temperature cycling

This type of test is known as PR2 level design validation and is covered by the API 6A standard. PR2F validation is applicable for valves that are subject to a sand slurry test as per API 6AV1. Normally, new designs require design validation through a qualification test or finite element analysis (FEA). The main question is, what changes in the design of a valve and actuator should be considered a new design? API 6DSS lists some of the parameter changes that lead to design change as follows:

- Change of the valve type: As an example, gate valves are typically used for subsea manifolds in sizes such as 5 1/8″ or any other sizes less than 8″. If the decision is to change the gate valve on the branch manifold in 5 1/8″ to a ball valve, most likely the ball valve requires a qualification test.

- Change in pressure class or rating;
- Valve configuration: As an example, changing a full-bore valve to a reduced bore means changing the valve configuration and most likely requires a new qualification test.
- Body style change: Ball valves could be side entry, top entry, and two-piece or three-piece body type. Changing a ball valve from side entry to top entry requires a qualification test if no validated design for top entry is available. If the top entry design has been validated through a field experience or previous validation test, then no qualification is required.
- Type of sealing change: Different sealing is available for valves, such as O-ring, lip seal, metallic seal, V-pack, etc.
- Change in the sizing of valve internals such as the seats, closure member, or stem;
- Change in the speed of valve operation requires new qualification;
- Change in the metallic materials to those that have less yield strength for critical parts such as the valve body and bonnet, stem and bolting, and controlling parts such as the seat and closure member;
- Change in the size and roughness of the seals;
- Change in the hard facing;
- Change in the hard-facing process or thickness.

In some cases, scaling could be performed instead of validation of the valve design through qualification tests. Scaling means that if a specific design of a valve has been validated through a test or field experience or any other means of validation, the same design of valve, including all similar parameters, that is just one size smaller or one larger more could be accepted as a qualified design without applying the test. Valve suppliers are sometimes asked to apply FEA for design validation of valves and actuators that have been approved through scaling.

Different pressure-temperature cycling procedures are given in API 6A and API 6DSS. The minimum and maximum temperature during the test should be specified in the test procedure. As an example, if the minimum and maximum design temperature values specified in the valve data sheet are −29°C and 121°C and the validation PR2 test will be performed for the whole temperature range, then the minimum and maximum test temperature would be −29°C and 121°C, respectively. The other important temperature during pressure-temperature range is the room temperature, which is defined as being between 4°C and 50°C. In addition, the test pressure should be defined in the procedure. The test pressure could be defined as the maximum working or design pressure of the valve, as examples. First, the test procedure given in API 6A is reviewed according to Fig. 14.22. The small letters in Fig. 14.22 represent the steps taken during the test. There are three thermal cycles in the test procedure: the first one is from step (a) to step (e), the

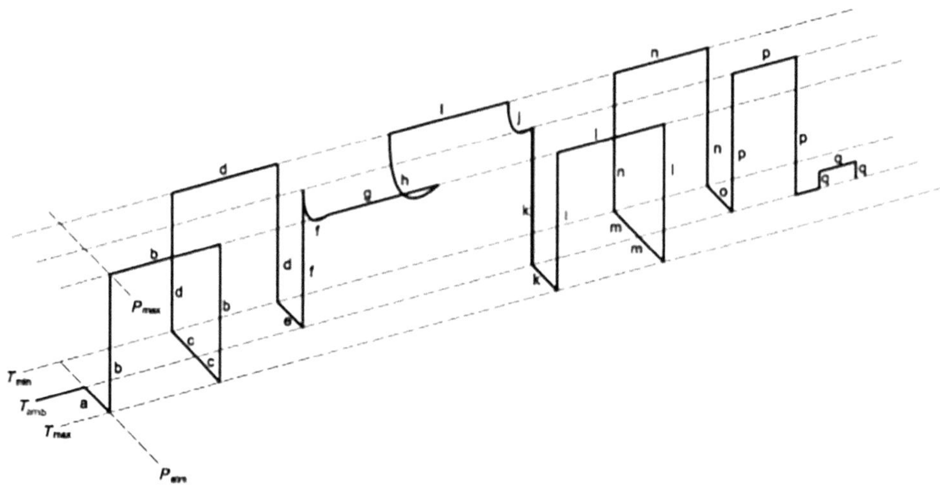

Fig. 14.22 Pressure-temperature testing steps according to API 6A.

second is from (i) to (j), and the third is from (k) through (o). A summary of the steps is as follows:

Step (a): Set the valve temperature to ambient temperature between 4°C and 50°C with atmospheric pressure and then raise the temperature to maximum (e.g., 121°C);

Step (b): Apply the test pressure, hold it for 1 h, and then release the pressure;

Step (c): Reduce the pressure to minimum;

Step (d): Apply the test pressure and hold it for a minimum of 1 h and then release the pressure;

Step (e): Return the temperature to ambient temperature;

Step (f): Apply the test pressure at ambient temperature between 4°C and 50°C and then change the pressure to a value between 50% and 100% of the test pressure while raising the temperature to the maximum;

Step (g): Hold the maximum temperature with 50%–100% of the test pressure for 1 h;

Step (h): Reduce the temperature to the minimum while maintaining the pressure between 50% and 100% of the test pressure;

Step (i): Hold the pressure and temperature given in step (h) for 1 h;

Step (j): Raise the temperature to ambient while the pressure is kept between 50% and 100% of the test pressure;

Step (k): Release the pressure and then raise the temperature to the maximum;

Step (l): Apply the test pressure and hold for a minimum of 1 h and then release the pressure;

Step (m): Reduce the temperature to minimum;

Step (n): Apply the test pressure, hold for a minimum of 1 h, and then release the pressure;

Step (o): Raise the temperature to ambient;

Step (p): Apply the test pressure, hold for a minimum period of 1 h, and then release the pressure;

Step (q): Apply 5%–10% of the test pressure, hold for a minimum period of 1 h, and then release the pressure.

The total number of cycles in PR2 pressure-temperature testing is 200. The first 160 cycles should be done at ambient temperature condition or room temperature in step (a) before raising the temperature. Twenty cycles are recommended to be done at high-temperature condition during step (b) as an example. The other 20 cycles should be done at low-temperature condition during step (c) as an example.

The pressure-temperature test is simpler and contains fewer steps in API 6DSS compared to API 6A. Fig. 14.23 illustrates the required steps for a PR2 test for subsea valves according to the API 6DSS requirements. The definitions of the pressure and temperature values are the same as those in API 6A. The description of each step is given as follows:

Step (a): Apply the test pressure at room temperature and maintain at 50%–100% of the test pressure while raising the temperature to the maximum;

Step (b): Hold the pressure at 50%–100% of the test pressure with the temperature at the maximum for 1 h;

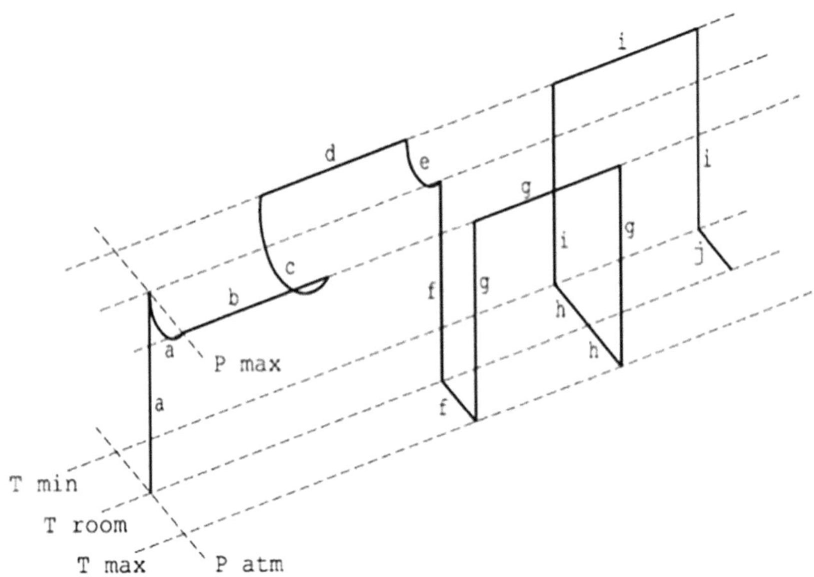

Fig. 14.23 Pressure-temperature testing steps according to API 6DSS.

Step (c): Reduce the temperature to the minimum when the pressure is between 50% and 100% of the test pressure;

Step (d): Keep the pressure and temperature in step (c) for 1 h;

Step (e): Raise the temperature to room temperature while the pressure is between 50% and 100% of the test pressure;

Step (f): Release the pressure and raise the temperature to the maximum;

Step (g): Apply the test pressure, hold for 1 h, and then release the pressure;

Step (h): Reduce the temperature to the minimum;

Step (i): Apply the test pressure, hold it for 1 h, and then release the pressure;

Step (j): Raise the temperature to room temperature.

Again, 200 pressure/load cycles and three temperature cycle tests should be performed during the pressure-temperature cycling: 160 cycles at ambient temperature, 20 cycles at low temperature, and 20 cycles at high-temperature conditions. A pressure-temperature cycle test is normally performed before the two other qualifications tests, namely hyperbaric and endurance testing. The important point is that API 6DSS lists different types of tests such as a hydrostatic body test, seat tests, etc., before and after pressure-temperature cycle tests. The sequence of the design validation tests before and after the PR2 test is given in Table 14.6, extracted from API 6DSS.

Note: Hyperbaric and endurance design validation tests are done after performing the tests provided in Table 14.6. A short description of each test is provided in this section:

1. *Hydrostatic body test*: This test should be performed as per the requirements given in Section 14.2.3.1 of this chapter. In general, 1.5 times of the design pressure should be applied for 1 h as a minimum and no leakage is allowed from the valve during the test.
2. *Seat test*: This test should be performed as per the requirements given in Section 14.2.3.1. The type of seat test in terms of hydrostatic or gas test has not been clarified in the relevant API and ISO standard. But the recommendation is to apply a hydrostatic seat test for a duration of 1 h as a minimum. The allowable leakage rate should be according to the valve data sheet. As an example, zero leakage or leak rate A as per ISO 5208 can be considered for a gate valve.
3. *Force/torque measurement*: The breakaway or break to open (BTO) torque should be measured during this test and compared with the torque value given by the valve manufacturer in the valve drawings and/or valve data sheets. The valve should be opened against maximum working pressure across the valve while one side of the valve is pressurized with water equal to the working pressure.
4. *Cycling at room temperature*: One hundred and sixty cycles (opening and closing) should be done and the operation speed during the cycles should be measured and should be according to the project documents. The test fluid is water at ambient temperature between 4°C and 50°C.
5. *High-pressure gas seat test*: A high-pressure gas seat test with design pressure or 1.1 times of the design pressure should be applied for 1 h and the leakage from the seat

Table 14.6 Design validation tests before and after pressure-temperature cycling according to API 6DSS.

Item #	*Design validation tests*
1	Hydrostatic body test
2	Seat test
3	Force/torque measurement
4	Open/close cyclic dynamic pressure test at room temperature (160 cycles)
5	Gas seat test
6	Open/close cycling dynamic pressure gas test at maximum temperature (20 cycles)
7	Gas body test at maximum temperature
8	Gas seat test at maximum temperature
9	Low-pressure seat test at maximum temperature
10	Open/close cycling dynamic pressure gas test at minimum temperature (20 cycles)
11	Gas body test at minimum temperature
12	Gas seat test at minimum temperature
13	Low-pressure seat test at minimum temperature
14	Raise the temperature to room temperature
15	*Pressure-temperature cycling*
16	Gas body test at room temperature
17	Gas seat test at room temperature
18	Low-pressure seat test at room temperature
19	Final force or torque measurement

should be according to the value given in the valve data sheet as per ISO 5208. The temperature is not mentioned in the standard but the interpretation of this author is that this test should be done at room temperature.

6. *Cycling at minimum and maximum temperature*: Twenty cycles at minimum and 20 cycles at maximum temperature are required. The test fluid is gas instead of water, which was used in the cycling at room temperature, test #4 in this section. The speed of operation is measured and compared with the values associated with cycling speed in the valve data sheet.
7. *Gas body test at minimum and maximum temperature*: Ball and gate valves should be partially open and check valves should be pressurized from the upstream side. The test pressure should be equal to the rated working pressure but can be increased to 1.1 times of the rated working pressure or design pressure. Rated working and design pressure are normally equal for subsea valves as per the API 6A and 17D standards. The duration of the test is 1 h for each temperature and no leakage is allowed.
8. *Low-pressure seat test at minimum and maximum temperature*: The valve should be subject to 5.5 or 20 bar or any pressure between 5% and 10% of the rated working pressure. The holding period for each test is 1 h and the allowable leakage should be based on the allowable leakage rate given in the valve data sheet (e.g., ISO 5208).
9. *Pressure and temperature cycles*: Refer to Section 14.3.1.

10. *High-pressure body gas test at room temperature*: The test implementation and requirements are similar to test #7, explained in this section, with one exception; the present test is performed at room temperature.
11. *Low-pressure gas seat test at room temperature*: The implementation and requirements of this test are similar to test #8, explained in this section, with one exception: it is performed at room temperature.
12. *Final force and torque measurement*: The tests may require only to measure the breakaway and running torques. But the recommendation is to measure all applicable torques or forces for operation of the valves as per author experience in the industry. The reason is that all the operating valve torques are essential. The other recommendation according to author experience is to measure the opening torque or force against the differential working pressure. The measurement of torque during closing could be done against the maximum operating pressure with zero pressure in the cavity of the valve. The measured torque during the test should not exceed the nominal torque values given by the valve supplier in the data sheets and drawings of the valves.

Fig. 14.24 illustrates an actuated gate valve undergoing PR2 temperature cycling at low temperature.

14.3.2 Endurance testing

Endurance testing is also called life cycle/endurance testing or endurance cycling testing. The main purpose of this test is to evaluate the possibility of long-term wearing and to assess the characteristics of the subsea equipment including valves and actuators. The temperature of the test can be selected from the operating temperature given by the manufacturer. The valve and actuator should undergo 200 operating cycles under full rated working pressure as the operational conditions.

14.3.3 Hyperbaric testing

Hyperbaric testing is very essential for subsea valves, since it is the only test that can demonstrate that the valve and actuator are suitable for installation and operation at a specific depth. In fact, this test can simulate deep-water depth conditions, especially water head pressure, by placing the assembled valve and actuator inside a hyperbaric chamber as illustrated in Fig. 14.25. Based on the information received from the valve suppliers, a maximum depth of 5000 m can be simulated in the hyperbaric chamber, which produces approximately 500 bar of pressure to the valve and actuator externals. The dimensions of hyperbaric chambers differ, but a hyperbaric chamber with a length of 7.5 m and a diameter of 2500 mm could be considered very large, maybe even one of the largest, according to current valve subsea technology. The hyperbaric chamber could be equipped with programmable logic control (PLC)—electronic devices to control

Fig. 14.24 Actuated gate valve at low temperature during a PR2 cycle test.

chamber process variants such as temperature and pressure. Calibrated pressure and temperature sensors are also installed on the chamber. A calibrated ROV torque tool is required for operation of the ROV bucket installed for manual operation of the valves or to override the hydraulic actuators. A high-pressure water-cooling system should be supplied to the chamber and maintained within a temperature range of 4°C, ambient

Fig. 14.25 Preparation for testing an assembled subsea valve and actuator in a hyperbaric chamber. *(Courtesy: Advanced Valve Technology.)*

temperature, according to the API 17D requirement. A subsea camera is required in the chamber to visually monitor the underwater conditions and the valve inside the chamber. The pressure of the chamber should be set at 1.1 times of the water depth head pressure as per the API 6DSS requirement.

14.3.3.1 Before hyperbaric testing

The assembly of the valve and actuator should be placed inside the chamber for hyperbaric testing together unless the chamber does not have enough space for the assembly. In that case, two separate tests should be performed on the valve and actuator individually. The other important point is to apply FEA by simulating the external water pressure on the valve and actuator before implementing the qualification hyperbaric test. The following steps should be taken before implementation of the hyperbaric test according to API 6DSS. Additionally, API 17D annex L provides a general guideline for the hyperbaric qualification test:

- The hyperbaric chamber shall contain atmospheric pressure and the valve and actuator assembly should be placed inside the hyperbaric chamber.
- The valve ends shall be closed and all the connections shall be connected to the valve. All test ports shall be connected.
- The valve should be kept in half open position and all the air removed from the valve internal and then the valve should be filled in with the test fluid. At this stage, the hyperbaric chamber should be empty of the test fluid (water).
- Perform the hydrostatic shell test at 1.5 times of the rated pressure typically equal to the design pressure. The ball and gate valves should be in partially open position during this test. The minimum test duration, also called holding time, should be 15 min. The acceptance criteria are no leakage from the valve or no pressure reduction during the test.
- Fill the chamber with the test fluid (water).
- Start with pressurizing the chamber in 5% and 10% of the water head pressure at depth before applying the complete water head test pressure.

14.3.3.2 During hyperbaric testing

Two hyperbaric tests are performed on valves and actuators: one is a hyperbaric ingress test and the other is a hyperbaric endurance test. Each of these tests is explained in this section. In total, four test steps are performed during the hyperbaric test as follows:

1. *Hyperbaric ingress test*: The hyperbaric ingress test is only applicable for subsea valves. Gate and ball valves should be in half open position during this test. The hyperbaric chamber should be filled in with water as the test fluid with the pressure equal to 1.1 times of the water depth design pressure. The density of the fluid should be taken into account for calculation of the pressure test. As an example, water test fluid with a

relative density of 1.03 provides higher pressure at a certain depth compared to water test fluid with a lower relative density. The valve internal pressure should be monitored continuously. Any increase inside the valve means that the valve should be rejected for the hyperbaric ingress test. The alternative way to monitor leakage to the valve is through a test port. A test port should be considered at the bottom part of the primary external stem seal. In case of having primary and secondary stem sealing, the test port should be designed between these two seals. Any leakage through/from the test port leads to rejection of the valve during the hyperbaric ingress test. The other important point is to fill in the compensation system with the compensation fluid to prevent damage of the actuator or gear box during the hyperbaric test. The procedure on how to fill the compensation system should be provided by the valve and actuator supplier in the installation, operation, and maintenance (IOM) manual.

2. *Hydrostatic shell test in hyperbaric condition*: The valve should be filled in with pressurized water equal to 1.5 times of the valve working pressure. At the same time, external hyperbaric pressure equal to 1.1 times of the water depth should be applied. The valve internal pressure shall be monitored constantly. The minimum duration of this test is 15 min. No leakage is allowed from the valve and the pressure change should be a maximum of 3% of the test pressure per hour. Any increase to this pressure reduction limit leads to rejection of the valve.
3. *Hydrostatic seat test*: The valve is kept in closed position and water test pressure equal to 1.1 times the valve working pressure or pressure rating (it could be more than 1.1, such as 1.25 times the working pressure rating, as per project specification) is applied sequentially to the seats of the valves. The amount of leakage from each seat is measured and should be less than the allowable leakage rate given in the valve data sheet. The valve is kept in hyperbaric condition, as before. The duration of each test should be at least 15 min. The seat test should be done in three stages as per Table 14.1, which means six tests in total (three sets of tests on two seats). The pressure should be reduced down to zero after each set of tests on both seats, and the valve should be cycled to fully open and closed after each test. The next set of tests is repeated after this cycling. Therefore, in total, three cycles shall be performed during the hydrostatic seat test.
4. *Hyperbaric endurance test*: This test requires 200 cycles of the subsea valve by means of an actuator or ROV torque tool or both against differential pressure equal to the rated working pressure in hyperbaric condition. Hyperbaric condition means that the valve is exposed to external pressure of water equal to 1.1 times of the water depth design pressure. The opening and closing force (applicable for gate valves) or torque (applicable for ball valves) should be recorded for every 10 cycles. If the actuated valve is equipped with ROV interface, 10% of the cycles (equal to 20) should be performed with the ROV torque tool and 180 cycles with the actuator. Hyperbaric endurance

testing is applicable to gate, ball, and choke valves, excluding check valves. The acceptance criterion for testing the valve is to have the measured torque and force within the limits given by the valve supplier.

14.3.3.3 After hyperbaric testing

Three test steps should be performed after the hyperbaric endurance test. These tests are a hydrostatic shell test at ambient condition, a hydrostatic seat test at ambient condition, and valve disassembly and visual inspection. Valve disassembly and visual inspection is explained in the next section.

1. *Hydrostatic shell test at ambient condition*: This test is a single stage shell test in which the valve internals are pressurized with 1.5 times of the working pressure while valves such as ball and gate are in half open position. The duration of the test is 15 min minimum and no leakage is allowed.
2. *Hydrostatic seat test at ambient condition*: This test is a typically one stage seat test with a minimum duration of 15 min.

14.3.3.4 Disassembly and inspection

The valve and operator (gear or actuator) as well as the position indicator are disassembled after performing all the qualification tests for a different type of inspection. The inspections could be visual or through nondestructive tests such as magnetic particle or liquid penetration tests. In addition, a dimensional check for some valve components such as seals may be performed. No galling or wearing is allowed to the metallic parts during the inspection. However, valves seals typically wear and should be changed. In some projects, the end user requires to receive the valve and actuator prototype that has undergone the qualification tests, so the valve manufacturer should refurbish them by changing the compensation oil and worn seals before delivery to the client. The figures provided in this section illustrate the disassembly and inspection of a valve and gear operator after the qualification tests. Fig. 14.26 illustrates the disassembly of a stem from a subsea ball valve. Fig. 14.27 illustrates the stem of a ball valve after removal. The pink component (light gray in print versions) on the top of the stem is a lifting lug for lifting purposes. The groove located on the top section of the stem is a place for the installation of a stem key that is parallel to the ball bore or hole and shows the position of the ball when the gear box or actuator and the position indicator are removed from the valve assembly. Fig. 14.28 illustrates a ball valve gear box that is pressure compensated after removal from the valve. Fig. 14.29 shows valve soft sealing such as lip seals and O-rings. A V-pack stem sealing belonging to a ball valve is shown in Fig. 14.30.

Figs. 14.31 and 14.32 illustrate the position indicator for a disassembled ball valve.

Fig. 14.33 shows a ball valve closure after disassembly.

Fig. 14.34 shows a ball and connected trunnion plate removal from a ball valve body during valve disassembly.

Fig. 14.26 Disassembly of the stem from a ball valve.

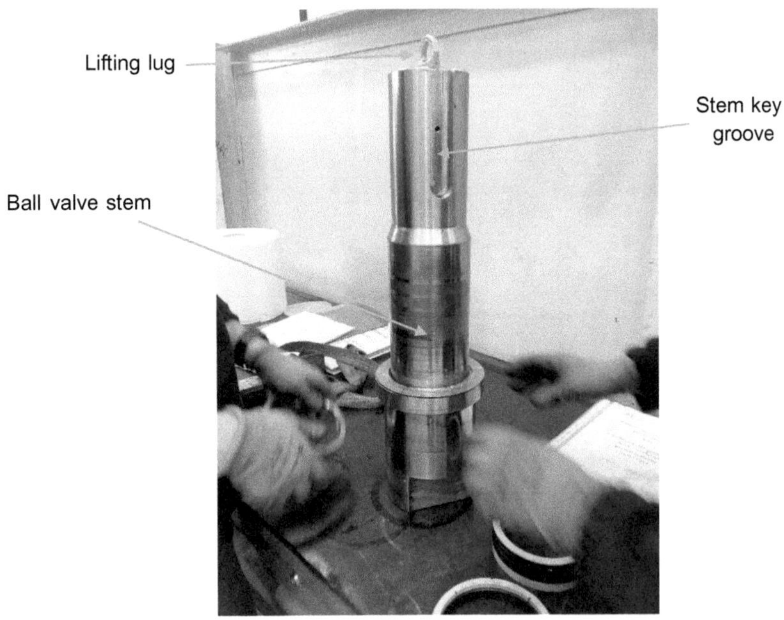

Fig. 14.27 Ball valve stem.

Fig. 14.28 Ball valve gear box after removal from the valve.

Fig. 14.29 Ball valve stem seals.

Fig. 14.30 V-Pack for a ball valve stem seal.

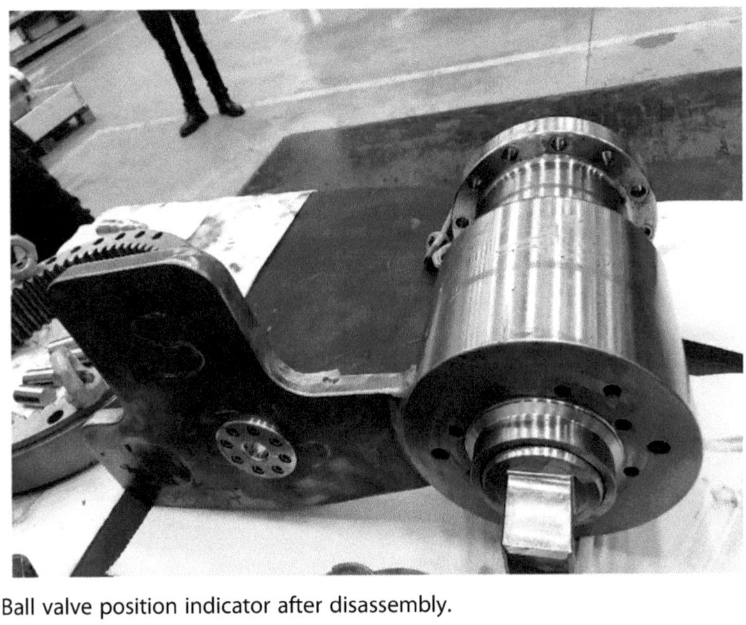

Fig. 14.31 Ball valve position indicator after disassembly.

Fig. 14.32 Ball valve position indicator after disassembly.

Fig. 14.33 Ball valve closure after disassembly.

Fig. 14.34 Ball and trunnion removal from a valve body during valve disassembly.

Fig. 14.35 shows the inside of a ball valve body after the removal of the ball and trunnion. Fig. 14.36 illustrates the ball and trunnion plate of a ball valve after disassembly from the valve. Fig. 14.37 illustrates the seats of a subsea ball valve after disassembly. Fig. 14.38 illustrates a manually ROV-operated slab gate valve after qualification tests that is ready for disassembly and inspection. Fig. 14.39 illustrates an ROV bucket removal from the top of a gate valve. The ROV bucket interface is according to API 17H class 4. The internal gears have been removed from the valve after ROV disassembly as illustrated in Fig. 14.40. The gears look fine after 600 cycles during the qualification program. Fig. 14.41 shows test port plugs and pressure relief valves on the bonnet of a slab gate valve after dismantling. The manual operator of the gate valve is pressure compensated with a closed compensation system in carbon steel material that is disconnected from the gate valve as illustrated in Fig. 14.42. Figs. 14.43 and 14.44 show the removal of the upper ROV gear and shaft. Different dimensions of the components should be measured to

Fig. 14.35 Ball valve internal after ball removal.

Fig. 14.36 Ball and trunnion of a ball valve after disassembly.

Fig. 14.37 Seats of a ball valve after disassembly.

check that there is no change in the original dimensions. No scratch or damage should be observed in the upper ROV gear and shaft. The stem nut removal from the valve operator is shown in Fig. 14.45. The next part that is disconnected from the valve operator assembly is override housing, illustrated in Fig. 14.46.

Additional components should be separated from ROV-operated valves, such as the yoke (see Fig. 14.47), upper ROV extension (see Fig. 14.48), and the override shaft (see Fig. 14.49). The dimensions of almost all of the components belonging to the manual override should be checked. Fig. 14.50 illustrates dimensions for a stem nut that should be checked after disassembly.

Fig. 14.51 illustrates disassembling the bonnet and stem of a slab gate valve from the body. Fig. 14.52 illustrates pulling out the bonnet and stem of a gate valve from the body, which in turn leads to movement of the disk to the outside of the valve body. Fig. 14.53 illustrates the gate or disk of a slab gate valve which is removed from the valve.

Fig. 14.54 illustrates the internals of a slab gate valve body after removal of the stem and disk. The seats of the gate valve are still inside the body. The seats of the valve have been removed from the body and are shown in Fig. 14.55. Visual examination should be

Fig. 14.38 ROV-operated gate valve after qualification and before disassembly.

performed on the metallic parts, especially the stem, disk, and seats, to make sure that there is no sign of damage or any scratch on these components. Any sign of damage to the metallic parts means that the valves fail the qualification tests. The pictures do not show any sign of damage to the valves' internals.

Fig. 14.56 illustrates the removal of the stem from the bonnet of a valve. The stem is disassembled from the bonnet as shown in Fig. 14.57.

Fig. 14.58 illustrates the body of a slab gate valve after disassembly of all the components. As mentioned before, a visual examination and dimensional check is performed on the valve components after disassembly. Defects of the components may not be observed visually. Alternatively, a liquid penetration test can be performed. This process is illustrated in Figs. 14.59–14.61.

The preparation of the disk for the test involves cleaning the surface of the disk to remove any grease, paint, dust, etc. Then the penetrant in red color (dark gray in print versions) is applied on the disk. The penetrant may need 5–30 min time. After that, a developer in white color is applied to the surface; this can show any possible defect such as a crack or damaged surface. The third liquid is the cleaner, which is used after the test to remove the penetrant and developer.

Fig. 14.39 ROV bucket removal (disassembly) from a gate valve.

Fig. 14.40 Slab gate operator internal gears after disassembly.

Fig. 14.41 Test port plugs and pressure relief valves after disassembly.

Fig. 14.42 Closed pressure compensation disassembled from a valve.

Fig. 14.43 Removal of ROV upper part gear and shaft from a slab gate valve.

Fig. 14.44 ROV upper part gear and shaft.

Fig. 14.45 Stem nut of an operator.

Fig. 14.46 Removal of the override housing from a gate valve operator.

Fig. 14.47 Yoke of a manual override for a gate valve.

Fig. 14.48 Upper ROV extension.

Fig. 14.49 ROV override shaft.

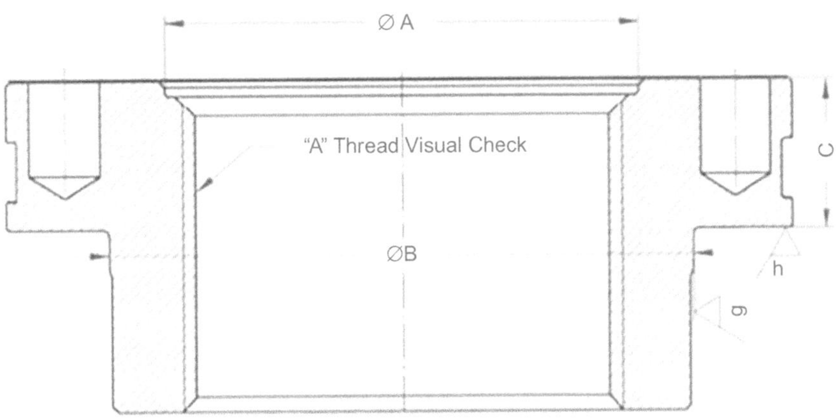

Fig. 14.50 Stem nut dimensions to be checked after disassembly.

Fig. 14.51 Disassembly of a gate valve bonnet and stem from the body.

Fig. 14.52 Pulling out the stem, bonnet, and disk from a slab gate valve.

Fig. 14.53 Gate (disk) of a slab gate valve after removal from the valve.

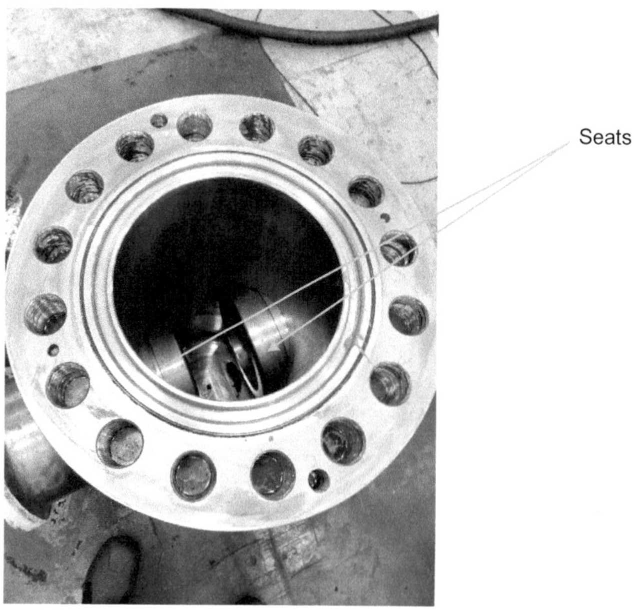

Fig. 14.54 Gate valve body internal after removal of the disk and stem.

Fig. 14.55 Seats of a slab gate valve after removal from the valve.

Fig. 14.56 Disassembling the stem from the bonnet of a slab gate valve.

Fig. 14.57 Stem of a gate valve.

Fig. 14.58 Slab gate valve body after disassembly of all the valve components.

Fig. 14.59 Preparation of a slab gate valve disk for a liquid penetration test.

Fig. 14.60 Slab gate valve disk after applying penetrant.

Fig. 14.61 Slab gate valve disk after applying developer.

Further reading

American Petroleum Institute (API) 6A, Specification for Wellhead and Tree Equipment, twenty-first ed., API, Washington, DC, 2018.

American Petroleum Institute (API) 6DSS, Specification for Subsea Pipeline Valves, third ed., API, Washington, DC, 2017.

American Petroleum Institute (API) 17D, Design and Operation of Subsea Production Systems, Subsea Wellhead and Tree Equipment, seond ed., API, Washington, DC, 2011.

Y. Bai, Q. Bai, Subsea Engineering Handbook, first ed., Elsevier, Atlanta, GA, 2012.

International Organization of Standardization (ISO) 10423, Petroleum and Natural Gas Industries—Drilling and Production Equipment—Wellhead and Christmas Tree Equipment, fourth ed., 2009. Geneva, Switzerland.

International Organization of Standardization (ISO) 13628-4, Petroleum and Natural Gas Industries—Design and Operation of Subsea Production Systems—Part 4: Subsea Wellhead and Tree Equipment, second ed., 2010. Geneva, Switzerland.

International Organization of Standardization (ISO) 14723, Petroleum and Natural Gas Industries—Pipeline Transportation System—Subsea Pipeline Valves, second ed., 2009. Geneva, Switzerland.

International Organization of Standardization (ISO) 5208, Pressure Testing of Metallic Valves, fourth ed., 2015. Geneva.

N. Placido, M. Franco, O. Alexandre, Failure distribution analysis of a novel subsea valve actuator based on reliability data base, IFAC-PapersOnLine 51 (24) (2018) 1247–1254.

J. Praveen, M. Pathan, K. Ansari, Hyperbaric pressure testing of a subsea valve to validate deep water condition, Int. J. Mech. Prod. Eng. Res. Dev. 8 (2) (2018) 1011–1022.

K. Sotoodeh, Valve preservation requirements, Valve World Mag. 21 (5) (2016) 53–56.

K. Sotoodeh, Valve failures, analysis and solution, Valve World Mag. 23 (11) (2018) 48–52. KCI publishing.

K. Sotoodeh, A review on subsea process and valve technology, J. Marine Syst. Ocean Technol. 14 (4) (2019), https://doi.org/10.1007/s40868-019-00061-4. Springer.

K. Sotoodeh, Actuator selection and sizing for valves, SN Appl. Sci. 1 (2019) 1207, https://doi.org/10.1007/s42452-019-1248-z.

CHAPTER FIFTEEN

High Integrity Pressure Protection System (HIPPS)

Contents

15.1 Introduction

High Integrity Pressure Protection Systems (HIPPS) have been utilized for many years in the oil and gas industry in order to prevent the overpressurizing of pressure facilities. Fundamentally, the main function of a HIPPS is to protect downstream equipment from overpressurization by continuously monitoring the line pressure and automatically closing the isolation valve(s) if the pressure rises above a certain level. A HIPPS is a specific application of a safety instrumented system (SIS) designed according to International Electrotechnical Commission (IEC) standards 61508 and 61511. These two international standards provide requirements for SIS specification, design, installation, operation, and maintenance to ensure that the system is situated and maintained in a safe state. Both standards are generic and do not address HIPPS specifically. All HIPPS components, which will be explained later, should follow IEC requirements to verify and validate their compliance with the functional safety requirements. The sensors and safety instrumentations should be in accordance with IEC 61511. SIS always contain three elements: sensors for

Subsea Valves and Actuators for the Oil and Gas Industry
https://doi.org/10.1016/B978-0-323-90605-0.00002-5

measurement, a logic solver for making decisions, and a final element that performs a task. Traditionally, American Petroleum Institute (API) standard 14C has been used for the implementation of safety systems in the offshore sector of the oil and gas industry; API 17O now directly addresses HIPPS in the subsea sector.

A HIPPS maybe referred to as a high integrity piping protection system because it deals with design and cost studies of piping and/or pipeline systems. API 521 defines HIPPS as an arrangement of instruments, final control elements (e.g., valves, switches, etc.), and logic solvers (see Fig. 15.1). Pressure transmitters monitor pipeline pressure against a predefined limit and send signals to the logic solver where an appropriate decision is taken based on the nature of the signal. The type of signal transmitted from the pressure sensors to the logic solver has different voting logic such as 2 out of 3 (2oo3). One of the key parameters in the design of SIS is the voting arrangement or architecture. The main concept behind using a voting system is redundancy—using extra components to tolerate the failure of one component. Redundancy consideration can reduce the risk of failure and improve the system's safety integrity level, which is explained in detail later in this chapter. There is a different SIS architecture for each voting arrangement such as 2oo3. There are two numbers used for voting: the first is the number of devices that should vote for a trip to occur; the second is the total number of installed devices or components. Therefore, in the case of a 2oo3 vote, two of the devices out of three agree on the shutdown of the system, so this will happen. 2oo2 voting indicates that both devices (two out of two) must agree to shut down the system.

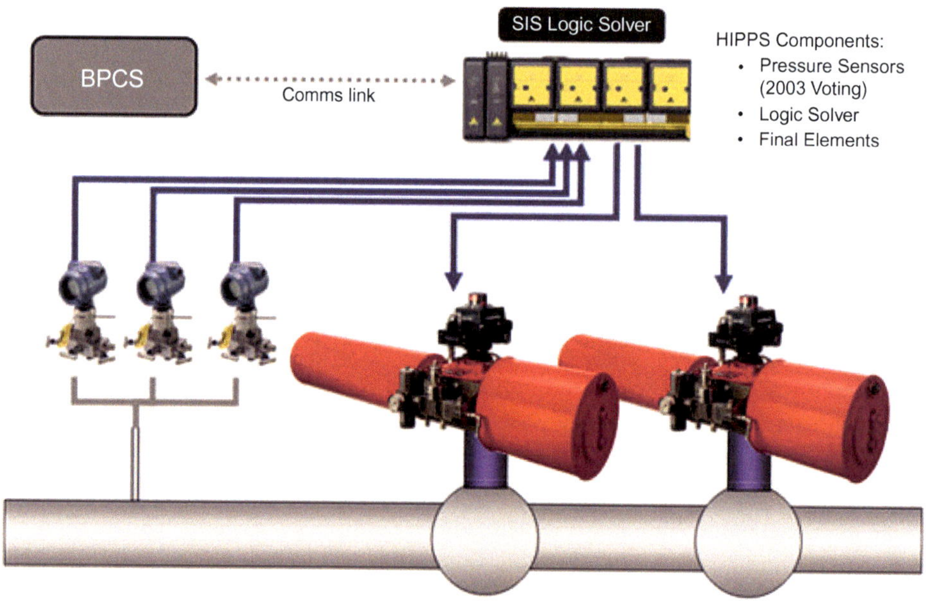

Fig. 15.1 HIPPS architecture. *(Courtesy: Emerson Automation.)*

The logic solver output in the form of an electrical signal results in an action taken by the final element, such as shutting down the process system by closing a valve. A HIPPS is designed to avoid overpressure incidents by removing the source of overpressure or by reducing the probability of an overpressure contingency to such a low level that it is no longer considered likely. HIPPS have been widely applied in both the subsea and topside sectors of the oil and gas industry. The main advantages of HIPPS can be summarized as protecting equipment, personnel and the environment, and reducing the total cost of capital and operation. The different essential challenges in HIPPS implementation can be categorized under the headings of design, performance, technology, and environment.

15.2 HIPPS architecture and layout

HIPPS is the outcome of evolution in the design of the mechanical and electronic safety devices used in process systems. Moving to deeper subsea fields leads to dealing with high(er) pressure and high(er) temperature applications, abbreviated as HPHT. This in turn leads to technical challenges such as the need for thicker, heavier, and more costly piping and valves inside the manifold and on flowlines and risers. One solution is to apply HIPPS in order to reduce the lower rated pressure of the piping and valves after the wellhead, compared to that of the wellhead and Christmas trees. Fig. 15.2 illustrates a field development schematic for a HIPPS module that could be placed between the wellhead and manifold; this would result in the use of lower pressure rated piping and the possibility of using smaller sized piping and a smaller manifold. Two valves connected to the

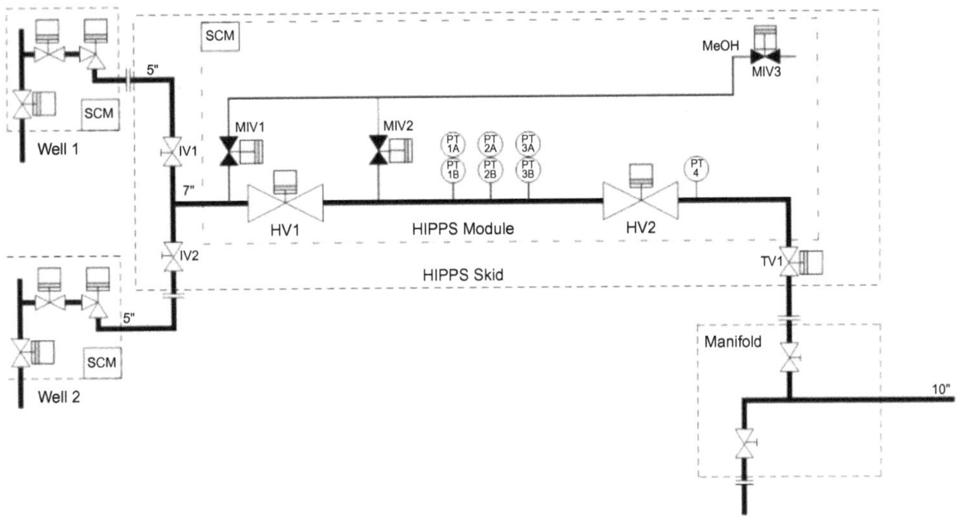

Fig. 15.2 HIPPS module between the wellhead and manifold.

HIPPS, HV1 and HV2, protect the low rating system which is the piping and valves inside the downstream manifold from the overpressure created in the wellhead and Christmas trees.

A HIPPS module can be placed between the Christmas tree and the platform; this set-up is called single well tieback and is shown in Fig. 15.3.

Alternatively, the HIPPS system can be placed after the manifolds and before the flowline; this is called pipeline HIPPS or manifold template tieback (see Fig. 15.4). In this case, the Christmas trees, jumpers, and manifolds should be pressure rated as full wellhead shut in pressure which is higher than the flowline and riser pressure rating. In the case of pipeline HIPPS, the following benefits are achieved:

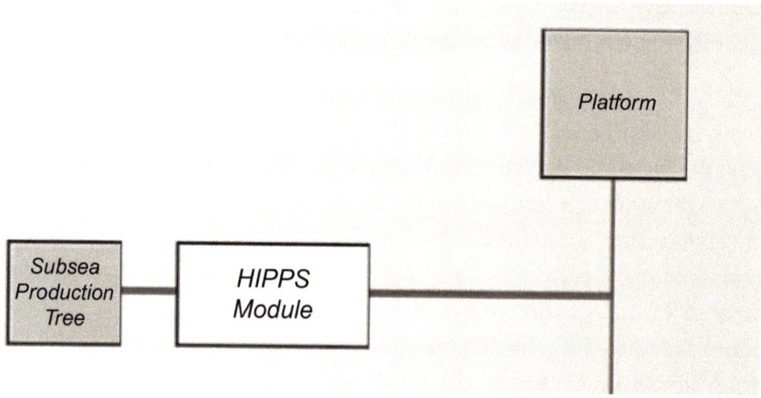

Fig. 15.3 HIPPS module between the Christmas tree and the host facility (platform).

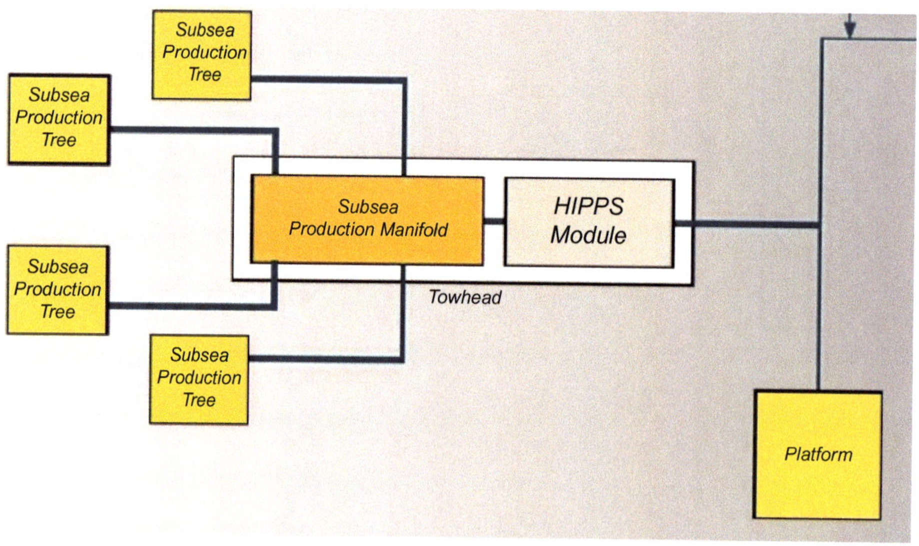

Fig. 15.4 HIPPS module between manifold and host facility (pipeline HIPPS).

- Reducing the pressure rating for the flowline, riser, and topside facilities;
- Reducing the wall thickness and weight of the riser and flowline;
- Less pressure drop in the flowlines and risers which leads to better reservoir recovery;
- Reducing the amount and duration of welding construction work on the subsea pipeline;
- Reducing the overall cost of flowlines and risers as well as their installation costs.

The structure of a HIPPS is like a control loop that includes a sensor (typically three pressure gauges) to measure the pressure. Pressure transmitters are typically combined with the pressure gauge and used to transmit the measured pressure value to the logic solver. Three different pressure transmitters are often installed to provide a higher safety function; since the three pressure transmitters each have a different design, the possibility of all of them failing at the same time under the same specific process condition is much less likely than if they all had a similar design. The combined failure of all of the pressure transmitters could jeopardize the proper functioning of the HIPPS system. Pressure transmitters send a signal to the next element, the logic solver, which compares the measured pressure value with the desired pressure value as per the SIS. The logic solver contains a central processing unit, which is normally software that can be programmed to take the proper action or respond based on the inlet signal. The logic solver sends a signal or response to the next element, which is a solenoid valve.

Solenoid valves are electromechanically operated valves and parts of a valve control system that receives an electrical signal and converts it to mechanical movement. There is a solenoid coil inside the valve that can produce a magnetic field after receiving electrical signals. The magnetic field is used to control the valve operation and provide valve cycling between open and closed positions. Solenoid valves could either be normally open or closed. The solenoid valves for a HIPPS system should normally be open and move to closed position after receiving signals from the logic solver. Solenoid valves are part of a control loop located between the final element (HIPPS valves) and the logic solver. There is a plunger and spring inside the solenoid valve. The plunger acts as the closure member of the valve and is kept upward by the spring force in normally open position (illustrated in Fig. 15.5). Therefore the valve is open and the fluid flows through the valve in normal process conditions. The fluid is hydraulic supply to the valves' actuator. When the electrical signal/current is sent to the solenoid from the logic solver and a magnetic field is produced by the coil, shown in brown color (dark gray in print versions) around the plunger in Fig. 15.6, the magnetic field compresses the spring and pushes the plunger down to close the valve.

The HIPPS valve is the final element. It should be closed very quickly (e.g., less than a second) by the signal coming from the logic solver to the solenoid in the event of high-pressure detection in the low-pressure section of piping after the HIPPS valves. The HIPPS control loop explained in this paragraph is illustrated in Fig. 15.7. The type of valve is normally ball or axial flow. Section 15.8 provides a systematic guideline to select the most suitable valve.

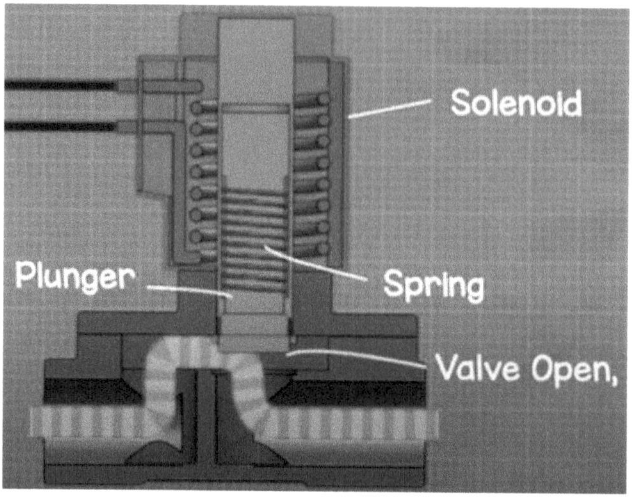

Fig. 15.5 Normally open solenoid valve in open position. *(Courtesy: The Engineering Mindset.)*

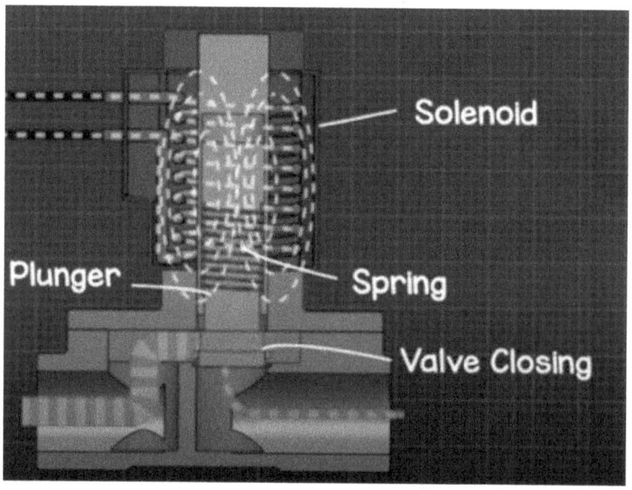

Fig. 15.6 Solenoid valve in closed position after receiving a signal from the logic solver. *(Courtesy: The Engineering Mindset.)*

Fig. 15.8 illustrates the detailed communication between the transmitter, logic solver, valve control system including solenoid valves and the final element, HIPPS valves and actuators. It should be noted that the HIPPS valve has an actuator that may be hydraulic or electrical. The source of power, whether hydraulic or electrical, is supplied from the topside through umbilicals to the subsea units and actuated valves. In addition, HIPPS valves are controlled by a HIPPS subsea control module (HSCM) which is part of a subsea

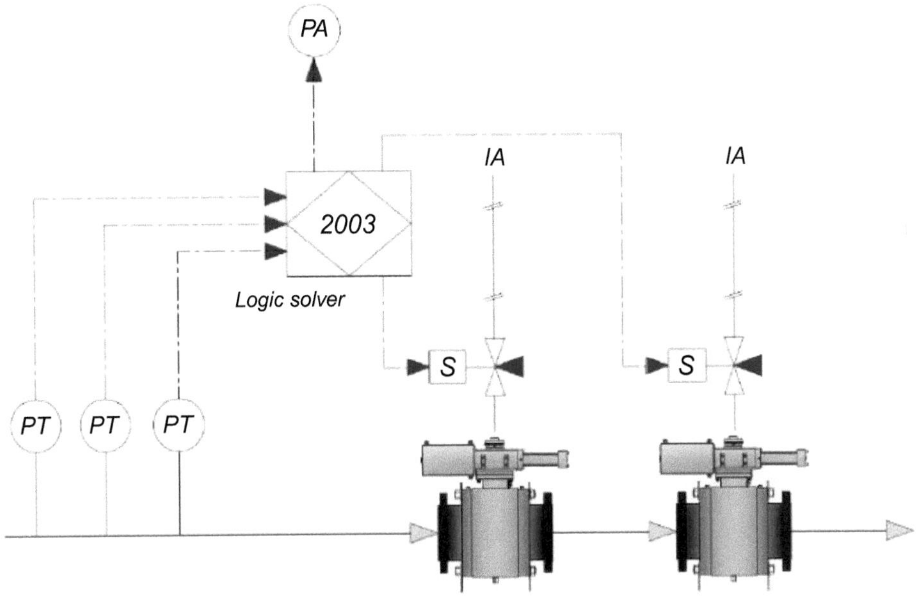

Fig. 15.7 HIPPS control philosophy. *(Courtesy: Ringo Volvulas.)*

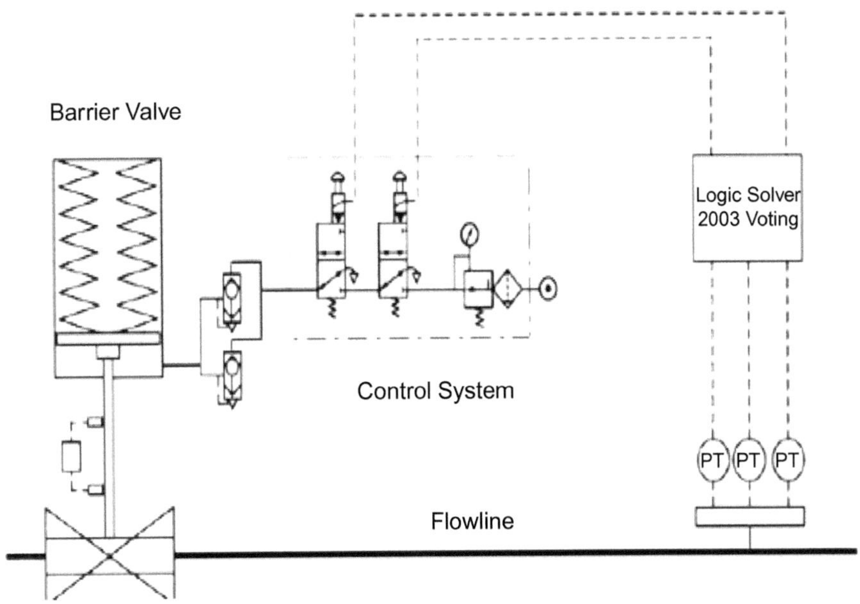

Fig. 15.8 Communication between logic control, control system, and HIPPS valve. *(Courtesy: Elsevier.)*

control module (SCM). HSCM controls all of the HIPPS module activities. HSCM is in contact with the topside master control station, the HIPPS valves and sensors, and the subsea control module. Thus the facilities and components that function in conjunction with the HIPPS system include the master control system, electrical and hydraulic power units, umbilicals, and the subsea control module as a minimum.

15.3 API 17O "HIPPS standard"

API 17O is a recommended practice (RP) and the only standard that addresses subsea HIPPS specifically. There was a concern early on that this standard would conflict with IEC standards. However, API 17O refers to both IEC 61508 and 61511 in several places, and the RP is completely compatible with these two standards.

15.3.1 HIPPS principle

Fig. 15.9, extracted from API 17O, provides a better overview of HIPPS implementation in conjunction with the reservoir, flowline, and host facilities. Using HIPPS brings new challenges to field development such as field architecture, offset distance, material, wall thickness and size of piping lines, pressure rating, etc. Fig. 15.10 shows the effect of a HIPPS module on pipeline wall thickness reduction.

The overpressure source is the reservoir in this case, but other sources of overpressure could be the subsea booster pumps, a connection to higher pressure pipeline, or a combination of factors. The overpressure fluid could be gas, liquid, or multiphase flow. Boarding shutdown valve (BSDV) is defined as an automatic valve assembly installed between an underwater production system and a surface facility that closes upon loss of power supply. The actuator for a BSDV valve is called a boarding shutdown actuator. More information about this valve is provided in Chapter 7.

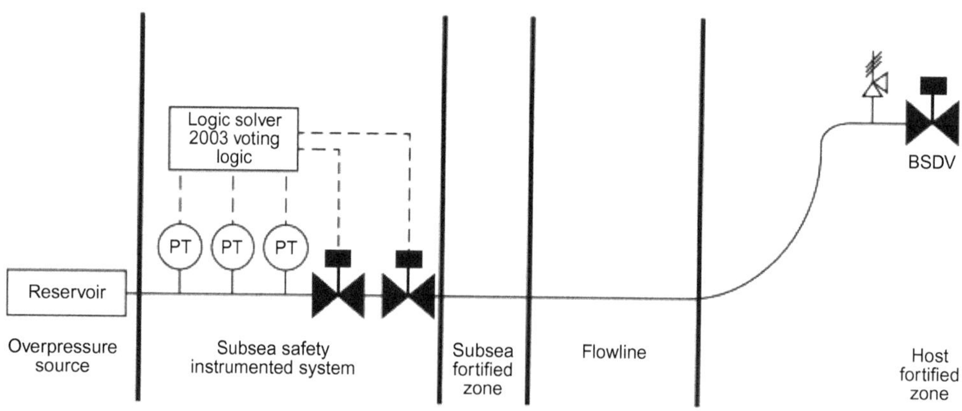

Fig. 15.9 Typical subsea production HIPPS valve diagram. *(Courtesy: API 17O.)*

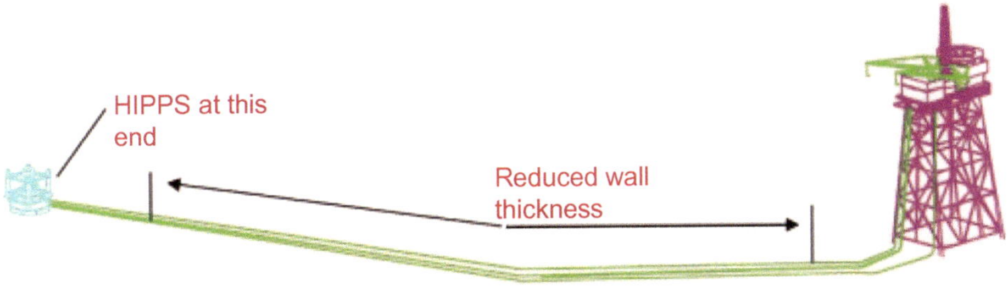

Fig. 15.10 Effect of HIPPS on reducing piping wall thickness. *(Courtesy: Vetco Gray and Elsevier.)*

Two HIPPS actuated valves with a fail-safe closed function are located after the reservoir and before the subsea fortified zone, which should ideally have zero leakage and a safety integrity level (SIL) of 3 as a minimum. SIL requirements, according to the IEC, are explained in more in detail later in this chapter. Some leakage may be tolerated from HIPPS valves under a special condition where the pressure safety valve (PSV) is on the surface. A PSV is illustrated in Fig. 15.9 before a BSDV on the host facility. HIPPS could be part of a Christmas tree, jumper, or manifold. The HIPPS module could be located after the Christmas tree or the manifold, as the production fluid is passing through the HIPPS module continuously. Different choices for HIPPS valves such as axial on-off, ball or gate valves are assessed later in this chapter. Choke valves are not a part of HIPPS, but more reliable performance is expected from chokes when HIPPS is implemented for subsea development. A subsea choke valve has a critical role: to control and adjust the flow from the well. Its failure to function can cause the release of high-pressure fluid to downstream sections such as the HIPPS and flowline; this is considered an unsafe and abnormal process condition and can trigger the alarm and trip for HIPPS valve closure. There are three important time units associated with HIPPS: response time, reaction time, and process safety time. HIPPS reaction time refers to the amount of time it takes HIPPS to shut down the system before a hazardous event takes place. HIPPS response time is typically shorter than HIPPS reaction time and depends on the response time of all three elements: the senor, logic solver, and final element (HIPPS valves(s)). Fig. 15.11 illustrates the process safety time, HIPPS response time, and reaction time according to IEC 61511. Process safety time defines as period of time between a failure occurring in the process or the process control system and the occurrence of the hazardous event if the safety function is not performed according to IEC 61508.

The subsea fortified zone, which is located downstream of the HIPPS valve, could be designed with the same pressure as the high-pressure section such as the subsea tree, or the maximum allowable working pressure of the downstream flowline. The reason why the fortified section should be designed for higher pressure than the upstream part is that it takes a short time for the HIPPS valve to close completely, so some high-pressure fluid can skip to the downstream section. The length of the fortified section after the HIPPS

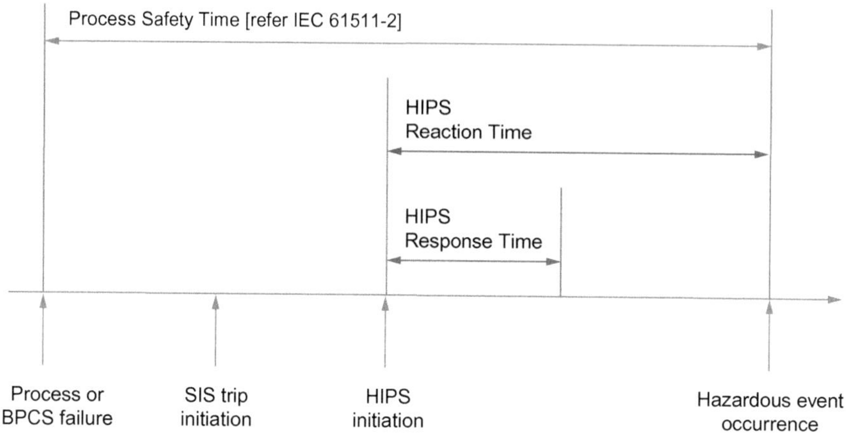

Fig. 15.11 Process safety time, HIPPS response time, and reaction time.

valve should be determined based on a flow assurance analysis. Normally, a subsea choke is installed on the wellhead to reduce the pressure of the fluid coming from the well. However, different undesirable events could lead to pressure buildup in the flowline such as malfunction of the choke to drop the pressure of the produced fluid from the well, unplanned shutting down of the boarding shutdown valve on the host facility (platform), or hydrate formation in the pipeline.

The flowline that transports the fluid from the subsea production units to the host facility is normally located on the unfortified zone and could be very long (10–20 km). In fact, a large portion of the budget for subsea development in the oil and gas industry is dedicated to pipeline cost. As illustrated in Fig. 15.9, the unfortified zone of the flowline is mainly located between the fortified zone and the host facilities such as a platform or ship. The starting point of the unfortified zone is determined according to hydraulic analysis and should be away from any possible leakage from the HIPPS valves. In addition, the length and location of the unfortified zone should be selected in such a way as to prevent injury to the topside personnel and damage to the topside facilities. The flowline pressure is monitored constantly and compared with the safe operating pressure. If the flowline pressure rises above normal conditions, the final element (the HIPPS valve) is triggered and shut down to stop the flow to the flowline. Blockage in the flowline, for example, could overpressure the flowline and lead to failure of the flowline as well. Thus the HIPPS valve should isolate the flow very quickly.

The connection of the riser to the host platform or ship should be designed and constructed in such a way as to prevent the release of hydrocarbon and protect personnel life. The area near the connection of the riser to the platform is a very critical point. As illustrated in Fig. 15.9, a blow down valve is placed on the platform to close the pipeline to the topside in the event of leakage from the HIPPS valves. The design of the whole system, illustrated in Fig. 15.9, should be implemented accurately by considering the different production chemical parameters such as water production rates, sand, carbon dioxide, hydrogen sulfide, gas-oil ratio, fluid velocity, etc. In fact, the fortified zone is designed to protect the host facility. A PSV is placed before the BDSV to release the overpressurizing fluid from the host facilities.

15.3.2 HIPPS implementation cost saving

A HIPPS system is located on the boundary between low- and high-pressure sections of pipe. Different architectures of HIPPS, such as a connection between a single well and the host platform, a connection between wells and the manifold, or between the manifold and host facilities were discussed before. Introducing HIPPS to a subsea system saves some costs while introducing some new challenges to the design of the pipeline system; these are summarized as follows:

- *Offset distance and wall thickness saving:* This distance refers to the section of flowline with less wall thickness due to HIPPS implementation upstream. Thinner walls mean lower capital expenditure (CapEx). The amount of saving on the wall thickness of the flowline could be as much as 30%.
- *Line size:* Implementation of HIPPS reduces the pressure on the downstream flowline section which leads to wall thickness reduction. Lowering the wall thickness of the pipe can cause line size reduction. Eq. (15.1) shows the relationship between wall thickness and the size of the piping.

  Pipeline wall thickness calculation

$$t = \frac{Pd_o}{2(H_s + P)} \tag{15.1}$$

where:

t: pipe wall thickness, in. (mm-metric);
H_s: maximum hoop stress, psi (MPa-metric);
P: internal pressure of pipe, psi (MPa-metric);
d_o: outside diameter of the pipe, in. (mm-metric).

- *Material selection:* The higher partial pressure of carbon dioxide and hydrogen sulfide increases the corrosion rate in the piping system and other facilities. The partial pressure of corrosive products is lowered by reducing the piping system pressure value. Minimizing the partial pressure of hydrogen sulfide and especially carbon dioxide can nullify the requirement for corrosion-resistant alloys. Thus HIPPS implementation minimizes the need for more exotic and expensive materials. Eqs. (15.2) and

(15.3) show the relationship between the pressure of the line and the partial pressure of the carbon dioxide and hydrogen sulfide.

CO_2 and H_2S partial pressure calculation

$$P_{CO_2} = P_{Operation} \times x_{CO_2} \tag{15.2}$$

$$P_{H_2S} = P_{Operation} \times x_{H_2S} \tag{15.3}$$

where:

P_{CO_2}: partial pressure of CO_2 (bar);
P_{H_2S}: partial pressure of H_2S (bar);
$P_{Operation}$: operation pressure (bar);
X_{CO_2}: mole fraction of CO_2; and
X_{H_2S}: mole fraction of H_2S.

As an example, for a 30 km section of 12″ pipeline in exotic corrosion-resistant alloys with a fluid pressure of 10,000 psi, the cost saving through HIPPS module implementation could be approximately 25 million USD.

15.3.3 HIPPS failure and risk analysis

Failure of HIPPS could lead to the overpressurization of downstream components such as the flowline. Thus the possibility of HIPPS failure should be taken into account during flowline design. API 17O lists three main types of flowline rupture that can result from HIPPS failure:

1. Uncontrolled: In this case, the flowline is ruptured and fails significantly so that all the internal fluid is released to the environment. The lives of the personnel at the topside facilities would be at high risk.
2. Controlled but uncontained: In this case, some quantities of overpressure could be released through a safety system. The risks and possibility of danger are lower in this case compared to an uncontrolled case.
3. Controlled and contained: In this case, the pressure-relieving system releases a large amount of overpressure so the catastrophic scenario due to HIPPS failure is less critical than in the two other cases.

Quantitative and qualitative risk analyses should be performed for the HIPPS system to identify the major risks associated with HIPPS failure and reduce the risks by reducing their probability and/or impact. The main risk is the failure of HIPPS, which would lead to overpressurizing the system. All of the factors contributing to this malfunction can be listed, evaluated, and given numbers to reflect their probabilities. The negative impact of occurrence through quantitative risk analysis and the total risk number can then be calculated by multiplying the probabilities by impact of occurrence. Higher risk scores are hazardous and require that some preventative measures be taken to reduce them. Tables

15.1 and 15.2 provide some examples of how to quantify the possibility and impact of different risks from 1 to 10 to calculate the final risk score from 100.

Based on the earlier tables, a risk factor with extremely high occurrence probability and hazardous impact without warning has a total score of 10 multiplied by 10 equal to 100, so this risk should be mitigated by means of suitable solutions.

One of the approaches to evaluating and estimating the reliability and the probability of failure for HIPPS is to determine the SIL according to IEC 61508. SIL provides a scientific and numeric approach to specifying and designing safety systems, thus enabling the risk of failure to be quantified. There are four SIL levels from 1 to 4 as presented in Table 15.3. SIL 4 has the least probability of failure or probability of failure on demand (PFD) and the highest reliability. The Probability of Failure on Demand (PFD) is a measure of the effectiveness of a safety function. It expresses the likelihood that the safety function does not work when required to. Conversely, SIL1 has the highest probability of failure and the least safety and reliability. SIL evaluation could include factors beyond

Table 15.1 Risk impact scoring guideline.

Rating	Effect	Severity (impact) effect
10	Hazardous without warning	Very high severity without warning
9	Hazardous with warning	Very high severity with warning
8	Very high	Destructive and unsafe failure
7	High	System is inoperable due to major equipment damage
6	Moderate	System is inoperable due to minor damage
5	Low	System is inoperable without damage
4	Very low	High degradation of performance
3	Minor	Less performance degradation
2	Very minor	Minimal system operability
1	None	No effect

Table 15.2 Risk probability scoring guideline.

Rating	Occurrence probability	Minimum probability percentage
10	Extremely high	50%
9	Very high	33%
8	Very high	10%–15%
7	High	5%
6	Marginal	1%
5	Marginal	0.25%
4	Unlikely	0.05%
3	Low	0.007%
2	Very low	0.0007%
1	Remote	0.000007%

Table 15.3 SIL Levels according to IEC 61508.

SIL	PFD	Risk Reduction Factor
SIL 1	0.1–0.01	10–100
SIL 2	0.01–0.001	100–1000
SIL 3	0.001–0.0001	1000–10,000
SIL 4[a]	0.0001–0.00001	10,000–100,000

[a]Not applicable in the process industry.

safety such as environmental and economical concerns. SIL 3 is normally required for subsea HIPPS valves as a minimum.

Failure rates (probabilities) are categorized into two types: safe failure ($\boldsymbol{\lambda}_S$) and dangerous failure ($\boldsymbol{\lambda}_D$) rates. Dangerous failures are those that will prevent achieving the required SIL. Safe failures put the safety function in its safe state, such as the emergency shutdown of a valve in case of failure. It is noticeable that SIL addresses only dangerous failures ($\boldsymbol{\lambda}_D$). Dangerous failures can be either detected or undetected. The possibility of detected dangerous failures and undetected dangerous failures is shown as $\boldsymbol{\lambda}_{DD}$ and $\boldsymbol{\lambda}_{DU}$. Dangerous detected failures can be detected by both diagnosis and proof testing, or they can be detected just after they occur. But dangerous undetected failures can only be revealed with proof testing. The classifications of different failure types are shown in Fig. 15.12. Eq. (15.4) shows the relationship between $\boldsymbol{\lambda}_D$, $\boldsymbol{\lambda}_{DD}$, and $\boldsymbol{\lambda}_{DU}$:

Relationship between dangerous failure rate and dangerous detected and dangerous undetected failure rates according to IEC 61508

$$\boldsymbol{\lambda}_D = \boldsymbol{\lambda}_{DU} + \boldsymbol{\lambda}_{DD} \tag{15.4}$$

The mean downtime of a system due to both dangerous detected and undetected failures $\mathbf{t}_{CE}$ includes the mean downtime of the system due to dangerous undetected failures $\mathbf{t}_{c1}$ plus the mean downtime of the system due to dangerous detected failures $\mathbf{t}_{c2}$ as shown in Fig. 15.13. It is important to pay attention that topology or system architecture 1oo1 is

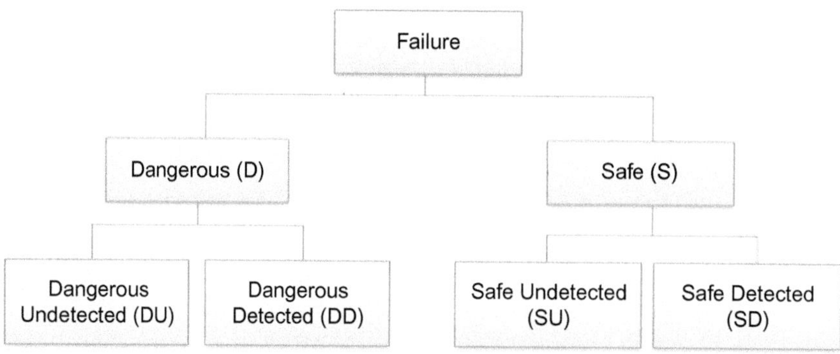

Fig. 15.12 Failure mode classification. *(Courtesy: NTNU.)*

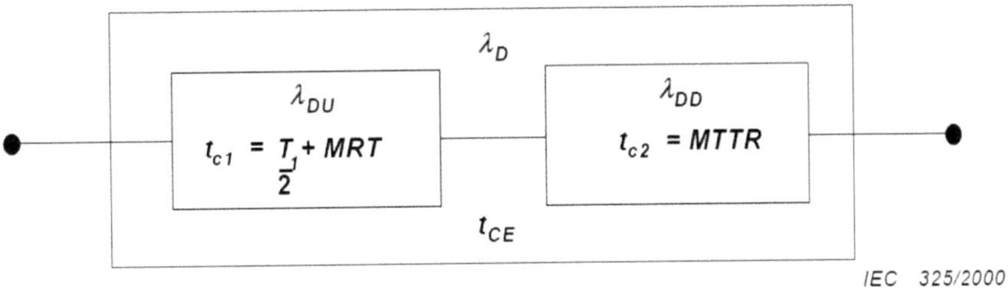

Fig. 15.13 1oo1 reliability block diagram.

considered for HIPPS reliability analysis and SIL study in this chapter. Topology 1oo1 means that all HIPPS elements including sensors, logic solver and final element are considered a single and unified system failure on each part leads to the failure of HIPPS.

$\mathbf{t}_{c1}$ and $\mathbf{t}_{c2}$ are calculated using Eqs. (15.5) and (15.6):

Calculation of the mean downtime of the system due to dangerous undetected failures as well as dangerous detected failures refer to IEC 61508

$$\mathbf{t}_{c1} = \frac{T_1}{2} + MRT \tag{15.5}$$

$$\mathbf{t}_{c2} = MTTR \tag{15.6}$$

where:

T_1: Proof test interval (h)
$MTTR$: Mean time to restoration (h)
MRT: Mean repair time (h)

$$\begin{aligned} MTTR = {} & \text{time to detect the failure} + \text{time spent before starting the repair} \\ & + \text{the effective time to complete the repair} \\ & + \text{the time before the component is put back into operation} \end{aligned} \tag{15.7}$$

$$MRT = MTTR - \text{time to detect the failure} \tag{15.8}$$

Fig. 15.13 shows the reliability block diagram for architecture 1oo1.

Placing values of $\mathbf{t}_{c1}$ and $\mathbf{t}_{c2}$ from Eqs. (15.5) and (15.6) into Eq. (15.9):

$$\mathbf{t}_{CE} = \frac{\boldsymbol{\lambda}_{DU}}{\boldsymbol{\lambda}_D}\left(\frac{T_1}{2} + MRT\right) + \frac{\boldsymbol{\lambda}_{DD}}{\boldsymbol{\lambda}_D} MTTR \tag{15.9}$$

The number of dangerous failures detected by automatic online diagnostic tests is called diagnostic coverage (DC). DC is the detected dangerous failure rate divided by the total rate of the failures, using Eqs. (15.10)–(15.12):

$$DC = \frac{\boldsymbol{\lambda}_{DD}}{\boldsymbol{\lambda}_D} \tag{15.10}$$

Table 15.4 Correlation between SIL and SFF for hardware fault tolerance—Type A (IEC).

Safe failure fraction	Hardware fault tolerance—Type A 0	1	2
SFF < 60%	SIL 1	SIL 2	SIL 3
60% < SFF < 90%	SIL 2	SIL 3	SIL 4
90% < SFF < 99%	SIL 3	SIL 4	SIL 4
SFF ≥ 99%	SIL 3	SIL 4	SIL 4

$$\text{Therefore,}\ \boldsymbol{\lambda}_{DD} = \boldsymbol{\lambda}_D \times \text{DC} \tag{15.11}$$

$$\boldsymbol{\lambda}_{DU} = \boldsymbol{\lambda}_D - \boldsymbol{\lambda}_{DD} = \boldsymbol{\lambda}_D - (\boldsymbol{\lambda}_D \times \text{DC}) -> \boldsymbol{\lambda}_{DU} = \boldsymbol{\lambda}_D(1 - \text{DC}) \tag{15.12}$$

The average PFD (PFD_G) is calculated using Eqs. (15.13) and (15.14):

$$PFD_G = (\boldsymbol{\lambda}_{DU} + \boldsymbol{\lambda}_{DD})\mathbf{t}_{CE} \tag{15.13}$$

$$PFD_G = \boldsymbol{\lambda}_D \times \mathbf{t}_{CE} = \boldsymbol{\lambda}_{DU}\left(\frac{T_1}{2} + MRT\right) + \boldsymbol{\lambda}_{DD} \times MTTR \tag{15.14}$$

The other important safety parameter is called safe failure fraction (SFF) which verifies the suitability of a device for usage in a system for a particular SIL. SFF is calculated as the ratio of the average failure rate of safe plus dangerous detected failures using Eq. (15.15):

$$\text{SFF} = \left(\frac{\boldsymbol{\lambda}_S + \boldsymbol{\lambda}_{DD}}{\boldsymbol{\lambda}_S + \boldsymbol{\lambda}_D}\right) = \left(\frac{\boldsymbol{\lambda}_{TOTAL} - \boldsymbol{\lambda}_{DU}}{\boldsymbol{\lambda}_{TOTAL}}\right) \tag{15.15}$$

It is possible to correlate SFF to SIL based on a table for different hardware fault tolerance. Hardware fault tolerance is $N = 0$, 1, or 2, and could be type A or B. Selection between type A or B depends on different system architectures such as 1oo2 or 2oo3. As an example, Table 15.4 presents the maximum allowable SIL for a safety function implemented by type A safety for different hardware systems.

15.3.4 Design process data

1. *Shut in pressure (SIP):* SIP is the full internal pressure contained by the HIPPS and the upstream piping when the HIPPS valve is closed and all the upstream valves are open.
2. *Flow properties:* The HIPPS system, including the valves, should be suitable for highly corrosive fluid over its design life as well as all possible changes in the fluid such as gas-oil ratio (GOR), flow rate, flow velocity, and change in chemical composition. Slug flow pattern formation in HIPPS valves is highly possible. In addition, the produced fluid could have a high likelihood of wax and hydrate formation. The fluid itself is corrosive with unwanted components such as carbon dioxide and hydrogen sulfide. Erosion is another concern due to the presence of sand and solids in the production fluid.

3. *Upstream and downstream pressure conditions:* Piping upstream of the HIPPS and some distance after the HIPPS valves should be designed based on the full pressure rating of the wellhead.
4. *Fortified section:* Explained earlier.
5. *Transition pressure due to blockage:* The possibility of sudden blockage of the flowline at various points downstream of the HIPPS shall be considered. Calculations of the transient pressure increase arising from the blockages should be developed.

15.3.5 Environmental data

Environmental data as a part of the design basis should be evaluated and taken into account for HIPPS analysis. This information is summarized as follows:

- Depth of water and water head pressure; this parameter affects the pressure class and wall thickness of the HIPPS valves.
- Seabed currents like tidal, eddy, hurricane, and other current information.
- Type of seabed soil.
- Seawater chemical and physical information such as density, salinity, etc.
- Site hazards that affect the risk of installation such as slope, earthquake risk, etc.
- Loads from connected pipeline including flowline; these loads should be considered in the design of the valve.

15.3.6 Process shutdown systems

Two main shutdown systems exist for process systems: process shutdown (PSD) and emergency shutdown (ESD). PSD is defined as the automatic isolation or activation of some or all parts of process systems. PSD typically includes sensors, valves, trip relays, and logic units for analyzing signals. ESDs are designed in plants to mitigate the negative consequences of emergency and uncontrolled events such as hydrocarbon release to the environment, high flow of fluid, etc. Therefore ESD is a safety layer and not a process layer, unlike PSD. PSD is part of a process plant safeguard designed and implemented to prevent or minimize the risk of conditions outside the normal operating conditions defined for the system. ESD is a process shutdown safeguard to prevent the occurrence of abnormal conditions in a process plant. HIPPS is also a safety system designed to prevent overpressurizing the pipeline system to prevent damage to the plant and facilities and to protect personnel; it is a different system from ESD. Fig. 15.14 illustrates the different layers of plant safety implementation and protection systems including PSD and ESD.

A HIPPS system is a separate system from ESD and PSD and is independent from the process control system (PCS). However, API 17O allows HIPPS and PCS to share electrical or hydraulic supply. PCS and HIPPS shall not use common solenoid valves for actuation or other control devices. But it should be noted that sharing HIPPS with control systems for subsurface safety valves and underwater control valves is not permitted

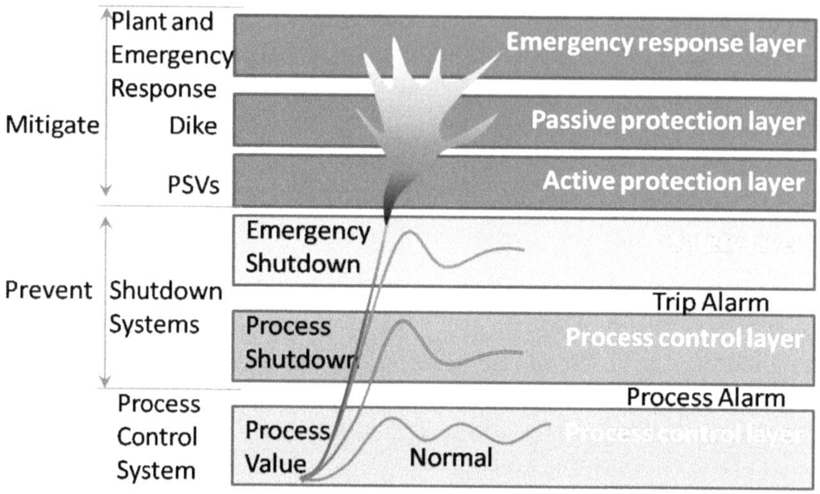

Fig. 15.14 Different layers of a plant safety implementation and protection system including PSD and ESD. *(Courtesy: Instrumentation Tools.)*

according to the API 17O standard. The important provision is that shared PCS system with HIPPS cannot give any credit to PCS regarding the higher process safety due to this immerging. HIPPS is not shown in the figure, but should be located above the ESD as the last layer of process protection safety system. HIPPS is comprised of process control devices, alarms, and trip systems.

15.4 HIPPS valve selection

The main aim of this section is to select the most suitable valve for HIPPS application based on some essential criteria. It should be noted that the approach in this section can be used as a guideline and that the selected parameters and given scores or priorities are subject to change in different projects. HIPPS valves have on/off or fluid stop/start application. Three types of subsea valves are suitable for on/off applications: ball, through conduit gate (TCG), and axial on-off valves. HIPPS valves are actuated; a linear actuator is the choice for gate and axial on/off valves, and rack and pinion or scotch yoke are the suitable actuators for ball valves. Different criteria are important for selecting a valve for HIPPS application such as fast operation (closing), reliability (feasibility), weight and space saving, flow capacity and pressure drop reduction, cost reduction, and sealability. These characteristics are explained further as follows:

1. *Fast closing:* Ball valves are quarter-turn valves that require only a 90-degree rotation of the ball between opening and closing. Thus this valve can be categorized as fast opening or closing. Like ball valves, axial valves require only a very short time for opening and closing due to their low mass disk and the short travel of the disk

between open and closed position. On the other hand, TCG valves are not as fast closing as ball and axial valves. That could be a reason why TCG valves are not considered a suitable option for HIPPS according to some valve engineering practices. However, TCG valves are used in the industry for HIPPS application. Fig. 15.15 illustrates two slab gate valves in 5 1/8″ and pressure class 15,000 psi for HIPPS application. The closing time for a HIPPS valve could be as short as 0.2 s.

2. *Reliability and feasibility:* TCG valves are arguably the least feasible choice for HIPPS valves. In contrast, axial valves have been used widely for HIPPS application.
3. *Weight and space saving:* Saving weight and space on an offshore platform is essential since offshore platforms have limited space and load capacity. The advantage of a horizontally oriented ball valves is that the actuator does not stand vertically, so in general less height is occupied by ball valves compared to TGC and axial valves on which the actuator stands vertically. Fig. 15.16 illustrates 5 1/8″ hydraulically operated slab gate valves in a pressure class of 10,000 psi with a fail-safe close linear actuator that is approximately 2.7 m in height. Fig. 15.17 illustrates a subsea ball valve with a rack and pinion actuator, which saves a lot of vertical space. Axial valves, like gate valves, have their actuators on the vertical line; however, some valve and actuator

Fig. 15.15 Two 5 1/8″ subsea slab gate valves in pressure class 15,000 psi for HIPPS. *(Courtesy: ATV.)*

Fig. 15.16 Subsea actuated 5 1/8″ TCG valve approximately 2.7 m in height (ATV).

Fig. 15.17 A ball valve with rack and pinion actuator.

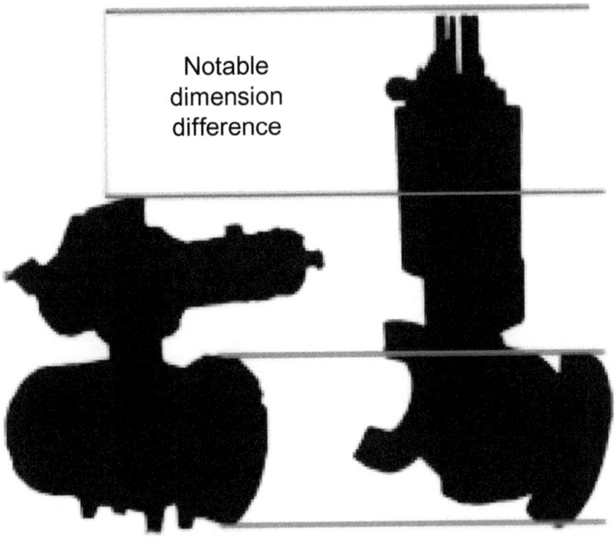

Fig. 15.18 Comparison between vertically and horizontally oriented actuators.

manufacturers can use horizontal actuators for HIPPS valves to save space vertically. Fig. 15.18 compares the actuated valves' height when the actuator is installed vertically and horizontally.

4. *Flow capacity and pressure drop:* TCG valves and full-bore ball valves provide the highest flow rate and the least flow restriction in the piping system. Full bore ball valves also have a very low amount of pressure drop. Axial valves have the least flow capacity compared to the two other options since the internals of the valve are located in front of the flow of the fluid inside the valve. In one case of HIPPS topside, the body of an axial valve was enlarged one size to compensate for the pressure drop in a recent offshore project.
5. *Cost:* A TCG valve could be the most expensive choice since a relatively large amount of metal is used for this valve, given its large bonnet and top entry design. Ball valves could be made in both top and side entry; the side entry choice weighs less and probably has a lower cost. An axial valve could be the cheapest choice. It should be noted that the best way to compare the costs of different valve choices is to contact the valve suppliers. Some manufacturers may produce ball valves that are heavier compared to TCG valves of the same size and pressure class.
6. *Seal-ability:* Ball valves can be made in soft seat which provides zero leakage possibility. However, for subsea operation and high-pressure classes, metal seat ball valves are more common; these could provide leak rate B according to ISO 5208 standard. The gate valve closure member, unlike that of the ball valve, is flat which can provide zero leakage equal to ISO 5208 leak rate A. Thus, in general, TCG valves can provide the

best sealing. Axial valves could provide leak rate B or C as per ISO 5208. The question is, what would be the leakage rate difference between a gate and ball valve in 8″ size in liquid service if the ball valve has an acceptable leak rate B and the gate valve has a leak rate A for seat and closure member sealing when the valve is closed? Leak rate A means no visible leakage. Leak rate B for a ball valve is calculated according to Eq. (15.16) given as follows (Table 15.5).

Leak rate B calculation as per ISO 5208 for a valve in liquid service

$$L = 0.00016^{*}\mathrm{DN} = 0.004^{*}\mathrm{DI} = 0.004 \times 8 = 0.032 \text{ bubble per second} \\ = 2 \text{ bubbles per minute} \tag{15.16}$$

where:

L: allowable leakage from the valve (drop/s);

DN: dimension nominal (mm);

DI: dimension in in..

The ranking of valves for each essential parameter discussed before is listed in Table 15.6.

The next step in value engineering is to evaluate the importance of each defined parameter. Each parameter has a weight W_j that is obtained through the implementation of a pair-wise comparison approach. The total weight of the parameters should be one as per Eq. (15.17).

Total weight of parameters used for valve selection for HIPPS application

$$\sum\nolimits_{i=1}^{j} W_i = 1, \text{ where} i = 1, 2, 3, \ldots j \tag{15.17}$$

A pair-wise comparison is used to determine the weight values for each parameter defined in the previous section (shown in Table 15.7) and to determine the weight values and hierarchy of each property presented in Tables 15.6 and 15.7. Regarding the determination of the weight values, a value of either 0 or 1 is assigned to the matrix for each compared pair, depending on which one is more important. After comparing all the pairs and assigning scores, the score of each parameter (creation) is calculated.

Referring to Table 15.7, reliability and fast operation (closing) are the most important performance requirements expected from HIPPS valves. Weight and space saving as well as flow capacity and pressure drop are the least important characteristics expected from valves used in HIPPS application according to the logic provided in the table.

The next step is to assign a score from 1 to 10 for each parameter defined in Table 15.7. The first ranking for each property is given 10 and the last ranking for each property is given 1. This logic of score assignment gives a value of 5.5 to second ranked properties. The next step is to create a quantitative performance matrix according to Table 15.8 for the four valve candidates based on the property and characteristic values listed in Tables 15.6 and 15.7. The performance of each parameter is calculated by

Table 15.5 Leakage rates based on ISO 5208 for gas and liquid services.

Test fluid	Unit leakage rates	Rate A	Rate AA	Rate B	Rate C	Rate CC	Rate D	Rate E	Rate EE	Rate F	Rate G
Liquid	mm^3/s	No visually detectable leakage for the duration of the test	0.006 × DN	0.01 × DN	0.03 × DN	0.08 × DN	0.1 × DN	0.3 × DN	0.39 × DN	1 × DN	2 × DN
	drops/s		0.0001 × DN	0.00016 × DN	0.0005 × DN	0.0013 × DN	0.0016 × DN	0.0048 × DN	0.0062 × DN	0.016 × DN	0.032 × DN
Gas	mm^3/s	No visually detectable leakage for the duration of the test	0.18 × DN	0.3 × DN	3 × DN	22.3 × DN	30 × DN	300 × DN	470 × DN	3000 × DN	6000 × DN
	bubbles		0.003 × DN	0.0046 × DN	0.0458 × DN	0.3407 × DN	0.4584 × DN	4.5837 × DN	7.1293 × DN	45.837 × DN	91.673 × DN

Table 15.6 Ranking the valves for HIPPS application.

Valve type	Fast closing operation	Feasibility/ reliability	Weight and space saving	Flow capacity/ pressure drop	Cost reduction	Sealability	Possible limitations
Ball	2	2	1	2	2	3	Soft seal ball valves are at risk of damage due to their fast closing operation action, so a metal seat should be used which reduces sealability
Through Conduit Gate	3	3	3	1	3	1	A TCG valve is not a quarter-turn valve like a ball or butterfly with a relatively longer distance for the disk to travel between open and closed positions.
Axial	1	1	2	3	1	2	Not Applicable

Table 15.7 Ranking the valves for HIPPS application.

Criteria	P_1	P_2	P_3	P_4	P_5	P_6	Total score	Normalized
P_1	P_1	0	1	0	1	1	3	0.22
P_2	1	P_2	1	1	1	1	5	0.36
P_3	0	0	P_3	1	0	0	1	0.07
P_4	0	0	0	P_4	0	0	1	0.07
P_5	0	0	1	1	P_5	1	2	0.14
P_6	0	0	1	1	0	0	2	0.14
Total							14	1

P_1: Fast closing operation; P_2 Feasibility and reliability; P_3: Weight and space saving; P_4: Flow capacity and pressure drop; P_5: Cost; P_6: Sealability.

multiplying each parameter rating given in Table 15.8 without considering the percentage by each parameter weight according to Table 15.7. Considering the fact that the rating is set between 1 and 10 and the weights are between 1 and 100%, the total performance value or score calculated on the bottom of Table 15.8 is between 100 and 1000.

Table 15.8 Quantitative performance matrix.

Mat.	Ball	TCG	Axial
P_1	2	3	1
P_1 Rate	5.5	1	10
P_1 Performance	121	22	220
P_2	2	3	1
P_2 Rate	5.5	1	10
P_2 Performance	198	36	360
P_3	1	3	2
P_3 Rate	10	1	5.5
P_3 Performance	70	7	38
P_4	2	1	3
P_4 Rate	5.5	10	1
P_4 Performance	38	70	7
P_5	2	3	1
P_5 Rate	5.5	1	10
P_5 Performance	77	14	140
P_6	3	1	2
P_6 Rate	1	10	5.5
P_6 Performance	14	140	77
Total Performance Score	518	289	842

Based on the total performance score provided in Table 15.8, the axial valve is the most suitable valve for HIPPS application. The ball valve has the second ranking and the TCG valve is the least suitable valve. It should be considered that other parameters such as a qualification requirement could be added to the matrix before, which would change the result of the evaluation.

15.5 HIPPS valve and actuator design considerations

Two special design considerations for HIPPS valves according to API 17O are piggability and the sand slurry test as per API 6AV1 requirements. HIPPS valves are installed on the pipeline and subject to pig running so the bore of the valve should be equal to the connected piping with some minor tolerances to facilitate pig passage through the valve. HIPPS valves are used in sandy services; testing the valve in sand slurry can reduce the possibility of valve failure due to sand in the fluid service during production. HIPPS valves should have ROV override as per API 17H/ ISO 13628-4. Temperature ratings provided in API 6A/17D could be used for HIPPS valves with the additional consideration of operational testing of the valve in a cold weather environment. The pressure of the valve follows standard pressures in API 6A/AP17D with the possibility of having an intermediate pressure between the standard pressure values. The external hydrostatic seawater head should be taken into account in the design pressure consideration. The hydraulic supply pressure to the valves' actuator could be standardized to 5000 psi equal to 345 bar. The actuator safety factor could be two. The integrity between the valve and the mounted actuator is very important. The valve should be capable of being opened against the maximum differential pressure on both sides of the valve and move from open to closed position in the defined required time which is normally less than a second.

15.6 HIPPS valve testing during operation

Correct functionality of HIPPS valves should be verified through qualification and factory acceptance testing. Typical qualification tests such as temperature-pressure cycles (PR2) and hyperbaric and endurance cycles are applicable for HIPPS valves. Scaling based on API 6A/17D is applicable for HIPPS valves. All of the factory acceptance tests listed in Chapter 14 for valves and actuators are applicable for HIPPS valves and actuators. HIPPS valves should be tested in the fabrication yard after installation and before commissioning; this is called site acceptance testing (SAT). One reason for SAT implementation is to make sure that the integrity of the HIPPS components is not jeopardized during transportation. Also, SAT is a way to make sure that the HIPPS components are installed correctly and separated from other systems such as ESD and PSD. SAT is a good approach to ensure proper communication between the valve actuator and the control system. In addition, API 17O standard proposes that proof testing be performed after the installation of HIPPS valves and during their operation. The main purpose of these tests is to make sure that HIPPS valves can maintain the required safety integrity level. Proof testing should also be performed as a part of IEC 61508 as a periodic test to detect hidden defects; that could require shutting down production. These tests are typically performed during the design process, but also should be performed during valve operation as per IEC 61508 requirements.

1. *Standard leak test:* HIPPS valves should be capable of shutting down the pipeline and provide sealing against shut-in pressure in closed position such that overall leakage from the valve does not exceed the given maximum leakage rate in the valve data sheet. The interval of these tests could be 6 months or 1 year, just as examples, depending on the regulation or end user technical requirements.
2. *Pressure sensor verification:* This test is performed to make sure that all the pressure sensors show the same value of piping pressure. The choke valve position can be changed to alter the pressure in the connecting pipeline. The pressure gauges should be monitored to make sure that all of them show the same value during fluid transition.

3. *Partial stroke testing:* Partial stroke testing (PST) is a type of online test performed on HIPPS valves and other emergency shutdown valves as a preventive failure diagnosis in a way that does not stop the production line. PST is an alternative to full stroke testing (FST), especially in cases when online testing should be performed more frequently. Applying FST stops the production line so it is not recommended. Introducing PST has two positive impacts: shortening test intervals and reducing system downtime. Reducing system downtime reduces the probability of failure on demand (PFD) and improves the SIL. The valve can be moved partially and slowly between fully open and 10%–20% open. The valve is probably designed this way because the majority of valve failures occur at closure or very close to closure, such as 10% or 20% open, while production continues. A controller connected to the valve can diagnose potential issues related to the valve. PST is a function test that can detect the possible failures fraction. According to the offshore installation data manual published by offshore reliability data, PST can detect 70% of frequent problems with the valve, actuator, and control system. The reliability of axial on-off valves is high for HIPPS application, so PST is not required for this type of valve based on the valve manufacturer recommendation. Typically, a screen shows the valve characteristics such as valve positioning during the PST.
4. *Full closure testing:* The valve should be tested to demonstrate that it moves to full closed position.

Some control system and data are implemented in the HIPPS system, including the pressure transmitter output signal to the master control station (MCS) and to the logic solver; the status of the HIPPS valve (open, closed, or partially open) is transmitted to the control station, alarms, and trip that are activated to close the valve as well as HIPPS controller status report which has self-diagnosis function with possibility of report to process or production control system. The alarm system is designed with the purpose of informing operators about malfunction of the system, which is overpressure in the case of HIPPS. The trip is a system designed to shut down the overpressure system by closing the HIPPS valve. Both the alarm and the trip are parts of the SIS.

Validation testing is not limited to HIPPS valves; it is also applicable to the directional control valves (DCVs) used in the control system for hydraulic actuators. DCVs are one of the most functional parts of the hydraulic actuator control system; they are used to change the direction of the hydraulic fluid to different destinations. Fig. 15.19 is used to explain the function of DCVs in the hydraulic system. Two DCVs are located in the subsea control module (SCM). The actuator in the figure is a double acting hydraulic actuator that contains two cylinders, both connected to one hydraulic line. The hydraulic supply passes the DCV on the left side and enters and opens the actuator through the orange color line (dark gray in print versions). Upon opening the actuator through hydraulic supply, the hydraulic fluid returns through the hydraulic line in blue color

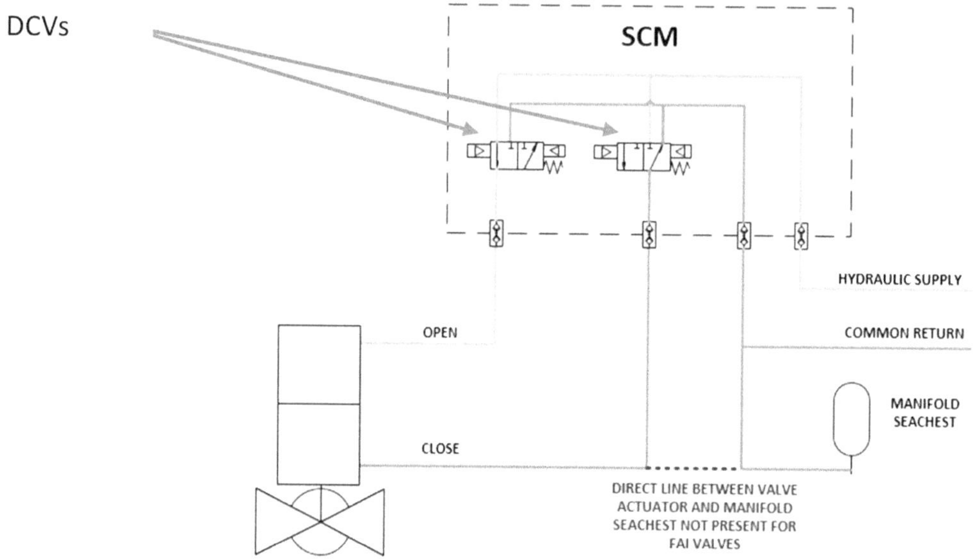

Fig. 15.19 Hydraulic system including DCVs.

(light gray in print versions). The return hydraulic on the blue line (light gray in print versions) passes through the DCV on the right side. Now suppose that the port in the DCV on the right changes its position from the blue line (light gray in print versions) to the orange line (dark gray in print versions). In this case the hydraulic supply enters and passes through the DCV on the right side and enters the cylinder of the actuator connected to the blue line (light gray in print versions). The actuator and the connected valve get closed in this case. Upon closing the actuator, the hydraulic from the "open" side or top side of the actuator leaves the actuator and enters the DCV on the left side. The hydraulic fluid that left the hydraulic actuator through the orange line (dark gray in print versions) will enter the common return line and flow back to the surface facilities.

The following tests shall be performed on DCVs according to the API 17O requirement:

1. *Cycle testing:* Ten thousand cycles of testing shall be performed on DCVs at maximum hydraulic supply pressure with a hydraulic fluid cleanliness level of SAE AS4059 class 6, B through F or better. The pressure should be monitored and dropped to zero for each cycle. After that, 1000 cycles should be applied with contaminated fluid with a cleanliness of SAE AS4059 class 10. As in the previous step, the pressure should be monitored and dropped to atmospheric pressure after each cycle.
2. *Hyperbaric testing at low temperature:* This test shall be done in a hyperbaric chamber at an external test pressure correspondent to the water depth head pressure. The test fluid should be clean based on SAE AS4059 class 6, B though F or better. The test pressure of the hydraulic fluid should be equal to the maximum rated hydraulic supply

pressure to the actuator. The water in the chamber should be cooled to a temperature below 2°C. The number of cycles is 100. Again, the pressure of the hydraulic fluid should be monitored and dropped to zero after each cycle.

The valve should pass these cycle tests successfully without any failure.

15.7 HIPPS valve main failures during operation

There are four main HIPPS failures:

1. *Fail to close:* This is known as the most severe failure of a HIPPS valve and can damage the downstream flowline and host facilities. Many factors could cause this type of failure such as malfunction of the actuator due to spring damage or lack of spring force to fully close the actuator and the connected HIPPS valve. Lack of spring force to close the valve could be due to blockage of the hydraulic fluid in the hydraulic return line of the actuator. High friction between the stem and stem sealings, or friction between the valve seat and the disk, could also cause the HIPPS valve to fail to close. This type of failure is typically detected by proof testing so this type of failure could be classified as an undetected dangerous failure.
2. *Leakage in closed position:* This type of failure is not as severe as fail to close and the severity or negative consequence of this problem depends on the amount of leakage from the valve. One of the main causes of leakage in closed position could be erosion and/or corrosion of the gate and/or seat(s). Galling or metal-to-metal friction between the disk and seats can be prevented by applying tungsten carbide hard facing on at least the contact surfaces of the disk and seats. Misalignment of the disk and/or seat could also be the cause of leakage in closed positions. Another reason for valve leakage in closed position could be vibration. This type of failure cannot be detected by partial stroke testing. The production line should be stopped and the valve closed for leak testing and measuring possible leakage from the valve seats in closed position.
3. *False trip:* This type of failure is associated with the closure of the valve without any requirement for closing or any demand from the logic unit. This type of failure does not have any negative safety consequence. The only concern is loss of production due to the closing of the valve. The reason for the sudden closing of the valve without any trip signal could be a failure in the supply of hydraulic fluid and pressure to the valve actuator, or failure of the solenoid valve in the control system. This type of failure is considered a safe detected failure.
4. *Fail to open:* This type of failure is also categorized as safe detected failure and can occur after valve closure due to trip. This type of failure can be detected easily by a maintenance team. This problem can be caused by friction between the gate and seats, a problem with the hydraulic supply to the actuator for opening the valve or accumulation of sand or hydrate in the valve cavity. If the HIPPS valve has an electrical actuator, failure of electrical supply to the actuator can cause fail to open.

15.8 Self-assessment questions

1. Which sentence is correct about HIPPS?
   - **A.** One of the challenges associated with HIPPS is that there is no standard to specifically address it.
   - **B.** HIPPS valves should have at least SIL3, but SIL3 is not a requirement for other HIPPS components such as the sensors and logic solver.
   - **C.** The HIPPS module is normally located on the low-pressure side of the pipeline or subsea development system.
   - **D.** HIPPS makes some of the oilfields located in deep-water depth economically suitable for development.

Answer: Option A is not correct because API 17O recommended practice addresses HIPPS specifically. Option B is not correct either because at minimum SIL 3 is required not only for the valve as a final element but also for the sensors and logic solver. Option C is not correct since HIPPS is normally mounted on the high-pressure side of the pipeline in a subsea development system. Option D is correct because subsea valves and piping are required to have a higher pressure rating due to the higher seawater pressure head. This makes subsea valves and piping more expensive, and subsea development very costly. Using HIPPS leads to pressure reduction in the pipeline and valves which could make the development of deep-water fields more economical.

2. Which sentence is wrong about HIPPS valves?
   - **A.** HIPPS valves are typically fail-safe closed actuated with a minimum SIL 3 requirement.
   - **B.** Axial on-off valves could be the most suitable type of valve for HIPPS application.
   - **C.** The closing time for a HIPPS valve could be 30 s.
   - **D.** Through conduit gate valves could be the least suitable option for HIPPS valves if they are selected for this purpose in the subsea sector of the oil and gas industry.

Answer: Options A, B, and D are correct. Option C is wrong because the closing time for a HIPPS valve is normally less than a second. A 30-s closing time for a HIPPS valve is quite long and would be unacceptable.

3. Which parameters are important for the selection of a HIPPS valve?
   - **A.** Cost, fast closing, reliability, sealability
   - **B.** Material, fast opening, space saving
   - **C.** Size, low friction stem sealing, flow capacity
   - **D.** Pressure class, temperature class, standard of the valve

Answer: Option A addresses all of the important parameters that affect valve selection for HIPPS. Option B is not correct since the material of the valve and fast opening are not important for a HIPPS valve. A HIPPS valve should have fast closing action to shut down the pipeline but fast opening is not a critical parameter. Option C is not correct since size and low friction stem sealing are not taken into account during the process of HIPPS valve selection. Option D is not correct since none of the given parameters are important for a HIPPS valve.

4. Which sentence is wrong about the testing of HIPPS?

A. Direct control valves (DCVs) are required to be qualified under cycle tests.

B. HIPPS valves typically require qualification tests such as PR2, hyperbaric, and endurance tests.

C. Full stroke testing instead of partial stroke testing is more common for HIPPS valves according to API 17O.

D. HIPPS valves should be tested on site after installation.

Answer: Option A is correct. DCVs are an essential part of the hydraulic control system and should be cycle tested with both clean and contaminated hydraulic fluid to verify their performance under severe operating conditions. Options B and D are also correct. HIPPS valves are typically subject to standard qualification tests such as PR2, hyperbaric, and endurance tests to validate the design, but scaling could be used for design validation as well. HIPPS valves are tested on site after installation for various reasons, such as to ensure that the HIPPS components are installed correctly, the integrity of the system is not jeopardized during transportation, and to test the integrity between the control system and the actuated valve. Option C is wrong because full stroke testing is not mentioned in API 17O; partial stoke testing is the proof testing mentioned in this standard.

5. Which sentence is wrong about HIPPS?

A. A HIPPS consists of pressure sensors and transmitters, a logic solver, solenoid valves, and an actuated valve.

B. Installation of the HIPPS after the Christmas tree and before the manifold is a less cost-saving approach compared to the installation of HIPPS after the manifold and before the flowline.

C. A voting system of 3oo5 means that 3 out of 5 pressure transmitters agree on the presence of a high-pressure value in the line and initiate a process for shutting down the process piping by a final element

D. Cost saving in HIPPS is related to reducing the thickness and weight of the pipeline and the possibility of using cheaper material and a smaller size of pipeline.

Answer: Options A, C, and D are correct. Option B is wrong since the installation of a HIPPS module after the Christmas tree and before the manifold leads to the selection of low-pressure piping and valves inside the manifold that saves the cost of the manifold and downstream pipeline. However, installation of HIPPS after the manifold only leads to less expensive flowline, and the manifold piping and valves would need to be designed according to the high-pressure value. Therefore installation of a HIPPS module after the Christmas tree and before the manifold leads to greater cost saving instead of placing a HIPPS between the manifold and the flowline.

Further reading

American Petroleum Institute (API) 17O, Standard for Subsea High Integrity Pressure Protection System (HIPPS), second ed., API, Washington, DC, 2014.

American Petroleum Institute (API) RP 14C, Analysis, Design Installation and Testing of Safety Systems for Offshore Production Facilities, eighth ed., API, Washington, DC, 2018.

American Petroleum Institute (API) 521, Pressure Relieving and De-Pressuring Systems, seventh ed., API, Washington, DC, 2020.

Y. Bai, Q. Bai, Subsea Engineering Handbook, first ed., Elsevier, Atlanta, GA, 2012.

C. Curran, API 17O—Recommended practice for subsea high-pressure protection systems (HIPPS), in: BSEE Workshop, 2014.

J. Davalath, H.B. Skeels, S. Corneliussen, Current state of the art in the design of subsea HIPPS systems, in: Offshore Technology Conference. Paper # OCT 14183. Houston, Texas, 2002.

J.G. Hoseth, B. Humphrey, Optimizing Pressure in Subsea Pipes With HIPPS, [online] Available at: https://library.e.abb.com/public/e2ed4866b60d409ec1256ef500457171/28-35%20-%20M136.pdf. (Accessed 10 September 2020).

V. Hutchings, Is the subsea high integrity pressure protection system (HIPPS) coming of age? Society of underwater technology, in: Subsea Control and Data Acquisition (SADA) Conference June 2–3 Newcastle, UK, 2010.

Inst Tools, Difference Between Process Shut Down and Emergency Shut Down, [online] Available at: https://instrumentationtools.com/process-shutdown-and-emergency-shutdown/. (Accessed 10 September 2020).

Inst Tools, Alarm and Trip Systems, [online] Available at: https://instrumentationtools.com/alarm-trip-systems/. (Accessed 10 September 2020).

International Association of Oil and Gas Producers (IOGP), High integrity protection systems—Recommended practice, 2015. Report # 443.

International Electrotechnical Commission (IEC), IEC 61508, Functional Safety of Electrical/Electronic/Programmable Electronic Safety-Related Systems, second ed., 2010. Geneva.

International Electrotechnical Commission (IEC), IEC 61511, Functional Safety-Safety Instrumented System for the Process Industry Sector, second ed., 2016. Geneva.

International Organization of Standardization (ISO) 5208, Industrial Valves—Pressure Testing of Metallic Valves, fourth ed., 2015. Geneva.

M. Luis, G. Garcia, C. Fialkowski, V. Sud, N. Christopher, HIPPS design, analysis, justification and implementation, in: Siemens Industry. Safety Control Systems Conference, 2015.

G. Nustad, Tackling corrosive challenges in the oil and gas industry, Valve World 20 (11) (2015) 56–60.

Ringo Valvulas, High Pressure Protection Integrity System, [online] Available at: http://www.ringospain.com/types/hipps-samson-ringo/. (Accessed 10 September 2020).

Smar Technology Compare, Introduction to Partial Stroke Testing (PST), [Online] Available at: https://www.smar.com/en/technical-article/introduction-to-pst-partial-stroke-test. (Accesseed 11 September 2020].

K. Sotoodeh, HIPPS application—a case study of suitable fast opening valve selection, Valve World Mag. 21 (10) (2016) 61–65.

K. Sotoodeh, Safety integrity level in valves, J. Fail. Anal. Prev. 19 (2019) 832–837, https://doi.org/10.1007/s11668-019-00666-2. Springer.

K. Sotoodeh, Failure mode and effect analysis (FMEA) of pipeline ball valves in the offshore industry, J. Fail. Anal. Prev. (2020), https://doi.org/10.1007/s11668-020-00924-8 Springer.

E.M. Stolen, Reliability Assessment of Subsea Hipps, Norwegian University of Science and Technology. Department of Production and Quality Engineering, 2014. [online] Available at: https://ntnuopen.ntnu.no/ntnuxmlui/bitstream/handle/11250/240965/713700_FULLTEXT01.pdf?sequence=1. (Accessed 10 September 2020).

Index

Note: Page numbers followed by *f* indicate figures and *t* indicate tables.

C

Q

R

T

U

V

W

Y